How to Manage Migratory Pests and Potential Food Crises: Locusts Plagues in the 2020's

How to Manage Migratory Pests and Potential Food Crises: Locusts Plagues in the 2020's

Editors

Michel Lecoq
Arianne Cease

MDPI • Basel • Beijing • Wuhan • Barcelona • Belgrade • Manchester • Tokyo • Cluj • Tianjin

Editors

Michel Lecoq
CBGP
CIRAD
Montpellier
France

Arianne Cease
School of Sustainability and
School of Life Sciences
Arizona State University
Tempe
United States

Editorial Office
MDPI
St. Alban-Anlage 66
4052 Basel, Switzerland

This is a reprint of articles from the Special Issue published online in the open access journal *Agronomy* (ISSN 2073-4395) (available at: www.mdpi.com/journal/agronomy/special_issues/locusts_plague).

For citation purposes, cite each article independently as indicated on the article page online and as indicated below:

LastName, A.A.; LastName, B.B.; LastName, C.C. Article Title. *Journal Name* **Year**, *Volume Number*, Page Range.

ISBN 978-3-0365-5752-6 (Hbk)
ISBN 978-3-0365-5751-9 (PDF)

Cover image courtesy of Michel Lecoq

Contents

About the Editors

Michel Lecoq

Dr. Michel Lecoq has been involved with pest locust population dynamics, ecology, surveillance and control during more than 50 years, mainly within the frame of CIRAD (Montpellier, France), where he was director of the research unit "Locust Ecology and Control" from 1997 to 2010. He has travelled extensively in tropical areas and worked in many countries affected by pest locusts. He contributes to the development of preventive control strategies for various species, including desert locust, migratory locust, red locust, Mato Grosso locust, and Senegalese grasshopper, aiming to better perceive the obstacles—mainly sociological and financial—to the implementation of effective controls. Michel Lecoq earned a PhD in entomology at Paris-Saclay University in 1975. His research activities have resulted in more than 110 peer-reviewed scientific publications, 60 books or book chapters, and numerous technical and consultation reports, as well as many popularization papers. Michel Lecoq was Consultant of the Food and Agriculture Organization of the United Nations on locust problems on numerous occasions; he served as president of the Orthopterists' Society from 2005 to 2009 and received the D.C.F. Rentz Award from the society in 2019 "in recognition of a lifetime dedicated to the study of Orthopteroid insects".

Arianne Cease

Dr. Arianne Cease has been studying locusts and grasshoppers since she was first exposed to locust swarms as a Peace Corps Volunteer in Senegal in 2005. Her research involves transdisciplinary approaches to understanding how human–plant–insect interactions affect the sustainability of agricultural systems, including laboratory and field studies. She works with collaborative teams to study locust plagues in response to agricultural practices in West Africa, South America, Australia, and China. A key goal of her research is to improve sustainable ecosystem management and rural livelihoods. For this work, she was named in the Brilliant 10 by *Popular Science* magazine in 2015. Her work has been supported by the US National Science Foundation, Foundation for Food & Agriculture Research, US Agency for International Development, and the US Department of Agriculture. She is currently an Associate Professor at Arizona State University. In 2018, she became the Founding Director of the Global Locust Initiative, which she continues to co-direct with Dr. Rick Overson.

Preface to "How to Manage Migratory Pests and Potential Food Crises: Locusts Plagues in the 2020's"

Locusts are a threat to agriculture and livelihoods in many countries globally. The economic, social, and environmental consequences of these highly migratory pests are substantial; they are treated as a national priority by many countries, and several international commissions have been established to unite efforts. Currently, one century after the revolutionary discovery of the phase phenomenon in locusts by Boris P. Uvarov, preventive strategies are increasingly being implemented as the most effective and economically viable control methods. They are based on the improved knowledge of locust biology and ecology, more efficient monitoring and control techniques, high-level technologies such as satellite imagery and geographic information systems, and the increased use of biological alternatives to chemical pesticides. As a result of the research efforts conducted over the past century, locust invasions are now better controlled and often shorter in duration and smaller in extent. Nevertheless, large-scale outbreaks persist, often arising unexpectedly. Considerable progress is still necessary in order to better understand locust behavior and key factors determining outbreaks, and to improve monitoring and forecasting methods. Critically, more focus is needed to understand the socio-political and economic aspects of locust governance to support sustained locust research and management capacity, as well as provide effective responses during a crisis. This book is a compilation of articles from a special issue of the *Agronomy* journal. It gathers the contributions of 54 authors from 19 countries in North and South America, Europe, Africa, Asia and Australia. It covers topics related to the main species of economic importance, in particular the Desert locust, Italian locust, Central American locust, Brown locust, Moroccan locust, and Migratory locust. It mobilizes diverse disciplines ranging from ecology, biology and biogeography to social sciences and political philosophy. Throughout the pages and articles, this book aims to shed light on some overarching questions: what have we learned from historical outbreaks; how serious is the threat; what research is ongoing and is needed to better manage these insects; how should the world respond to plagues today, especially in the context of climate change; are recommended preventive strategies really effective and what are the constraints to their application; and is there a possibility to make better use of biological alternatives to chemical pesticides? We hope that this book will contribute by highlighting recent research and management advancements, and stimulate new activities to improve management strategies for these dangerous pests that have plagued humanity for millennia.

Michel Lecoq and Arianne Cease

Editors

agronomy

MDPI

Editorial

What Have We Learned after Millennia of Locust Invasions?

Michel Lecoq [1,2,*] and Arianne Cease [3]

1 CIRAD, UMR CBGP, F-34398 Montpellier, France
2 CBGP, CIRAD, INRAE, IRD, SupAgro, University Montpellier, Montpellier, France
3 School of Sustainability and School of Life Sciences, Arizona State University, Tempe, AZ 85287, USA; acease@asu.edu
* Correspondence: mlecoq34@gmail.com

Abstract: Locust outbreaks have long been a very serious problem for agriculture and livelihoods in many countries globally. This article is an introduction to a Special Issue of the journal Agronomy devoted to the management of these pests. Although not exhaustive, the nineteen articles herein cover a variety of species, many regions of the world and many aspects of pest locust management and research in the early 21st century. This book is a source of information and reflection, as well as a resource, to support new areas of investigation and practice contributing to the process of developing sustainable solutions for locust invasions.

Keywords: locust plagues; control; management; preventive strategy; locust biology; locust ecology

Citation: Lecoq, M.; Cease, A. What Have We Learned after Millennia of Locust Invasions? *Agronomy* **2022**, *12*, 472. https://doi.org/10.3390/agronomy12020472

Received: 29 January 2022
Accepted: 10 February 2022
Published: 14 February 2022

1. Introduction

Since time immemorial, pest locusts in their gregarious phase have been a very serious problem for agriculture and livelihoods of human populations in many countries around the world. The economic and social consequences of locusts are so substantial that these highly migratory pests are treated as a national priority by many countries, and several international commissions have been established to unite efforts. The start of 2020 was marked by the continued South American Locust upsurge—the first major upsurge of this species in 60 years, as well as the onset of a dangerous Desert Locust invasion extending from Kenya to India—the first many of these countries have seen in decades. This Special Issue aims to shed light on the following overarching questions: What have we learned from historical outbreaks? What research is ongoing? What action is needed? How serious is the threat? How should the world respond to plagues today?

Around twenty-five of the estimated 6700 Acrididae (Order: Orthoptera) species [1] are considered to be major pest locusts to varying degrees [2–6]. When populations of these insects increase under favorable ecological conditions, they acquire gregarious and migratory behavior, leading to the formation of spectacular swarms that can migrate vast distances and leave behind devastated crops and desolation (Figure 1). However, these insects are also a source of food for various human populations [7], and swarms certainly have beneficial but largely unknown ecological roles [8].

Figure 1. Example of locust swarms and hopper bands: the Migratory Locust, *Locusta migratoria*, in Madagascar (**a**,**b**) and the Mato Grosso Locust, *Rhammatocerus schistocercoides*, in Brazil (**c**,**d**) (photos M. Lecoq). The Migratory Locust swarm is approximately 5 × 2 km, measured from a helicopter, and it completely covers the landscape due to its high density. Its color varies according to the position of the observer in relation to the sun.

2. History of Locusts and People

Due to the extent of their devastation and the spectacular nature of their migratory swarms, locusts have left an indelible mark since ancient times in the minds, beliefs, literature and arts of many cultures all over the world [9,10]. The first mention of their damage—albeit conjectural—dates to Assyrian times, ca. 3200 BC, and was noted on a stone tablet in pre-cuneiform writing [11,12]. Since then, the evidence has multiplied, in numerous chronicles, books, religious texts, and testimonies of travelers, missionaries and naturalists throughout the ages [13–15]. Some cultures include attribution to locusts in the creation of the world and mastery of fire [16,17]. Locusts have also been used for political and military purposes [18–20].

Locust management has been ongoing for centuries, probably since the beginning of agriculture, as testified by ancient Egyptian, Chinese, Hebrew, Roman and Greek texts [9,21,22]. For example, China has 3000 years of recorded Migratory Locust outbreaks, with >800 outbreak episodes. The frequency of invasions led ancient emperors and their governments to impose treatment programs: full-time locust control officers were appointed in the Tang Dynasty in the eighth century [23]. Laws were also passed in the 11th, 12th and 15th centuries requiring local governors to conduct control, which included paying people to catch locusts and bury them in pits [23,24].

Over the centuries, more and more resources have been deployed on all continents. At the beginning of the 20th century, concerted international cooperation was developed (see, for example, [25–30]). Since then, research, surveillance and control efforts have steadily increased and been refined. In the present day, the Food and Agriculture Organization of the United Nations plays a centralized role in coordinating survey and control activities, strengthening national capacities and providing emergency assistance in response to locust upsurges and plagues [31–36].

3. 100 Years since Pivotal Phase Theory Published

This Special Issue coincides with the 100th anniversary of the phase theory Sir Boris Petrovitch Uvarov published in 1921 in a famous article [37–39]. The article demonstrated that locusts have two phases and that a single individual can shift phases in response to population density. The solitarious phase is prevalent at a low density, avoids conspecifics, does not form swarms and is typically a cryptic green or tan color. The gregarious phase is prevalent when favorable ecological conditions support high-density populations; it then aggregates and migrates *en masse*, and typically has a combination of dark and bright markings. Prior to this discovery, the two phases were thought to be different species. This article corresponds to a pivotal, founding moment, from which the phenomenon of locust invasions began to be better understood. Phase theory became the major paradigm within which locust research has been able to flourish and develop effectively for a century.

After Uvarov's discovery, research was aimed at locating the so-called "outbreak areas"—areas of origin of the outbreaks, which were previously unknown—and studying the ecological conditions that supported the development of gregarious populations in these areas. Additional focal areas over the century included efficient and environmentally safer control methods, the real-time monitoring of field populations, and general biology and ecology, with a particular focus on the mechanisms of phase change. Several recent articles have masterfully reviewed these various topics [36,40–42].

The result of this research and scientific and technical progress is that outbreaks have now become less frequent and smaller in scale. If they do occur, they are now controlled quicker, although the damage caused can still jeopardize the food security of rural populations in many Global South countries [36]. Recent outbreaks of Desert Locust in Africa and Asia [43–45] and the South American Locust *Schistocerca cancellata* [46,47] are indeed reminders that the problem remains, albeit to a lesser extent than in the past. Each resurgence is a surprise to many—most often through unawareness—that this ancient and spectacular problem is still around. Beyond the immense scientific and technical progress attested by the reviews mentioned above, many lessons have been learned over the century.

4. A Few Key Lessons

(1) **The catastrophic impact of major locust outbreaks is a reality.** Despite denials and challenges in recent decades [48–50], the impact is real, devastating and documented, affecting not only crops but many aspects from family to national economics. Crop and pasture losses can lead to severe food shortages, large price fluctuations in markets, insufficient availability of grazing areas, the sale of animals at very low prices to meet household subsistence needs, high tensions between pastoralists and local farmers, and large human migrations to urban areas [36,51]. Impacts can reverberate for many years. Children born during plague years in impacted communities are less likely to ever start or to perform well in school [52,53]. Importantly, conventional techniques and simple cost–benefit ratios based on the market value of affected crops fail to capture true costs where >90% of crops are for subsistence. Furthermore, in the case of migratory pests such as locusts, the burden of control may fall on one region, while the benefits may be recorded completely elsewhere [54].

(2) **Prevention strategies have been shown to be effective where they have been applied** [55–57]. Advancements in locust biology since Uvarov's 1921 article revealed how plagues develop from pockets of gregarious locust outbreaks. These advancements created opportunities for preventative management involving the regular monitoring of high-risk areas and early treatment, in contrast to reactive control once a crisis has evolved. While prevention has been considered an unattainable goal [58], delayed control interventions lead to a locust situation that is overwhelming and exponentially more difficult and costly to manage [59]. Preventive management strategies for locusts have thus been developed with dedicated organizations set up in many regions globally; however, the sustainability of these management systems remains an important concern [55,60].

(3) **Current prevention strategies may not be widely feasible** since they require consistent resources and training and can be weakened by the erosion of infrastructure in the time between outbreaks, insecurity and war; however, community partnerships can help. If preventative management is successful, it inevitably leads to a vicious cycle with decreased investment when the problem is perceived to be solved [60,61]. Outbreaks then occur unfettered when ecological conditions become favorable. Regional insecurity and conflict can also preclude preventative management in key regions [62]. One aspect that has a clearer solution is the better engagement of communities in locust-prone regions. Prevention plans focused exclusively on the permanent surveillance of smaller outbreak areas that neglect surrounding rural communities during quiet periods and fail to inform them about the ins and outs of preventive control may prove to be largely counter-productive in the long term. Management plans that are better able to include local communities and learn from their traditional ecological knowledge through periods of calm as well as in the event of failure of prevention may be more apt to maintain their effectiveness over many years.

(4) **During major outbreaks, there continues to be pressures requiring the massive use of fast-acting insecticides.** The various shortcomings of preventative strategies continue to result in the development of large-scale outbreaks that directly threaten crops, thus requiring the massive use of fast-acting insecticides. This situation regularly raises controversy about the quantities used and their harmfulness. It habitually calls into question all preventive control strategies, forgetting—most often due to lack of information—their largely positive aspects. However, there are still no synthetic chemical pesticide alternatives that are viable on a large scale. Some biopesticides have been developed, most notably, the fungus *Metarhizium acridum*. Though, perhaps apart from China (Zhang and Hunter p.c. in [40]), many logistical challenges remain [36]. A targeted treatment option with limited secondary environmental impacts that takes into account the many inherent logistical challenges could make great strides in improving sustainable locust management.

(5) **Climate change is likely to alter the probability, severity and location of outbreaks** [63–65]. However, the level of understanding of the magnitude and directionality of impacts and responses of various locust species to climate change remains low [66]. In fact, in view of the many uncertainties, there is an urgent need to coordinate research and monitoring actions on the topic of global change [67].

(6) **With each new major outbreak, a similar cycle of events repeats itself.** As a locust crisis hits international headlines, consistent themes arise. The initial response is concern for the lack of action and food security, followed by the questioning of control strategies, profitability of control and environmental impacts. Emphasis on safer techniques leads to a multitude of miracle solutions being proposed (e.g., massive locust vacuums). Articles in the media flourish, many opportunistic scientific publications are born, and declarations of good intentions multiply, including developing new research in various fields, training people, reforming institutions and reinforcing their resources. Each major crisis has its share of recommendations, always more or less the same, but between two major outbreaks, the locust problem is quickly considered to be solved, the memory of the problem and of past outbreaks fades, vigilance is reduced and qualified personnel retire or are fired, resulting in a devastating loss of knowledge. In this way, major outbreaks succeed one another, regularly followed by collective amnesia. In this sense, locust plagues are no exception in the list of calamities that have affected humanity for millennia. This can be likened to the myth of Sisyphus, the hero of Greek mythology. Everything appears to be an eternal and absurd restart [68].

(7) **Transdisciplinary approaches are critical.** The situation of the repetitive cycle has long been denounced [60,61,69]. Despite the immense improvements in monitoring and control methods, if outbreaks remain—albeit at a more modest level than in the past—it indicates it is no longer only a problem of limited biological and technical advancements. It is also, and above all, a question of removing the obstacles linked to the way we, as a global community, approach the problem, and of realizing that we have always forgotten major components: people, human behavior, decision-making processes, and the hazards

of any human action. A better integration of social sciences in the process of developing sustainable solutions seems increasingly necessary, in addition to the more traditional approaches, which are obviously indispensable [61]. There is a need to develop a more comprehensive research, response, and resilience framework by engaging actors across disciplines, sectors, cultures, and boundaries. Supporting a global network to enable transdisciplinary approaches and continuity between outbreaks is part of the mission of the Global Locust Initiative at Arizona State University.

5. Summary of Articles Included in This Special Issue

This Special Issue covers a variety of species, many regions of the world, and many different aspects of locust management and research in the early 21st century. It discusses the Desert Locust—a major pest and model species—but also the Central American Locust, the Brown Locust in southern Africa, the South American Locust and the Italian Locust in the Siberian steppes. The aspects discussed are diverse: ecology and behavior; control strategies and methods; pesticides, their impact and substitutes; possible contributions from social sciences to improve the management of locust outbreaks, and even the positive aspects of locust invasions, both in terms of human nutrition and ecosystem health. Here, we present briefly the various contributions that constitute this Special Issue.

First, concerning the behavioral aspects, Maeno et al. [70] studied the previously poorly understood nocturnal behavior of swarms of the Desert Locust, *Schistocerca gregaria*. This nocturnal behavior appears to be predictable and dependent on plant size, and the authors propose that it can be used to facilitate locust swarm management and to adopt a general strategy of nocturnal locust control.

Using time series of abundance indices for the Desert Locust and the Oriental Migratory Locust (*Locusta migratoria*), Cheke et al. [71] analyzed them independently and in relation to measures of solar activity and ocean oscillation systems. The results suggest that solar activity can be used to predict locust abundance, offering hope that information on these phenomena might enable a better early forecasting of Desert Locust upsurges.

Liu et al. [72], for the first time, studied the behavior of Desert Locust swarms in southern Tibet at very high altitudes of up to 5400 m. They show that low temperatures, high humidity and low atmospheric oxygen put the locusts under severe stress and that the Himalayan mountains provided an important natural barrier that limited the northward expansion of Desert Locust populations.

Concerning management strategies, Showler et al. [59] discuss the three major approaches to Desert Locust population control: reaction, proaction and outbreak prevention. They suggest that while research on the biology and behavior of this species is no longer a highly urgent requirement for improving control efficacy, new research priorities have emerged for developing epidemic prevention capability (and for improving proactive management). Salient needs presently include long residual tactics for prophylactic (preventive) control in breeding areas; intervention thresholds; and improved, sustainable coordination among stakeholders at national, regional and international levels. The most recent Desert Locust episode of 2020 illustrates how prophylactic control could have prevented the entire upsurge, and how proactive management in some countries contained the spread of swarms.

Despite the high level of progress in recent years, Showler and Lecoq [62] show that locust surveillance and control are impaired by many obstacles, perhaps the most intractable of which is insecurity. After a historical analysis of insecurity in the various Desert Locust range countries, the 2020 upsurge is used to show how direct insecurity still contributes to the genesis and expansion of locust outbreaks. The possible mitigation of the effects of direct insecurity on some Desert Locust operations is discussed.

Sergeev [73] covers the eco-geographic distribution and long-term dynamics of the Italian Locust, *Calliptamus italicus*, in Asia, especially in Western Siberia, and analyzes their consequences for the management of pest species and rare forms. He discusses issues arising from insecticide experimental treatments, as well as ecological associations between the Italian Locust and rare Orthoptera. Suggestions are provided regarding the need for the improved monitoring, supervision, control and forecasting of Italian Locust distribution and population dynamics.

Ciplak [74] provides a historical background for locust outbreaks in Southwest Asia (mainly Arabian Peninsula, Cyprus, Iraq, Israel, Jordan, Lebanon, Palestine, Syria and Turkey), assesses the potential for outbreaks of local species, and defines pathways for future actions, especially with regard to global change and agricultural expansion that may increase locust activities.

Barrientos et al. [75] present a review of the Central American Locust, *Schistocerca piceifrons piceifrons*, an important transboundary pest distributed from Mexico to Panama. The paper covers life history, habitat, ecology, permanent breeding areas, band and swarm formation, phase polyphenism, gregarization and migration. They provide a history of recent outbreaks and consider the current management strategy for this locust and recommended control measures, as well as current and future research.

In South Africa, the regular and often intense outbreaks of the Brown Locust, *Locustana pardalina*, are a formidable pest control problem. Price [76] discusses the operational constraints associated with the traditional ground control strategy in the Karoo outbreak area of this species. He makes recommendations for a modernized and technologically equipped integrated management strategy for the Brown Locust, combining ground and aerial tactics with the flexibility and capacity to effectively deal with outbreaks.

Trumper et al. [47] present a very interesting case of preventive management in the South American Locust, which has been successful for many years but has recently experienced a major resurgence. Their paper provides a review of this species' biology, management history and perspectives on navigating a plague period after a 60-year recession.

Monitoring and control methods have made many advances in recent decades, but challenges remain. Matthews [77] discusses the difficulties of regularly monitoring locust populations, often in remote and unpopulated areas, and the recent contribution of new technologies, particularly drones and mycopesticides. Drones would facilitate more efficient surveys, determining where sprays need to be applied at an early stage, and minimize the risk of swarms developing and migrating to feed on large crop areas. They could also spray groups of hoppers and adults. The use of biopesticides, which have been available for many years, is encouraged due to their effectiveness and greater acceptability to the environment and non-target fauna.

Control operations are naturally a crucial tool for managing locust crises. Retkute et al. [78] use the example of the recent Desert Locust upsurge in 2019–2021 to analyze the nature of control operations conducted to contain the crisis. These treatments were used both for the immediate protection of crops and to reduce overall locust numbers. Without the latter, the upsurge might have lasted many years as they did in the first half of the 1900s.

The harmful effects of pesticides used in locust control are illustrated by Peveling's article [79]. His studies in Madagascar show a long-term decline of harvester termites following multiple barrier treatments with fipronil carried out to control an outbreak of the Migratory Locust in the late 1990s. The main outcome of this research is a strikingly low resilience of harvester termite populations, which did not recover within eight years, with likely repercussions on food webs. The environmental benefit of barrier treatments is lost if the same areas are treated repeatedly during the same campaign. Recommendations are given as to how to mitigate those risks.

As we have seen above, biopesticides appear to be a promising and readily available solution to control locusts, though there are currently logistical barriers to their widespread use. Abdelatti and Hartbauer [80] propose an alternative treatment whose main component is linseed oil. The authors describe the impact of this formulation on the aggregation behavior of Desert Locusts. Their results suggest that linseed oil may act as a bioactive agent capable of disrupting swarm formation.

Protozoa may also be used to control locust and grasshopper populations. Zhang and Lecoq [81] reviewed the role of *Nosema locustae*, a biological agent developed in the 1980s for locust and grasshopper control. The authors review the many studies that have focused on pathogenicity, the host spectrum, mass production, epizootiology, applications, genomics and molecular biology. This entomopathogen has many benefits: lower environmental impacts, vertical transmission to offspring through eggs and long-term persistence in populations of locusts and grasshoppers for more than 10 years, and it is well adapted to a variety of ecosystems. However, some limitations still need to be overcome.

While locusts have been at the heart of the natural sciences for over a century, the social sciences remain vastly under-represented. The potential contributions of the social sciences to improve the management of locust plagues are explored by Therville et al. [61]. Organizational, economic and cultural variables clearly have a considerable impact on the management strategies of these pests, and the social sciences are an important means to better understand these questions. This article examines the scope and purpose of different subfields of the social sciences and explores how they can be applied to different issues facing entomologists and practitioners to implement sustainable locust research and management.

In a similar vein, Lockwood and Sardo [82] use the Desert Locust as an iconic case from a humanities perspective. They provide a summary of standard moral theories and examine their shortcomings in developing a framework for understanding the socio-economic complexity of locust management. They address some of the models of global justice and focus on two fundamental questions: Who is a moral agent in Desert Locust management? How can responsibilities be equitably distributed between agents in preventive and reactive modes? After identifying the agents, they use a quadruple set of principles to construct a Desert Locust management framework consistent with global justice and apply this conceptual system to two hypothetical scenarios.

While major locust outbreaks exacerbate food insecurity, locusts can also be a food source for many populations. Samejo et al. [83] conducted a field study in India to investigate whether locust collection could be an attractive control method to protect crops in the event of an invasion, as well as an accepted food resource for poor rural communities. The authors conclude that locust consumption could be an effective practice to prevent malnutrition and protein deficiency and, to some extent, a mitigating measure to help communities better protect themselves and their crops from locust invasions. They argue that the collection and consumption of locusts should be encouraged while being realistic about their actual impact on locust control. They also advocate that this should be carried out in concert with local authorities to minimize risks to human health by avoiding the consumption of insects treated with pesticides.

Finally, Kietzka et al. [8] further explore the ecological and human food value of locusts. They note that while locusts have been considered since the Antiquity to be very serious pests of sedentary agriculture in many parts of the world—and as such combated with increasingly sophisticated means—they can be a nutritious food source for people and of great ecological importance as major components of ecosystem nutrient cycling. The authors estimated their potential benefits for human nutrition and ecosystem function using calculations based on a 1 km^2 area of swarming and breeding Desert Locusts, *Schistocerca gregaria*.

Although not fully comprehensive, this Special Issue covers most major current issues in locust management. We hope that it will be a source of information and reflection for many already working in the field, as well as serving as a resource to support new areas of investigation and practice to aid in the process of developing sustainable solutions for locust outbreaks.

Conflicts of Interest: The authors declare no conflict of interest.

References

1. Song, H.; Mariño-Pérez, R.; Woller, D.A.; Cigliano, M.M. Evolution, diversification, and biogeography of grasshoppers (Orthoptera: Acrididae). *Insect Syst. Divers.* **2018**, *2*, 3. [CrossRef]
2. COPR. *Locust and Grasshopper Manual*; Centre for Overseas Pest Research: London, UK, 1982.
3. Lecoq, M.; Zhang, L. (Eds.) *Encyclopedia of Pest Orthoptera of the World*; China Agricultural University Press: Beijing, China, 2019; 310p, ISBN 978-7-5655-2232-1.
4. Le Gall, M.; Overson, R.; Cease, A.J. A global review on locusts (Orthoptera: Acrididae) and their interactions with livestock grazing practices. *Front. Ecol. Evol.* **2019**, *7*, 263. [CrossRef]
5. Song, H. Phylogenetic perspectives on the evolution of locust phase polyphenism. *J. Orthoptera Res.* **2005**, *14*, 235–245. [CrossRef]
6. Song, H. Density-dependent phase polyphenism in nonmodel locusts: A minireview. *Psyche A J. Entomol.* **2011**, *2011*, 741769. [CrossRef]
7. Van Huis, A. Harvesting desert locusts for food and feed may contribute to crop protection but will not suppress upsurges and plagues. *J. Insects Food Feed* **2021**, *7*, 245–248. [CrossRef]
8. Kietzka, G.J.; Lecoq, M.; Samways, M.J. Ecological and human diet value of locusts in a changing world. *Agronomy* **2021**, *11*, 1856. [CrossRef]
9. Kevan, D.K.M.E. *Land of the Locusts*; Parts I-Before 450 A.D. Memoir n°6; Lyman Entomological Museum and Reserch Laboratory, McGill University: Quebec, QC, Canada, 1978; p. 530.
10. Dominy, N.J.; Fannin, L.D. The sluggard has no locusts: From persistent pest to irresistible icon. *People Nat.* **2021**, *3*, 542–549. [CrossRef]
11. Barton, G.A. A Text from the Oldest Period of Babylonian Writing. *Orient. Lit.* **1913**, *1*, 6–12. Available online: https://archive.org/details/orientalistische16deutuoft/page/5/mode/2up?view=theater (accessed on 19 December 2021). [CrossRef]
12. Horne, C.F. The Locust Charm. In *The Sacred Books and Early Literature of the East. Babylonia and Assyria*; Parke, Austin, and Lipscomb: New York, NY, USA, 1917; Volume 1, pp. 34–35. Available online: https://archive.org/details/sacredbooksearly01hornuoft/page/n9/mode/2up (accessed on 26 January 2022).
13. Buj, A. *Plagas de langosta—De la Plaga Bíblica a la Ciencia de la Acridología [Locust Plagues—From the Biblical Plague to the Science of Acridology]*; Serbal: Barcelona, Spain, 2016.
14. Desanges, J. Témoignages antiques sur le fléau acridien [Ancient Evidence on the Locust Plague]. In L'homme face aux calamités naturelles dans l'Antiquité et au Moyen Âge. In Proceedings of the 16th Symposium of the Villa Kérylos, Beaulieu-sur-Mer, France, 14–15 October 2005; Cahiers de la Villa "Kérylos" N° 17; Académie des Inscriptions et Belles-Lettres: Paris, France, 2006; pp. 221–235.
15. Plauchut, E. Les Invasions des sauterelles (The locust invasions). *Rev. Mondes* **1895**, *132*, 406–429.
16. Pereira, A.H. O pensamento mítico de Paresí. *Pesqui. Antropol.* **1986**, *41*, 1–441. Available online: http://www.anchietano.unisinos.br/publicacoes/antropologia/antropologia.html (accessed on 11 January 2022).
17. Lecoq, M.; Pierozzi, I., Jr. *Rhammatocerus schistocercoides* locust outbreaks in Mato Grosso (Brazil): A long-standing phenomenon. *Int. J. Sustain. Dev. World Ecol.* **1995**, *2*, 45–53. [CrossRef]
18. Péloquin, C. Unruly Nature and Technological Authority: Governing Locust Swarms in the Sahel. Ph.D. Thesis, University of Arizona, Tucson, AZ, USA, 2014. Available online: http://hdl.handle.net/10150/321308 (accessed on 28 January 2022).
19. Peloquin, C. Locust swarms and the spatial techno-politics of the French Resistance in World War II. *Geoforum* **2013**, *49*, 103–113. [CrossRef]
20. Forestier-Peyrat, E. Fighting locusts together: Pest control and the birth of Soviet development aid, 1920–1939. *Glob. Environ.* **2014**, *7*, 536–571. [CrossRef]
21. Uvarov, B.P. The Locust Plague. *J. Econ. Entomol.* **1944**, *37*, 93–99. [CrossRef]
22. Zadoks, J.C. *Crop Protection in Medieval Agriculture*; Sidestone Press: Leiden, The Netherlands, 2013. Available online: https://www.sidestone.com/openaccess/9789088901874.pdf (accessed on 26 January 2022).
23. Wang, Z.; He, K.; Zhao, J.; Zhou, D. Integrated pest management in China. In *Integrated Pest Management in the Global Arena*; Maredia, K.M., Dakuou, D., Mota-Sanchez, D., Eds.; CABI Publishing: Oxon, UK, 2003; pp. 197–207. [CrossRef]
24. Tian, H.; Stige, L.; Cazelles, B.; Kausrud, K.; Svarverud, R.; Stenseth, N.; Zhang, Z. Reconstruction of a 1,910-y-long locust series reveals consistent associations with climate fluctuations in China. *Proc. Natl. Acad. Sci. USA* **2011**, *108*, 14521–14526. [CrossRef] [PubMed]
25. Buj, A. International experimentation and control of the locust plague. Africa in the first half of the 20th century. In *Les Sciences Hors d'Occident au XXe Siècle*; Nature et Environnement; Chatelin, Y., Bonneuil, C., Eds.; ORSTOM: Paris, France, 1995; Volume 3, pp. 93–105. Available online: http://www.documentation.ird.fr/hor/fdi:010008856 (accessed on 11 February 2022).
26. International Institute of Agriculture. *La Lutte Contre les Sauterelles dans les Divers Pays [Locust Control in Various Countries]*; International Institute of Agriculture: Rome, Italy, 1916.
27. International Institute of Agriculture. *Proceedings of the International Conference for the Organization of Locust Control, Rome, Italy, 28–31 October 1920*; International Institute of Agriculture: Rome, Italy, 1921.
28. Künckel d'Herculais, J. *Invasion des Acridiens, Vulgo Sauterelles, en Algérie (1893–1905) [Locusts Invasion in Algeria, 1893–1905]*; Imprimerie Administrative et Commerciale Giralt: Algiers, Algeria, 1893–1905; Volumes 1–3.

29. Schemper, L. Years of the Locust—What history can tell us about international locust control. Fondation Pierre du Bois pour l'histoire du temps présent: Pully, Switzerland. *Curr. Aff. Perspect.* **2020**, *3*, 1–7. Available online: https://www.researchgate.net/publication/339943779_Years_of_the_Locust_-_What_History_Can_Tell_Us_about_International_Locust_Control (accessed on 28 January 2022).
30. Vayssière, P. Le problème acridien et sa solution internationale. *Matériaux Pour L'étude Calam.* **1924**, *2*, 122–158.
31. Cressman, K. Desert locust. In *Biological and Environmental Hazards, Risks, and Disasters*; Sivanpillai, R., Ed.; Elsevier: Amsterdam, The Netherlands, 2016; pp. 87–105. [CrossRef]
32. FAO. *Desert Locust*; Food and Agriculture Organization of the United Nations: Rome, Italy, 2022. Available online: https://www.fao.org/locusts/en/ (accessed on 26 January 2022).
33. FAO. *Locust Watch*; Food and Agriculture Organization of the United Nations: Rome, Italy, 2022. Available online: https://www.fao.org/ag/locusts/en/info/info/index.html (accessed on 26 January 2022).
34. FAO. *Locust Hub*; Food and Agriculture Organization of the United Nations: Rome, Italy, 2022. Available online: https://locust-hub-hqfao.hub.arcgis.com (accessed on 26 January 2022).
35. Lecoq, M. Desert Locust threat to agricultural development and food security and FAO/ international role in its control. *Arab J. Pl. Prot.* **2003**, *21*, 188–193. Available online: https://www.academia.edu/49263970/ (accessed on 26 January 2022).
36. Zhang, L.; Lecoq, M.; Latchininsky, A.; Hunter, D. Locust and grasshopper management. *Annu. Rev. Entomol.* **2019**, *64*, 15–34. [CrossRef]
37. Uvarov, B.P. A Revision of the genus Locusta, L. (= Pachytylus, Fieb.), with a new theory as to the periodicity and migrations of locusts. *Bull. Entomol. Res.* **1921**, *12*, 135–163. [CrossRef]
38. Latchininsky, A.V.; Sergeev, M.V.; Fedotova, A.A. The centennial of Sir Boris Uvarov's locust phase theory. *Metaleptea* **2021**, *41*, 11–13. Available online: https://orthsoc.org/2021/10/03/new-metaleptea-issue-413/ (accessed on 1 January 2022).
39. Pflüger, H.J.; Bräunig, P. One hundred years of phase polymorphism research in locusts. *J. Comp. Physiol. A* **2021**, *207*, 321–326. [CrossRef] [PubMed]
40. Cullen, D.A.; Cease, A.J.; Latchininsky, A.V.; Ayali, A.; Berry, K.; Buhl, J.; De Keyser, R.; Foquet, B.; Hadrich, J.C.; Matheson, T.; et al. Chapter Seven—From molecules to management: Mechanisms and consequences of locust phase polyphenism. In *Advances in Insect Physiology*; Verlinden, H., Ed.; Academic Press: Cambridge, MA, USA, 2017; Volume 53, pp. 167–285. [CrossRef]
41. Pener, M.P.; Simpson, S.J. Locust phase polyphenism: An update. *Adv. Insect Phys.* **2009**, *36*, 1–272. [CrossRef]
42. Wang, X.; Kang, L. Molecular mechanisms of phase change in locusts. *Annu. Rev. Entomol.* **2014**, *59*, 225–244. [CrossRef]
43. FAO. *Desert Locust Bulletin, Nos. 484–495*; Food and Agriculture Organization of the United Nations: Rome, Italy, 2019. Available online: http://www.fao.org/ag/locusts/en/archives/archive/1823/index.html (accessed on 26 January 2022).
44. FAO. *Desert Locust Bulletin, Nos. 496–507*; Food and Agriculture Organization of the United Nations: Rome, Italy, 2020. Available online: http://www.fao.org/ag/locusts/en/archives/archive/2521/index.html (accessed on 26 January 2022).
45. Sultana, R.; Samejo, A.A.; Kumar, S.; Soomro, S.; Lecoq, M. The 2019–2020 upsurge of the Desert Locust and its impact in Pakistan. *J. Orthoptera Res.* **2021**, *30*, 145–154. [CrossRef]
46. Medina, H.; Cease, A.; Trumper, E. The Resurgence of the South American Locust (*Schistocerca cancellata*). *Metaleptea* **2017**, *37*, 17–21.
47. Trumper, E.V.; Cease, A.J.; Cigliano, M.M.; Copa Bazán, F.; Lange, C.E.; Medina, H.E.; Overson, R.P.; Therville, C.; Pocco, M.E.; Piou, C.; et al. A Review of the biology, ecology, and management of the South American Locust, *Schistocerca cancellata* (Serville, 1838), and future prospects. *Agronomy* **2022**, *12*, 135. [CrossRef]
48. Krall, S.; Herok, C. Economics of Desert Locust control. In *New Strategies in Locust Control*; Krall, S., Peveling, R., Diallo, D.B., Eds.; Birkhäuser Verlag: Basel, Switzerland, 1997; pp. 401–413. [CrossRef]
49. Joffe, S. *Economic and Policy Issues in Desert Locust Management: A Preliminary Analysis*; Desert Locust Technical Series n° 27; Food and Agriculture Organization of the United Nations: Rome, Italy, 1998; 27p. Available online: https://www.fao.org/ag/locusts/common/ecg/1290/en/327_en_TS27p1.pdf (accessed on 19 December 2021).
50. Joffe, S. *Desert Locust Management: A Time for Change*; World Bank: Washington, DC, USA, 1995. [CrossRef]
51. Brader, L.; Djibo, H.; Faye, F.G.; Ghaout, S.; Lazar, M.; Luzietoso, P.N.; Babah, M.A.O. *Towards a More Effective Response to Desert Locusts and Their Impacts on Food Security, Livelihoods and Poverty—Multilateral Evaluation of the 2003–05 Desert Locust Campaign*; Food and Agriculture Organization of the United Nations: Rome, Italy, 2006; Available online: http://www.fao.org/ag/locusts/common/ecg/1913/en/DesertLocustEvalReportE.pdf (accessed on 3 November 2020).
52. De Vreyer, P.; Guilbert, N.; Mesplé-Somps, S. *The 1987–1989 Locust Plague in Mali: Evidences of the Heterogeneous Impact of Income Shocks on Education Outcomes*; No. DT/2012/06, Working Papers; DIAL—Développement, Institutions et Mondialisation: Paris, France, 2012; 48p. Available online: https://hal-pse.archives-ouvertes.fr/hal-00961739 (accessed on 11 February 2022).
53. De Vreyer, P.; Guilbert, N.; Mesple-Sompsa, S. Impact of natural disasters on education outcomes: Evidence from the 1987–89 locust plague in Mali. *J. Afr. Econ.* **2014**, *24*, 57–100. [CrossRef]
54. Lomer, C.J.; Bateman, R.P.; Johnson, D.L.; Langewald, J.; Thomas, M. Biological control of locusts and grasshoppers. *Annu. Rev. Entomol.* **2001**, *16*, 667–701. [CrossRef]
55. Lecoq, M. Recent progress in desert and migratory locust management in Africa. Are preventative actions possible? *J. Orthopt. Res.* **2001**, *10*, 277–291. [CrossRef]
56. Magor, J.I.; Lecoq, M.; Hunter, D.M. Preventive control and desert locust plagues. *Crop Prot.* **2008**, *27*, 1527–1533. [CrossRef]

57. Sword, G.A.; Lecoq, M.; Simpson, S.J. Phase polyphenism and preventative locust management. *J. Insect Physiol.* **2010**, *56*, 949–957. [CrossRef] [PubMed]
58. Symmons, P. A Critique of "Preventive control and Desert Locust plagues". *Crop Prot.* **2009**, *28*, 905–907. [CrossRef]
59. Showler, A.T.; Ould Babah Ebbe, M.A.; Lecoq, M.; Maeno, K.O. Early intervention against Desert Locusts: Current proactive approach and the prospect of sustainable outbreak prevention. *Agronomy* **2021**, *11*, 312. [CrossRef]
60. Lecoq, M. Le Criquet pèlerin. Enseignements de la dernière invasion et perspectives offertes par la biomodélisation. In *La Lutte Anti-Acridienne*; AUPELF-UREF, John Libbey Eurotext: Paris, France, 1991; pp. 71–98.
61. Therville, C.; Anderies, J.M.; Lecoq, M.; Cease, A. Locusts and people: Integrating the social sciences in sustainable locust management. *Agronomy* **2021**, *11*, 951. [CrossRef]
62. Showler, A.T.; Lecoq, M. Incidence and ramifications of armed conflict in countries with major Desert Locust breeding areas. *Agronomy* **2021**, *11*, 114. [CrossRef]
63. Meynard, C.N.; Gay, P.E.; Lecoq, M.; Foucart, A.; Piou, C.; Chapuis, M.P. Climate-driven geographic distribution of the desert locust during recession periods: Subspecies' niche differentiation and relative risks under scenarios of climate change. *Glob. Chang. Biol.* **2017**, *23*, 4739–4749. [CrossRef]
64. Kayalto, M.; Idrissi Hassani, M.; Lecoq, M.; Gay, P.E.; Piou, C. Cartographie des zones de reproduction et de grégarisation du criquet pèlerin au Tchad (Mapping of breeding and outbreak areas of the Desert Locust in Chad). *Cah. Agric.* **2020**, *29*, 14. [CrossRef]
65. Wang, B.; Deveson, E.D.; Waters, C.; Spessa, A.; Lawton, D.; Feng, P.; Liu, D.L. Future climate change likely to reduce the Australian Plague Locust (*Chortoicetes terminifera*) seasonal outbreaks. *Sci. Total Environ.* **2019**, *668*, 947–957. [CrossRef]
66. Meynard, C.N.; Lecoq, M.; Chapuis, M.P.; Piou, C. On the relative role of climate change and management in the current Desert Locust outbreak in East Africa. *Glob. Chang. Biol.* **2020**, *26*, 3753–3755. [CrossRef]
67. Kapuka, A.; Hlásny, T. Climate change impacts on ecosystems and adaptation options in nine countries in southern Africa: What do we know? *Ecosphere* **2021**, *12*, e03860. [CrossRef]
68. Camus, A. *Le Mythe de Sisyphe*; Gallimard: Paris, France, 1942.
69. Lecoq, M. Desert locust management: From ecology to anthropology. *J. Orthoptera Res.* **2005**, *14*, 179–186. [CrossRef]
70. Maeno, K.O.; Ould Ely, S.; Ould Mohamed, S.; Jaavar, M.E.H.; Ould Babah Ebbe, M.A. Adult desert locust swarms, *Schistocerca gregaria*, preferentially roost in the tallest plants at any given site in the Sahara desert. *Agronomy* **2020**, *10*, 1923. [CrossRef]
71. Cheke, R.A.; Young, S.; Wang, X.; Tratalos, J.A.; Tang, S.; Cressman, K. Evidence for a Causal Relationship between the Solar Cycle and Locust Abundance. *Agronomy* **2021**, *11*, 69. [CrossRef]
72. Liu, J.; Lecoq, M.; Zhang, L. Desert locust stopped by Tibetan highlands during the 2020 upsurge. *Agronomy* **2021**, *11*, 2287. [CrossRef]
73. Sergeev, M.G. Ups and downs of the Italian Locust (*Calliptamus italicus* L.) Populations in the Siberian steppes: On the horns of dilemmas. *Agronomy* **2021**, *11*, 746. [CrossRef]
74. Çiplak, B. Locust and grasshopper outbreaks in the Near East: Review under global warming context. *Agronomy* **2021**, *11*, 111. [CrossRef]
75. Barrientos-Lozano, L.; Song, H.; Rocha-Sánchez, A.Y.; Torres-Castillo, J.A. State of the art management of the Central American Locust *Schistocerca piceifrons piceifrons* (Walker, 1870). *Agronomy* **2021**, *11*, 1024. [CrossRef]
76. Price, R. Alternative strategies for controlling the Brown Locust, *Locustana pardalina* (Walker). *Agronomy* **2021**, *11*, 2212. [CrossRef]
77. Matthews, G.A. New technology for Desert Locust control. *Agronomy* **2021**, *11*, 1052. [CrossRef]
78. Retkute, R.; Hinton, R.G.K.; Cressman, K.; Gilligan, C.A. Regional differences in control operations during the 2019–2021 desert locust upsurge. *Agronomy* **2021**, *11*, 2529. [CrossRef]
79. Peveling, R. Long-term decline in harvester termites in Madagascar following multiple barrier treatments with fipronil against Migratory Locust. *Agronomy* **2022**, *12*, 310. [CrossRef]
80. Abdelatti, Z.A.S.; Hartbauer, M. Linseed oil affects aggregation behaviour in the Desert Locust *Schistocerca gregaria*—A potential swarm disruptive agent. *Agronomy* **2020**, *10*, 1458. [CrossRef]
81. Zhang, L.; Lecoq, M. *Nosema locustae* (Protozoa, Microsporidia), a biological agent for locust and grasshopper control. *Agronomy* **2021**, *11*, 711. [CrossRef]
82. Lockwood, J.A.; Sardo, M.C. A swarm of injustice: A sociopolitical framework for global justice in the management of the Desert Locust. *Agronomy* **2021**, *11*, 386. [CrossRef]
83. Samejo, A.A.; Sultana, R.; Kumar, S.; Soomro, S. Could entomophagy be an effective mitigation measure in Desert Locust management? *Agronomy* **2021**, *11*, 455. [CrossRef]

agronomy

Editorial

Are Mycopesticides the Future of Locust Control?

Michel Lecoq [1,*] **and Arianne J. Cease** [2]

[1] CIRAD, UMR CBGP, F-34398 Montpellier, France
[2] School of Sustainability and School of Life Sciences, Arizona State University, Tempe, AZ 85287, USA
* Correspondence: mlecoq34@gmail.com

Locusts are a very serious problem for agriculture and for the livelihoods of populations around the world. Many lessons have been learned from decades of controlling these pests [1]. After the intensive use of chemical insecticides, new perspectives that are more respectful of people and the environment have emerged over the last twenty years. Protozoa, such as *Nosema locustae*, have been used and have many benefits; however, some limitations still need to be overcome [2]. Currently, entomopathogenic fungi are the most promising alternative, at least partially, to traditional chemical insecticides. The article from Wakil et al. [3] in this Special Issue dedicated to locust management is the first field study to test the efficacy of four different entomopathogenic fungal formulations together—*Metarhizium acridum* (Green Muscle® and Green Guard®), *Metarhizium anisopliae*, and *Beauveria bassiana*—against nymphs and adults of the desert locust *Schistocerca gregaria*. The study shows that all of these formulations have the potential to control locusts under field conditions. In addition, there are sublethal effects on their reproductive ability and behavior, with reductions in diet consumption, frass production, and weight observed. The authors conclude that future research needs to explore the combination of fungus with other control agents—including chemical insecticides, microsporidia, and botanical extracts—as a means of integrated pest management of locusts in the field.

Chemical pesticides have many side effects, and these have been increasingly elucidated, including their impact on human health, the environment, nontarget organisms, and biodiversity [4]. They have been shown to contribute to insect decline worldwide [5]. In addressing this issue, the use of biopesticides as a possible important component of locust management programs is a most significant recent development [6]. The benefits of using biopesticides include specificity to locusts and grasshoppers and the preservation of natural enemies [7]. The credibility of biopesticides as part of locust management programs is increasingly recognized, and programs that have included biopesticides have found them invaluable for treating locusts and grasshoppers wherever they occur [8]. As restrictions on the use of chemical pesticides rightly increase, treatment programs in the future will need to ensure their effectiveness by including biopesticides and by putting mechanisms in place to facilitate their use. Research and development supported by governments and donors will be critical in providing pathways to navigate logistical challenges such as manufacturing, storing, applications, and environmental limitations, especially the low temperature range over which locust mycopesticides are effective. It seems that we are at a turning point in locust control and that substitutes to chemical insecticides—notably mycopesticides—will become increasingly important in the future, not only in prevention—as originally thought—but also in periods of invasion, as shown by the mycopesticide-based treatments carried out during the last desert locust invasion in 2019–2020 [9–11]. *Metarhizium anisopliae* has already been listed for some years as one of the few products recommended by the Locust Pesticide Referee Group of the FAO to control the desert locust and any other locust species [12].

Citation: Lecoq, M.; Cease, A.J. Are Mycopesticides the Future of Locust Control? *Agronomy* **2022**, *12*, 2344. https://doi.org/10.3390/agronomy12102344

Received: 15 July 2022
Accepted: 26 September 2022
Published: 28 September 2022

Publisher's Note: MDPI stays neutral with regard to jurisdictional claims in published maps and institutional affiliations.

Author Contributions: M.L. wrote the first draft of the manuscript; A.J.C. reviewed and edited. All authors have read and agreed to the published version of the manuscript.

Funding: This research received no external funding.

Conflicts of Interest: The authors declare no conflict of interest.

References

1. Lecoq, M.; Cease, A. What Have We Learned after Millennia of Locust Invasions? *Agronomy* **2022**, *12*, 472. [CrossRef]
2. Zhang, L.; Lecoq, M. *Nosema locustae* (Protozoa, Microsporidia), a biological agent for locust and grasshopper control. *Agronomy* **2021**, *11*, 711. [CrossRef]
3. Wakil, W.; Ghazanfar, M.U.; Usman, M.; Hunter, D.; Shi, W. Fungal-Based Biopesticide Formulations to Control Nymphs and Adults of the Desert Locust, *Schistocerca gregaria* Forskål (Orthoptera: Acrididae): A Laboratory and Field Cage Study. *Agronomy* **2022**, *12*, 1160. [CrossRef]
4. Zhang, L.; Lecoq, M.; Latchininsky, A.; Hunter, D. Locust and grasshopper management. *Annu. Rev. Entomol.* **2019**, *64*, 15–34. [CrossRef]
5. Sánchez-Bayo, F.; Wyckhuys, K.A.G. Worldwide decline of the entomofauna: A review of its drivers. *Biol. Conserv.* **2019**, *232*, 8–27. [CrossRef]
6. Lomer, C.J.; Bateman, R.P.; Johnson, D.L.; Langewald, J.; Thomas, M. Biological control of locusts and grasshoppers. *Annu. Rev. Entomol.* **2001**, *46*, 667–701. [CrossRef] [PubMed]
7. Zhang, L.; Hunter, D.M. Laboratory and field trials of Green Guard® Metarhizium anisopliae var. acridum (Deuteromycotina: Hyphomycetes) against the oriental migratory locust (*Locusta migratoria manilensis*) (Orthoptera: Acrididae) in China. *J. Orthoptera Res.* **2005**, *14*, 27–30. [CrossRef]
8. Hunter, D.M. Credibility of an IPM approach for locust and grasshopper control: The Australian example. *J. Orthoptera Res.* **2010**, *19*, 133–137. [CrossRef]
9. FAO. *Desert Locust Bulletin, Nos. 484–495*; Food and Agriculture Organization of the United Nations: Rome, Italy, 2019. Available online: https://www.fao.org/ag/locusts/en/archives/archive/2521/2570/index.html (accessed on 12 July 2022).
10. FAO. *Desert Locust Bulletin, Nos. 496–507*; Food and Agriculture Organization of the United Nations: Rome, Italy, 2020. Available online: https://www.fao.org/ag/locusts/en/archives/archive/2521/2618/index.html (accessed on 12 July 2022).
11. CABI. Green Muscle Providing Strength against Devastating Locusts in the Horn of Africa. 2020. Available online: https://www.cabi.org/news-article/green-muscle-providing-strength-against-devastating-locusts-in-the-horn-of-africa/ (accessed on 12 July 2022).
12. FAO. *Evaluation of Field Trials Data on the Efficacy and Selectivity of Insecticides on Locusts and Grasshoppers*; Report to FAO by the Locust Pesticide Referee Group, Eleventh Meeting (Virtual); Food and Agriculture Organization of the United Nations: Rome, Italy, 2021. Available online: https://www.fao.org/ag/locusts/common/ecg/2614/en/PRG11.pdf (accessed on 12 July 2022).

agronomy

Article

Adult Desert Locust Swarms, *Schistocerca gregaria*, Preferentially Roost in the Tallest Plants at Any Given Site in the Sahara Desert

Koutaro Ould Maeno [1,2,*], Sidi Ould Ely [2,3], Sid'Ahmed Ould Mohamed [2], Mohamed El Hacen Jaavar [2] and Mohamed Abdallahi Ould Babah Ebbe [2,4]

1 Japan International Research Center for Agricultural Sciences (JIRCAS), Livestock and Environment Division, Ohwashi 1-1, Tsukuba, Ibaraki 305-8686, Japan
2 Centre National de Lutte Antiacridienne (CNLA), Nouakchott 665, Mauritania; sidiouldely@yahoo.com (S.O.E.); sidmd2002@yahoo.fr (S.O.M.); mjaavar@yahoo.fr (M.E.H.J.); maouldbabah@yahoo.fr (M.A.O.B.E.)
3 National Centre of Agricultural Research and Development (CNRADA), Kaedi, Mauritania
4 Institute of Sahel, Bamako 1530, Mali
* Correspondence: kmaeno@affrc.go.jp; Tel.: +81-29-838-6622

Received: 23 October 2020; Accepted: 27 November 2020; Published: 7 December 2020

Abstract: The desert locust, *Schistocerca gregaria*, is a major migratory pest that causes substantial agricultural damage. Flying adult swarms disperse widely during the daytime, but they densely roost on plants at night. Swarm control operations are generally conducted during the daytime, but night-time control is a significant potential alternative. However, the night-roosting behavior of swarms is poorly understood. We determined night-roosting plant preferences of migrating sexually immature swarms of *S. gregaria* at four different sites in the Sahara Desert in Mauritania during winter. The night-roosting sites were divided into two types based on presence or absence of large trees. Swarms tended to roost on the largest trees and bushes at a given site. Swarms used medium-sized plants when large trees were not locally available, but the same medium-sized plant species were hardly used when large trees were available. Plant choice influenced roosting group size—large locust groups roosted on larger plants. Night-roosting locusts rarely fled from approaching observers. These results suggest that swarms of *S. gregaria* exhibit plasticity in their utilization patterns of night-roosting plants depending on the plant community encountered and they selectively use larger plants. We propose that this predictable plant-size dependent night-roosting can be used to particularly ease locust swarm control and to generally adopt anti-locust night control strategy.

Keywords: aggregation; density-dependent phase polyphenism; migration; night-roosting site choice; *Schistocerca gregaria*

1. Introduction

The desert locust, *Schistocerca gregaria* (Forskål, 1775), is one of the most destructive migratory pests in the world [1–3]. Locust outbreaks are irregular and can cause serious agricultural damage over a wide range across West Africa, the Middle East, and southwest Asia [4–8]. Understanding the biology and ecology of wild locusts can help improve locust management [9,10].

Locusts are polyphenic grasshopper species that show density-dependent phase polyphenism in behavioral, morphological, and physiological characteristics [11–13]. Solitarious locusts occurring at low densities are usually dispersed across the landscape, while gregarious locusts at high densities aggregate together and migrate over long distances as a group [14,15]. Nymphal and adult groups are traditionally called "bands" and "swarms", respectively. Swarm sizes vary greatly and sometimes cover several hundred square kilometers [7,15]. Preventive control has been promoted, where the nymphal stages are treated before they reach adulthood and can fly as swarms to agricultural areas, because control of highly mobile, gigantic swarms is costly and ineffective [1,5,9,10,16]. Although desert locust management systems have been developed by the Food and Agriculture Organization of the United Nation (FAO), locust affected countries, and researchers by using chemical pesticides and biopesticides, swarms often occur at remote and inaccessible habitats due to geography and insecurity issues and then invade cultivated areas near human activity [17]. This may be one of the reasons why we still face desert locust problems in 2020 [18,19]. Therefore, improving control techniques and methods against swarms is important.

Surveys and control operations are usually conducted during the daytime and consider environmental factors such as wind, temperature, plant community, time of day, and locust behavior [20]. Swarms show various diel-cycling behavioral activities such as migratory flying, settling on the ground for basking, feeding, and resting during the daytime, and roosting on plants at night [15]. Aerial spraying using aircrafts is one of the few control techniques against flying swarms [21–23], but the flying locusts become hazardous to low-flying aircraft. When swarms settle on the ground, ground control measures can be used, which involve less pesticides than control measures against actively flying swarms [23]. However, the settling period is relatively short and irregular during the daytime [20,22]. Therefore, the efficacy of night-time control as an additional management strategy should be explored. As a first step, knowledge of night-roosting behaviors of swarms is essential.

Swarms roost for the night in trees or bushes. Locust densities in roosting swarms are much higher than that in migrating swarms [15]—the ratio between the area taken up by a swarm in flight and that by a swarm when settled can range from 12:1 to 900:1 [22]. Therefore, night-roosting swarms may be a suitable target for control. However, little is known about the night-roosting plant preference of swarms at a fine scale [24]. Our previous studies observed that both solitary adults and small-grouped gregarious adults selectively roosted on large trees [25,26]. Habitats of *S. gregaria* are diverse in terms of topography and plant communities, including vast plains with a variety of soil types such as rocky dry soils, dunes, and playas [27–29]. Some areas lack large trees and only have small bushes [30,31]. Highly mobile swarms of *S. gregaria* could encounter a variety of plant communities during migration [6,21,32], thus it is likely that they sometimes cannot reach favorable plant communities until night. However, they can access favorable plants by flying within the local areas. Accordingly, we hypothesized that migratory swarms of *S. gregaria* exhibit plasticity in their night-roosting plant utilization patterns depending on local plant communities and preferentially use relatively large plants. Previous studies have suggested that night-roosting plant choice influenced locust group size—gregarious nymphs selectively roosted on the largest plants and formed large groups [33]. Accordingly, we also tested the hypothesis that night-roosting swarms of *S. gregaria* form larger groups on larger plants.

Mauritania is a major breeding and recession area for the desert locust [27,30,34,35]. In 2013, swarms invaded the Banc d'Arguin National Park in Northwestern Mauritania, where plants are abundant and natural conditions are preserved. This ecological Park is nationally strictly protected from any chemical pesticides or alternative pest control strategies, only mechanical methods are allowed. We used this opportunity to test our hypotheses. We investigated the night-roosting plant preference and group formation of migratory swarms of *S. gregaria* at a fine scale. This work will contribute to developing better locust swarm survey and control methods.

2. Materials and Methods

2.1. Study Area

Mauritania, in West Africa, is an important area for desert locust outbreaks [27,30,34]. Our study sites were located in the Banc d'Arguin National Park, midway between Nouakchott and Nouadhibou, in north-western Mauritania. The area is a vast plain with a variety of soil types including rocky dry soils, dunes, playas, and small hills along the coast of the Atlantic Ocean. The primary plant community of the survey sites were identified according to Duranton et al. [28]. Sparse, low-growing desert annuals (grasses, herbs, vines, etc.) grow between the bushes and trees. We conducted field surveys from December 22–25 and 29–31, during the winter of 2013, when we encountered migrating sexually immature swarms. At this time, rain was rare and desert annuals were still partially green but starting to dry. Sunrise and sunset occurred at about 07:30 and 18:30 h local time, respectively. The mean temperature during the observation period (21:00–07:00) was 13.6 °C (SE: 0.2 °C, range: 8.6–19.5 °C) and the mean humidity was 44.4% (SE: 0.6%, range: 24.1–71.4%) at 50 cm above the ground, measured using a thermo-hygro recorder (TR-72wf, ONDOTORI, Tokyo, Japan).

2.2. Study Species

We studied migratory swarms of sexually immature adult desert locusts, *Schistocerca gregaria* (Forskål, 1775). Classical morphometric characteristics (hind femur length/head width) of locusts [36] confirmed their gregarious-phase status. During our study, locusts moved and fed in a daily cyclical rhythm—during the colder night and dawn period they roosted in trees and large bushes. After dawn they warmed by solar basking, moved to the ground, and began alternatively feeding, migrating, and resting, often on relatively bare ground. Near dusk, they flew or climbed into trees and large bushes for nocturnal roosting. They mainly fed on low-growing annuals and used the bushes and trees as refuges. Sexually mature gregarious locusts become yellow in body coloration [37]. Our population were dark red in body coloration and no sexual behaviors were observed, indicating that our population were sexually immature.

2.3. Sampling Regime

We followed migratory swarms encountered within the survey site without disturbing them until they roosted at night (17:00–19:00). Field surveys were conducted at night (21:00–07:00). We established 25 × 2-m belt transects and recorded information on night-roosting plants (species and abundance) and locust group size within each transect. We recorded the aggregation level of locusts on each bush and tree based on visual estimation. At least 30 transects were surveyed during each sample period and transects were separated by at least 10 m.

2.4. Night-Roosting Plants

Plants were patchily distributed at the study site. In our previous studies we measured all the plants within transects to examine plant size and locust group levels [31,33], but this was time consuming. In the present study, we determined the maximum length, width, and height of individuals of the dominant plant species at four survey sites within a strait belt transect until obtaining enough sample size, according to Maeno et al. [31]. At least 35 individuals or, whenever possible, all plants were measured and all data were pooled and analyzed to calculate mean size. The volume (m^3) of each plant species was calculated as maximum length × width × height. The abundance of each plant species was calculated from the transects (50 m^2). Maeno and Ould Babah Ebbe [33] reported that maximum plant height rather than width and volume was the most important factor for night-roosting site choice by gregarious late instar nymphs. Accordingly, we categorized relative plant size based on mean plant height: small, <1.0 m; medium, 1.0–2.0 m; large, >2.0 m. At Site 1, some trees *Euphorbia balsamifera* and bushes *Nucularia perrini* had started to dry out. Therefore, we also recorded plant states (green vs. dry) based on leaves from the two plant species to test whether green plants attracted more locusts and formed larger groups than dry ones.

2.5. Locust Group Size

We estimated the number of adults roosting on each plant by directly counting them after 20:00 when locusts were inactive. Each plant was given a score based on the number of roosting locusts (i.e., group size): 0 (0 locusts), 1 (<10), 2 (10–100), 3 (100–1000), 4 (1000–10,000) and 5 (>10,000), following the estimation method described by Maeno et al. [38].

2.6. Defensive Response

To determine defensive behaviors of the swarms at night, their response to an approaching observer carrying a light was recorded according to modified methods by Maeno et al. [39]. We carried a low-intensity wide-angle headlight (Gentos, DPX-233H, Delta Peak) and walked within 1 m from roosting locusts, filming their behavior with a video camera (Panasonic, HC-V520, Tokyo, Japan). This was only done for locusts in groups of size 5 and was repeated for seven night-roosting groups (i.e., >70,000 individuals). This observation was conducted on 30 December 2013.

2.7. Statistical Analysis

The effect of plant species on the percentage of plants roosted by locusts were determined using a post hoc Fisher's exact test after Bonferroni correction. To analyze differences in plant size (maximum length, height, and volume) and abundance between different plant species, Tukey–Kramer HSD tests were conducted. Percentages of plants roosted by locusts were arc-sine transformed and analyzed using Tukey–Kramer HSD tests. Two-way analysis of variance (ANOVA) was used to analyze the effects of plant species and presence of trees on locust group size. Differences in plant sizes between green and dry plants were analyzed using *t*-tests. The proportions of different group sizes were analyzed using a χ^2-test. Statistical analyses were conducted using the software package R, version 4.0.1 [40] and JMP (SAS Institute, Cary, NC, USA).

3. Results

3.1. Plant Characteristics

In the present study, only Site 4 had three large tree species (Figure 1a,b, Table 1). Large trees were relatively scarce (Tables 1 and 2: Tukey–Kramer HSD test, $p < 0.05$). Smaller plants were more abundant than trees (Tukey–Kramer HSD test, $p < 0.05$).

3.2. Daily Cyclical Movement and Night-Roosting Plant Preference

During our study, swarms migrated, fed, settled and roosted on plants in a daily cyclical rhythm. During the colder night and dawn periods, they roosted in relatively large plants including trees and bushes (Figure 1c–e). After dawn they displayed solar basking on the plants or ground by orientating perpendicular to the sun's rays to maximize the body surface (Figure 1f,g). They began alternating between feeding and resting, often on relatively bare ground (Figure 1h). At mid-day they usually migrated downstream (Figure 1i) and sometimes settled on the ground (Figure 1j). When swarms reached the coast, they avoided the ocean and returned inland. As a result, migrating locust density apparently increased (Figure 1k). Near dusk, they flew or climbed into trees and larger bushes for nocturnal roosting. No locusts were observed on the bare ground at night during any of the 150 belt surveys conducted, suggesting that locusts avoid habitats that lack plants. Locusts roosting on large trees rarely flew away from approaching observers with lights (i.e., when we tested their defensive response) (Figure 1l).

Plant utilization patterns by swarms varied. Figure 2 summarizes the percentage of each plant species roosted by locusts at our four survey sites. Large trees, such as *Capparis decidua*, *Acacia tortilis* and *Maerua crassifolia*, were relatively scarce, but most of them were roosted by locusts (post hoc Fisher's exact test after Bonferroni correction, $p < 0.00091$). More than 50% of *Euphorbia balsamifera* trees (medium-sized), *Panicum turgidum* bushes (medium-sized), and *Salsola imbricata* bushes (small-sized) were also utilized by locusts. The other small-sized plants (*Stipagrostis plumosa*, *Crotalaria saharae* and *Hyoscyamus muticus*) were rarely used, except for *Nucularia perrini*.

Table 1. Abundance of 11 dominant plant species in four survey sites. Mean (±SE) number of plants per belt transect (50 m^2).

			ID	Site 1	Site 2	Site 3	Site 4
			Data Location	**2013/12/22 19°55′ N, 16°14′ W**	**2013/12/23 19°53′ N, 16°18′ W**	**2013/12/24 19°53′ N, 16°18′ W**	**2013/12/29 19°24′ N, 16°13′ W**
Plant Species	**Family**	**Types of Plant**	**Plant Form**				
Capparis decidua	Capparaceae	Perennial	Tree	-	-	-	0.2 ± 0.3 a
Acacia tortilis	Fabaceae	Perennial	Tree	-	-	-	0.2 ± 0.3 a
Maerua crassifolia	Capparaceae	Perennial	Tree	-	-	-	0.2 ± 0.3 a
Euphorbia balsamifera	Euphorbiaceae	Perennial	Tree	4.2 ± 0.9 b	-	-	1.2 ± 0.3 a
Calotropis procera	Poaceae	Perennial	Bush & tree	-	-	-	0.2 ± 0.3 a
Panicum turgidum	Apocynaceae	Perennial	Bush	-	2.0 ± 0.9 a	6.6 ± 0.8 b	-
Stipagrostis plumosa	Poaceae	Annual	Bush	-	-	-	6.8 ± 0.3 b
Salsola imbricata	Amaranthaceae	Annual	Bush	2.7 ± 0.9 ab	10.5 ± 0.9 b	3.7 ± 0.8 ab	-
Nucularia perrini	Amaranthaceae	Perennial	Bush	17.0 ± 0.9 c	10.8 ± 0.9 b	6.3 ± 0.8 b	0.3 ± 0.3 a
Crotalaria saharae	Fabaceae	Perennial	Bush	-	2.0 ± 0.9 a	1.2 ± 0.8 a	-
Hyoscyamus muticus	Solanaceae	Annual	Bush	0.7 ± 0.9 a	-	-	-
			No. of transects	30	30	30	60

Different letters after values indicate significant differences within each survey site (Tukey–Kramer HSD test, $p < 0.05$). "-" indicates the absence of the plant species at the study site.

Night-roosting plant choice varied depending on plant community structure, i.e., the presence of relatively large trees (Figure 3). Our survey sites could be roughly divided into either areas with trees (Site 4) or without trees (Sites 1–3). For example, at Site 1, *E. balsamifera* trees, which are medium-sized, were locally the largest plants and most of them were roosted by locusts, while the percentage of *E. balsamifera* trees roosted by locusts was lower at Site 4. A similar tendency was also observed for *N. perrini* between Site 2 and 4. The relatively small-sized plant, *C. saharae*, was rarely used by locusts even at sites without trees. We could not analyze *H. muticus* due to a small sample size.

Figure 1. Habitats used by sexually immature swarms of *Schistocerca gregaria* for night-roosting sites and various diel behaviors in a population in Northwestern Mauritania during winter. (**a**) Habitats without trees and (**b**) with trees. (**c**) At night, locusts roosted on large trees (*Capparis decidua*), (**d**) medium-sized trees (*Euphorbia balsamifera*) and (**e**) small bushes (*Salsola imbricata*). (**f**) In the morning, locusts moved to the top of night-roosting trees or (**g**) to the open bare ground for basking. (**h**) Swarms feeding on low vegetation, (**i**) flying swarms, (**j**) settled swarms on the bare ground, (**k**) flying swarms along a coast, and (**l**) an observer staying near night-roosting swarms on a tree at night.

Table 2. Mean (±SE) plant sizes of 11 dominant plant species in the four survey sites.

Plant Species	Relative Size	Maximum Width (m)	Maximum Height (m)	Volume (m^3)	*n* Plants Measured
Capparis decidua	Large	9.34 ± 0.23 h	3.71 ± 0.08 h	331.5 ± 13.2 d	18
Acacia tortilis	Large	7.08 ± 0.20 g	2.84 ± 0.07 g	117.5 ± 11.2 c	25
Maerua crassifolia	Large	5.03 ± 0.25 f	2.29 ± 0.08 f	55.7 ± 14.0 b	16
Euphorbia balsamifera	Medium	2.20 ± 0.11 e	1.26 ± 0.04 e	5.9 ± 6.0 a	87
Panicum turgidum	Medium	2.10 ± 0.12 de	1.18 ± 0.04 e	4.9 ± 6.7 a	70
Calotropis procera	Medium	0.71 ± 0.26 abc	1.03 ± 0.09 de	0.6 ± 15.0 ab	14
Stipagrostis plumosa	Small	1.53 ± 0.17 cd	0.74 ± 0.06 dc	1.5 ± 9.5 ab	35
Salsola imbricata	Small	1.23 ± 0.09 bc	0.54 ± 0.03 bc	0.8 ± 5.3 a	110
Nucularia perrini	Small	1.00 ± 0.08 abc	0.43 ± 0.03 ab	0.5 ± 4.5 a	157
Crotalaria saharae	Small	0.76 ± 0.12 ab	0.32 ± 0.04 a	0.2 ± 6.6 a	71
Hyoscyamus muticus	Small	0.41 ± 0.22 a	0.34± 0.08 ab	0.1 ± 12.5 ab	20

Different letters after values indicate significant differences within each survey site (Tukey–Kramer HSD test, $p < 0.05$). Relative plant size was based on mean plant height: small, <1.0 m; medium, 1.0–2.0 m; large, >2.0 m.

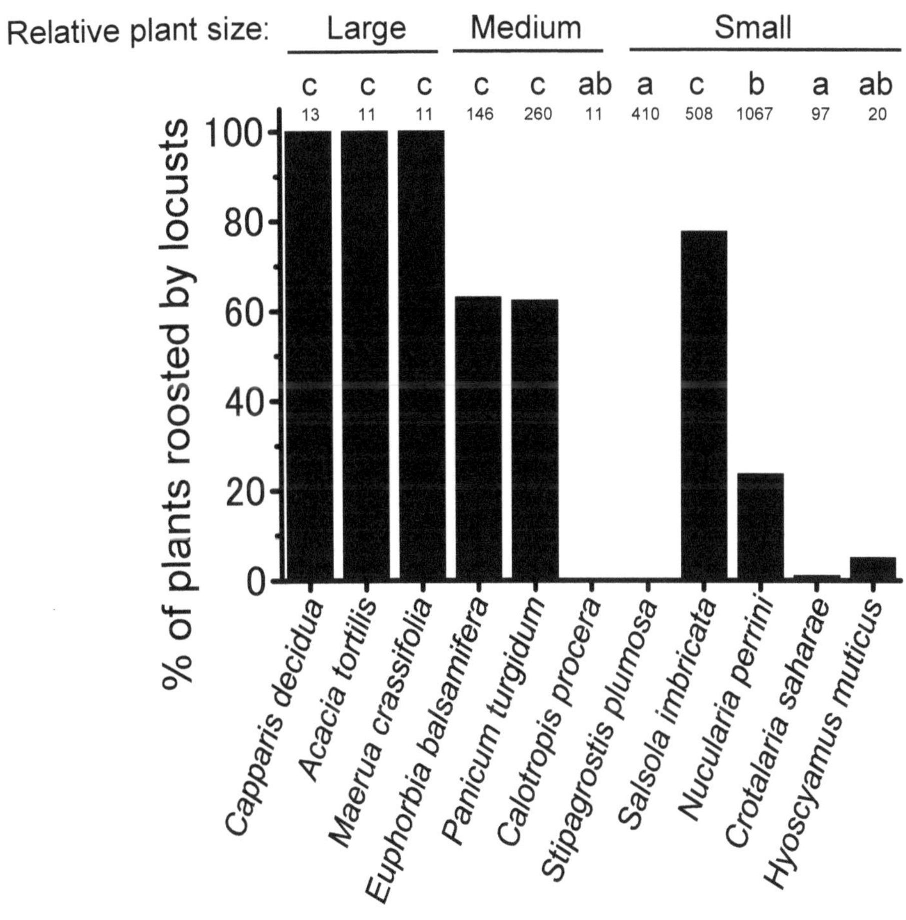

Figure 2. Percentage of plants from 11 plant species roosted by swarms of *Schistocerca gregaria* at four survey sites at night. Numbers above bars indicate sample sizes. Different letters above bars indicate significant differences at $p < 0.00091$ (post hoc Fisher's exact test after Bonferroni correction).

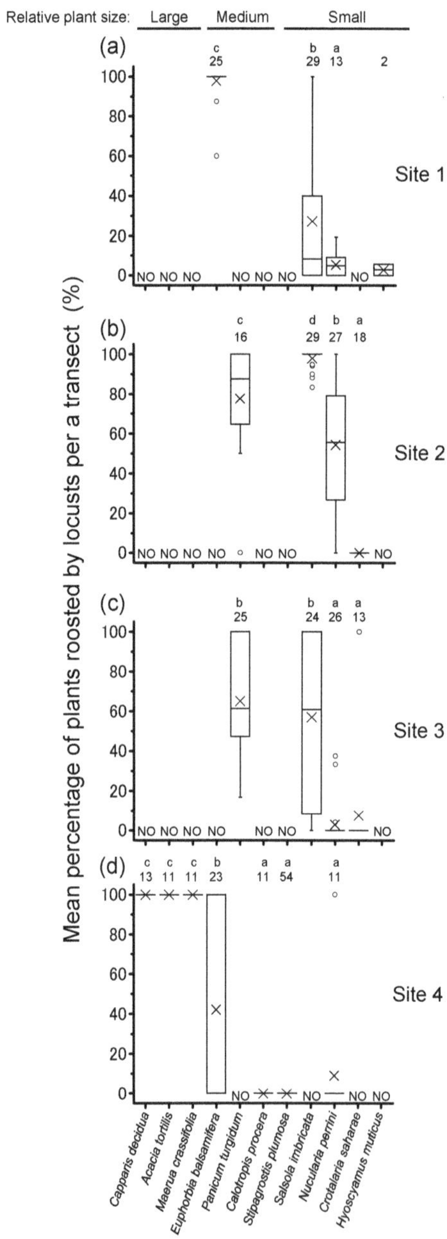

Figure 3. Mean percentage of plants from 11 plant species roosted by swarms of *Schistocerca gregaria* at night at four survey sites [Site 1: (**a**), Site 2: (**b**), Site 3: (**c**), and Site 4: (**d**)]. Each box plot displays the median value with the ends of the boxes representing the 25th and 75th percentiles and the ends of the lines representing the 10th and 90th percentiles. Crosses indicate mean values and open circles are outliers. Different letters above each bar indicate significant differences at $p < 0.05$ based on a Tukey-Kramer HSD test after arc-sine transformation. Numbers above bars indicate sample sizes (numbers of transects which certain plant species were recorded). "NO" in the figure indicates absence of the plant species at the study site.

3.3. *Group Size on Roosting Plants*

At night, locust groups were rarely observed on the ground or on grass; they mainly aggregated on relatively large plants. Figure 4 summarizes the different group sizes observed for 11 plant species across the four survey sites. Group sizes ranged from size 0 to 5 for each plant species. Larger groups (>10,000) tended to be formed on large trees, and medium-sized groups (3, 100–1000 locusts; 4, 1000–10,000 locusts) on medium-sized trees and bushes, while large-sized group was not formed on small bushes. The 11 plant species differed significantly in the proportion of group sizes that roosted on them (Figure 4: $\chi^2 = 3861.814$, d.f. $= 10$, $p < 0.01$).

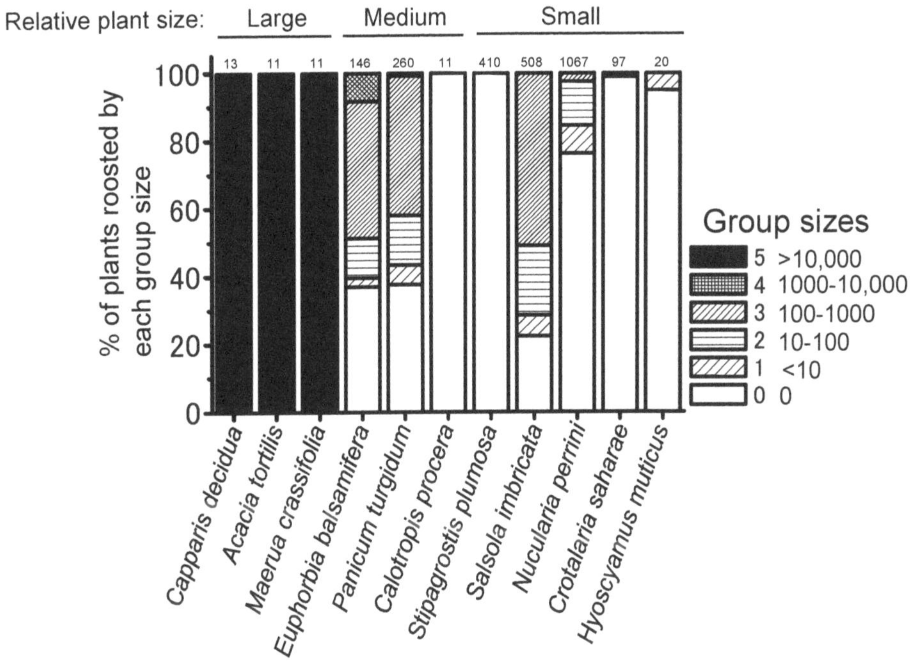

Figure 4. Percentage of plants roosted by each group size of *Schistocerca gregaria* swarms for the 11 dominant plant species at the four study sites. Numbers above bars indicate sample sizes (numbers of plants observed).

Plant community structure also influenced group size (Figure 5). The largest groups (>10,000 locusts per single plant) were observed at Site 4, where there were trees, but not at Sites 1, 2 and 3.

Group sizes on night-roosting plants varied depending on local plant community structure. For example, the size of *E. balsamifera* plants did not differ significantly between Site 1 and 4 (Figure 6a; Wilcoxon rank sum test, $z = -0.598$, $p > 0.05$), but significantly more large locust groups were observed on this species in Site 1 than in Site 4 (Figure 6b: $\chi^2 = 121.170$, d.f. $= 4$, $p < 0.001$).

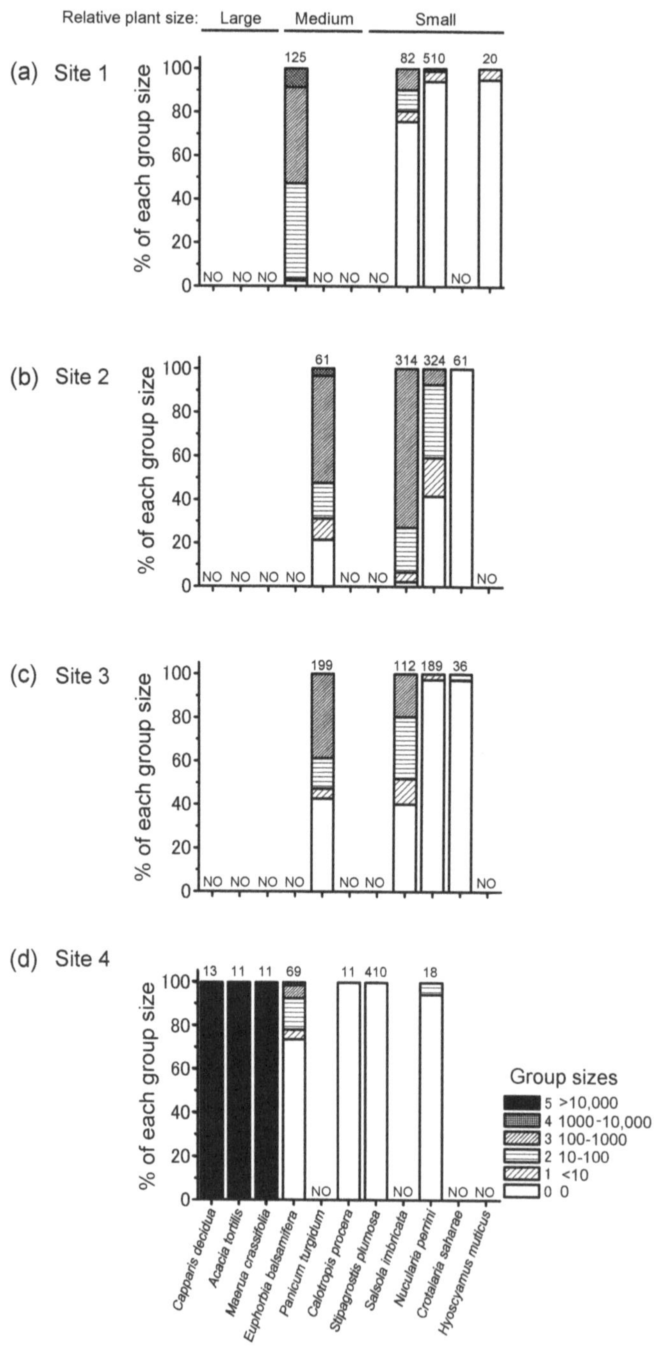

Figure 5. Percentage of plants roosted by each group size of *Schistocerca gregaria* swarms for the 11 dominant plant species at the four study sites [Site 1: (**a**), Site 2: (**b**), Site 3: (**c**), and Site 4: (**d**)]. Numbers above bars indicate sample sizes (numbers of plants observed). "NO" in the figures indicates absence of the plant species at the study site.

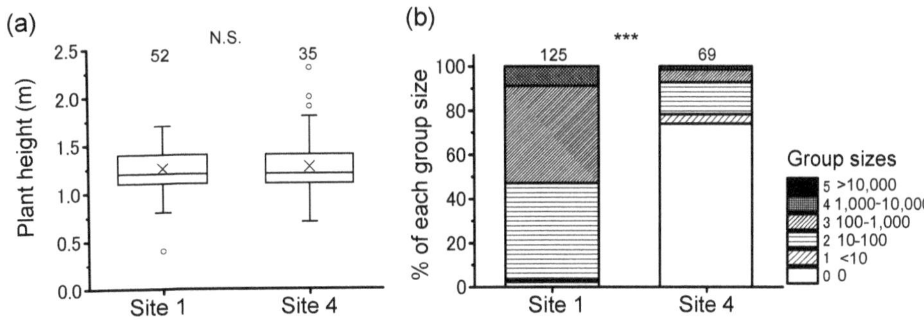

Figure 6. Height of *Euphorbia balsamifera* plants (**a**) and the percentage of *E. balsamifera* plants roosted by each group size of *Schistocerca gregaria* swarms (**b**) at a site without trees (Site 1) and one with trees (Site 4). Each box plot displays the median value with the ends of the boxes representing the 25th and 75th percentiles and the ends of the lines representing the 10th and 90th percentile values. Crosses indicate mean values and open circles are outliers. N.S. indicates no significant differences at $p > 0.05$ according to a Wilcoxon rank sum test. Numbers in figures indicate sample sizes. ***, significant differences between the two groups (χ^2-test, $p < 0.001$).

3.4. Plant Conditions: Green vs. Dry

At Site 1, some *E. balsamifera* and *N. perrini* plants were still green but others were dry. We examined whether greenness of plants influenced night-roosting plant choice of swarms. The utilization of green and dry plants by swarms was compared for each species. Plant size did not differ between green and dry plants for either plant species (Wilcoxon rank sum test, $p > 0.05$ for both). For *E. balsamifera*, greenness did not affect roosting. Conversely, for *N. perrini*, green plants were roosted by locusts significantly more than dry plants (Figure 7a; Wilcoxon rank sum test, $z = -3.573$, $p < 0.01$). Plant greenness did not influence group size composition on *E. balsamifera* (Figure 7: $\chi^2 = 0.8731$, d.f. = 4, $p > 0.05$). Conversely, for *N. perrini*, there were significantly more large locust groups on green plants than on dry plants (Figure 7: $\chi^2 = 24.5$, d.f. = 4, $p < 0.01$).

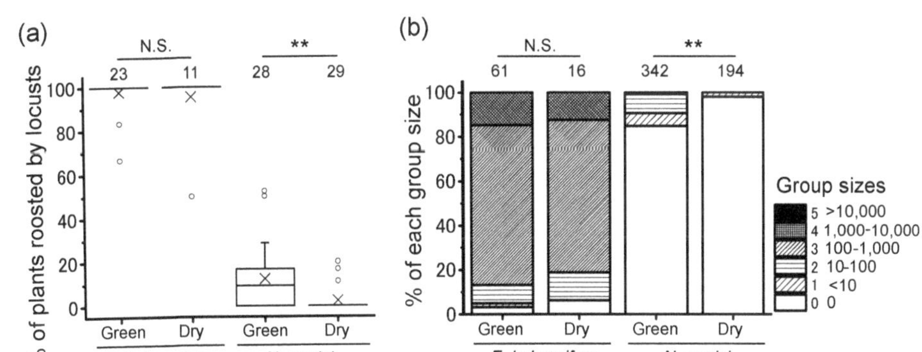

Figure 7. Percentage of green and dry plants roosted by *Schistocerca gregaria* swarms for *Euphorbia balsamifera* and *Nucularia perrini* at Site 1 (**a**) and the percentage of each locust group size on each plant (**b**). Each box plot displays the median value with the ends of the boxes representing the 25th and 75th percentiles and the ends of the lines representing the 10th and 90th percentiles. Crosses indicate mean values and open circles are outliers. Numbers in figures indicate sample sizes. (**a**) **, $p < 0.01$; N.S. indicates no significant differences between the two groups at $p > 0.05$ according to a Wilcoxon rank sum test. (**b**) **, $p < 0.01$; N.S. indicates no significant differences between the two groups at $p > 0.05$ according to a χ^2-test.

4. Discussion

The present results support our hypothesis that migratory swarms of *S. gregaria* exhibit plasticity regarding their utilization patterns of night-roosting plants depending on the plant community encountered and they selectively use larger plants. Furthermore, we found that their night-roosting plant choice influenced locust group size—larger groups were formed on larger trees, and night-roosting locusts rarely fled from approaching observers. This knowledge not only aids our understanding of night-roosting behaviors of swarms, but is also important for effective and easier locust control by night. In the following, we will discuss night-roosting plant preference and group formation by swarms of *S. gregaria* and how to apply this knowledge for improving locust control operations.

4.1. Night-Roosting Plant Preference

Many grasshopper species are known to roost on plants overnight [15]. In *S. gregaria*, swarms are known to roost on large trees or bushes at night [15,22,24]. The present study confirmed this trend for sexually immature swarms in Northwestern Mauritania during a winter season. Further, we found plasticity in their night-roosting plant utilization—they preferentially used relatively larger plants within local plant communities, but they used medium-sized trees and bushes as refuges when large trees were not available. For example, swarms were observed on most of the *E. balsamifera* trees, which are medium-sized, at Site 1, where this plant species was locally the largest. Conversely, the percentage of *E. balsamifera* trees roosted by locusts at Site 4 was less than half that at Site 1, as large trees were present in Site 4. Height of *E. balsamifera* trees was not significantly different between the two sites, thus swarms might use relative plant size to select night-roosting plants. Swarms also occupied medium-sized bushes at Site 2 and 3 where large trees were not available. To our best knowledge, we are the first to report that swarms of *S. gregaria* change their night-roosting plant preference depending on the local plant community, preferentially using the largest plants within the local area. We have previously observed that marching gregarious nymphs of *S. gregaria* also selectively roosted on the largest trees in the local plant community and plant height rather than width and volume was the primary cue for night-roosting plant choice [33]. Solitarious adults and gregarious adults at relatively low densities, and even solitarious nymphs, used relatively larger plants as their night-roosting sites [25,26,31]. These results suggest that *S. gregaria* tend to use relatively larger night-roosting plants irrespective of population density, developmental stage, and phase.

Plant height seems to be a useful criterion not only for ground-dwelling locust nymphs but also for flying adults, likely because they can visually recognize tall plants from all directions and they are comparable and available in all locust habitats. In fact, plant height is frequently used by insects as a cue for selecting aggregation sites, known as a "hill-topping" behavior [41]. At Site 3 where the ground was almost flat and there were no large trees, our car (TOYOTA, Land Cruiser, 5 m × 2 m × 2 m) was the largest object at the site. We observed that some flying locusts tried to land on our car around dusk (18:20–18:50). This observation also suggested that locusts preferred relatively large plants (or objects) within the local area and provided a hint for developing an artificial trapping system which can be used as an environmentally friendly control option. The simple criterion of plant height seems to be useful for migratory locusts as they encounter various habitats characterized by different vegetation cover, plant species, and plant sizes [24,27,28].

Plants are frequently used as a refuge by prey animals [42]. Greenness, i.e., the amount of leaves on refuge plants, may influence detectability by predators [42]. In the present survey sites where trees were not available, greenness of medium-sized plants such as *E. balsamifera* did not influence night-roosting site choice by swarms of *S. gregaria*, but more green plants of small-sized plants (*N. perrini*) were used than dry ones. Plant size of *N. perrini* did not differ depending on greenness, thus this may suggest that greenness is an important factor for small-sized plants. These plants are mainly used as a shelter rather than as food during a night. However, larger dry *E. balsamifera* plants attracted more locusts than green *N. perrini* plants, thus attraction of greenness could be masked by plant size. Although plant height was almost similar between the small bushes *S. plumosa* and *S. imbricata*, the former were rarely used by locusts. The stem and leaves of *S. plumosa* are thin and soft, thus they seem to have an unstable structure. This observation suggests that morphological characteristics of plants could be another factor for night-roosting plant choice.

Anti-predatory strategies are usually associated with microhabitat selection [42]. In the present survey area, ambient temperature can fall below 20 °C at night. Because escaping performance is temperature-dependent in grasshoppers [43,44], adult locusts cannot escape quickly from approaching predators below 20 °C [39]. This is most likely why locusts roost in large trees and bushes at night, away from nocturnal ground-foraging mammals. Morphological structures of medium-sized *P. turgidum* bushes and small-sized *N. perrini* bushes, which locusts used in the present study, were complex and apparently prevented access of relatively large predators. Some smaller plants such as *S. plumosa* and *C. saharae* had enough space for many locusts to roost, but the percentages used by swarm were small, probably because they did not provide shelter. Furthermore, aggregation with conspecifics could reduce the risk of predation through a dilution effect, or the "selfish herd effect" [45], as suggested for lubber grasshoppers, *Romelea guttata* [46], and Mormon crickets, *Anabrus simplex* [47]. Therefore, using refuges and grouping on the night-roosting plants might jointly serve as an anti-predator strategy. Interestingly, night-roosting locusts sometimes responded to an approaching observer by dropping and moving their hind legs, but they usually remained in place. It remains unclear how darkness (i.e., intensity of moon light) and low temperature influence escaping behaviors of night-roosting locusts, but this reduction of escaping performance may be useful, because sprayers can closely approach swarms.

4.2. *Local Locust Group Size on Night-Roosting Plants*

The present results supported our hypothesis that migratory swarms of *S. gregaria* would form larger groups on larger night-roosting plants. We found that larger locust groups (>10,000 individuals) roosted on large trees (>2 m tall), while medium- and smaller-sized locust groups (10–10,000 individuals) roosted on medium- and small-sized trees and bushes. This variation cannot be explained by plant size as a capacity of roosting space alone. Of course, large trees have more roosting space than small bushes, but the presence of large trees influenced the locust group size on other roosting plants. Locust group size on medium-sized plants significantly decreased when large trees were nearby, as observed for the medium-sized tree *E. balsamifera* at Site 1 and 4. This suggests that larger trees attracted more locusts from surrounding smaller plants. This spatial pattern was similar to that observed in gregarious late instar nymphs of *S. gregaria*—marching bands formed one large group on the largest plant with several smaller groups scattered at the local level [33]. This trend could be associated with plant distribution patterns and sizes in the semi-arid area. In the desert plant community, resource competition for water often results in heterogeneous patchy distributions [48], as was observed in the present survey area. The largest trees in a local area were conspicuous, because no other large plants existed near them. This may allow swarming locusts to see the roosting tree from afar without the need to evaluate other potential roosting plants.

In *S. gregaria*, it has been reported that heterogeneous plant distributions promote local crowding leading to gregarization, whereas a uniform distribution promotes scattering [29,30,49–53]. Cisse et al. [29] reported that vegetation cover and greenness influenced the threshold of gregarization in adults—low vegetation cover and dry vegetation led to a low density threshold of gregarization. Drying or dry vegetation forces locusts to meet on the few usable resources of a given area. Ould Babah and Sword [30] showed that specific plant communities promoted gregarization in nymphs. The present results suggest that not only plant distribution, communities, greenness, and cover, but also relative plant size, particularly the presence of large trees, and plant community structure played an important role in aggregation and local locust density.

Physical contact with conspecifics is the primary factor inducing gregarization in locusts [54–57], but flying locusts stay at a distance from conspecifics [15]. This raises the questions of when and where swarms aggregate and maintain gregariousness during migration. Microhabitat preferences are critical to recruit scattered conspecifics to a particular site [58], and individual attraction, group basking behavior and microhabitat preferences jointly lead conspecifics to concentrate within a limited area [49]. In *S. gregaria*, some scattered swarming members were also attracted to the large roosting trees, suggesting that scattered swarms can fuse to form an aggregation via night-roosting behaviors. Local crowding in the habitat has been observed on discrete resources such as host plants, basking, or shelter sites [49,59]. It is reasonable to assume that night-roosting plant preference of gregarious locusts may concentrate them in limited areas where they would receive physical contact while roosting and ground basking, leading to the maintenance of gregariousness.

4.3. Applications and Future Directions

The present study suggests that future locust control methods could also target night-roosting swarms for the following reasons: (1) night-roosting was a regular event and predictable, (2) locusts densely aggregated on large plants, which were patchily distributed, instead of on the ground or in the air, (3) roosting locusts were relatively immobile and escaping performance was low, thus sprayers could approach roosting locusts even using lights, (4) large trees or bushes are easily located in the habitat. Conversely, it may be difficult to conduct aerial spraying of night-roosting locusts due to the following: (1) a large proportion of the locusts were often inside trees and so would be sheltered if spraying was done from the air, (2) locusts roosted so densely that they would shield one another from aerial spraying, and (3) night-roosting locust swarms are not easily visible from aircraft in addition to the spatial dispersion of trees which make full coverage spraying not rational. To overcome these disadvantages, we suggest the following: (a) ground survey teams should follow migratory swarms until roosting time to determine roosting areas, (b) spraying should be done from the ground by car or by man with backsprayers, (c) sprayers should target night-roosting locusts and control them directly. (d) an eventual trapping system based on high similar object of the trees could be created to attract swarm before they roost, (e) although night-time control operations have been poorly documented, we are planning to test the efficacy of spraying from the ground at night. In 2013, the Mauritanian National Anti-locust centre (CNLA) have collected adults in bags by hands in the early morning within the natural park.

In addition, although the present study did not show any data about locust density of migratory swarms, their density was apparently high when they migrated along the coast. Migratory swarms avoid the polarization of oceans [60], thus members of swarms could be locally concentrated due to such a natural barrier. Therefore, remote sensing techniques and historical survey data could be used in the future to predict night-roosting areas [27,34,61–63]. As suggested by multi-agent models [19,64], the spatial conditions and accessibility of locusts should be considered to improve preventive control system.

In the future, if we can develop techniques to repel and attract swarms, a combination of these two techniques can concentrate swarms in a limited area. At the present, unfortunately, we have not developed such techniques, but integrating various information such as night-roosting site preference, natural barriers such as water, and meteorological and topographical information can help to inform which areas are suitable for extensive control operations. The present study identified factors potentially influencing locust group size on night-roosting plants such as site, plant type, plant conditions (green/dry), plant relative size, plant abundance and plant community. We should determine major parameters by using appropriate statistical models. It will be necessary to repeat this exercise with data from other countries and seasons in order to broaden our understanding.

Local people and field workers have empirically known that swarms roost on large trees at night, but this has yet to be sufficiently reported in the scientific literature. It is important to share field and local observations, not only to inspire scientists but also to improve control methods.

Author Contributions: Conceptualization, formal analysis, and investigation, K.O.M., S.O.E., S.O.M., M.E.H.J. and M.A.O.B.E. organized field surveys; original draft preparation, K.O.M. All authors have read and agreed to the published version of the manuscript.

Funding: This study was funded by the Ministry of Agriculture, Forestry and Fisheries (MAFF) of Japan through the Japan-CGIAR Fellowship Program implemented by the Japan International Research Center for Agricultural Sciences (JIRCAS).

Acknowledgments: We thank Ahmed Salem Bennahi, Cheikh Tidjani Babe Taleb and members of the Centre National de Lutte Antiacridienne for their assistance with the field survey. We are grateful to Editage (www.editage.com) for English language editing. We greatly appreciate the constructive and valuable comments on the manuscript from two anonymous reviewers.

Conflicts of Interest: The authors declare no conflict of interest. The funders had no role in the design of the study; in the collection, analyses, or interpretation of data; in the writing of the manuscript, or in the decision to publish the results.

References

1. Sword, G.A.; Lecoq, M.; Simpson, S.J. Phase polyphenism and preventative locust management. *J. Insect Physiol.* **2010**, *56*, 949–957. [CrossRef]
2. Le Gall, M.; Overson, R.; Cease, A. A global review on locusts (*Orthoptera: Acrididae*) and their interactions with livestock grazing practices. *Front. Ecol. Evol.* **2019**, *7*, 263. [CrossRef]
3. Brader, L.; DJibo, H.; Faya, F.G.; Ghaout, S.; Lazar, M.; Luzietoso, P.N.; Ould Babah, M.A. *Towards a More Effective Response to Desert Locusts and Their Impacts on Food Security, Livelihood and Poverty*; Multilateral Evaluation of the 2003-05 Desert Locust Campaign; FAO: Rome, Italy, 2006; p. 114.
4. Cressman, K. Desert Locust. In *Biological and Environmental Hazards, Risks, and Disasters*; Shroder, J.F., Sivanpillai, R., Eds.; Elsevier: Amsterdam, The Netherlands, 2016; ISBN 9780123948472.
5. Van Huis, A.; Cressman, K.; Magor, J.I. Preventing desert locust plagues: Optimizing management interventions. *Entomol. Exp. Appl.* **2017**, *122*, 191–214. [CrossRef]
6. Waloff, Z. The upsurges and recessions of the desert locust plague: An historical survey. *Anti Locust Mem.* **1966**, *8*, 1–111.
7. Steedman, A. *The Locust Handbook*; Natural Resources Institute: Chatham, UK, 1990.
8. Showler, A.T. Desert locust control: The effectiveness of proactive interventions and the goal of outbreak prevention. *Am. Entomol.* **2019**, *65*, 180–191. [CrossRef]
9. Lecoq, M. Desert locust management: From ecology to anthropology. *J. Orthoptera Res.* **2005**, *14*, 179–186. [CrossRef]
10. Zhang, L.; Lecoq, M.; Latchininsky, A.; Hunter, D. Locust and grasshopper management. *Annu. Rev. Entomol.* **2019**, *64*, 15–34. [CrossRef]
11. Cullen, D.A.; Cease, A.J.; Latchininsky, A.V.; Ayali, A.; Berry, K.; Buhl, J.; De Keyser, R.; Foquet, B.; Hadrich, J.C.; Matheson, T.; et al. From molecules to management: Mechanisms and consequences of locust phase polyphenism. *Adv. Insect Phys.* **2017**, *53*, 167–285. [CrossRef]
12. Pener, M.P.; Simpson, S.J. Locust phase polyphenism:an update. *Adv. Insect Phys.* **2009**, *36*, 1–286. [CrossRef]
13. Ayali, A. The puzzle of locust density-dependent phase polyphenism. *Curr. Opin. Insect Sci.* **2019**, *35*, 41–47. [CrossRef]
14. Rogers, S.M.; Cullen, D.A.; Anstey, M.L.; Burrows, M.; Despland, E.; Dodgson, T.; Matheson, T.; Ott, S.R.; Stettin, K.; Sword, G.A.; et al. Rapid behavioural gregarization in the desert locust, *Schistocerca gregaria* entails synchronous changes in both activity and attraction to conspecifics. *J. Insect Physiol.* **2014**, *65*, 9–26. [CrossRef]
15. Uvarov, B.P. *Grasshoppers and Locusts*; Centre for Overseas Pest Research: London, UK, 1977; Volume 2.
16. Magor, J.I.; Lecoq, M.; Hunter, D.M. Preventive control and Desert Locust plagues. *Crop Prot.* **2008**, *27*, 1527–1533. [CrossRef]
17. Krall, S.; Peveling, R.; Diallo, D.B. Economics of Desert Locust Control. In *New Strategies in Locust Control*; Birkhäuser Verlag: Basel, Switzerland, 1997.
18. Kimathi, E.; Tonnang, H.E.Z.; Subramanian, S.; Cressman, K.; Abdel-Rahman, E.M.; Tesfayohannes, M.; Niassy, S.; Torto, B.; Dubois, T.; Tanga, C.M.; et al. Prediction of breeding regions for the desert locust *Schistocerca gregaria* in East Africa. *Sci. Rep.* **2020**, *10*, 1–10. [CrossRef]
19. Gay, P.E.; Lecoq, M.; Piou, C. The limitations of locust preventive management faced with spatial uncertainty: Exploration with a multi-agent model. *Pest Manag. Sci.* **2020**, *76*, 1094–1102. [CrossRef]
20. Dobson, H.M. *Desert Locust Guidelines 4: Control*; FAO: Rome, Italy, 2001.

21. Rainey, R.C.; Sayer, H.J. Some recent developments in the use of aircraft against flying locust swarms. *Nature* **1953**, *172*, 224–228. [CrossRef]
22. Gunn, D.L.; Perry, F.C.; Seymour, W.G.; Telford, T.M.; Wright, E.N.; Yeo, D. Behaviour of the Desert Locust (*Schistocerca gregaria* Forskål) in Kenya in relation to aircraft spraying. *Anti Locust Bull.* **1948**, *3*, 1–70.
23. Joyce, R.J.V. The evolution of an aerial application system for the control of Desert Locusts. *Philos. Trans. R. Soc. Lond. B Biol. Sci.* **1979**, *287*, 305–314. [CrossRef]
24. Symmons, P.M.; Cressman, K. *Desert Locust Guidelines 1: Biology and Behavior*; FAO: Rome, Italy, 2001.
25. Maeno, K.O.; Ould Ely, S.; Nakamura, S.; Abdellaoui, K.; Cissé, S.; Jaavar, M.E.; Ould Mohamed, S.; Atheimine, M.; Ould Babah, M.A. Daily microhabitat shifting of solitarious-phase Desert locust adults: Implications for meaningful population monitoring. *Springerplus* **2016**, *5*, 107. [CrossRef]
26. Maeno, K.O.; Ould Ely, S.; Ould Mohamed, S.; Jaavar, M.E.H.; Nakamura, S.; Ould Babah Ebbe, M.A. Defence tactics cycle with diel microhabitat choice and body temperature in the desert locust, *Schistocerca gregaria*. *Ethology* **2019**, *125*, 250–261. [CrossRef]
27. Babah Ebbe, M.A. *Biogéographie du Criquet pèlerin en Mauritanie*; Hermann: Paris, France, 2010.
28. Duranton, J.-F.; Foucart, A.; Gay, P.-E. *Florule des Biotopes du Criquet Pèlerin en Afrique de l'Ouest et u Nord-Ouest à L'usage des Prospecteurs de la Lutte Antiacridienne*; FAO-CLCPRO: Montpellier, France, 2012.
29. Cisse, S.; Ghaout, S.; Mazih, A.; Ould Babah Ebbe, M.A.; Benahi, A.S.; Piou, C. Effect of vegetation on density thresholds of adult desert locust gregarization from survey data in Mauritania. *Entomol. Exp. Appl.* **2013**, *149*, 159–165. [CrossRef]
30. Ould Babah, M.A.; Sword, G.A. Linking locust gregarization to local resource distribution patterns across a large spatial scale. *Environ. Entomol.* **2004**, *33*, 1577–1583. [CrossRef]
31. Maeno, K.O.; Piou, C.; Ould Ely, S.; Ould Mohamed, S.; Jaavar, M.E.H.; Ould Babah, M.A.; Nakamura, S. Field observations of the sheltering behavior of the solitarious phase of the Desert locust, *Schistocerca gregaria*, with particular reference to antipredator strategies. *Jpn. Agric. Res. Q.* **2012**, *46*, 339–345. [CrossRef]
32. Rainey, R.C. Meteorology and the migration of desert locusts. *Anti Locust Mem.* **1963**, *7*, 1–115.
33. Maeno, K.O.; Ould Babah Ebbe, M.A. Aggregation site choice by gregarious nymphs of the desert locust, *Schistocerca gregaria*, in the Sahara Desert of Mauritania. *Insects* **2018**, *9*, 99. [CrossRef]
34. Piou, C.; Jaavar Bacar, M.E.H.; Ould Babah Ebbe, M.A.; Chihrane, J.; Ghaout, S.; Cisse, S.; Lecoq, M.; Ben Halima, T. Mapping the spatiotemporal distributions of the Desert Locust in Mauritania and Morocco to improve preventive management. *Basic Appl. Ecol.* **2017**, *25*, 37–47. [CrossRef]
35. Piou, C.; Gay, P.E.; Benahi, A.S.; Babah Ebbe, M.A.O.; Chihrane, J.; Ghaout, S.; Cisse, S.; Diakite, F.; Lazar, M.; Cressman, K.; et al. Soil moisture from remote sensing to forecast desert locust presence. *J. Appl. Ecol.* **2019**, *56*, 966–975. [CrossRef]
36. Uvarov, B.P. *Grasshoppers and Locusts*; Cambridge University Press: Cambridge, UK, 1966; Volume 1.
37. Norris, M.J. Sexual maturation in the desert locust with special reference to the effects of grouping. *Anti Locust Bull.* **1954**, *18*, 1–43.
38. Maeno, K.O.; Piou, C.; Ould Ely, S.; Ould Babah, M.A.; Pélissié, B.; Ould Mohamed, S.; Jaavar, M.E.H.; Etheimine, M.; Nakamura, S. Plant size-dependent escaping behavior of gregarious nymphs of the Desert locust, *Schistocerca gregaria*. *J. Insect Behav.* **2013**, *26*, 623–633. [CrossRef]
39. Maeno, K.O.; Ould Ely, S.; Ould Mohamed, S.; Jaavar, M.E.H.; Nakamura, S.; Ould Babah Ebbe, M.A. Behavioral plasticity in anti-predator defense in the desert locust. *J. Arid Environ.* **2018**, *158*, 47–50. [CrossRef]
40. R Development Core T. *R: A Language and Environment for Statistical Computing*; R Foundation for Statistical Computing: Vienna, Austria, 2020.
41. Alcock, J.; Dodson, G. The diverse mating systems of hilltopping insects. *Am. Entomol.* **2008**, *54*, 80–87. [CrossRef]
42. Cooper, W.E., Jr.; Blumstein, D.T. *Escaping from Predators: An Integrative View of Escape Decisions*; Cambridge University Press: London, UK, 2015.
43. Whitman, D.W. Function and evolution of thermoregulation in the desert grasshopper *Taeniopoda eques*. *J. Anim. Ecol.* **1988**, *57*, 369–383. [CrossRef]
44. Chappell, M.A.; Whitman, D.W. Grasshopper Thermoregulation. In *Biololgy Grasshoppers*; Chapman, R.F., Joern, A., Eds.; John Wiley: New York, NY, USA, 1990; pp. 143–172.
45. Hamilton, W.D. Geometry for the selfish herd. *J. Theor. Biol.* **1971**, *31*, 295–311. [CrossRef]

46. Stauffer, T.W.; Hegrenes, S.G.; Whitman, D.W. A laboratory study of oviposition site preferences in the lubber grasshopper, *Romalea guttata* (Houttuyn). *J. Orthoptera Res.* **1998**, *1998*, 217–221. [CrossRef]
47. Sword, G.A.; Lorch, P.D.; Gwynne, D.T. Migratory bands give crickets protection. *Nature* **2005**, *433*, 703. [CrossRef] [PubMed]
48. Fonteyn, P.J.; Mahall, B.E. An experimental analysis of structure in a desert plant community. *J. Ecol.* **1981**, *69*, 883–896. [CrossRef]
49. Kennedy, J.S. The behaviour of the desert locust *Schistocerca gregaria* (Forsk.)(*Orthoptera*) in an outbreak centre. *Trans. R. Entomol. Soc. Lond.* **1939**, *89*, 385–542. [CrossRef]
50. Ellis, P.E.; Ashall, C. Field studies on diurnal behaviour, movement and aggregation in the desert locust (*Schistocerca gregaria* Forskål). *Anti Locust Bull.* **1957**, *25*, 1–103.
51. Collett, M.; Despland, E.; Simpson, S.J.; Krakauer, D.C. Spatial scales of desert locust gregarization. *Proc. Natl. Acad. Sci. USA* **1998**, *95*, 13052–13055. [CrossRef]
52. Despland, E.; Collett, M.; Simpson, S.J. Small-scale processes in desert locust swarm formation: How vegetation patterns influence gregarization. *Oikos* **2000**, *88*, 652–662. [CrossRef]
53. Van der Werf, W.; Woldewahid, G.; van Huis, A.; Butrous, M.; Sykora, K. Plant communities can predict the distribution of solitarious desert locust, *Schistocerca gregaria. J. Appl. Ecol.* **2005**, *42*, 989–997. [CrossRef]
54. Simpson, S.J.; Despland, E.; Hägele, B.F.; Dodgson, T. Gregarious behavior in desert locusts is evoked by touching their back legs. *Proc. Natl. Acad. Sci. USA* **2001**, *98*, 3895–3897. [CrossRef]
55. Cullen, D.A.; Sword, G.A.; Dodgson, T.; Simpson, S.J. Behavioural phase change in the Australian plague locust, *Chortoicetes terminifera*, is triggered by tactile stimulation of the antennae. *J. Insect Physiol.* **2010**, *56*, 937–942. [CrossRef]
56. Maeno, K.; Tanaka, S.; Harano, K.-I. Tactile stimuli perceived by the antennae cause the isolated females to produce gregarious offspring in the desert locust, *Schistocerca gregaria. J. Insect Physiol.* **2011**, *57*, 74–82. [CrossRef]
57. Maeno, K.O.; Piou, C.; Ghaout, S. The desert locust, *Schistocerca gregaria*, plastically manipulates egg size by regulating both egg numbers and production rate according to population density. *J. Insect Physiol.* **2020**, *122*, 104020. [CrossRef]
58. Beauchamp, G. *Social Predation: How Group Living Benefits Predators and Prey*; Academic Press: Amsterdam, The Netherlands, 2014; ISBN 978-0-12-407228-2.
59. Roffey, J.; Popov, G. Environmental and behavioural processes in a desert locust outbreak. *Nature* **1968**, *219*, 446–450. [CrossRef]
60. Shashar, N.; Sabbah, S.; Aharoni, N. Migrating locusts can detect polarized reflections to avoid flying over the sea. *Biol. Lett.* **2005**, *1*, 472–475. [CrossRef]
61. Piou, C.; Lebourgeois, V.; Benahi, A.S.; Bonnal, V.; Jaavar, M.E.H.; Lecoq, M.; Vassal, J.-M. Coupling historical prospection data and a remotely-sensed vegetation index for the preventative control of Desert locusts. *Basic Appl. Ecol.* **2013**, *14*, 593–604. [CrossRef]
62. Latchininsky, A.; Piou, C.; Franc, A.; Soti, V. Applications of Remote Sensing to Locust Management. In *Land Surface Remote Sensing, Environment and Risks*; Baghdadi, N., Zribi, M., Eds.; ISTE Press: London, UK; Elsevier: Oxford, UK, 2016; pp. 263–293, ISBN 9780081012659.
63. Cressman, K. Role of remote sensing in desert locust early warning. *J. Appl. Remote Sens.* **2013**, *7*, 075098. [CrossRef]
64. Gay, P.E.; Lecoq, M.; Piou, C. Improving preventive locust management: Insights from a multi-agent model. *Pest Manag. Sci.* **2018**, *74*, 46–58. [CrossRef]

agronomy

MDPI

Article

Evidence for a Causal Relationship between the Solar Cycle and Locust Abundance

Robert A. Cheke [1,*], Stephen Young [1], Xia Wang [2], Jamie A. Tratalos [3], Sanyi Tang [2] and Keith Cressman [4]

1 Natural Resources Institute, University of Greenwich at Medway, Central Avenue, Chatham Maritime, Kent ME4 4TB, UK; Stephen.Young@greenwich.ac.uk
2 School of Mathematics and Information Science, Shaanxi Normal University, Xi'an 710119, China; xiawang@snnu.edu.cn (X.W.); sytang@snnu.edu.cn (S.T.)
3 Centre for Veterinary Epidemiology and Risk Analysis, School of Veterinary Medicine, University College, Belfield, Dublin 4, Ireland; jamie.tratalos@ucd.ie
4 Food and Agriculture Organization of the United Nations, Viale delle Terme di Caracalla, 00153 Rome, Italy; Keith.Cressman@fao.org
* Correspondence: r.a.cheke@greenwich.ac.uk

Abstract: Time series of abundance indices for Desert Locusts *Schistocerca gregaria* (Forskål 1775) and Oriental Migratory Locusts *Locusta migratoria manilensis* (Meyen 1835) were analysed independently and in relation to measures of solar activity and ocean oscillation systems. Data were compiled on the numbers of territories infested with swarms of the Desert Locust from 1860–2015 and an inferred series that compensated for poor reporting in the 1860 to 1925 period. In addition, data for 1930 to 2014, when reports are considered to have been consistently reliable were converted to numbers of 1° grid squares infested with swarms and separated according to four different geographical regions. Spectral analysis to test the hypothesis that there are cycles in the locust dynamics revealed periodicities of 7.5 and 13.5 years for the inferred series that were significant according to the Ornstein-Uhlenbeck state-space (OUSS) test. Similar periodicities were evident in the 1° grid square data and in each of the regions but even though these were significantly different from white noise, they were not significant according to the OUSS criterion. There were no significant peaks in the Oriental Migratory Locust results with the OUSS test, but the data were significantly different from white noise. To test hypotheses that long term trends in the locust dynamics are driven by solar activity and/or oceanic oscillation systems (the Southern Oscillation Index (SOI), the North Atlantic Oscillation Index (NAO) and the Indian Ocean Dipole (IOD)), the original locust data series and their Kalman-filtered low frequency (LF) components were tested for causality using both spectral coherence tests and convergent cross mapping. Statistically significant evidence was found that solar activity measured by numbers of sunspot groups drive the dynamics, especially the LF components, of both species. In addition, causal links were inferred between both the SOI and NAO data and Desert Locust dynamics. Spectral coherence was also found between sunspot groups and the NAO, the IOD and LF SOI data. The data were also analysed showing that the LF SOI had causal links with the LF inferred Desert Locust series. In addition, the LF NAO was causally linked to the LF 1° grid square data, with the NAO for December-March being most influential. The results suggest that solar activity plays a role in driving locust abundance, but that the mechanisms by which this happens, and whether they are mediated by fluctuations in oceanic systems, is unclear. Furthermore, they offer hope that information on these phenomena might enable a better early warning forecasting of Desert Locust upsurges.

Keywords: Desert Locust *Schistocerca gregaria* (Forskål 1775); Oriental Migratory Locust *Locusta migratoria manilensis* (Meyen 1835); spectral analysis; Kalman filter; spectral coherence; convergence cross mapping; sunspot groups; ENSO; SOI; IOD; NAO

Citation: Cheke, R.A.; Young, S.; Wang, X.; Tratalos, J.A.; Tang, S.; Cressman, K. Evidence for a Causal Relationship between the Solar Cycle and Locust Abundance. *Agronomy* **2021**, *11*, 69. https://doi.org/10.3390/agronomy11010069

Received: 30 November 2020
Accepted: 28 December 2020
Published: 31 December 2020

Publisher's Note: MDPI stays neutral with regard to jurisdictional claims in published maps and institutional affiliations.

1. Introduction

The Desert Locust *Schistocerca gregaria* (Forskål 1775) is a major agricultural pest in Africa, the Middle East and Asia. As with all locusts, the species is polyphenic, occurring in different phases. When low density solitarious phase insects become sufficiently numerous they begin to congregate and transform into gregarious phase insects, via a *transiens* intermediate phase. Triggers for such population increases are rainfall and consequent vegetation growth, with the cover and status of the vegetation affecting threshold densities that elicit gregarisation in both hoppers and adults [1,2]. Once in the gregarious phase, the nymphs occur in hopper bands and later, as adults, they may form swarms which migrate long distances by day. In contrast, solitarious insects do not swarm and migrate by night. For details of the biology and control of the Desert Locust see [3–8].

Desert Locusts require adequate rain at sites where their egg-pods have been laid to allow hatching and, if the rainfall has been sufficiently high, the hatching will coincide with flushes of green vegetation for them to feed on. Consequently, there has been extensive research on the causes of outbreaks such as relations between rainfall patterns and Desert Locust upsurges from which we know that changes in rainfall determine locust breeding (proximate causes) [9,10] and regional studies have sought patterns that could be used for forecasting [11–13]. In this study we examined whether drivers of these changes in rainfall—the effects of sunspot activity or oceanic cooling—are themselves correlated with locust abundance (ultimate causes). Although it has been shown that outbreaks of the Brown Locust *Locustana pardalina* (Walker 1870) are influenced by the Pacific *El Niño/La Niña* Southern Oscillation (ENSO) system, with generally more locusts associated with cold *La Niña* events [14], no published attempt to investigate if this is also the case for the Desert Locust has been made. Here we make good this omission by analysing long-term data sets on numbers of territories and on numbers of 1° grid squares in which swarms of the Desert Locust have been reported in relation to ENSO and North Atlantic Oscillation (NAO) data.

As the genesis of locust outbreaks may vary according to the timing and locations of upsurges, we further hypothesise that any links between such upsurges and oceanic systems will vary with season and geographic region and so we have also analysed the 1° grid square locust data, which are available as monthly series, according to quarterly periods for four different regions in addition to our analyses of the annual data.

Swinton [15] proposed that periodicities in the now extinct Rocky Mountain Locust *Melanoplus spretus* (Walsh 1866) were linked to variations in solar phenomena, an idea supported by others in relation to sunspot minima and the Desert Locust [16–20]. However, Ramchandra Rao [16] and Uvarov [17] questioned how the proposed link could be causal, except indirectly via changes in rainfall and temperature. It is such an indirect link that we re-examine here with more up-to-date data, given that there is now renewed interest in how solar activity might cause temperature changes on earth [21,22] and hence influence weather patterns. We also examine whether any such link with sunspot activity may be mediated by fluctuations in oceanic oscillation systems, using indices of oceanic systems such as the Southern Oscillation Index (SOI) and the NAO. When the NAO index is positive, conditions are colder and drier than average over the north-western Atlantic and Mediterranean regions and when it is negative conditions are warmer and wetter than average in northern Europe, the eastern United States of America and parts of Scandinavia. However, of particular relevance for this study, the NAO also has influences on the weather patterns remote from the Atlantic across parts of Asia and the Middle East [23]. Another index likely to influence weather patterns in the eastern part of the Desert Locust's geographical range is the Indian Ocean Dipole (IOD).

In addition, in a preliminary explorative study, we examined possible effects of sunspot activity on populations of the Oriental Migratory Locust. The biology of the Migratory Locust *Locusta migratoria* (Linnaeus 1758) is similar to that of the Desert Locust and other polyphenic locusts. The form that occurs in China and neighbouring countries, the Oriental Migratory Locust *L. m. manilensis* (Meyen 1835), was monitored by monks for many years and a very long time series of 1910 years for this insect exists [24]. Its dynamics are determined by precipitation and temperature, with there being more locusts under dry and cold conditions and when locust abundance was high in the year or years before [24], so this locust's dynamics may warrant further investigations of how they might be influenced by fluctuations in oceanic systems, a topic to be considered elsewhere.

2. Materials and Methods

2.1. Data Sets

Details of the data sets used are summarised in Table 1. Data on numbers of territories infested with swarms of Desert Locusts between 1860 and 2015 inclusive were collated from reports held at the Food and Agricultural Organisation of the United Nations (FAO) to provide a yearly time series (DL) (Figure 1). A second time series, known as the inferred time series of the numbers of territories infested with swarms (IT), was prepared according to the methods described by Waloff [25]. This includes information on probable or deduced occurrences of Desert Locust swarms for the period 1860 to 1925. The IT series is considered to be a more accurate reflection of events than the original data series, given the vagaries of reporting rates and the efficiencies of recording surveys which varied during the above lengthy period. The IT was added to the 1926 to 2015 data and the two data sets are depicted in Figure 1, which also distinguishes periods of recession, plague onset, plague peak and plague decline. Given the variation in sizes of the territories used to compile these data sets, data from the more reliable recording period from 1930 to 1987 were converted in a previous study to numbers of 1° grid squares infested with locust swarms and analysed in relation to rainfall at monthly intervals [10]. Here we also use these data, but we have updated them to provide an annual series for 1° grid squares during the 1930 to 2014 period. The 1° grid square data were also separated according to recording regions (Western, North Central, South Central and Eastern; see maps in [26] in which it is also shown that the numbers of reported swarms are cross-correlated).

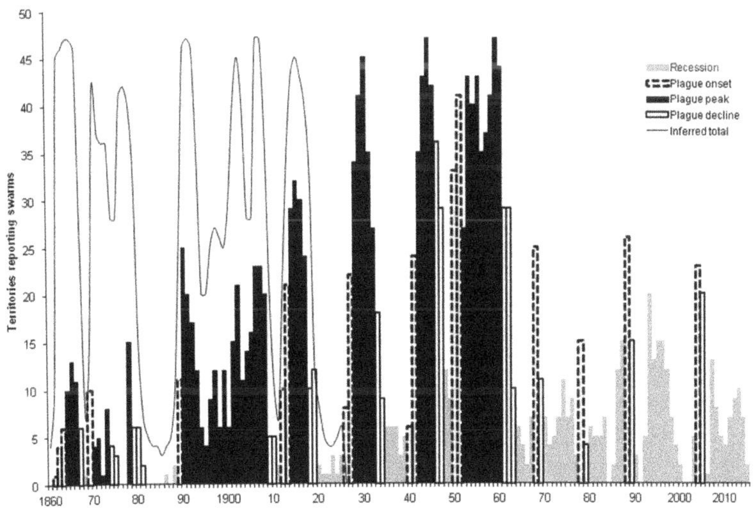

Figure 1. Numbers of territories as defined by Waloff (1976) [24] infested with swarms of the Desert Locust 1860 to 2015 inclusive. The solid line represents the inferred time series for 1860 to 1925 (see text), grey areas are recession periods, the dashed line depicts periods of plague onset, solid black lines indicate plagues and open bars show periods of plague declines.

Data on the Oriental Migratory Locust *L. m. manilensis* for the period 1700 to 1911 were analysed using data published by Tian et al. [24].

Although monthly data for the locust, sunspot and oceanic data were available, annual data were analysed as being most appropriate because the locusts need more than a month to complete their life cycles.

The sunspot data used were the sunspot group number series compiled by Hoyt and Schatten [27,28] updated by Vaquero et al. [29]. These data comprised daily counts which we converted to yearly mean data from 1610 to 2010. The sunspot group numbers refer to counts of active regions rather than the number of, individual, small spots.

Possible effects of the North Atlantic Oscillation (NAO) on locusts were examined. Positive NAO index values are associated with stronger-than-average westerly winds over the middle latitudes, more intense weather systems over the North Atlantic, and wetter/milder weather over western Europe (https://climatedataguide.ucar.edu/climate-data/hurrell-north-atlantic-oscillation-nao-index-pc-based). The NAO data were available as annual totals and for different periods: December to February inclusive (DJF); March to May (MAM); June to August (JJA); September to November (SON) and also for December to March (DJFM) as it is known that the NAO is most noticeable during the boreal winter (December to March inclusive) [30], when its effects may reach across Europe to Asia [23]. Another index likely to influence weather patterns in the eastern part of the Desert Locust's geographical range is the Indian Ocean Dipole (IOD). As precise measurements for the IOD are only available for short periods, relationships between it and locust abundance were investigated using a proxy series, the mean unfiltered dipole mode index, which is based on measurements taken from marine corals [31].

Table 1. Data sets used in the analyses.

Data Set	Unit	Interval	Start Date	End Date	Source
Oriental Migratory Locust series	Counties with locusts multiplied by outbreak intensity (graded 1–3) adjusted for recording effort (*Ladj* in [9])	Annual	1610	1911	Tian et al. (2011)
Total Desert Locust (DL)	Territories	Annual	1866	2015	FAO
Total Desert Locust Inferred Series (IT)	Territories	Annual	1866	2015	FAO
Total Desert Locust 1° grid square series (DLSA)	1° grid square	Annual	1930	2014	Converted by JAT from FAO data
Regional versions of DLSA for the Western (DLSAW), North Central (DLSANC), South Central (DLSASC) and Eastern Regions (DLSAE)					
Southern Oscillation Index (SOI)		Annual	1866	2015	http://www.esrl.noaa.gov/psd/gcos_wgsp/Timeseries/SOI/
North Atlantic Oscillation (NAO)		Annual, DJF, MAM, JJA, SON & DJFM series	1899	2016	https://climatedataguide.ucar.edu/climate-data/hurrell-north-atlantic-oscillation-nao-index-pc-based. (Principal Components based time series of the leading Empirical Orthogonal Function of Sea Level Pressure anomalies over the Atlantic sector, 20°–80° N, 90° W–40° E.)
Indian Ocean Dipole	Coral reconstruction of mean unfiltered dipole mode index	Annual	1846	1994	https://www.ncdc.noaa.gov/paleo-search/study/8607; [31]
Sunspot Group Numbers		Annual	1610	2010	http://www.sidc.be/silso/groupnumberv3

2.2. Statistical Analyses

Spectral analyses were performed on the original unadjusted series (DL), on the inferred series (IT) and on the 1° degree gridded data series (DLSA) using the *peacots* package (version 1.3) run in R version 3.6.1 (downloaded from the University of Bristol, UK, CRAN) [32] that calculates the periodogram using a Fast Fourier Transform (FFT) and tests for significance against the null hypothesis of the Ornstein-Uhlenbeck state-space (OUSS) model [33]. Given results of the spectral analyses, for additional analyses the data were separated into high (HF) and low frequency (LF) components by Kalman filtering [34]. Kalman filtering was achieved by using the integrated random walk method (function irwsm in the CAPTAIN toolbox programmed for MATLAB, available from: http://captaintoolbox.co.uk/Captain_Toolbox.html/Captain_Toolbox.html). The LF component is derived first and this is subtracted from the original series to provide the HF component. The LF component of variability in the time series can be thought of as representing decadal variability, with the HF component illustrating interannual variability.

To compare pairs of time series to seek causal relationships, two tests were used (a) spectral coherence with the function *myspec* in the R package *astsa* [35] and (b) convergent cross mapping (CCM; [36]). The spectral coherence test used calculates the squared spectral coherence. This is a measure that picks out those frequencies for which the strongest degree of covarying by two variables occurs and for which significance values can be assigned. Prior to testing for spectral coherence, the data were smoothed with a Daniell kernel, with parameter $m = 4$, which is a centred moving average that creates a smoothed value at time t by averaging all values between times $t - m$ and $t + m$ (inclusive). For example, the smoothing formula for a Daniell kernel with $m = 2$ is $x_t = (x_{t-2} + x_{t-1} + x_t + x_{t+1} + x_{t+2})/5$. The smoothed data were then tapered 10% (meaning application of a function, here a cosine taper, to split the series into short windows to minimise the effect of discontinuities between the beginning and end of the time series, thereby reducing the standard deviation of each spectral estimate to 10%; i.e., the centre of the data window is enhanced relative to the extremities) and made stationary by removing the mean and linear trends. Significant spectral coherence for ergodic, stationary, linear systems indicates causal links between pairs of variables [37].

An alternative test for causality is the Granger test [38] but, as Tsonis et al. [22] pointed out, the Granger test is inappropriate for nonlinear dynamic systems, so convergent cross mapping was carried out using the R function *CCM_boot* from the *multispatialCCM* package produced for Multispatial Convergent Cross Mapping [39]. CCM is suitable for non-linear coupled systems and tests the ability of lagged versions of one series to predict the dynamics of another series. It is also appropriate for cases in which the sign of correlations changes with time or is not significant for some sections of the series. The method's essentials are described by Chang et al. [40]. Based on applications of Takens's theorem [41], CCM first uses simplex projection [42] to test predictabilities by using data from numerous lagged data points of a process that is part of a bigger system to predict the dynamics of the process in question. The relationship is then tested for causality based on the assumption that interacting processes have information on each other if the information from the first process can be used to predict the second one. To achieve this, it is first necessary to estimate the optimal embedding dimension (E), i.e., the number of temporal steps needed for predictions, for each series. Next, confirmation that the system is non-linear is obtained from showing that the predictive power measured by Pearson's correlation coefficient (ρ) decreases with increasing prediction time steps. Finally, causality can be confirmed only if the predictive power (ρ) increases with the number of historical data points (L, the library length) needed to make predictions. To prevent the order of sampling affecting the results, a bootstrapping method is used with the number of iterations needed determined by when the mean and standard deviation of the estimates of (ρ) stabilise. For the analyses reported here 1500 iterations were run. For statistical significance testing, if (ρ) is larger at the longest L available than at the shortest L and is also greater than zero at the longest L, then causal forcing is concluded. Results were only accepted as significant if, in addition to significance for a potential driver causing a particular response, the impossible reverse relation of the response causing the potential driver was not significant. In the examples of Chang et al. [40], the data were first normalised to zero mean and unit variance, but here we test unfiltered and Kalman-filtered data (see above) using the method of Clark et al. [39]. In all of these tests, values of time delays (tau) were taken as 1, as we were interested in seeking relationships rather than developing forecasting models, in contrast to methods including lags used by Ye et al. [43] and Chen et al. [44].

3. Results

3.1. Spectral Analysis: Oriental Migratory Locusts

There were no significant peaks in the Oriental Migratory Locusts results with the OUSS test, but the data were significantly different from white noise (random signals with equivalent intensities at different frequencies), according to the less stringent Kolmogorov–Smirnov test.

3.2. Spectral Analysis: Numbers of Territories Infested with Desert Locust Swarms, 1866–2010

Figure 2 shows the results of spectral analyses of the inferred Desert Locust series (IT), the sunspot groups and the SOI for the 1866 to 2010 data (truncated from 2014 as the available sunspot groups data only extend to 2010). For the Desert Locusts (1866 to 2014) there was a significant peak at a periodicity of 13.5 years ($p = 0.02$; based on the OUSS test) and another at a periodicity of 7.5 years ($p = 0.02$). The sunspot groups show a strong peak corresponding to a periodicity of 11.19 years ($p < 0.0001$). The SOI has a series of peaks at various periodicities but even the highest at a periodicity of 3.5 years was not significant ($p = 0.62$). The strongest peak in the annual NAO data for the 1899–2016 period was at a periodicity of 7.9 years, but this was only just significant with the OUSS test ($p = 0.05$).

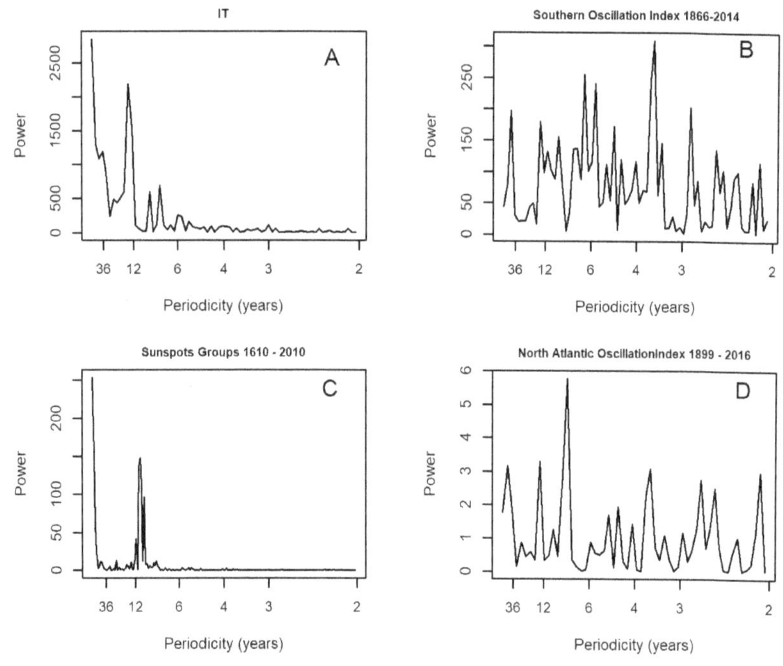

Figure 2. Periodograms of (**A**) the inferred Desert Locust series (IT), (**B**) the Southern Ocean oscillation Index (SOI), (**C**) the sunspot groups; and (**D**) the North Atlantic Oscillation Index (NAO).

3.3. Spectral Analysis: Numbers of 1° Grid Squares Infested with Desert Locust Swarms, 1930–2014

Figure 3 shows the results of spectral analyses of the 1° grid square locust data total and according to region, together with the sunspot groups data for the shorter period (1930–2010). All of the locust series, analysed up to 2014 inclusive, show peaks for periodicities between 7 and 14 years. The peaks for the whole area's data and for the Western and North Central regions corresponding to periodicities of 14 years are not significant with the OUSS test ($p = 0.61$) but are significantly different from white noise ($p = 0.02$). However, the peak corresponding to a periodicity of 7 years in the North Central Region's data is significant with the OUSS test ($p = 0.04$). The highest peak in the Eastern region data corresponds to a periodicity of 10.6 years, not significant with the OUSS test ($p = 0.2$) but significantly different from white noise ($p = 0.01$). This peak is similar to the sunspot groups' peak for this shorter series (1930–2010) which has a periodicity of 10.12 years, which is highly significant (OUSS test, $p < 0.0001$). Thus, although many of the peaks in the 1° grid square data are not significant with the OUSS test they do correspond in range with those for the longer IT series (see Section 3.2. above) which are significant at periodicities of about 7 and 14 years. No significant peaks were found by spectral analysis of the SOI data for 1930–2014.

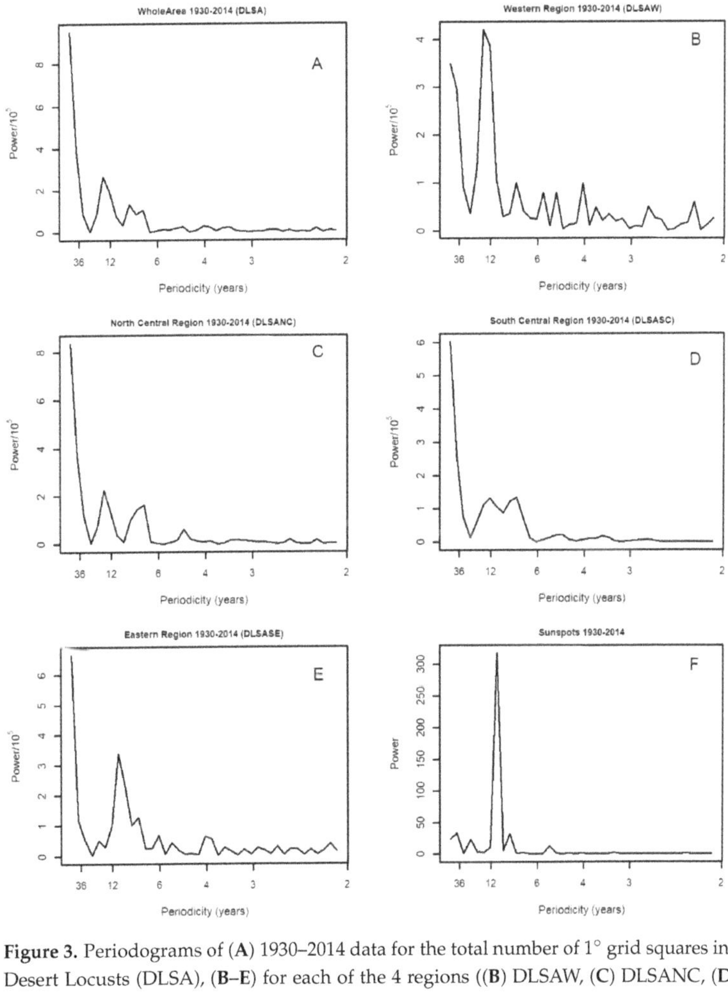

Figure 3. Periodograms of (**A**) 1930–2014 data for the total number of 1° grid squares infested with Desert Locusts (DLSA), (**B–E**) for each of the 4 regions ((**B**) DLSAW, (**C**) DLSANC, (**D**) DLSASC, (**E**) DLSAE) and (**F**) the sunspot groups.

3.4. *Kalman Filtering*

Figure 4 shows the inferred numbers of territories series (IT) split into high and low frequency (ITlf) components by Kalman filtering and Figure 5 shows the SOI and NAO and their low and high frequency Kalman-filtered series. Figure 6 shows the numbers of 1° grid squares infested with swarms of Desert Locusts from 1930 to 2014 and the results of Kalman filtering and Figure 7 show the data arranged according to region.

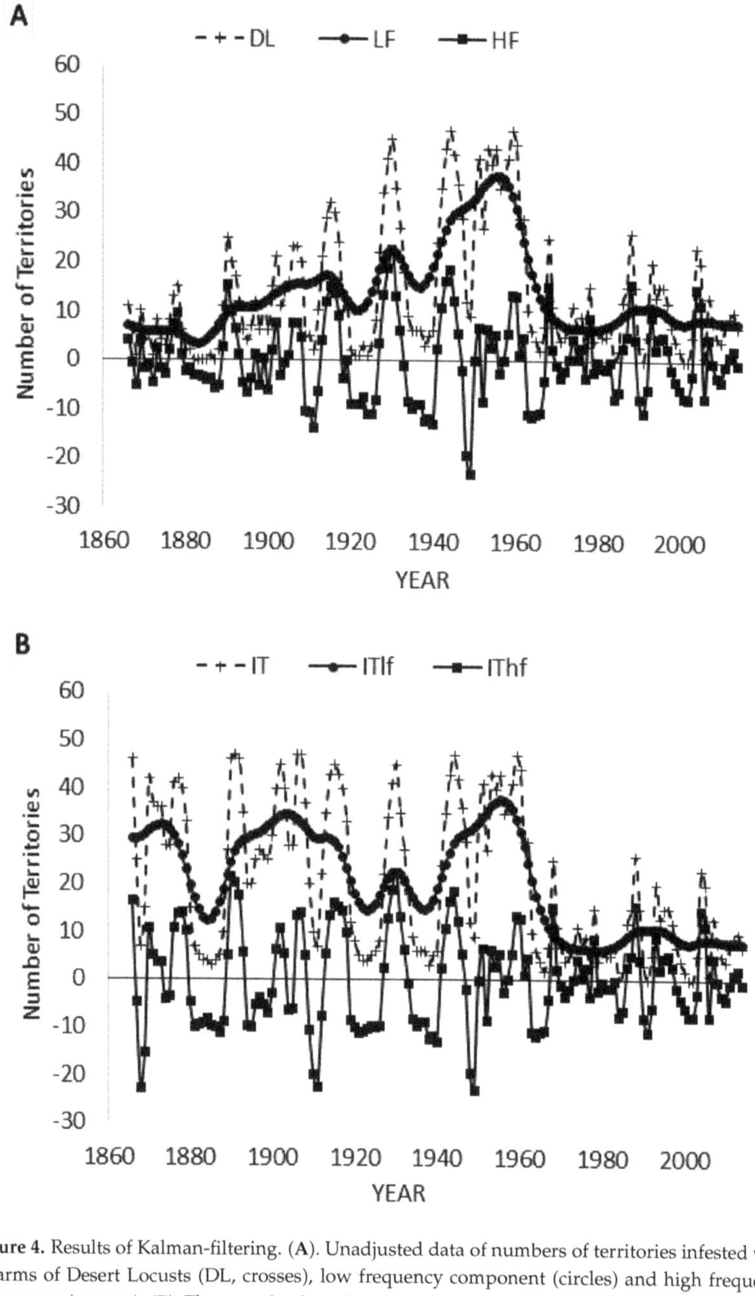

Figure 4. Results of Kalman-filtering. (**A**). Unadjusted data of numbers of territories infested with swarms of Desert Locusts (DL, crosses), low frequency component (circles) and high frequency component (squares). (**B**). The same for the inferred numbers of territories infested with swarms (IT).

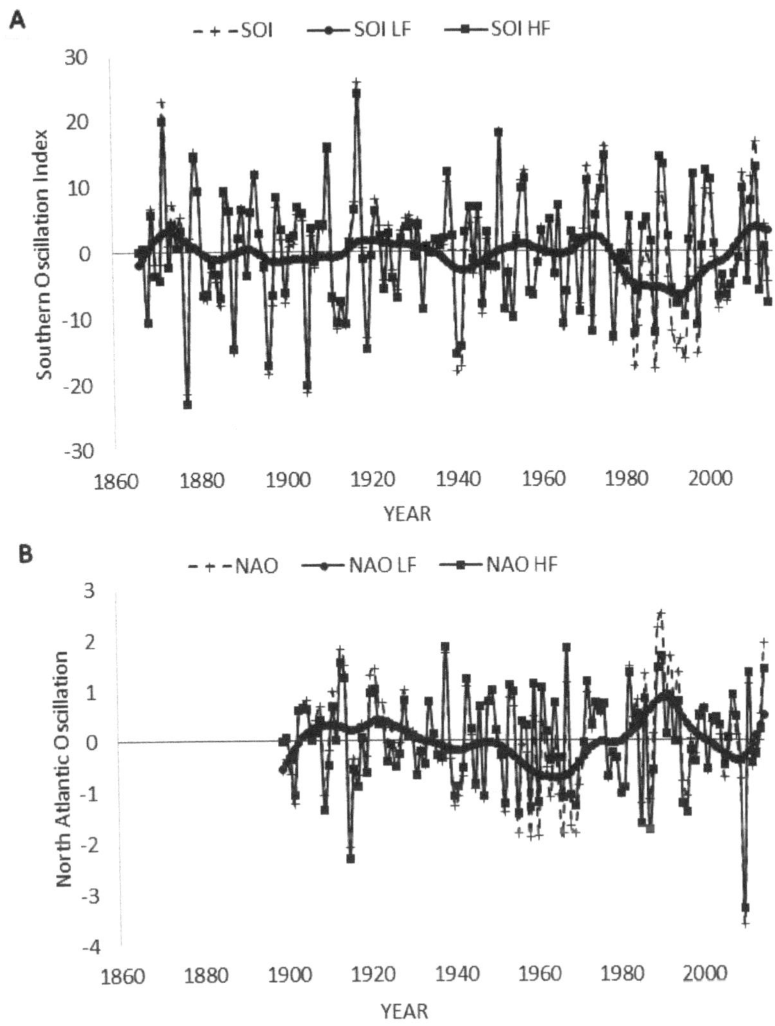

Figure 5. Results of Kalman-filtering. (**A**). Southern Oscillation Index unfiltered data 1866–2014 (SOI, crosses), low frequency component (circles) and high frequency component (squares). (**B**). The same for the North Atlantic Oscillation Index (NAO) 1899–2016.

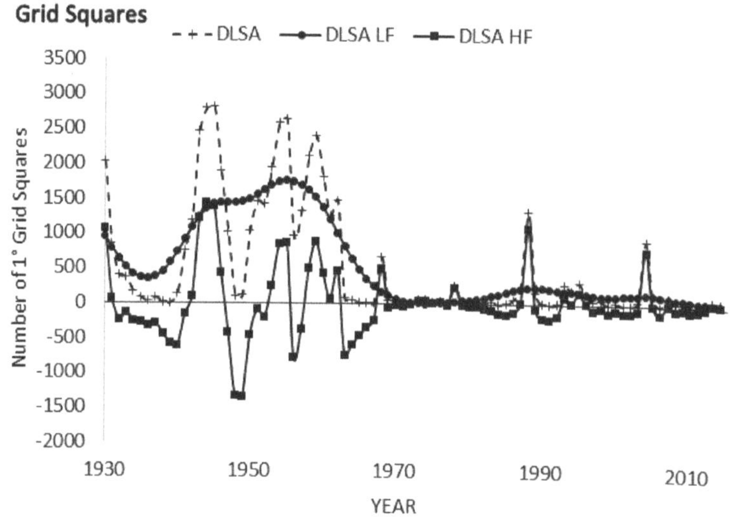

Figure 6. Numbers of 1° grid squares infested with swarms of Desert Locusts 1930–2014 (DLSA, crosses), together with results of Kalman filtering to provide low frequency components (circles) and high frequency components (squares).

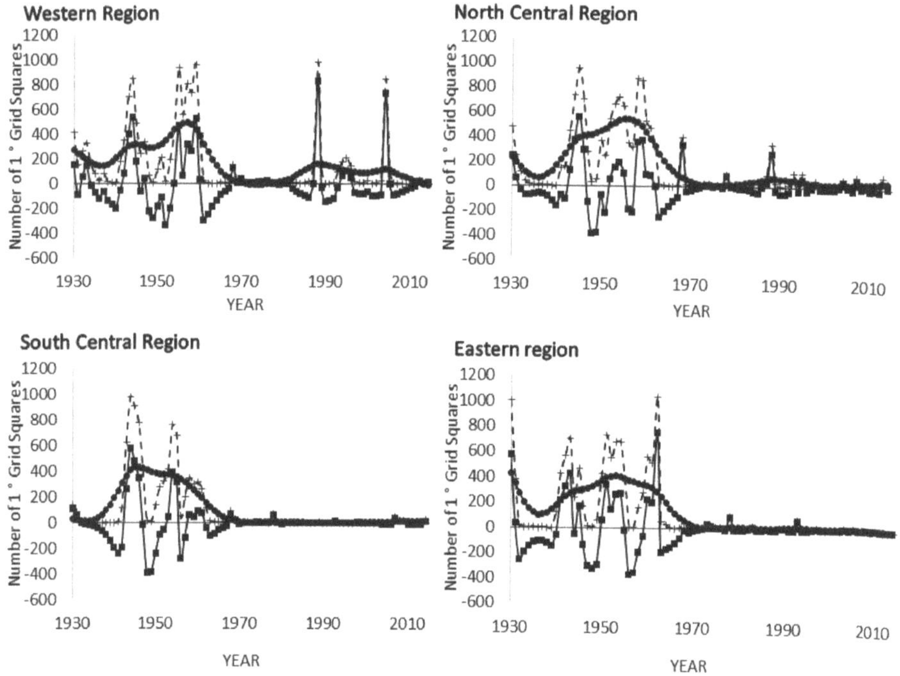

Figure 7. Numbers of 1° grid squares infested with swarms of Desert Locusts 1930–2014 according to region (crosses), together with results of Kalman filtering to provide low frequency components (circles) and high frequency components (squares).

3.5. Spectral Coherence: Sunspots and Oceanic Systems

Table 2 presents results of spectral coherence tests for the sunspot groups data and various oceanic indices. Although the coherence between the sunspots and the unfiltered SOI is not significant there is a significant result ($p < 0.01$) for the sunspots and the low frequency SOI data at cycles of 7.89 years (Figure 8). There are also significant coherences between the sunspot groups and both the NAO and the low frequency component of the NAO at cycles of 4.05 years and with the IOD at 4.5 years.

Table 2. Spectral coherence and convergent cross mapping CCM results between sunspot groups and oceanic indices. For spectral coherence significant coherent frequencies are denoted * for $p < 0.05$; ** for $p < 0.01$; *** for $p < 0.001$. Frequencies common to at least two pairs of time series in a row are highlighted in bold. NAO = North Atlantic Oscillation; NAOlf = low frequency component of NAO; IOD = Indian Ocean Dipole; SOI = Southern Oscillation Index; SOIlf = low frequency component of SOI.

	Time Series 2 (Response)				
	NAO	NAOlf	IOD	SOI	SOIlf
Time Series 1 (Driver)					
Spectral Coherence					
Sunspot Groups	**4.05** **	**4.05** ***, 5.79 *	4.50 *, 2.37 **	NS	7.89 **
Convergence Cross Mapping					
Sunspot Groups	NS	$p < 0.0001$	NS	NS	$p = 0.04$

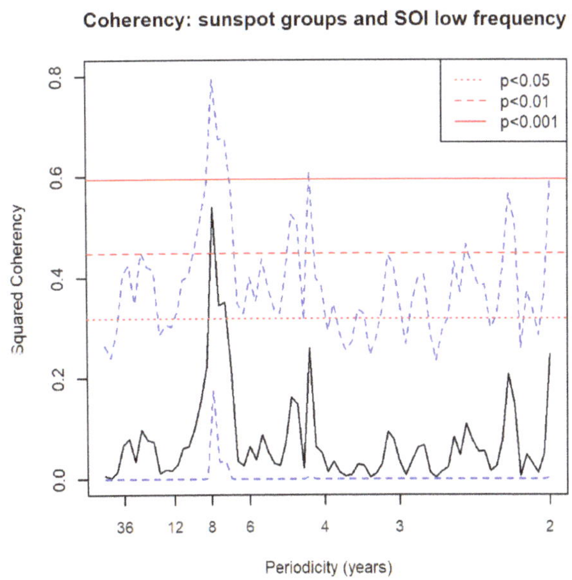

Figure 8. Spectral coherence between sunspot groups and the low frequency component of the SOI (SOIlf). The solid horizontal red line indicates significance at $p = 0.001$, dashed line $p = 0.05$, dotted line $p = 0.05$. The blue dashed lines represent 95% confidence intervals.

3.6. Spectral Coherence: Oriental Migratory Locusts

Significant spectral coherence was found between the 1610–1911 series of sunspot groups and the Oriental Migratory Locusts, with the main peak at 12.3 years ($p < 0.01$) and subsidiary peaks at 11.4, 5.6 and 4.2 years (all $p < 0.05$, Figure 9).

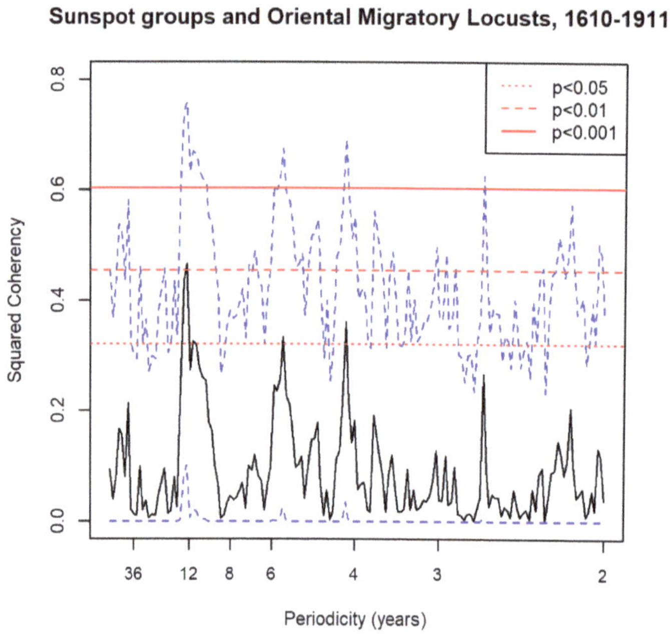

Figure 9. Spectral coherence between sunspot groups and Oriental Migratory Locusts, 1610–1911. The solid horizontal red line indicates significance at $p = 0.001$, dashed line $p = 0.05$, dotted line $p = 0.05$. The blue dotted lines represent 95% confidence intervals.

3.7. Spectral Coherence: Numbers of Territories Infested with Desert Locust Swarms, 1866–2014 (1866 to 2010 for Sunspot Groups Data)

Table 3 shows the results of spectral coherence tests for different potential drivers and each of the Desert Locust series. There were significant coherences between all potential drivers tested and both IT (inferred Desert Locust series) and ITlf (low frequency (lf) component of IT) at many different frequencies, but these significant frequencies differed amongst both drivers and locust series making interpretation difficult. Furthermore, these cycle lengths differed slightly as the Daniell smoothing factor varied (e.g., by increasing it from 4 to 7), although the same combinations of drivers and locust series were still found to be significant. Nevertheless, given the results of spectral analyses indicative of cycles of between 7 and 14 years, coherences within this range may be the most informative. Those of interest include the link between the sunspot groups and the ITlf at 7.14 years ($p < 0.05$), a frequency recurring for the effects of the low frequency component of the SOI (SOIlf) on both IT ($p < 0.05$) and ITlf ($p < 0.01$). The coherences between the SOI and both IT and ITlf at 8.8 years (both $p < 0.01$; Figure 10), a result repeated for the coherence between SOIlf and ITlf ($p < 0.001$), are also of interest. In the table, frequencies repeated along a row have been highlighted in bold; these include the coherences between SOIlf and both IT ($p < 0.05$) and ITlf ($p < 0.01$) at 10.7 years. The low frequency component of NAO (NAOlf) has highly significant ($p < 0.001$) coherences with IT and ITlf at 9.23 years, a frequency repeated for both IT and ITlf with all of the seasonal versions of NAOlf except that for June–August.

Table 3. Spectral coherence results between Desert Locust time series and oceanic indices and sunspots. All statistically significant values of coherent cycle lengths are presented. Significant results are denoted * for $p < 0.05$; ** for $p < 0.01$; *** for $p < 0.001$. IT = Inferred Desert Locust series (1866–2014 [or 2010 for link with sunspots]); ITlf = low frequency component of IT; DLSA = 1° gridded data (1930–2014); DLSAlf = low frequency component of DLSA, followed by the same for regional data (W = West; NC = North Central; SC = South Central; E = East). SOI = Southern Oscillation Index; NAO = North Atlantic Oscillation; IOD = Indian Ocean Dipole; NAOlf = low frequency component of NAO, followed by seasonal data. Frequencies common to at least two pairs of time series in a row are highlighted in bold.

	Time Series 2 (Response)							
	IT	**ITlf**	**DLSA**	**DLSAlf**	**DLSAWlf**	**DLSANClf**	**DLSASClf**	**DLSAElf**
Time Series 1 (Potential Driver)								
Sunspot groups	22.86 *, 2.32 **	7.14 *, 2.27 *	6.23 *	**6.75** ***, 4.26*	**6.75** *	**6.75** **, 4.26 *	**6.75** *	**6.75** ***
SOI	**8.82** **, 7.89 *, 4.29 *, 3.95 *, **2.68** ***, **2.54** **	**8.83** **, 8.33 *, 3.41 *, **2.67** **, **2.54** *	NS	NS	NS	NS	NS	NS
SOIlf	**10.71** *, **7.14** *, 5.17 *, 4.0 **, 3.75 *, **2.68** **, **2.54** **	15.0 *, **10.71** *, 8.82 ***, **7.14** **, 5.0*, 3.41 *, 3.09 * **2.68** **, **2.54** *	NS	**6.43** **, 4.74 ***, **2.90** *	12.86 *, 6.92 ***, 4.09 **, **3.21** **	**6.42** *, **5.0** ***, 3.10 *, **2.90** *	**5.0** ***, **3.21** *, **2.90** *	**6.43** **, **5.0** **, **2.90** *, 2.14 **
NAO	**2.10** **	3.24 *, **2.10** **	NS	**4.5** *	**4.5** *	**4.5** *, 4.09 *, 3.75 *, 3.33 *	NS	NS
NAOlf	**9.23** ***, 6.32 ***, **2.10** ***	60.0 *, 17.14 **, **9.23** ***, **6.0** **, 3.16 *, **2.10** ***	NS	**15.0** *, **8.18** ***, **4.5** ***	18.0 *, **8.18** *, **6.0** *, **4.74** **	**15.0** *, **8.18** ***, **4.5** *	**8.18** ***, 5.29 *, **4.74** *, **4.5** *, **2.19** *, 2.0 *	20 *, 7.5 *, 6.92 **, **4.5** ***, 3.12 *, **2.18** *
NAODJFlf (winter)	**9.23** *, **6.0** **	17.14 ***, **9.23** ***, **6.0** ***, 5.22 *	NS	15.0 *, 8.18 ***, 4.5 ***	**15.0** *, **8.18** *, 6.0 *, **4.74** **, 4.29 *	**15.0** *, **8.18** *** **4.74** ***	**8.18** ***, 5.29 *, **4.74** *	**15.0** *, 6.92 **, 4.5 ***, 3.21 *, 2.90 *
NAOMAMlf (spring)	**9.23** *, **6.0** *	**9.23** ***, **6.0** ***, 3.24 *	3.21*	**15.0** *, **10.0** *, **6.43** *, **4.74** ***	**18.0** *, **9.0** **, **6.43** ***, 4.5 *	**15.0** **, **10.0** *, **6.43** **, **4.74** ***	**15.0** **, **9.0** *, 7.5 *, **4.74** ***	**18.0** **, **6.43** ***, **4.74** ***, 2.81 *
NAOJJAlf (summer)	**20** *,**10** *, **6.67** *, **4.8** *	**20** **, 13.33 **, **10** **, 8 **, **6.67** **, **4.8** **, 3.53 ***	NS	**30.0** **, **15.0** **, **9.0** ***, 6.43 *, **5.29** ***, 4.28 *, **3.10** *	**30.0** **, 15.0 **, 11.25 **, **9.0***, 6.92 **, **5.29** *, **4.29** **, **3.21** *, **2.09** *	**15.0** **, **9.0** ***, **5.29** **	**30.0** **, **15.0** **, **9.0** ***, **5.29** ***, **3.21** **	**15.0** **, **9.0** **, 6.0 **, **5.29** **, **4.29** ***, **3.10** **, **2.09** **
NAOSONlf (autumn)	**9.23** *, 3.64 *, **2.40** **	15.0 *, **9.23** ***, 6.0 *, 4.8 *, **3.33** * **2.40** ***	**9.0** *, 5.62 *	**9.0** ***, **6.43** **, **5.0** **	12.86 *, **9.0** ***, 6.92 ***, **3.33** *	**9.0** ***, **6.43** **, **5.0** ***	11.25 ***, **9.0** ***, **5.0** ***	**9.0** *, **6.43** ***, **5.0** *
IOD (Mean DMI)	12.27 *, 3.21 *	3.37 **	NS	7.20 *	NS	NS	NS	NS

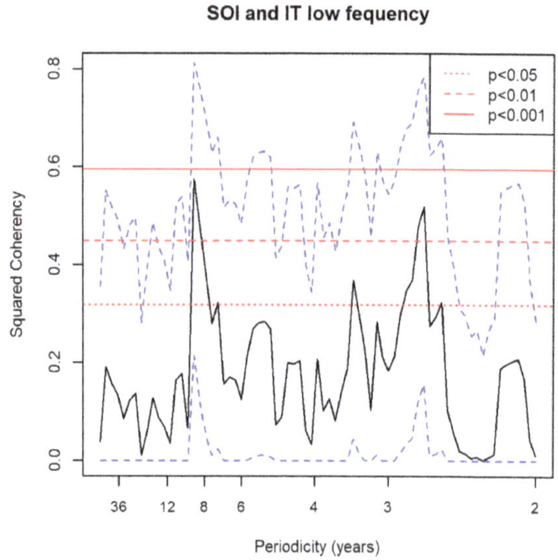

Figure 10. Squared coherency of SOI and ITlf. The solid horizontal red line indicates significance at $p = 0.001$, dashed line $p = 0.05$, dotted line $p = 0.05$. The blue dotted lines represent 95% confidence intervals.

3.8. Spectral Coherence: Numbers of 1° Grid Squares Infested with Desert Locust Swarms, 1930–2014

The sunspot groups have significant spectral coherence with the unfiltered total number of 1° grid squares infested with desert locusts (DLSA) series at a frequency of 6.23 years ($p < 0.05$) but a consistent coherence at 6.75 years for all of the low frequency gridded data series regardless of region ($p < 0.001$ for DLSAlf (Figure 11) and DLSAElf, $p < 0.01$ for DLSANClf and $p < 0.05$ for DLSAWlf and DLSASClf).

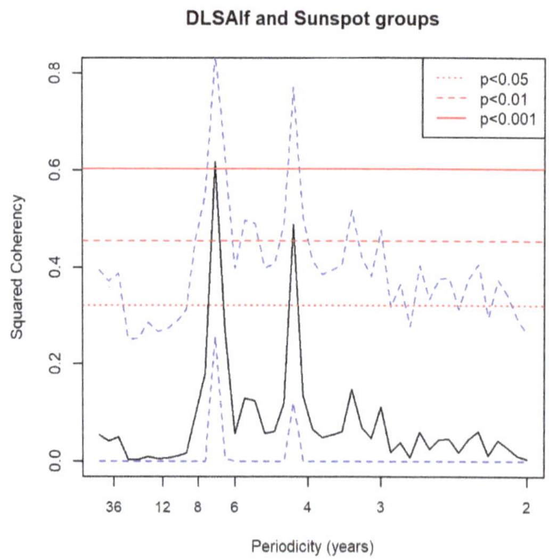

Figure 11. Squared coherency of the sunspot groups and DLSAlf (the low frequency component of DLSA, the 1° gridded data (1930–2014)). The solid horizontal red line indicates significance at $p = 0.001$, dashed line $p = 0.05$, dotted line $p = 0.05$. The blue dotted lines represent 95% confidence intervals.

No significant spectral coherence was detected between the SOI and the DLSA series (Table 3), but the SOIlf had coherences with all low frequency gridded data sets with a frequency of 6.4 years recurring in DLSAlf and the South Central and Eastern regions.

The unfiltered NAO data showed consistent coherence at 4.5 years with DLSAlf and the low frequency regional data sets for the Western and North Central regions (all $p < 0.05$). The NAOlf and the December to January version of it were consistently coherent with all low frequency versions of the locust gridded data at 8.18 years, except in the Eastern region. The link at 9 years between the September-November version of the NAOlf and all versions of the locust gridded data, which also tallied with the 9.2 years noted for the same potential driver with the IT and the ITlf, was also remarkable.

The IOD showed weak coherence with DLSAlf at 7.2 years ($p < 0.05$), but not with the regional data, but as this oceanic system is linearly correlated with the SOI ($p < 0.001$) it was not surprising that results were similar for both systems.

3.9. Convergent Cross Mapping: Sunspot Groups and Oceanic Systems

Although there was no evidence of a causal relationship between the sunspot groups and the ocean oscillation systems when unfiltered data were used, significant causal relationships were found between sunspot groups and the low frequency components of both the SOI ($p = 0.04$) and the NAO ($p < 0.0001$), which is consistent with the spectral coherence results (Table 2).

3.10. Convergent Cross Mapping: Oriental Migratory Locusts

Although there was no evidence of a causal relationship between the sunspot groups and the unfiltered Oriental Migratory Locust data, a significant effect of the sunspot groups causing its low frequency was detected ($p = 0.001$) but this result is invalid as the reverse, impossible, result (OML causing sunspots) was also significant ($p < 0.04$).

3.11. Convergent Cross Mapping (CCM): Desert Locusts

Remarkably, CCMs gave strong evidence for sunspot groups causing changes in Desert Locust abundance as measured in all of the locust series ($p < 0.0001$ in every case; Table 4; Figure 12), except for the unfiltered data of the inferred series (IT). No potential driver was found to be causing changes in the IT but, in addition to the sunspot groups result, there was also evidence of the ITlf being driven by the SOIlf ($p < 0.005$), and by the low frequency versions of the NAO for March to May ($p < 0.0001$), June to August ($p < 0.01$) and September to November ($p < 0.0001$).

The low frequency component of the Kalman-filtered numbers of 1° grid squares infested with Desert Locust swarms for the 1930 to 2014 period (DLSAlf) was also found to be causally linked to the low frequency component of the SOI (SOIlf, $p < 0.0001$). This relationship also held true when the SOIlf data were analysed at the regional level for the South Central ($p < 0.001$) and Eastern regions ($p = 0.002$), but not for the North Central and Western regions.

The low frequency component of the NAO for the 1930–2014 period was consistently shown by CCM to drive the low frequency components of the gridded data (DLSAlf, $p < 0.0001$; DLSAWlf, $p < 0.001$; DLSANClf, $p < 0.005$; DLSASClf, $p < 0.0001$; DLSAElf $p < 0.003$). These results were to a large degree matched when the seasonal versions of the NAOlf data were used, with strong evidence ($p < 0.01$) of a causal link between the December to March version and all of the low frequency gridded data sets, including the entire (whole year, all regions aggregated) series (DLSAlf) (Figure 13).

Table 4. Results of analyses using CCM tests of causality of various time series (drivers) causing the low frequency inferred Desert Locust series 1866–2014 (ITlf) and the gridded Desert Locust series (1930–2014). p values given if relationships statistically significant (NS = not significant). Note that the results for the unfiltered sunspot groups data (1866–2010) reveals significance for all relationships with the gridded Desert Locust data. DLSA = 1° gridded data (1930–2014); DLSAlf = low frequency component of DLSA, followed by the same for regional data (W = West; NC = North Central; SC = South Central; E = East). SOI = Southern Oscillation Index; NAO = North Atlantic Oscillation; NAOlf = low frequency component of NAO, followed by seasonal data.

	Time Series 2 (Response)						
	ITlf	**DLSA**	**DLSAlf**	**DLSAWlf**	**DLSANClf**	**DLSASClf**	**DLSAElf**
Time Series 1 (Driver)							
Sunspot Groups	<0.0001	<0.0001	<0.0001	<0.0001	<0.0001	<0.0001	<0.0001
SOIlf	<0.005	NS	<0.0001	NS	NS	0.001	0.002
NAOlf	NS	NS	<0.0001	0.001	<0.005	<0.0001	0.003
NAODJFlf	NS	NS	0.034	NS	NS	0.0007	NS
NAOMAMlf	<0.0001	NS	<0.0001	NS	0.009	<0.0001	0.0007
NAOJJAlf	0.01	NS	0.038	NS	0.021	<0.005	<0.0001
NAOSONlf	<0.0001	NS	0.048	0.05	<0.003	<0.003	NS
NAODJFMlf	NS	NS	0.0007	0.002	0.004	<0.0001	<0.003

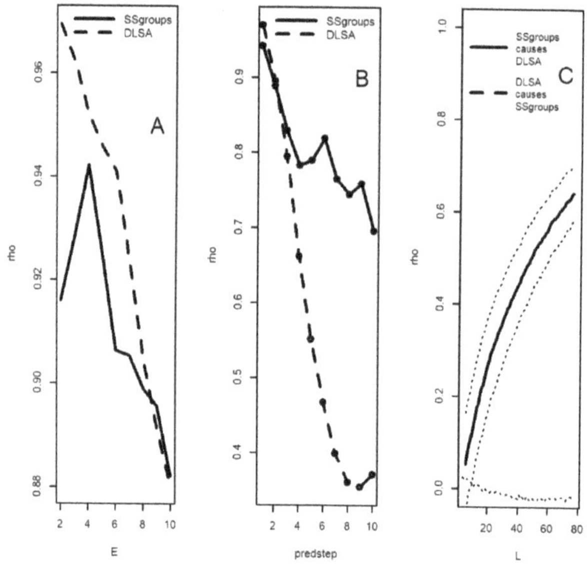

Figure 12. Convergent cross mapping results for sunspot groups and the 1° gridded Desert Locust series 1930–2014. (**A**). Showing use of maximum values of correlation coefficient rho (ρ) to estimate the embedding dimension (E) of each series; (**B**) showing decreasing rho with increasing number of prediction steps to confirm non-linearity and (**C**) showing increasing rho with increasing library length (L) for the hypothesis that sunspot groups cause DLSA ($p < 0.0001$) and for the converse (DLSA causes sunspot groups, $p = 0.69$).

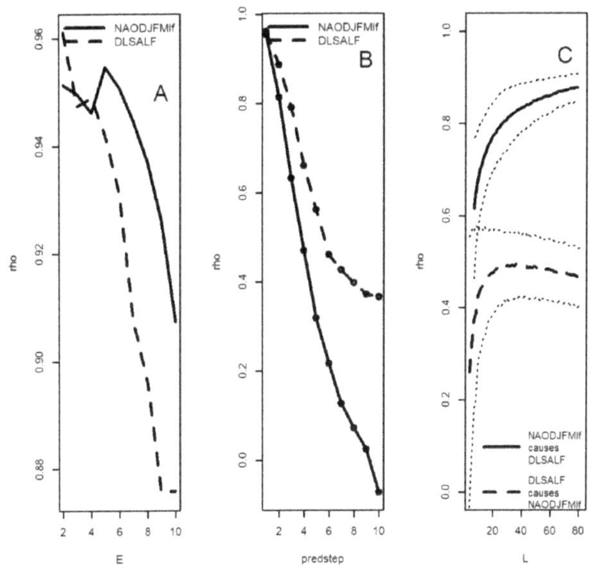

Figure 13. Convergent cross mapping results for the low frequency NAO series for December-March (NAODJFMlf) and the low frequency 1° gridded Desert Locust series 1930–2014 (DLSAlf). (**A**). Showing use of maximum values of correlation coefficient rho (ρ) to estimate the embedding dimension (E) of each series; (**B**) showing decreasing rho with increasing number of prediction steps to confirm non-linearity and (**C**) showing increasing rho with increasing library length (L) for the hypothesis that NAODJFMlf causes DLSAlf (p = 0.0007) and for the converse (DLSAlf causes NAODJFMlf, p = 0.29).

4. Discussion

The analyses described above have provided evidence for causal links between the solar cycle, measured by numbers of sunspot groups, and low frequency components of the NAO and the SOI (Table 2) and between both unfiltered NAO and SOI data and their low frequency components with indices of population variation in both Desert Locusts (Tables 3 and 4) and Oriental Migratory Locusts (Sections 3.6 and 3.10). From these results it might appear that the solar activity is driving the oceanic systems and that these are, in turn, driving the locusts. However, such an explanation is simplistic since the links between the sunspot data and the oceanic systems are weak and the cycle lengths detected are not consistent with each other. Indeed, a recent study failed to find a significant link between solar activity and the NAO [45], so our finding of spectral coherence between sunspot group numbers and both the unfiltered series and the low frequency components of the NAO (and with the LF SOI) may be of more interest to climatologists than to entomologists. Nevertheless, what is clear from our results is that there are causal relationships between solar activity and locusts, but this may not necessarily be via the oceanic systems, directly, but perhaps via some other means whereby the sun influences weather patterns.

We have also demonstrated for the first time that there is cyclicity in the dynamics of Desert Locust populations (Section 3.2). The latter result contrasts with the conclusion of Waloff [25] that there are no regular cycles in Desert Locust outbreaks, but lends support to the findings of Cheke and Holt [9] that there were was some evidence for 7 and 13 year cycles in a series of data up to 1990, although these were not statistically significant. Spectral analysis confirmed the existence of a statistically significant 13.5 year cycle, a near harmonic of the similarly significant peak at 7.5 years, in the 1866 to 2015 inferred series of numbers of territories infested with Desert Locust swarms. Similar periodicities were evident in the shorter 1930–2014 1° grid square data sets but these were not significant according to the stringent OUSS test, although they were significantly different from white noise according to the Kolmogorov-Smirnov test. Although Waloff and Green [46] argued that there was no evidence of periodicity in locust plagues, their data on the frequency distribution of intervals between onsets of successive regional plagues were highest (9 instances) at 5–7 years and 14–16 years (8 instances). With the above evidence for cycles of 7.5 and 13.5 years, it is of considerable interest that the significant and most consistent coherent frequencies between different drivers and the locust dynamics reported in Table 3 include 8.8 (SOI) and range from 6.23 to 7.14 (sunspot groups) and from 8.2 to 15 (low frequency NAO). A possible explanation for some of these results, as far as the sunspot groups is concerned given their well-known approximately 11 year cycle (confirmed here, Figure 2c), could be forthcoming if there is a certain low value of solar activity responsible for triggering appropriate meteorological events conducive to locusts. Such a value would be reached both when the solar activity is rising towards its peak and when it is descending to a trough and, depending on the magnitude of such a certain value in relation to the period of the cycle, the interval between these events could be about every 6–8 years.

Although there were no significant cycles detected by spectral analysis of the Oriental Migratory Locust data, there was significant spectral coherence between the sunspot groups and the Oriental Migratory locusts at 12.3 years, evidence of a link between them. In this context it is of interest that statistically significant ($p < 0.05$) peaks at 17.3, 3.7 and 2.9 years have been reported previously for the Brown Locust [14].

The different values for the spectral coherences associated with different drivers (Table 3) are difficult to interpret and may reflect inaccuracies resulting from the methods used (e.g., see Section 3.7. for comment on effects of varying the Daniell smoothing span in spectral coherence) but it needs to be borne in mind that Desert Locust outbreaks do not spring from a single source outbreak area, but from a variety of gregarisation zones located in each of the four regions (Western, North Central, South Central and Eastern, see maps in [26,47]) within a huge 10,000 km wide geographical area from West Africa to India. Therefore, it is to be expected that different oceanic drivers or their seasonal variants could be responsible for causing outbreaks in different regions. Thus, Waloff [25] reported three upsurges originating in the Western region, three in the North Central region, one in the South Central region and four in the Eastern region. Therefore, different oceanic drivers or seasonal variants of them could be responsible for causing outbreaks in different regions at different frequencies, as suggested by some of the results in Table 3. In addition, once an outbreak has started the locusts will often move out of their region of origin into another region and flourish there, resulting in a detectable signal from an oceanic driver being linked to the second rather than the first region. Such considerations may explain why the low frequency series of the NAO for September-November shows consistent coherence with all of the Desert Locust series tested (Table 3) at nine year intervals, whereas they might be expected to be most influential on the Western or North Central regions. It should also be borne in mind that an increase in the low frequency component of the locust dynamics will not always be reflected in a corresponding high frequency increase if weather conditions or control activities curtail a locust upsurge.

Both conventional statistical methods and the convergent cross mapping revealed relations between the SOI and the Desert Locust time series. The results for the inferred series and the shorter 1° grid square series were not always consistent and there are some regional variations to be discussed below. The SOI has high values during cold *La Niña* conditions and low values during warm *El Niño* circumstances (See https://www.ncdc.noaa.gov/teleconnections/enso/indicators/soi/). The short-term high frequency components of Brown Locust numbers have previously been found to be high during *La Niña* periods [14], but here we have concentrated on the long-term low frequency components. The rationale for the association is that there will be higher rainfall, essential for locust success, in many of the Desert Locust outbreak areas when *La Niña* cold conditions lead to more rain than the quantities that fall during the warm *El Niño* periods. This is supported by the CCM analyses suggesting causal links with the low frequency SOI driving the low frequency DLSA data set for the total number of infested 1° grid squares. However, at the regional level the links only held for the low frequency SOI with the South Central and Eastern regions' locust data, confirming our hypothesis that there are regional variations.

The NAO is a key factor that is responsible for inter-annual variability over extensive geographical zones, but especially of the north African climate [48]. When the NAO is low in winter, the north African climate tends to be warm and wet, and there is an inverse relation between the December to February NAO and November to April rainfall over Morocco [49]. Across western Africa, annual variations in seasonal climatic conditions are determined primarily by the Atlantic Ocean, although the rest of the world's oceans also play important roles (http://www.ipcc.ch/ipccreports/tar/wg2/index.php?idp=380). The temperature in the Sahara is affected by the NAO, with warmer winter temperatures during negative NAO events and cooler winters with more frosts when the NAO is positive [30]. Thus, the finding that the low frequency DJFM NAO data have the strongest causal links with the low frequency locust data according to the CCM tests (Table 4) is consistent with such NAO variability. Indeed, as already mentioned above, the NAO has correlations in DJFM that reach across Europe into the Middle East and Asia (see Figure 1 of [30]).

Inspection of Figure 1 reveals that, until the current upsurge began in 2018, in recent years there has been reduced locust activity. An exception was the Western region, where many upsurges begin. It is likely that the lack of major plagues until very recently was attributable to modern control practices. However, it may also explain why some of the cycles and trends were not as clear in the 1930–2014 1° grid square series in comparison with the long 1866–2014 inferred territories series, but this could also be explained by the higher likelihood of any patterns in the data being demonstrated to be statistically significant as the number of observations increases [50].

Given the number of statistical tests carried out (Tables 2–4), it is likely that some statistically significant results could have been found by chance alone but there are sufficient consistencies in the results for us to consider that it is unlikely that they are spurious. For instance, many of the CCM results are consistent with the spectral coherence tests (e.g., for sunspot groups) and the evidence suggests that there are links between sunspots and the locusts, probably mediated by the influence of the sunspots on oceanic systems (Table 2 shows the low frequency SOI to be coherent at a frequency of 7.89 years with the sunspots), and hence through weather systems, although exactly how this is achieved remains unclear and requires further research. In addition, given the evidence for cyclicity in the sunspots and likely delays in a similar response in the insect dynamics, it is possible that with further research a method of forecasting long-term trends in Desert Locust outbreaks may become feasible to supplement the short term methods [12] currently in use. As emphasised in the Introduction, the analyses presented here sought trends for the long term dynamics and ultimate causative factors, with results suggesting recurrences every 7–15 years, whereas the proximate factor, rainfall, is linked to the short-term dynamics with lags of 4 to 12 months [10]. Future research will seek to use the results presented here to develop models capable of providing early warning forecasts, perhaps by developing models based on the techniques described by Ye et al. [43] and Chen et al. [44]. For this, it will be necessary to examine detailed effects of lags (all of the CCM results discussed here assumed that tau = 1) to devise predictive models that could be tested to see if they anticipate the Desert Locust outbreak that began in 2018 (incidentally during a solar minimum) and to test the hypothesis that some outbreaks are driven by the SOI and others by the NAO in relation to seasonal and geographic variations in rainfall and locust abundance.

5. Conclusions

Statistically significant cyclicity exists in the population dynamics of the Desert Locust *Schistocerca gregaria* at periods of 7.5 and 13.5 years, according to results of spectral analysis of the well-known inferred time series. Tests using the methods of spectral coherence and convergent cross mapping revealed statistically significant evidence that dynamics of both Desert Locusts and Oriental Migratory Locusts *Locusta migratoria manilensis* and, in particular their low frequency dynamics, are causally linked with numbers of sunspot groups. In addition, causal links were inferred between both the Southern Oscillation Index (SOI) and the North Atlantic Oscillation (NAO) and Desert Locust dynamics, with the NAO for December-March being most influential. Solar activity seems to play a role in driving locust abundance, but by exactly what means and the extent to which fluctuations in oceanic systems are determinants remains unclear, although weak causal links between the sunspots and both the SOI and the NAO were also found. Nevertheless, it is possible that further research on the phenomena identified may eventually enable better early warning forecasting of desert locust upsurges.

Author Contributions: R.A.C., K.C. and S.T. conceived the study. K.C. provided the locust data in GIS format that were converted to time series by J.A.T., X.W. calculated the Kalman filtered series. R.A.C. and S.Y. conducted the analyses. R.A.C. wrote the paper with input from all authors. All authors have read and agreed to the published version of the manuscript.

Funding: This study was supported by the National Natural Science Foundation of China (NSFC, 11471201 and NSFC, 12031010), the Fundamental Research Funds for the Central Universities (grant number: GK202007039) and research funds of the University of Greenwich.

Informed Consent Statement: Not applicable.

Data Availability Statement: Data on the sunspots and oceanic oscillation systems are available in the publicly accessible repositories listed in Table 1. Other data presented in this study are available on request from the corresponding author.

Acknowledgments: We are grateful to the late Joyce Magor for her compilations of, and corrections to, the FAO Desert Locust data set, without which this study would not have been possible, and to two anonymous referees for their helpful comments.

Conflicts of Interest: The authors declare no conflict of interest.

References

1. Cisse, S.; Ghaout, S.; Mazih, A.; Ebbe, M.A.O.B.; Piou, C. Effect of vegetation on density thresholds of adult desert locust gregarization from survey data in Mauritania. *Entomol. Exp. Appl.* **2013**, *149*, 159–165. [CrossRef]
2. Cisse, S.; Ghaout, S.; Mazih, A.; Ebbe, M.A.O.B.; Benahi, A.; Piou, C. Estimation of density threshold of gregarization of desert locust hoppers from field sampling in Mauritania. *Entomol. Exp. Appl.* **2015**, *156*, 136–148. [CrossRef]
3. Uvarov, B. *Grasshoppers and Locusts. A Handbook of General Acridology*; Cambridge University Press: Cambridge, UK, 1966; Volume 1.
4. Uvarov, B. *Grasshoppers and Locusts. A Handbook of General Acridology*; Centre for Overseas Pest Research: London, UK, 1977; Volume 2.
5. Simpson, S.J.; McCaffery, A.R.; Hägele, B.F. A behavioural analysis of phase change in the desert locust. *Biol. Rev. Camb. Philos. Soc.* **1999**, *74*, 461–480. [CrossRef]
6. Simpson, S.J.; Despland, E.; Hägele, B.F.; Dodgson, T. Gregarious behavior in desert locusts is evoked by touching their back legs. *Proc. Natl. Acad. Sci. USA* **2001**, *98*, 3895–3897. [CrossRef] [PubMed]
7. *Desert Locust Guidelines*, 2nd ed.; FAO: Rome, Italy, 2001.
8. Cullen, D.A.; Cease, A.J.; Latchininsky, A.V.; Ayali, A.; Berry, K.; Buhl, J.; De Keyser, R.; Foquet, B.; Hadrich, J.C.; Matheson, T.; et al. From molecules to management: Mechanisms and consequences of locust phase polyphenism. *Adv. Insect Physiol.* **2017**, *53*, 167–285. [CrossRef]
9. Cheke, R.A.; Holt, J. Complex dynamics of desert locust plagues. *Ecol. Entomol.* **1993**, *18*, 109–115. [CrossRef]
10. Tratalos, J.A.; Cheke, R.A.; Healey, R.; Stenseth, N.-C. Desert locust populations, rainfall and climate change: Insights from phenomenological models using gridded monthly data. *Clim. Res.* **2010**, *43*, 229–239. [CrossRef]
11. Pedgley, D.E. (Ed.) *The Desert Locust Forecasting Manual*; Centre for Overseas Pest Research: London, UK, 1981.
12. Cressman, K. Information and forecasting. In *Desert Locust Guidelines*, 2nd ed.; Food and Agriculture Organization of the United Nations: Rome, Italy, 2001; Chapter 3.
13. Vallebona, C.; Genesio, L.; Crisci, A.; Pasqui, M.; Di Vecchia, A.; Maracchi, G. Large-scale climatic patterns forcing desert locust upsurges in West Africa. *Clim. Res.* **2008**, *37*, 35–41. [CrossRef]
14. Todd, M.C.; Washington, R.; Cheke, R.A.; Kniveton, D. Brown locust outbreaks and climate variability in southern Africa. *J. Appl. Ecol.* **2002**, *39*, 31–42. [CrossRef]

15. Swinton, A.H. Data obtained from solar physics and earthquake commotions applied to elucidate locust multiplication and migration. In *Third Report of the United States Entomological Commission, Relating to the Rocky Mountain Locust, the Western Cricket, the Armyworm Canker-Worms, and the Hessian-Fly*; Riley, C.V., Packard, A.S., Thomas, C., Eds.; Government Printing Office: Washington, DC, USA, 1883; pp. 65–85.
16. Ramchandra Rao, Y. A preliminary note on a study of locust infestations in North-West India since 1863 in relation to sunspot cycles. In Proceedings of the 5th International Locust Conference, Brussels, Belgium, 25 August–1 September 1938; pp. 252–257.
17. Ramchandra Rao, Y. *The Desert Locust in India*; No. 21; Indian Council for Agricultural Research: New Delhi, India, 1960; p. 721.
18. Pasquier, R. Les sauterelles pèlerines. L'invasion actuelle, les recherches, la lutte. *L'Agria* **1942**, *99*, 1–12.
19. Shcherbinovskiĭ, N.S. *The Desert Locust, Schistocerca gregaria. The Problem of Defending the Southern Territories of the U.S.S.R. Against Invasions by Swarms of Schistocerca*; Gosudarstvennoe Izdatel'stvo Sel'skohozâjstvennoj Literatury: Moscow, Russia, 1952.
20. Shumakov, E.M. *Study of Locust Ecology in the U.S.S.R. with Special Reference to the Development of Opinions on Phase Variation*; No. 114; Colloques Internationaux du Centre National de la Recherche Scientifique: Paris, France, 1962; pp. 217–228.
21. Meehl, G.A.; Arblaster, J.M.; Matthes, K.; Sassi, F.; van Loon, H. Amplifying the Pacific climate system response to a small 11-year solar cycle forcing. *Science* **2009**, *325*, 1114–1118. [CrossRef] [PubMed]
22. Tsonis, A.A.; Deyle, E.R.; May, R.M.; Sugihara, G.; Swanson, K.; Verbeten, J.D.; Wang, G. Dynamical evidence for causality between galactic cosmic rays and interannual variation in global temperature. *Proc. Natl. Acad. Sci. USA* **2015**, *112*, 3253–3256. [CrossRef] [PubMed]
23. Visbeck, M.H.; Hurrell, J.W.; Polvani, L.; Cullen, H.M. The North Atlantic Oscillation: Past, present, and future. *Proc. Natl. Acad. Sci. USA* **2001**, *98*, 12876–12877. [CrossRef] [PubMed]
24. Tian, H.; Stige, L.C.; Cazelles, B.; Kausrud, K.L.; Svarverud, R.; Stenseth, N.-C.; Zhang, Z. Reconstruction of a 1910-y-long locust series reveals consistent associations with climate fluctuations in China. *Proc. Natl. Acad. Sci. USA* **2011**, *108*, 14521–14526. [CrossRef]
25. Waloff, Z.; Green, S.M. Some temporal characteristics of Desert Locust plagues with a statistical analysis. In *Anti-Locust Memoir*; Centre for Overseas Pest Research: London, UK, 1976.
26. Cheke, R.A.; Tratalos, J.A. Migration, patchiness and population processes illustrated by two migrant pests. *BioScience* **2007**, *57*, 145–154. [CrossRef]
27. Hoyt, D.V.; Schatten, K.H. Group Sunspot Numbers: A New Solar Activity Reconstruction. *Sol. Phys.* **1998**, *179*, 189–219. [CrossRef]
28. Hoyt, D.V.; Schatten, K.H. Group Sunspot Numbers: A New Solar Activity Reconstruction. *Sol. Phys.* **1998**, *181*, 491–512. [CrossRef]
29. Vaquero, J.M.; Svalgaard, L.; Carrasco, V.M.S.; Clette, F.; Lefèvre, L.; Gallego, M.C.; Arlt, R.; Aparicio, A.J.P.; Richard, J.-G.; Howe, R. Revised collection of the number of sunspot groups from 1610 to 2010. *Sol. Phys.* **2016**, *291*, 3061–3074. [CrossRef]
30. Meehl, G.A.; van Loon, H. The sea-saw in winter temperatures between Greenland and Northern Europe, Part 3: Teleconnections with lower latitudes. *Mon. Weather Rev.* **1979**, *107*, 1095–1106. [CrossRef]
31. Abram, N.J.; Gagan, M.K.; Cole, J.E.; Hantoro, W.S.; Mudelsee, M. Recent intensification of tropical climate variability in the Indian Ocean. *Nat. Geosci.* **2008**, *1*, 849–853. [CrossRef]
32. R Core Team. The R Project for Statistical Computing. Available online: https://www.R-project.org/ (accessed on 9 August 2020).
33. Louca, S.; Doebeli, M. Detecting cyclicity in ecological time series. *Ecology* **2015**, *96*, 1724–1732. [CrossRef]
34. Chatfield, C. *The Analysis of Time Series: An Introduction*; Chapman & Hall: New York, NY, USA, 1992.
35. Shumway, R.H.; Stoffer, D.S. *Time Series: A Data Analysis Approach Using R*; CRC Press: Boca Raton, FL, USA, 2019.
36. Sugihara, G.; May, R.; Ye, H.; Hsieh, C.; Deyle, E.; Fogarty, M.; Munch, S. Detecting causality in complex ecosystems. *Science* **2012**, *338*, 496–500. [CrossRef] [PubMed]
37. Bendat, J.S.; Piersol, A.G. *Random Data: Analysis and Measurement Procedures*; John Wiley and Sons, Inc.: New York, NY, USA, 2000.
38. Granger, C.W.J. Investigating causal relations by econometric models and cross-spectral methods. *Econometrica* **1969**, *37*, 424–438. [CrossRef]
39. Clark, A.T.; Ye, H.; Isbell, F.; Deyle, E.R.; Cowles, J.; Tilman, G.D.; Sugihara, G. Spatial convergent cross mapping to detect causal relationships from short time series. *Ecology* **2015**, *96*, 1174–1181. [CrossRef]
40. Chang, C.-W.; Ushio, M.; Hseih, C.-H. Empirical dynamic modeling for beginners. *Ecol. Res.* **2017**, *32*, 785–796. [CrossRef]
41. Takens, F. Detecting strange attractors in turbulence. In *Dynamical Systems and Turbulence*; Rand, D.A., Young, L.-S., Eds.; Springer: Berlin, Germany, 1981; pp. 366–381.
42. Sugihara, G.; May, R.M. Nonlinear forecasting as a way of distinguishing chaos from measurement error in time series. *Nature* **1990**, *344*, 734–741. [CrossRef]
43. Ye, H.; Deyle, E.R.; Gilarranz, L.J.; Sugihara, G. Distinguishing time-delayed causal interactions using convergent cross mapping. *Sci. Rep.* **2015**, *5*, 14750. [CrossRef]
44. Chen, D.; Xiao, Y.; Cheke, R.A. Early warning signals for an environmental driver of respiratory infections. *Bull. Math. Biol.* **2021**. In Review.
45. Chiodo, G.; Oehrlein, J.; Polvani, L.M.; Fyfe, J.C.; Smith, A.K. Insignificant influence of the 11-year solar cycle on the North Atlantic Oscillation. *Nat. Geosci.* **2019**, *12*, 94–99. [CrossRef]
46. Waloff, Z.; Green, S.M. Regularities in duration of regional desert locust plagues. *Nature* **1975**, *256*, 484–485. [CrossRef]

47. Waloff, Z. The upsurges and recessions of the Desert Locust plague: An historical survey. In *Anti-Locust Memoir no. 8*; Anti-Locust Research Centre: London, UK, 1966.
48. Lamb, P.J. Case studies of tropical Atlantic circulation patterns during recent sub-Saharan weather anomalies: 1967 and 1968. *Mon. Weather Rev.* **1978**, *106*, 482–491. [CrossRef]
49. Lamb, P.J.; Peppler, R.A. North Atlantic oscillation: Concept and an application. *Bull. Am. Meteorol. Soc.* **1987**, *68*, 1218–1225. [CrossRef]
50. Kühberger, A.; Fritz, A.; Lermer, E.; Scherndl, T. The significance fallacy in inferential statistics. *BMC Res. Notes* **2015**, *8*, 84. [CrossRef] [PubMed]

agronomy

Article

Desert Locust Stopped by Tibetan Highlands during the 2020 Upsurge

Jun Liu [1], Michel Lecoq [2] and Long Zhang [3,*]

1 College of Grassland Science and Technology, China Agricultural University, Beijing 100193, China; liujun2637096337@163.com
2 French Agricultural Research Centre for International Development, Center for Biology and Management of Populations, F-34398 Montpellier, France; mlecoq34@gmail.com
3 Shandong Academy of Agricultural Sciences, Jinan 250100, China
* Correspondence: locust@cau.edu.cn

Abstract: Desert locust is an important pest to agriculture. In 2019–2020, a major upsurge originated in the southern Arabian Peninsula and gradually spread to east Africa, then to south-west Asia, as far as Pakistan and India, even reaching Nepal, resulting in major agricultural losses. For the first time, a few swarms entered southern Tibet. Using field observations and experiments, we studied their path to the Tibetan plateau and their behavior at these very high altitudes. The locusts moved up the Tibetan valleys from low-lying areas (1700 m) to much higher elevations (5400 m). The low temperatures and high humidity put them under severe stress; their activities were limited, and they did not survive long or produce local offspring. It is clear that the high-altitude environmental conditions in the Himalayan mountains provided an important natural barrier that limited the northward expansion of the desert locust populations.

Keywords: *Schistocerca gregaria*; Tibet; high altitude plateau; invasion route; survival; natural barrier

Citation: Liu, J.; Lecoq, M.; Zhang, L. Desert Locust Stopped by Tibetan Highlands during the 2020 Upsurge. *Agronomy* **2021**, *11*, 2287. https://doi.org/10.3390/agronomy11112287

Academic Editor: Andrea Sciarretta

Received: 27 September 2021
Accepted: 2 November 2021
Published: 11 November 2021

1. Introduction

Desert locust, *Schistocerca gregaria* (Forskål, 1775) is an important pest to agriculture for many countries in Africa, the Middle East, and SW Asia. This polyphagous insect can cause damage to almost all types of vegetation and crops [1–3]. It is endowed with a high reproductive potential under favorable rainfall conditions [1,4,5], as well as a great capacity for flight, frequently migrating hundreds of kilometers, sometimes even over oceans [6,7]. The transition from a solitary to a gregarious phase is another characteristic of the desert locust that depends on population density, and locusts in the gregarious phase can migrate long distances in large swarms and cause huge crop losses [8–11].

In general, the desert locust is mainly distributed in arid and semi-arid regions from the Sahara through the Arabian Peninsula to southwest Asia in lowland areas [3,4,12–14]. This recession area covers around 19M km^2. During invasions, desert locusts can spread over larger territories (an invasion area of 31M km^2) and reach the limits of the areas they can colonize, often unfavorable for breeding, but where they can still cause significant damage. This is the case of the humid woodlands of Africa to the south, or the too cold and mountainous regions to the north. Until now, in the Asian part of its range, the northern mountains (such as Himalayas and Kunlun Mountains with and average altitude of 4000 m) seemed to constitute an uncrossable barrier. Particularly in the southern end of the range, if northern India and Nepal have sometimes been colonized, no swarms have ever been reported from Tibet. Yet, this is what has happened recently.

In 2019 and 2020, the desert locust again threatened large areas of east Africa, the near east, and south-west. The upsurge originated in the desert regions of Arabia that received exceptional rainfall. The Food and Agriculture Organization of the United Nations (FAO) described this as the most severe locust situation in decades [15–17]. To the south, swarms invaded much of eastern Africa, reaching Kenya, Uganda, and Tanzania, which have not faced a threat of this magnitude in 70 years. To the east, from March 2019, the invasion spread to Iran, Pakistan, and India, even reaching Nepal, and for the first time Tibet [15–18]. In particular, in June 2020, locust swarms invaded Nepal, and some of them reached the foot of the Himalayas [17]. Until now, there had been no reports of locust swarms in the Tibetan highlands. It was considered that swarms cannot cross high mountain ranges such as the Atlas Mountains, the Hindu Kush, or the Himalayas and that the temperature was too cold for the desert locust at these high altitudes. During past major desert locust invasions, and even though northern India was widely contaminated, no desert locusts had been reported in Tibet. While Nepal is included by Waloff in the range of the desert locust, Tibet is not [19–21]. Only a solitary phase adult female has been reported in Tibet, collected by Zhang Xue Zhong, on 29 April 1974, in the Zhangmu area of Nyalam County (about 28°20′ N, 86° E, altitude 2250 m) [22,23]. According to Cai [24], the desert locust is also present in Yunnan (China), but there is no specific information to support this assertion. We believe that this is mainly an assumption based on Yunnan's proximity to neighboring areas of Burma known to be part of the desert locust invasion area. However, in early July 2020, during this new upsurge, as swarms advanced from Iran, Pakistan, India, and Nepal, seven counties or cities in Tibet reported light swarms of locusts, clearly originating in Nepal. Some were recorded at very high altitudes, around 4000 m or more.

Although the risk of desert locusts invading the high-altitude Tibetan plateaus and causing damage was low, the Tibet Forestry and Grass Bureau, Lhasa Customs, and other departments have initiated prevention and control monitoring work and issued specific prevention and control and emergency plans [25]. However, even with a near-zero risk, this situation offered the first opportunity to study locusts in these cold, high-altitude areas. How could these swarms invade Tibet and enter the Himalayan range in 2020? What was their fate in these high-altitude areas? Using field observations and experiments, we studied for the first time the path of the desert locust to the Tibetan plateau, as well as its behavior, feeding, maturation, and survival in this region, providing valuable information for eventual forecasting.

2. Material and Methods

Between early July and late September 2020, field surveys were conducted in seven localities in southern Tibet where desert locusts were reported: Pulan, Zhongba, Saga, Nyalam, Chentang, Jilong, and Dingri counties (Figure 1). These sites benefit from a wide range of climates: semi-arid in the intense cold zone of the Tibetan plateau (mean annual temperature, T = 6 °C, and annual rainfall, R = 280 mm: Saga) or sub-cold (T < 20 °C and R = 173 mm to 280 mm: Pulan and Zhongba, respectively); temperate semi-humid climate, with low temperature and R = 618 mm (Nyalam); monsoon climate, whether subtropical (mild climate and heavy rainfall in summer and autumn: Chentang), mountain subtropical (T = 10–13 °C, R = 1000 mm: Jilong), or semi-arid temperate (T = 12 °C, R = 319 mm: Dingri).

We focused, in particular, on Nyalam County (ranging from 1700 to 4800 m) and the Boqu River valley, flowing from the north to the south and opening towards Nepal at the level of Zhangmu town (Figure 1). This valley is one of the areas where locust swarms appear to have entered Tibet, obviously from Nepal where locusts have been reported in abundance and where the greatest number of swarm reports have been made. This valley was monitored daily in July and August 2020, except on rainy or foggy days.

Study sites included rugged and steep mountainous areas, riverbanks, sandy lands, and a wide variety of plant communities (Figure 2). Field monitoring focused on locust density per square meter, sex ratio, maturity, and ovary development status within desert locust populations. At least five locust samples were collected from each sampling site. The sex ratio was estimated on a larger number of individuals collected by hand in the morning when the locusts were still inactive. The maturity status of the population was determined based on the appearance of ovaries and ovarioles (immature: white ovarioles, small and not clearly visible; mature 1: white ovarioles, larger, oocytes starting to develop; mature 2: yellow ovarioles, larger with clearly developing oocytes; mature 3: ovarioles containing well-formed eggs, almost ready for laying; laying: distended, enlarged abdominal segments, no eggs in abdomen).

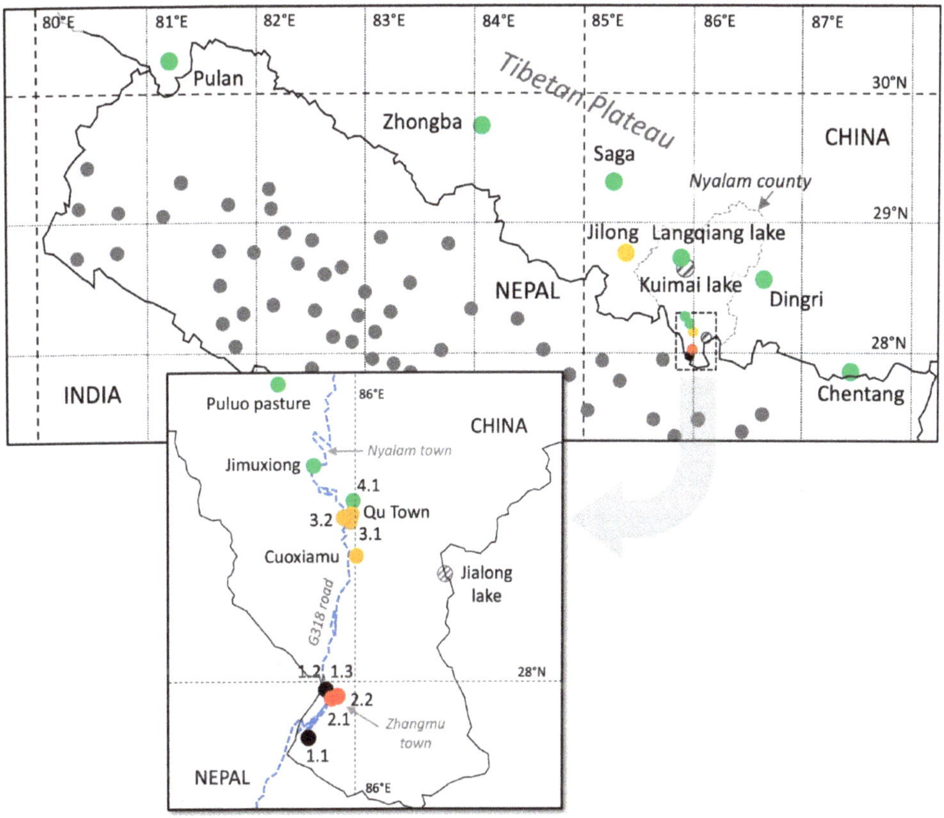

Figure 1. Desert locust records in Tibet in July 2020. Colored dots indicate the abundance of desert locust populations in Tibet: gray striped, none; green,very light/light; yellow, middle; red, heavy; black, very heavy. The gray dots in Nepal indicate desert locust reports (swarms or scattered populations) between June 27 and July 16 (modified from Shrestha et al., 2021). The box below the general map details the situation in southern Nyalam County where the degree of infestation was the most serious, along the G318 road (blue dotted line), for about 26 km along the Boqu River valley (1700 to 4200 m).

Caging experiments were conducted in the field to observe and record daily life habits, mating, breeding, and longevity of desert locusts. They were conducted in two sites: Zhangmu (2300 m) and Nyalam (3800 m) (Figure 1). Gregarious locust adults were collected in the field for rearing in cages with *Lolium perenne* (Gramineae) leaves. The cages measured 70 × 70 × 70 cm, had a wooden frame, and were covered with wire mesh. In Zhangmu, 6 cages contained 10 adults each (5 males and 5 females) while in Nyalam, 2 cages were used, each containing 27 locusts (18 females and 9 males).

In addition, food preference tests were implemented. Healthy and active adult locusts were collected from Zhangmu in July 2020 and reared in cages under natural conditions. The tested locusts were starved for 12 h and then placed individually in a transparent plastic cylindrical bottle (60 mm diameter, 200 mm length). Leaves, freshly collected on the morning of the test, were inserted into the bottles, with all leaves of the same plant being the same size as far as possible. Thirty-four plant species, collected in the region, were tested. At least 10 adults (5 males and 5 females) were fed each plant species. Leaves were removed after 24 h and the area consumed was measured to assess feeding preferences according to the following scale: high >1700 mm^2, medium >650 mm^2, low >100 mm^2; very low <100 mm^2.

Figure 2. Some views of desert locusts in southern Tibet, in 2020, not far from the border with Nepal. (**a**) A valley having been invaded. (**b**,**c**) Light swarms of locusts flying in the clouds at about 3000 m. (**d**) The Boqu River valley at 3360 m. (**e**) Landscape at 4670 m altitude with lakes, sandy beaches, and a rare vegetation. (**f**) Desert locusts found dead in the field at an altitude of 3350 m. (**g**) basking in the sun on stones, or (**h1**–**h3**) feeding on wormwood and willow leaves.

We used Graphpad Prism 8.0 for statistical analyses. For food preference tests, Graphpad Prism 8.0 was used to draw the leaf area consumed, imageJ 1.8.0 to measure this area, and unpaired *t*-test or one-way ANOVA for comparisons.

3. Results

3.1. Desert Locust Penetration in the Tibetan Highlands

From July 1 to 9, 2020 at least 7 counties in Tibet (Chentang, Dingri, Nyalam, Jilong, Pulan, Saga, Zhongba) reported adult locusts near the border with Nepal. These counties are characterized by valleys opening at a lower altitude at the border with Nepal. The largest swarm appeared in Zhangmu District (Nyalam County), and another in Jilong County at an altitude of about 1700 m. However, small groups of locusts have also appeared at altitudes above 4000 m (Figure 1, Table 1). How did they manage to climb to such places in the Himalayas? To answer this question, we mainly surveyed Nyalam County, from the lower elevations around 1700 m, up to 4200 m, following the G318 road for about 26 km along the Boqu River valley, passing through Zhangmu and Nyalam towns. This valley opens to the south at the China–Nepal border (Figure 2a) and gradually rises northward from 1700 to 3800 m.

Table 1. Monitoring sites and assessment of desert locust numbers along some valleys in South Tibet.

Sites	Altitude (m)	Locust Density	Latitude	Longitude
Saga (Jilong valley)	5400	Light	29°19′48″	85°13′56″
Zhongba (Zhongba County)	4800	Light	29°46′12″	84°01′53″
Jialong Lake	4750	None	28°03′54″	86°03′43″
Kuimai lake	4725	None	28°40′52″	85°56′27″
Langqiang Lake	4670	Very light	28°43′26″	85°53′31″
Puluo pasture	4200	Very light	28°10′39″	85°56′47″
Jimuxiong	3750	Very light	28°08′33″	85°58′17″
Pulan (Pulan valley)	3600	Light	30°16′05″	81°10′06″
Jilong (Jilong valley)	3530	Middle	28°45′01″	85°21′47″
4.1	3380	Light	28°06′27″	85°59′55″
Qu Town	3365	Middle	28°05′49″	85°59′46″
3.2	3350	Middle	28°05′59″	85°59′50″
3.1	3340	Middle	28°05′45″	85°59′46″
Chentang (Chentang valley)	3300	Light	27°51′39″	87°25′13″
Dingri (Gama valley)	3200	Light	28°34′00″	86°38′00″
Cuoxiamu	3150	Middle	28°04′37″	85°59′54″
2.2	2500	Heavy	27°59′37″	85°59′8″
2.1	2350	Heavy	27°59′34″	85°58′57″
1.3	2330	Very heavy	27°59′50″	85°58′47″
1.2	2320	Very heavy	27°59′50″	85°58′47″
1.1	1950	Very heavy	27°58′09″	85°57′54″

Estimated degree of desert locust density (number of locusts seen per site): Very heavy >10,000 (black), heavy >100 (red), middle 50–100 (brown), light 10–50 (green), very light <10 (light green). Unless otherwise indicated, all sites belong to the Boqu River valley in the Nyalam County.

The largest swarms were observed in Bang and Lixin villages (1700 m) in Zhangmu District, downstream of the Boqu River valley near its opening to Nepal (Figure 1 and Table 1, sites 1.1, 1.2, 1.3). The population size was estimated to be over 0.4 million adults. Locusts rested on both trees (mainly bamboo, but also willow, pine, and fir) and herbaceous vegetation at the base of valley cliffs (Figure 2a,h1–h3). On sunny or slightly cloudy days, locusts formed a large, cumuliform swarm. Crows and other birds were observed catching locusts in flight (Figure 2c). Some locusts formed into stratiform swarms migrating upward along the valley midway up the slope tops (Figure 2b). The size of each migrating group varied from several hundred to several thousand adults. During flight, many individuals fell into the river or landed on the road or mountain slopes.

Along the valley, at three sites between 2500 and 3200 m (Table 1, sites Cuoxiamu, 2.1, 2.2) locusts were observed forced to stop their flight by strong winds. Hundreds of individuals were found gathered on trees or in the grass. At four other sites located in the same valley at an altitude above 3300 m (Table 1, sites Qu town, 3.1, 3.2, 4.1), at a place where the river makes many bends, locusts were observed grouped on the ground (Figure 1). In Qu town (at 3365 m), a few locusts were found on the banks of the river as well as at the foot of the high cliffs, covered with lush vegetation, that overlook it (Figure 2d,f). Some were dead, others alive and basking on the rocks on sunny days (Figure 2g). The same situations were observed at the three other sites (Sections 3.1, 3.2 and 4.1).

The Boqu River valley ends near Nyalam, at about 3800 m elevation (Figure 1 and Table 1, Jimuxiong site). Several scattered locusts were observed flying in this area, with flight speed estimated at about 6 m/s. A few adults were also collected locally in potato fields. At Puluo pasture (4200 m), about 25 km northwest of Nyalam, 6 locusts were found in early July (Figure 1 and Table 1). On the shores of Langqiang (4700 m) and Kuimai (4725 m) lakes, about 30–40 km north of Nyalam, a few locusts in flight were observed as well as a few dead locusts in late July (Figure 2e, Table 1). It is interesting to note that while locusts were observed at Cuoxiamu, none were seen at Jialong Lake, although both sites are located at almost the same latitude (about 28°4′), but the former in the Boqu River valley at 3150 m, while the latter, at 4750 m, is outside this valley about 10 km to the east.

3.2. *Population Dynamics of the Desert Locust in Tibet in 2020*

3.2.1. Field Survey

Field surveys were conducted in areas at elevations from of 2300 m to 4750 m, from mid-July to early October 2020.

In Zhangmu, at 2300 m, from 13 to 27 July, the average density of locusts at surveyed sites gradually decreased from 3.1 locusts/m^2 to zero from 27 July. After this date, desert locusts were no longer observed in this area (Figure 3a). Twenty-five live females were collected on 25 July, and 38 and 177 dead females on 1 and 11 August. Females were dominant and the sex ratios on these dates were 60%, 68.4%, and 71.2%, respectively (Figure 3b, column graph). On August 1, all females were immature and on August 11, only 1.3% of the females were at maturity stage 1 (Figure 3b, pie chart).

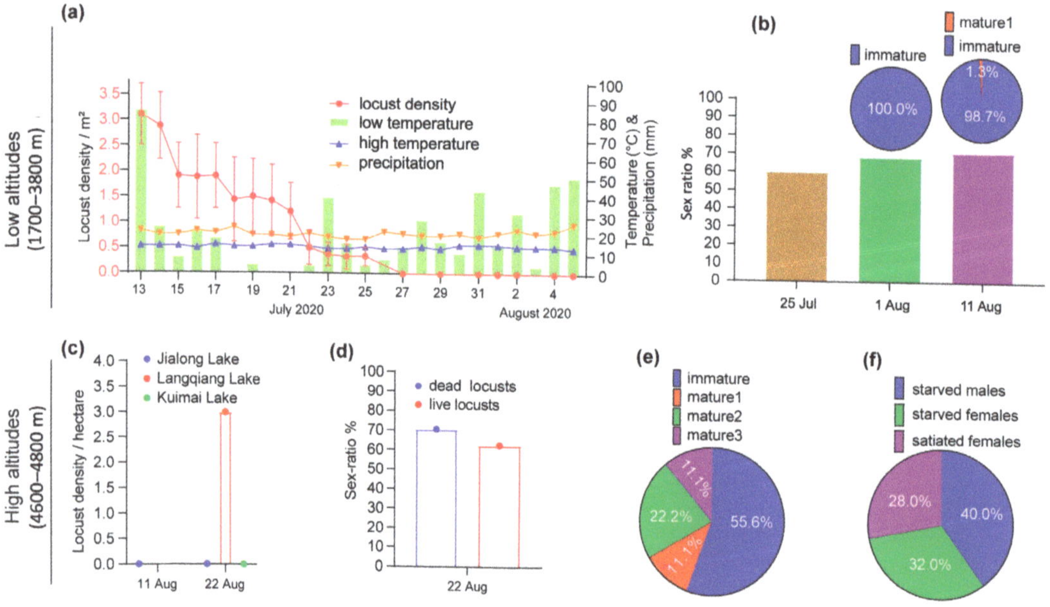

Figure 3. Desert locust field monitoring in low and high altitudes in Nyalam County, Tibet, in July–August 2020. Lower altitudes: changes in locust average density as a function of temperature and rainfall (**a**), locust sex ratio and ovarian development (**b**). Higher altitudes: locust density at three sites (**c**), sex ratio (**d**), ovarian development (**e**). Filling degree of the digestive tract of locusts collected in the field (**f**).

Weather conditions must have been unfavorable for the desert locusts. In Zhangmu, the temperature was relatively low and stable during the survey period, with the average ranging from 14.0 °C to 20.8 °C, and the lowest temperature being about 10 °C. In Qu City, in the middle of the Boqu River valley, the temperature was even lower, with an average of 5 to 8 °C due to heavy rainfall, higher elevation, and strong wind. It rained almost every day and only three days had no rain. The highest rainfall was recorded on July 13 with 85.2 mm. On July 14, 16, and 17, more than 18 mm of rain fell, and on July 23 more than 40 mm. Since desert locusts prefer higher temperatures and drier weather, the low temperatures and sustained rainfall are expected to have been a major stress on these insects, resulting in high mortality and population loss in this area in late July.

At higher elevations, Jialong (4750 m), Langqiang (4670 m), and Kuimai (4725 m) lakes were monitored (Figure 1). No locusts were found near Jialong (11 and 22 August) and Kuimai lakes (22 August). Only at Langqiang lake were locusts observed on 22 August, at a low density of only 3 locusts/ha (Figure 3c). On the lake shore, some locusts were feeding or resting on shrubby plants such as acacia trees. A few dead locusts were also observed. On a sunny day, when the maximum temperature of the day was above 20 °C, a few locusts were observed flying 2–3 m above the ground in various directions at a speed of about 10 m/s. The distance flown in a single flight was about 20–50 m. The sex ratio was 70.6% and 62.5% for dead (n = 38) and live (n = 16) locusts, respectively (Figure 3d). Live females were mostly immature (55.6%), with the remainder in various stages of maturity (Figure 3e). No females had yet laid eggs. It appears that locusts lived longer in these drier, higher areas (above 3700 m) than in the lower (2300 m), very rainy areas.

Examination of the digestive system of locusts (n = 25) collected from Qu town area revealed that 72.0% of individuals were starved, including 53.3% of females and 100.0% of males (Figure 3f). Locust feces were liquid, probably due to the high humidity throughout the valley.

As of late July 2020, no swarms were observed in all previously infested areas and only dead locusts were found throughout the Boqu River valley.

3.2.2. Field Caging Experiments

At the same time, in Zhangmu and Nyalam, cage field studies were conducted. In Zhangmu (20 July to 4 August) locust life span averaged 26 days, ranging from 22 to 35 days. Probably under the influence of heavy rains, survival rate dropped sharply from 95% to 38% between July 22 and 25, and from 37% to 10% from July 26 to 28. On 4 August, no locusts were alive (Figure 4a). These figures are consistent with field observations made at the same time in this area (Figure 3a). This is despite the fact that locusts in cages were provided with sufficient food each day. In addition, 90% of the females were immature in the caged field studies, and no mating behavior was observed (Figure 4b), which is also very similar to the field study. These results confirm that low temperatures and high precipitation are critical limiting factors for locust survival in the southwestern Himalayan foothills.

In Nyalam (where the experiments were conducted from July 26 to October 18), the average lifespan of locusts was 70.9 days, with the longest being 110 days. The survival rate in the cages gradually decreased from 100% to 52% in 56 days from 26 July to 19 September. After the heavy rains on 23 and 24 September, survival continued to decline and then dropped rapidly after 10 October until the locusts disappeared completely on 18 October (Figure 4c). During the first 59 days, low precipitation occurred on 38 days (64.4%), and the mean daily temperature ranged from 8.7 °C to 18.8 °C. On September 23 and 24, heavier precipitation occurred (22.3 and 31.3 mm) followed by a sharp drop in temperature (daily average ranging from 10.5 °C and 1.5 °C between September 19 and October 18). Precipitation ceased after September 25 (Figure 4c).

Thus, locust mortality in the field cages in both areas was also clearly influenced by temperature and rainfall. Heavy precipitation seems a key factor, but low average temperatures over a long period and low minimum temperatures can also be critical.

All arriving locusts were pinkish and immature. Under the low temperatures prevailing from July to October, ovarian development of females was very slow. Twenty-six females could be dissected to examine the ovaries. On August 1, the females were immature. Early maturation stage 1 was observed on August 12, stage 2 on August 15, stage 3 on

September 10, and the only female (over 26) that laid eggs was observed on October 5. The egg-pod contained 78 eggs (Figure 4f), but none hatched after incubation in the laboratory (L:D = 12:12, T = 35 °C, RH = 30%).

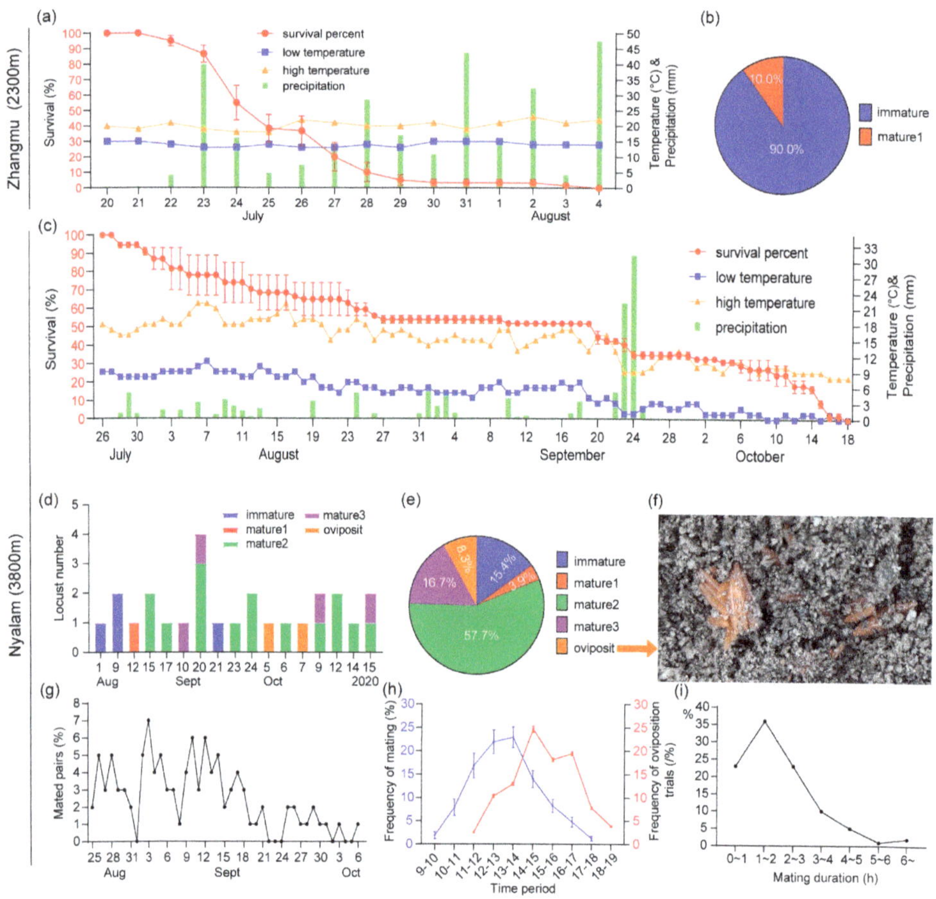

Figure 4. Cage field studies in Zhangmu (2300 m) and Nyalam (3800 m). Desert locust survival and weather conditions in Zhangmu (**a**) and Nyalam (**c**) in 2020 (the % of survival is the mean of survival percentages per cage). Percentage of maturation of females in Zhangmu (**b**). In Nyalam: Number (**d**) and percentage (**e**) of maturation status of females; eggs in the ootheca (**f**); daily mating rate of locusts (**g**); mating (left) and pre-oviposition behavior (right) rates (**h**); mating duration (%) (**i**).

In the cage study, adults mated many times during their lifetime, especially on sunny days and when the temperature was above 10 °C. However, no mating occurred on cloudy and rainy days. The period during which mating pairs were observed was very long, ranging from late August to early October, or 43 days. The vast majority (91%) of matings occurred from 25 August to 18 September, or 25 days (Figure 4g). During the day, matings could be observed from 09:00 to 18:00, but most (75.7%) occurred between 11:00 and 15:00 (Figure 4h). The duration of mating varied from a few minutes to several hours; 36% lasted 1–2 h; the longest being 19 h, probably due to the low evening temperatures that made the locusts not very active (Figure 4i). At the time of oviposition, the female makes regular movements of opening and closing the ovipositor valves and the abdomen elongates significantly. The whole oviposition process usually lasted a few minutes but could sometimes last several tens of minutes. Overall, oviposition lasted about 2 h and eggs were deposited in sandy soil at a depth of about 12 cm. Pre-oviposition behavior (females using their abdomen to test whether the soil is suitable for oviposition) were observed from 11:00 to 18:00, but 75.3% occurred from 14:00 to 17:00 (Figure 4h). When the temperature reached 15 °C at noon in August and September, locusts flew into the cages but could still be hit easily.

3.2.3. Food Preference Tests

Most of the desert locusts (72%) collected on the Tibetan Plateau were starving, with an empty digestive tract (Figure 3f). In addition to the fact that low temperatures must have reduced feeding activity, we questioned whether or not local plants might be favorable and conducted food preference tests. A total of 34 local plant species (belonging to 19 families and 32 genera) were collected from Zhangmu area and presented to adult desert locusts (Figure 5). Only three plants were clearly consumed: *Lophatherum gracile* (LG), *Mentha canadensis* (MC), and *Fagopyrum esculentum* (FE) with, respectively, 2600, 1850, and 1400 mm^2 of leaf area consumed in 24 h. For the other plants, the consumed surface was lower or largely lower than 1200 mm^2 and ten plants were not or very little consumed.

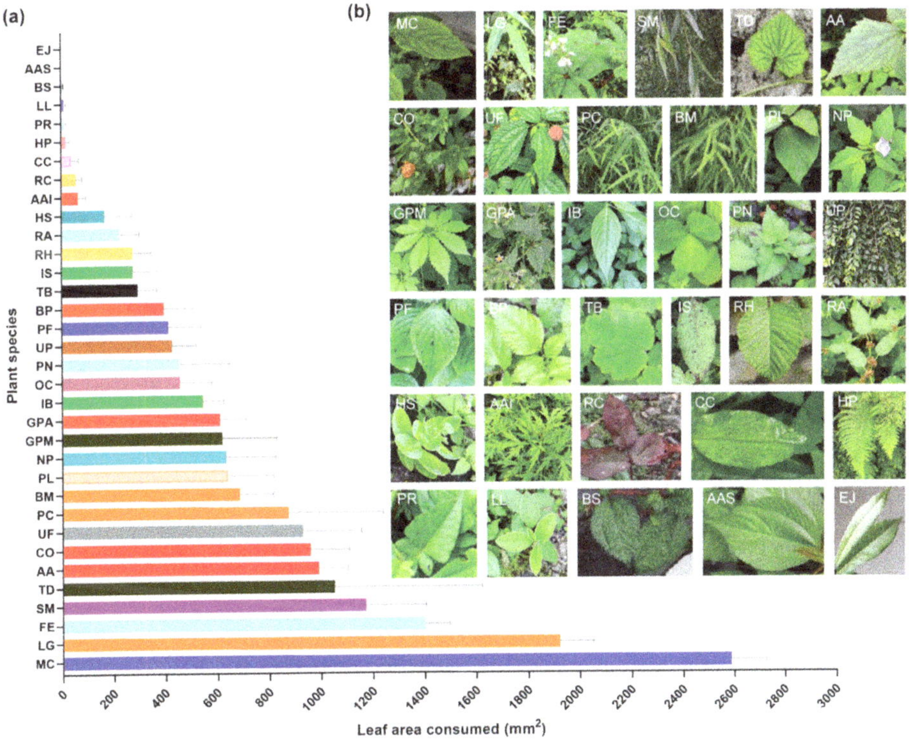

Figure 5. Desert locusts' food preferences for some Tibetan plants. (**a**) Leaf area consumed by locusts (n = 10, female:male = 1:1; same color bars represent plants of the same family). (**b**) Leaf aspect of the 34 plant species tested: MC, *Mentha canadensis* (syn. *M. haplocalyx*); LG, *Lophatherum gracile*; FE, *Fagopyrum esculentum*; SM, *Salix matsudana*; TD, *Thladiantha dubia*; AA, *Ageratina adenophora* (syn. *Eupatorium adenophorum*); CO, *Calendula officinalis*; UF, *Urtica fissa*; PC, *Pogonatherum crinitum*; BM, *Bambusa multiplex*; PL, *Pueraria lobata*; NP, *Nicandra physalodes*; GPM, *Gynostemma pentaphyllum*; GPA, *Galinsoga parviflora*; IB, *Impatiens balsamina*; OC, *Oxalis corniculata*; PN, *Polygonum nepalense*; UP *Ulmus pumila*; PF, *Perilla frutescens*; BP, *Bidens pilosa*; TB, *Thalictrum baicalense*; IS, *Impatiens siculifer*; RH, *Rubus hirsutus*; RA, *Rumex acetosa*; HS, *Hylotelephium spectabile* (syn. *Sedum spectabile*); AAI, *Artemisia argyi*; RC, *Rosa chinensis*; CC, *Commelina communis*; HP, *Hypolepis punctata*; PR, *Polygonum runcinatum*; LL, *Ligustrum lucidum*; BS, *Boehmeria spicata*; AAS, *Acalypha australis*; EJ, *Eriobotrya japonicu*.

4. Discussion

4.1. Desert Locust Routes in Tibet in July 2020

The Himalayas, with their high altitudes and low temperatures, may seem an insurmountable natural barrier for locusts despite the great flight ability of these insects. However, some swarms have managed to penetrate it and our observations show that some isolated locusts have even been able to advance to very high areas. In fact, the valleys—thanks to the associated aerological systems—were the gateways of the locusts to the Tibetan highlands.

The highest concentrations of locusts in Tibet in 2020 were recorded along the Boqu River valley, with the highest densities observed south of the valley near the Nepalese border and at low elevations of only 1700 m (Figure 2a). Based on the observations, the flight path of the swarms mainly followed the river valley and gradually rose toward the Tibetan high plateau to an altitude of about 3200 m. Outside the valley floor, the density was much lower. Density decreased as one moved northward and elevation increased. A few isolated locusts were able to progress, still following the valley, to much higher areas between 3200 and 5400 m. They did not migrate in large swarms, but in small groups with several stops in the climbing route. These locusts do not seem to have been able to benefit from thermal lift that would have allowed them to rise to high altitudes, sometimes to the top of the convective layer at heights of 1 to 2 km [26]. The aerological conditions forced them to fly at low altitude, at a few tens of meters from the ground, while going up very gradually the bottom of the valley, until reaching too high altitudes where the thermal conditions must have ended up preventing them from flying. The observed flight speed along the Boqu River was only about 6 m/s, which is not much different from the average speeds of 7–18 m/s reported in the locust's usual habitat range [27,28]; however, the flights were only of short duration. The progression of these locusts up the valley must have been aided by slope and valley breezes, blowing from the bottom up during the day, at the time of locust flight activity. It is known that transport by such anabatic winds can account for the frequent occurrence of grasshoppers in mountainous areas at altitudes well above their area of residence [29], as well as for the movement of desert locust swarms to and on mountain slopes [30] (see Appendix A).

This Boqu River valley is surely only one of the entry routes of the desert locust into Tibet in July 2020. At the same time, desert locusts are widely reported throughout Nepal, immediately to the south. This country was invaded in three successive waves, on 25–29 June, 12 July, and 16 July. The swarms then spread over much of Nepal in about 10 days [31,32] (Figure 1). This Himalayan region on the southern border of Tibet has four valleys (ChenTang, Gama, Zhangmu, and Jilong) leading into Nepal and of relatively low altitude. All these valleys could thus have been, like the Boqu River valley studied here, an access route to invade southern Tibet from the beginning of July.

4.2. High Altitude Climate as a Major Obstacle for Desert Locusts

Once in Tibet, the high-altitude climate must have been a major obstacle to entry, reproduction, and local survival for locust individuals. The desert locust's usual range, its recession area, covers about 19 million km^2 from western Africa to western India. These are generally low-lying areas (<1000 m) with a tropical desert or semi-desert climate. The average annual temperature is around 20–30 °C, the average monthly temperature from July to October is above 30 °C, and the annual rainfall is very low and below 250 mm [2]. Rainfall over 25 mm in two consecutive months is usually assumed to be enough for locust breeding and development [27]. In the highlands of Tibet, in areas where locusts have been observed, climatic conditions are quite different and a priori not very conducive to the survival of new arrivals. The average annual temperature is about 8 °C in southern Tibet. Zhangmu area, at relatively low elevations of 1700–3800 m, has a humid or semi-humid subtropical monsoonal plateau climate. Rainfall is abundant with an annual average of 2820 mm and a rainy season extending from June to October. The average daily temperature during this season is less than 20 °C or even 10 °C in the highest areas such as Qu City [33,34]. Areas at elevations above 3800 m, such as Nyalam, have a plateau mountain climate where the highest daily temperature in August is mostly below 20 °C and the nighttime temperature is close to 0 °C [35–37].

The lowest temperature at which locusts can migrate and propagate over a long distance is 20 °C [27,38–40] and this limiting temperature is even higher (about 23 °C) if the sky is overcast, with insufficient sunlight rendering locusts unable to absorb sufficient heat energy [26,27]. Therefore, temperatures in Tibet make it nearly impossible for locusts to fly uninterrupted over long distances. Furthermore, above 4000 m, the oxygen content of the air is only about 14.6%, well below the 21.0% recorded in Africa at sea level [41–43]. Decreased oxygen availability (and decreased temperature) make life at such altitudes challenging. The physiological effects of environmental hypoxia may indeed contribute to limiting the ability of insects to persist at high altitudes [44]. This lack of oxygen at high altitudes must therefore have been another limiting factor in locust flight activity in Tibet. This reduced flight ability is probably responsible for the large number of drowned locusts found in small ponds and lakes in the Qu town area (3365 m) and at Langqiang Lake (4670 m), where individuals have been observed falling into the water, even on a sunny day, and although adults are accustomed to avoiding water when flying.

These extreme climatic conditions for desert locusts in the Tibetan highlands appear to have affected not only the flight ability of individuals but also their survival and reproductive ability. In a field cage, with sufficient food and no natural enemies, the average life span of desert locust adults was 26 days in Zhangmu (at 2300 m elevation) and 71 days in Nyalam (at 3800 m). Such life spans seem short even considering that these are immigrant individuals with only part of their imaginal life observed, with the average duration normally ranging from 2.5 to 5 months [27]. This longevity appears to have been significantly shortened even though many individuals remained immature, a state in which locusts can normally remain quiescent for several months until suitable conditions return.

Other evidence suggests that the locusts that entered Tibet were placed in unfavorable conditions.

First of all, they had a low food intake since 72% of the locusts collected in the field had an empty digestive tract. In the cages, the locusts also showed a reduced feeding activity. The majority of local plants tested appeared to have low palatability. Even though they represent a small proportion of the flora of Tibet, one third of these plants were not consumed or were consumed only slightly, which is surprising considering the very polyphagous nature of the desert locust. It is interesting to note that of the 34 plants tested, 15 belong to genera that are not represented in the usual range of this insect.

Secondly, sexual maturation was disturbed. In Zhangmu area, 99.4% of the females were immature. In the field cage only 10.0% of the females were observed in early maturation. In the Langqiang Lake area, 55.6% of the females were immature, the rest were in various stages of ovarian development, but none had laid eggs. In Nyalam, in a cage, only one female laid eggs while the other 35 tested died before laying eggs. In the end, most of the locusts died quickly in the immature state. For sexually matured females, the time from adult emergence to first matings was 43 days at 3800 m, whereas it may be as little as 15 days under optimal thermal conditions [27,45,46].

Finally, no eggs were observed except for one, containing 78 eggs, none of which hatched after incubation in the laboratory. At best, the embryonic development time was strongly slowed down by the low temperatures. Desert locust eggs require an air temperature of 20–35 °C and a soil temperature of 15.5–35 °C, low temperatures make egg survival impossible [27,47,48]. On the Tibetan plateau, the winter temperature is very low and the snow is deep. Even if locusts were successful in laying eggs, they would probably not withstand the low temperatures.

In conclusion, the combined influences of low temperatures, high precipitation, high altitude and low oxygen content, and unfavorable food supply weakened the immigrant locust populations, reduced their flight ability, disrupted sexual maturation, and made egg laying almost impossible. In addition, various natural enemies must have contributed to rapidly limiting the populations, including various birds and entomopathogenic fungi actually observed. In the end, these locusts died quickly without any local offspring and without having been able to cross the Himalayan barrier.

Author Contributions: L.Z. and J.L. conceived the study, performed the field work, and collected data; J.L., M.L. and L.Z. wrote the first draft of the manuscript; all authors contributed substantially to data analysis, writing, review and editing. All authors have read and agreed to the published version of the manuscript.

Funding: This work was supported by grants from The Ministry of Agriculture and Rural Affairs of China (15206030 and 12210030), and the grant for innovation project from Shandong Academy of Agricultural Sciences (CXGC2021B12).

Conflicts of Interest: Authors declare no competing interests.

Appendix A

Desert locust invasion pathways, survival, and reproduction on the Tibetan Plateau in China. The desert locust swarms cannot pass the high-altitude but enter through the lower parts of the Himalayan river valleys. A limited number of locusts climb along the river valleys, step by step. When the altitude exceeds 3000 m, only a few locusts can fly and some may fall suddenly in flight. They are not adapted to the low temperatures and high humidity of the plateau climate and eventually die abnormally and are unable to reproduce.

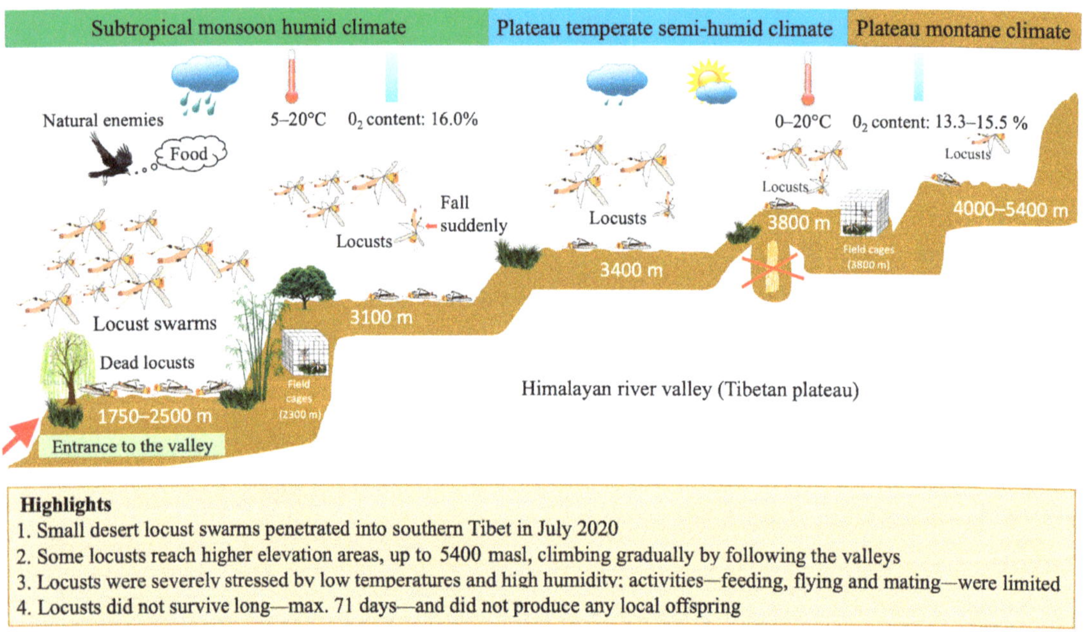

Figure A1. Desert locust invasion pathways, survival, and reproduction on the Tibetan Plateau in China.

References

1. Bennett, L.V. The development and termination of the 1968 plague of the desert locust, *Schistocerca gregaria* (Forskål) (Orthoptera, Acrididae). *Bull. Entomol. Res.* **1976**, *66*, 511–552. [CrossRef]
2. Steedman, A. *Locust Handbook*, 3rd ed.; Overseas Development Natural Resource Institute: London, UK, 1990.
3. Despland, E. Diet breadth and anti-predator strategies in desert locusts and other Orthopterans. *J. Orthoptera Res.* **2005**, *14*, 227–233. [CrossRef]
4. Roffey, J.; Popov, G. Environmental and behavioural processes in a desert locust outbreak. *Nature* **1968**, *219*, 446–450. [CrossRef]

5. Magor, J.I.; Lecoq, M.; Hunter, D.M. Preventive control and Desert Locust plagues. *Crop Prot.* **2008**, *27*, 1527–1533. [CrossRef]
6. Kennedy, J.S. The migration of the desert locust (*Schistocerca gregaria* Forsk). I. The Behaviour of Swarms. II. A theory of long-range migrations. *Philos. Trans. R. Soc. Lond. B Biol. Sci.* **1951**, *235*, 163–290.
7. Lorenz, M.W. Migration and trans-Atlantic flight of locusts. *ScienceDirect* **2009**, *196*, 4–12. [CrossRef]
8. Uvarov, B.P. A revision of the genus Locusta, L. (= Pachytylus, Fieb.), with a new theory as to the periodicity and migrations of locusts. *Bull. Entomol. Res.* **1921**, *12*, 135–163. [CrossRef]
9. Uvarov, B.P. Locust as a world problem. In Proceedings of the Première Conférence Internationale Pour la Protection Contre les Calamités Naturelles, Paris, France, 13–17 September 1937; Commission Française d'études des Calamités with the Support of Union Internationale de Secours: Paris, France, 1938; pp. 376–382.
10. Uvarov, B.P. *Grasshoppers and Locusts: A Handbook of General Acridology. II: Behaviour, Ecology, Biogeography, Population Dynamics*; Centre for Overseas Pest Research: London, UK, 1977.
11. Magor, J.I.; Ceccato, P.N.; Dobson, H.M.; Pender, J.; Ritchie, L. *Preparedness to Prevent Desert Locust Plagues in the Central Region, an History Review, No. AGP/DL/TS/35*; Food and Agriculture Organization of the United Nations: Rome, Italy, 2007. Available online: http://www.fao.org/ag/locusts/common/ecg/1288/en/TS35ePart1text.pdf (accessed on 17 June 2021).
12. Symmons, P.M.; Cressman, K. *Desert Locust Guidelines 1. Biology and Behavior*; Food and Agriculture Organization of the United Nations: Rome, Italy, 2001. Available online: http://www.fao.org/ag/LOCUSTS/common/ecg/347_en_DLG1e.pdf (accessed on 17 June 2021).
13. Cressman, K. Desert Locust. In *Biological and Environmental Hazards, Risks, and Disasters*; Shroder, J.F., Sivanpillai, R., Eds.; Elsevier: Amsterdam, The Netherlands, 2016; pp. 87–105.
14. Zhang, L.; Lecoq, M.; Latchininsky, A.; Hunter, D. Locust and Grasshopper Management. *Annu. Rev. Entomol.* **2019**, *64*, 15–34. [CrossRef]
15. FAO. *Desert Locust Bulletin. No. 486, General Situation during March 2019*; Food and Agriculture Organization of the United Nations: Rome, Italy, 2019. Available online: http://www.fao.org/ag/locusts/common/ecg/2471/en/DL486e.pdf (accessed on 17 June 2021).
16. FAO. *Desert Locust Bulletin. No. 488, General Situation during May 2019*; Food and Agriculture Organization of the United Nations: Rome, Italy, 2019. Available online: http://www.fao.org/ag/locusts/common/ecg/2475/en/DL488e.pdf (accessed on 17 June 2021).
17. FAO. *Desert Locust Bulletin. No. 501, General Situation during June 2020*; Food and Agriculture Organization of the United Nations: Rome, Italy, 2020. Available online: http://www.fao.org/ag/locusts/common/ecg/2556/en/DL501e.pdf (accessed on 17 June 2021).
18. FAO. *Desert Locust Bulletin. No. 502, General Situation during July 2020*; Food and Agriculture Organization of the United Nations: Rome, Italy, 2020. Available online: http://www.fao.org/ag/locusts/common/ecg/2559/en/DL502e.pdf (accessed on 17 June 2021).
19. Waloff, Z. The upsurges and recessions of the Desert Locust plague: An historical survey. *Anti-Locust Mem.* **1966**, *8*, 1–111.
20. Waloff, Z. Some temporal characteristics of Desert Locust plagues. *Anti-Locust Mem.* **1976**, *13*, 1–36.
21. COPR. *The Locust and Grasshopper Agricultural Manual*; Centre for Overseas Pest Research: London, UK, 1982.
22. Chen, Y.L. A new record of the desert locust, *Schistocerca gregaria* (Forskål) in Tibet, China. *Acta Entomol. Sin.* **1982**, *25*, 67.
23. Chen, Y.L. Watch out for the outbreaks of desert locusts. *Entomol. Knowl.* **2002**, *39*, 335–339.
24. Cai, B.H. *Taxonomy of Insects*; Financial and Economic Publishing House: Beijing, China, 1956; Volume 1.
25. Zhang, W. Will Desert Locust Invade Tibet? It's not just the Himalayas that Stand in Front of It. Available online: https://www.tellerreport.com/life/2020-03-13---will-desert-locust-invade-tibet---it-\T1\textquoterights-not-just-the-himalayas-that-stand-in-front-of-it-.BJDLn1YH8.html (accessed on 1 September 2021).
26. Rainey, R.C. Biometeorology and insect flight: Some aspects of energy exchange. *Annu. Rev. Entomol.* **1974**, *19*, 407–439. [CrossRef]
27. Cressman, K.; Stefanski, R. *Weather and Desert Locusts*; WMO-No. 1175; WMO: Geneva, Switzerland; FAO: Rome, Italy, 2016. Available online: https://library.wmo.int/doc_num.php?explnum_id=3213 (accessed on 17 June 2021).
28. Schaefer, G.W. Radar detection of individual locusts and swarms. In Proceedings of the International Study Conference on the Current and Future Problems of Acridology, London, UK, 6–16 July 1970; Hemming, C.F., Taylor, T.H.C., Eds.; Centre for Overseas Pest Research: London, UK, 1972; pp. 379–380.
29. Alexander, G. Occurrence of grasshoppers as accidentals in the Rocky Mountains of northern Colorado. *Ecology* **1964**, *45*, 77–86. [CrossRef]
30. Rainey, R.C. Meteorology and the migration of desert locusts. In *Applications of a Synoptic Meteorology in Locust Control*; Technical Notes, No. 54; World Meteorological Organization: Geneva, Switzerland, 1963. [CrossRef]
31. Shrestha, S.; Thakur, G.; Gautam, J.; Acharya, N.; Shrestha, J. Desert locust and its management in Nepal: A review. *J. Agric. Nat. Resour.* **2021**, *4*, 1–28.
32. Karki, S. *Locust: Winged-Terror Declines Gradually in Nepal*; Khabarhub: Kathmandu, Nepal, 2020. Available online: https://english.khabarhub.com/2020/07/110094/ (accessed on 1 September 2021).
33. Jia, S.L.; Wang, Q.; Li, L.; Fang, X.K.; Shi, Y.H.; Xu, W.G.; Hu, J.H. Comparative study on PCDD/F pollution in soil from the Antarctic, Arctic and Tibetan Plateau. *Sci. Total Environ.* **2014**, *497*, 353–359. [CrossRef]

34. Liu, D.X.; Tong, B.; Yang, Y.S.; Yao, X.Y.; Ding, Y.; Xia, Z.Y. Basic Characteristics and Proneness Evaluation about Debris Flow in Zhangmu Gully of Tibet. *Electron. J. Geotech. Eng.* **2016**, *21*, 7433–7446.
35. Liu, X.D.; Chen, B.D. Climatic warming in the Tibetan Plateau during recent decades. *Int. J. Climatol.* **2000**, *20*, 1729–1742. [CrossRef]
36. Wu, G.X.; Duan, A.; Liu, Y.; Mao, J.; Ren, R.; Bao, Q.; He, B.; Liu, B.; Hu, W. Tibetan Plateau climate dynamics: Recent research progress and outlook. *Natl. Sci. Rev.* **2015**, *2*, 100–116. [CrossRef]
37. Wu, S.H.; Yin, Y.H.; Zheng, D.; Yang, Q.Y. Climate changes in the Tibetan Plateau during the last three decades. *Acta Geogr. Sin.* **2005**, *60*, 3–11.
38. Neville, A.C.; Weis-Fogh, T. The effect of temperature on locust flight muscle. *J. Exp. Biol.* **1963**, *40*, 111–121. [CrossRef]
39. Roffey, J. Observations on gliding in the desert locust. *Anim. Behav.* **1963**, *11*, 359–366. [CrossRef]
40. Weis-Fogh, T. Biology and physics of locust flight II. Flight performance of the desert locust (*Schistocerca gregaria*). *Philos. Trans. R. Soc. Lond. B Biol. Sci.* **1956**, *239*, 459–510.
41. Harrison, J.F.; Kaiser, A.; VandenBrooks, J.M. Atmospheric oxygen level and the evolution of insect body size. *Philos. Trans. R. Soc. Lond. B Biol. Sci.* **2010**, *277*, 1937–1946. [CrossRef]
42. Hoback, W.W.; Stanley, D.W. Insects in hypoxia. *J. Insect Physiol.* **2001**, *47*, 533–542. [CrossRef]
43. Rascón, B.; Harrison, J.F. Oxygen partial pressure effects on metabolic rate and behavior of tethered flying locusts. *J. Insect Physiol.* **2005**, *51*, 1193–1199. [CrossRef]
44. Dahlhoff, E.P.; Dahlhoff, V.C.; Grainger, C.A.; Zavala, N.A.; Otepola-Bello, D.; Sargent, B.A.; Roberts, K.T.; Heidl, S.J.; Smiley, J.T.; Rank, N.E. Getting chased up the mountain: High elevation may limit performance and fitness characters in a montane insect. *Funct. Ecol.* **2019**, *33*, 809–818. [CrossRef]
45. Duranton, J.F.; Lecoq, M. *Le Criquet pèlerin au Sahel*; Collection Acridologie Opérationnelle n°6; Comité Inter-Etats de Lutte contre la Sécheresse dans le Sahel, Département de Formation en Protection des Végétaux: Niamey, Niger, 1990. Available online: http://locust.cirad.fr/ouvrages_pratiques/pdf/DFPV6.pdf (accessed on 2 June 2021).
46. Norris, M.J. Reproduction in the desert locust *Schistocerca gregaria* (Forsk.) in relation to density and phase. *Anti-Locust Bull.* **1952**, *13*, 1–49.
47. Gómez, D.; Salvador, P.; Sanz, J.; Casanova, C.; Taratiel, D.; Casanova, J.L. Desert locust detection using Earth observation satellite data in Mauritania. *J. Arid Environ.* **2019**, *164*, 29–37. [CrossRef]
48. Huque, H.; Jaleel, M.A. Temperature-Induced Quiescence in the Eggs of the Desert Locust. *J. Econ. Entomol.* **1970**, *63*, 1398–1400. [CrossRef]

agronomy

MDPI

Article

Early Intervention against Desert Locusts: Current Proactive Approach and the Prospect of Sustainable Outbreak Prevention

Allan T. Showler [1,*], Mohammed Abdallahi Ould Babah Ebbe [2], Michel Lecoq [3] and Koutaro O. Maeno [4]

1 USDA-ARS Knipling-Bushland U.S. Livestock Insects Research Laboratory, 2700 Fredericksburg Road, Kerrville, TX 78028, USA
2 Sahel Institute, Bamako, BP 1530, Mali; mohamedabdellahi.ebbe@cilss.int
3 CIRAD, UMR CBGP, F-34398 Montpellier, France; mlecoq34@gmail.com
4 Japan International Research Center for Agricultural Sciences, Livestock and Environment Division, Ohwashi 1-1, Tsukuba 305-8686, Japan; kmaeno@affrc.go.jp
* Correspondence: Allan.Showler@usda.gov

Abstract: The desert locust, *Schistocerca gregaria* (Forskål) (Orthoptera: Acrididae), a major Old World pest, is associated with agricultural losses and undesirable societal effects. There are three broad approaches to its control: reaction, proaction, and outbreak prevention. Reaction protects crops from swarms but it is costly and disruptive. Proaction involves early intervention during outbreaks to avert further development to plague status; it is in current use because it is effective, relatively inexpensive, and it is the best available option for now. Outbreak prevention, largely unavailable since the 1970s, at least on a regional scale, will require highly sensitive surveillance to detect the onset of gregarization. Sufficiently early intervention can, hypothetically, extend desert locust recession indefinitely. While research on desert locust biology and behavior is, almost, no longer an urgent requirement to improve the efficacy of control, new priorities have arisen for developing outbreak prevention capability (and for enhancing proaction). Salient needs presently include long residual tactics for prophylactic (preventive) control in breeding areas, intervention thresholds, and improved, sustainable coordination among stakeholders at national, regional, and international levels. The most recent desert locust episode of 2020 provides an illustrative example of how prevention might have averted the entire upsurge, and how proaction in some countries contained the spread of swarms. The initial outbreak in Saudi Arabia escaped control due to unpreparedness, and impacts of armed conflict in Somalia and Yemen, which weakened surveillance and control, further contributed to the invasion of ≥22 countries, and the spraying of ≈4.9 million ha, by the end of 2020.

Keywords: control; plague; proaction; *Schistocerca gregaria*; surveillance; threshold; upsurge

Citation: Showler, A.T.; Ould Babah Ebbe, M.A.; Lecoq, M.; Maeno, K.O. Early Intervention against Desert Locusts: Current Proactive Approach and the Prospect of Sustainable Outbreak Prevention. *Agronomy* **2021**, *11*, 312. https://doi.org/10.3390/agronomy11020312

Academic Editor: Imre J. Holb
Received: 8 January 2021
Accepted: 7 February 2021
Published: 10 February 2021

1. Introduction

Basic scientific principles for controlling the desert locust, *Schistocerca gregaria* (Forskål) (Orthoptera: Acrididae), were established during the first half of the 20th century. Following the elucidation of phase theory (whereby the pest can change between solitary and gregarious morphologies and behaviors) by Uvarov [1] and the location of major desert locust breeding areas in which outbreaks tend to originate, the Fifth International Locust Conference, held in Brussels, 1938, promulgated the need for international coordination as well as the possibility of developing a preventative strategy against swarm formation [2]. Since that time, scientific and technical advances have been substantial, greatly improving approaches to surveillance and control [3,4]. Desert locust control strategy, however, has remained a topic of considerable debate. In recent years there has been extensive discussion of early intervention strategies [3–16]. The desert locust is a biologically and behaviorally complex insect [11,17] that, when gregarious and uncontrolled, is often fraught with adverse economic, environmental, and social consequences [3,8,10,11,16,18–21].

2. Prevention Is Relative

Constructive discussion of desert locust control is often confounded by different perspectives on common terminology [11]. Regarding "prevention", confusion arises from disharmonious views on what we intend to prevent. Do we permit the occurrence of upsurges before intervening to prevent plague development? Or is the ultimate goal to prevent outbreaks, thereby maintaining recession status indefinitely [9]? To prevent is to avert an event or condition, precluding its occurrence. Outbreak prevention is gregarization management; it entails intervention prior to, or during, phase transformation from solitary to gregarious [9,10,12,21]. Intervention during outbreaks, on the other hand, is aptly described as "proaction", aimed at halting an outbreak's expansion toward plague status [9]. There are, broadly, three desert locust control approaches: reaction, proaction, and outbreak prevention [4,8–11].

2.1. Reaction

The reactive approach involves intervention after economically and socially threatening desert locust upsurges and plagues are underway [11], a result of unpreparedness to take action earlier. Reaction primarily focuses on defensively protecting croplands from destruction instead of offensively intervening in the main breeding areas (commonly in remote regions) [11,17,19]. Major upsurges and plagues, unfortunately, are associated with serious economic losses and social upheaval [11,20].

Some observers from countries unafflicted by desert locusts have espoused allowing plagues to build and run their course without intervention; they propose that farmers who incur losses will later receive compensation through insurance policies [22], and that reactive campaigns are, on average, less expensive [13,14] than early intervention. Both views are cavalier, after Europe and United States took strong measures to decimate their own locust populations (e.g., Rocky Mountain locusts, *Melanoplus spretus* [Walsh]) [11,23], and because insurance is impractical where it is largely an unfamiliar concept and where subsistence agriculture predominates. Calculating costs associated with gregarious desert locust episodes is challenging [8,9,11,19,24] because it involves the monetary value of crops; the social implications of subsistence agriculture; negative impacts on pasturage, rangeland, and livestock; food aid; assistance for displaced people; and long-term impact on populations and education outcomes [20,21,25]. Economic assessments are available [17,18,26–31], but they are not necessarily complete. Agricultural losses were suggested as being insufficient to warrant investment in control [22], but that was founded on the cash value of selected crop systems without considering the array of other factors. The 2003–2005 upsurge, for example, was conservatively assessed to have inflicted an average crop loss of ≈30%, but that estimate failed to incorporate losses incurred by small Sahelian farms and to consider areas suffering from particularly severe food shortages [11,20,21]. The 2003–2005 upsurge further resulted in displacement of farmers and pastoralists to urban areas; conflicts between farmers, nomads, and pastoralists over limited resources; declines in market commodities that offset agricultural product prices; and food aid from international agencies (donors), and multiple cascades of indirect socio-economic impacts [20,29–34]. Belayneh [20] conservatively estimated crop losses of 10–20%, while loss of anticipated cereal harvest reached 80%, 90%, and 100% during 2004 in areas of Burkina Faso, Mali, and Mauritania, respectively [21]. Legume crop reductions were as much as 90% in the same three countries; further, ≈30% of pasture vegetation was destroyed in Mali and Burkina Faso, and fodder production was reduced by up to 80% in Mauritania, resulting in a total of ≈8.4 million affected sahelian farmers and pastoralists [21].

While it can be argued that reaction is, technically, preventive by aiming to avert crop losses under some circumstances, reaction is not regarded as being a preventive strategy. Reaction is instead a default response when desert locust activity becomes overwhelming and threatens agricultural production [9–12]. Examples of reactive control include the 1986–1989, 1992–1995, and the 2003–2005 campaigns [5,8,10,11,20]. Each of those campaigns was expensive in terms of international aid (US$ millions ≈300, ≈19, and

≈35, respectively) and involved 18–23 desert locust afflicted countries in Africa and Asia (Table 1) [5,8,10,11,20].

Table 1. Desert locust episode and control effort statistics 1986–2016 [11].

Desert Locust Episodes	No. Countries	Ha Sprayed	Duration	Cost to Donors [b]	Intervention Timing
1986–1989	23	25 million	4 years	$310 million	Late
1992–1994	18	4 million	2 years	$18.8 million	Moderate
1997–1998	7	430,000	7 months	$30,000	Early
2003–2005	20	13 million	2.5 years	$35 million	Late
2007–2016 [a]	22	1.8 million	10 years	<$1 million	Early

[a] Ten-year period involved numerous outbreak episodes. [b] US$.

2.2. Proaction

Proaction is an intermediate response between reaction and outbreak prevention [9,12,26,35,36]. The progression of desert locust populations from recession to plague status is a continuum; outbreak prevention occurs at the initial stage, while later intervention during outbreaks is proactive. Although both terms refer to early interventions for averting plagues, the difference is in the timing. Conflation of outbreak prevention with proaction (i.e., [13,14]) obscures the important distinction between them [4,9,11]. Historical records show that the longer gregarious populations breed, the greater the risk of being overwhelmed by the locust situation and the greater the likelihood of reaching plague status [4,11,25,28,37].

Proactive intercession against desert locusts is initiated based on an often ad hoc blend of estimated gregarizing populations, local control capabilities, political pressure, and gestalt [9]; there is no empirically calculated intervention threshold. The default is killing all gregarious nymphal bands and swarms [4], representing zero-tolerance. The best available approach is proaction (Figure 1) and, following the 1986–1989 plague, it has had different degrees of success [4,8–11,20,25,38,39].

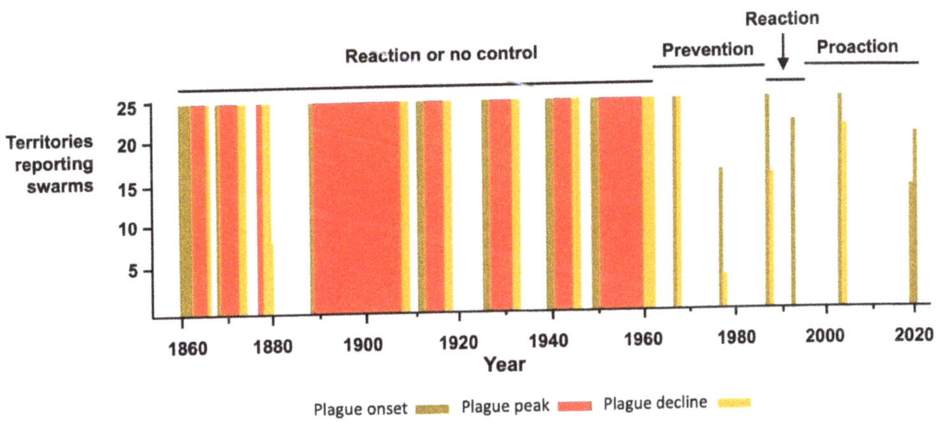

Figure 1. Bars represent years during which major desert locust activity (major upsurges and plagues) occurred (adapted and modified from 3), 1860–2020 (most recent upsurge extended past 2020); prevalent control approaches are indicated, prevention involved use of long-residual organochlorinated insecticides.

Desert locust episodes from 1986 to the present (2020) are relatively well-documented. Post-1970s, campaigns discontinued use of highly persistent, broad spectrum organochlorinated insecticides (e.g., DDT, dieldrin) due to environmental concerns [19,40]. Long-residual insecticides, however, were useful (Figure 1) for applying in limited swathes, or bands, on vegetation in breeding areas that killed nymphs crawling into those areas by direct contact and ingestion of treated foliage [25]. Early control of desert locust outbreaks was obviously weakened by the banning of dieldrin [25,41–43]. Control since the 1970s has relied almost exclusively on relatively short-residual insecticides, including malathion, propoxur, and some pyrethroids [19]. Contemporary inability to control nymphal bands using long residual prophylactic methods means that spraying must be spatially and temporally accurate. Reliance on short residual insecticides has resulted in the need to take more individual control actions against nymphal bands, as well as highly mobile adult swarms. This has increased areas sprayed and numbers of afflicted countries, and intensified dependence on international aid [11]. In terms of insecticide treated areas, the 1997–1998 outbreak, involving proaction, was only 1.7% and 3.3% the magnitude of the 1986–1989 plague and the 2003–2005 upsurges, respectively [11]. During the ten-year period 2007–2016, when there were no plagues (Figure 1) but numerous outbreaks, proaction likely contributed to limiting the magnitude of sprayed area, across all ten years, to 7.2% and 13.8% of the 1986–1989 and 2003–2005 episodes, respectively (Table 2) [11]. International aid supporting proaction against the 1997–1998 outbreak was 0.01% and 0.3% of costs incurred during reactive 1986–1989 and 2003–2005 campaigns, respectively [11]. Accrued donor expenditures for 2007–2016 constituted only 0.1% and 2.9% of the 1986–1989 and 2003–2005 campaigns, respectively [11]. Furthermore, the proactive 1997–1998 campaign took just seven months, whereas the reactive campaigns persisted for 2.5–4 years [5,11,20].

Table 2. Numbers of countries infested, percentage of months in which proactive control occurred, area sprayed with insecticides, and number of times FAO Bulletin used "alarm" words to describe the desert situation, 2007–2016 (adapted from 11).

Year	No. Countries Infested	% Months with Proactive Intervention	Hectares Sprayed	Number of Alarm Words [a]
2007	14	100	255,887	18
2008	9	75	64,744	0
2009	9	92	37,613	12
2010	10	75	48,163	3
2011	11	83	197,380	11
2012	9	92	215,660	26
2013	11	100	482,159	21
2014	13	100	373,231	11
2015	5	50	108,162	0
2016	11	92	54,916	15

[a] Alarm words: critical, extremely alarming, extremely/very serious, outbreak, potentially dangerous, potentially threatening, potentially very dangerous, serious, and threatening.

Belayneh [20] concluded that lack of proactive intervention in 2003–2005 allowed desert locust populations to breed until attaining upsurge status; the dynamic was mainly curbed by unfavorable climate-driven conditions. Had proactive control been conducted sufficiently early to spare 10% of the 2003–2005 agricultural losses, US$ 226 million might have been conserved [20]. The cost of control during the 2003–2005 upsurge equaled projected costs of 170 years of proactive control [37].

Effective proaction hinges on the efficiency of outbreak suppression in breeding areas [19,25,37,44]. During the decade 2007–2016, outbreaks were usually controlled within five months [11]. Proactive interventions helped to hasten return to recession status before outbreaks could expand into upsurges (Figure 1) [11]. Although not every outbreak during the ten years might have become an upsurge, some of them produced swarms that moved beyond breeding areas and reproduced in other regions (Table 2) [45] before being controlled. The Food and Agriculture Organization (FAO) monthly bulletins warned, commonly in ominous terms (Table 2), that desert locust populations and habitat conditions favored imminent development toward upsurges [11,45]. Claims that ten years of proactive strategy is riskier than reactive campaigns [13] have not been supported by events [11].

2.3. Outbreak Prevention

Outbreak prevention refers to intervention that is sufficiently early to avert the onset, and initial development, of gregarization before and when nymphs first amass in small patches and adults gather in loose groups [9,10,12]. The approach will eliminate a sufficient proportion of the population to suppress pheromone cues associated with crowding (crowding elicits positive feedback for building fully gregarious populations) that trigger gregarization [4,46–54]. Sustainable interference with semiochemical gregarization triggers can, hypothetically, maintain recession status indefinitely [8,9,11,12]. Although proaction is an improvement over reaction, an earlier, preventive, approach offers an even more efficient control option [4,10,11,55]. While outbreak prevention is the logical ultimate goal of contemporary research efforts, it has not yet been implemented without using long-residual organochlorinated insecticides or through sufficiently early detection and intervention, as was accomplished in Mauritania in 2009, and again several years later [56]. On the other hand, preventive locust control strategies had achieved success against some swarming species, such as the red locust, *Nomadacris septemfasciata* (Serville), in southern Africa [57], and the South American locust, *Schistocerca cancellata* (Serville) [17,58]. Assuming that outbreak prevention becomes possible, it is likely that proactive interventions will continue intermittently until prevention is sustainable. A proactive contingency capability will likely remain prudent, even when outbreak prevention is sustainable, to counter unforeseen gregarization events.

3. Proaction: A Technological and Social Accomplishment

Surveillance locates desert locust populations for control. Both surveillance and control have benefitted from technological advances, as has communication, which is essential for gathering data from afflicted countries, disseminating desert locust forecasts from the FAO, and intra- and interregional coordination.

3.1. Surveillance

Proactive intervention relies upon well-targeted, accurate, and quickly communicated surveillance information, particularly in areas prone to producing gregarious populations. The geographic foci of surveillance efforts can change over time depending upon several factors, including weather and greenness, and it is accomplished by multiple means involving different levels of technological sophistication [5,8,11]. Surveillance must be sustainable in order to produce routine reports during times of calm as well as site-specific, detailed reports contributing to national, regional, and international data bases to better understand population dynamics [7,10,11,26,59]. In general, information gathered by technologically sophisticated methods requires relatively low-technology ground verification in order to detect and characterize actual desert locust activity.

Technological improvements to surveillance began during the mid-1980s, including use of remote sensing imagery (particularly to assess cover by green vegetation and soil moisture) and climatic and weather data collection methods, both of which help to anticipate desert locust activity [11,25,38,44,60–66]. These factors, combined with historical data and field-based reports, identify where conditions are conducive to outbreaks [8,10,11,15,59]. Suspect areas can be monitored for incidence of gregarizing populations [9,10,59] by scouts in terrestrial vehicles, and by obtaining reports from farmers, nomads, administrative authorities, militaries, and sometimes fire lookouts in the Atlas Mountains and rebel groups in contested territories [5,8,11,59]. Aerial surveillance (and control) is conducted, when possible, by the regional Desert Locust Control Organization for Eastern Africa (DCLO-EA), which has aircraft [67] (Saudi Arabia has its own aircraft). Field information is radioed to national ministries of agriculture that transmit reports by email and other electronic means to regional coordination centers and to the desert locust information service of FAO in Rome. Recently developed technology (elocust 3) [68] is used to transmit, in parallel, information to national locust units and to the Food and Agriculture Organization. The information is collated, synthesized, interpreted, and reported back to the afflicted and donor countries accompanied by desert locust activity forecasts. The value of technological advances for enhancing surveillance, communication, and forecasting is corroborated by successful proaction from 2007 through 2016 [11].

3.2. Control

Following the shift away from long-residual organochlorinated insecticides during the 1980s [3,40], use of conventional, synthetic, relatively short residual insecticides sprayed directly onto nymphal bands and settled swarms has been the only available control tactic [10,25], excluding recent limited use of entomopathogens. Short residual insecticides are effective when desert locust populations are detected early and with sufficient accuracy to halt outbreaks. Currently used insecticides can also be applied as "barriers" against nymphal bands [66,69,70], but this is not particularly efficacious for long-term prophylactic control. Selective insecticides, including insect growth regulators [25], might also be efficacious as barrier treatments, but they are unlikely to be more persistent than conventional insecticides. Entomopathogens are commercially available, and others are being evaluated [45,66,71]. In 2020, entomopathogens were applied to at least 10,845 ha in Somalia against desert locusts [72,73]. Other alternative tactics have not been reported [10,25].

Additional technological advances include ultra-low volume (ULV) spray equipment, now in routine aerial and terrestrial use against desert locusts [44,66]. ULV formulations are oil-based, reducing droplet evaporation such that only 0.5–1.0 L of the insecticide is needed per hectare [44], decreasing environmental contamination and costs of transportation, handling, and storage [11]. In addition, global positioning systems (GPS) technology improves precision of insecticide applications, enhances navigation to targets that are often located in vast, remote wilderness [25,66], and reduces spurious environmental exposure to, and wastage of, insecticides. Research is underway to test the utility of unmanned aerial vehicles (UAV) or drones for desert locust detection and control.

4. Outbreak Prevention: Salient Requirements

The most important research goal now is arguably the development of a reasonably long-residual and environmentally benign control tactic that, whether chemical, biological, or other, is useful in a prophylactic context. Furthermore, the length of time constituting a reasonably long effective residual (days, weeks, months) should be established.

Because early intervention occurs within limited windows of opportunity, insecticide application timing is important even for reasonably long-residual tactics; hence, surveillance must be able to sustainably detect the onset of gregarization. It is not yet clear how this can be achieved, but it will rely, at least in part, on global information systems (GIS) imagery and weather information [4,11]. Outbreak prevention will require study of geo-historical desert locust survey and control to refine our knowledge of breeding and phase transformation in salient areas, such as has been accomplished in Mauritania and a few other countries [15,59,74–76]. Outbreak prevention will also entail use of empirical measurements that serve as triggers, or thresholds, for interventions [4,46]. It is likely that thresholds will have to be tailored for local conditions across time (e.g., seasonally) and in different areas [4]. An intervention threshold might be based on integrations of population density, food source abundance, and food source clumping which, in certain configurations, are associated with the initiation of gregarization [4,46]. In addition to using visual assessments of potentially gregarious desert locust activity as intervention threshold tools, it might be possible to develop a sensor for detecting semiochemicals associated with phase transformation as a cue to intervene.

Surveillance of population densities alone is inadequate for predicting gregarization because of differences between discrete desert locust habitats [4], which extend from Mauritania to India [11,17]. In western Africa, limited north-south movement of the intertropical convergence zone (ITCZ), which can be tracked, affects amounts of precipitation, and wind speed and direction, across the Sahel [15]. This provides a broad notion of where conditions are likely to favor desert locust population intensification, and, because swarms move downwind, projections about areas at risk of invasion [15]. A study in Mauritania and Morocco determined that a displacement cycle for swarms, largely based on ITCZ position, has a seasonal association, but this has not been confirmed for solitarious populations [15]. Population increases during recessions were associated with areas that received sufficient rainfall, which changes spatially over time because the ITCZ moves [15]. Following rainfall, persistence of water beneath the soil surface in wadis and other low-lying areas support flushes of vegetational growth [15]. The foliage supplies enough food to increase desert locust populations that, when conditions dry and the vegetated areas recede, results in crowding, semiochemical release, and gregarization [11,49]. Anticipation of gregarization based upon suitable breeding conditions has helped to protect northwestern Morocco from swarms developing south of the Atlas Mountains (e.g., the Sahel) since 1988 [15], a dividend of proaction [11]. Improved surveillance by ground teams can be attributed to superior map-based orientation, the effort invested, and training [11,15]. Heightened ability of ground surveillance teams to find desert locusts; however, might produce skewed information unless scouts also report where desert locusts were not observed [15]. Failure to record nonpopulated locations can result in inaccurate estimates of area-wide desert locust activity (it can also bias historical desert locust activity data) [15].

Habitat differences involve variables that might affect the propensity of a desert locust population to initiate gregarization. Understanding how food source clumping, community and individual plant structure and phenology, nutritional value, and availability influence population density is important for assessing the risk of gregarization [4,46,52,75,77]. To streamline scouting, areas that need to be searched are reduced by use of timely forecasts [5,12], and although currently gathered information is sufficient to support proactive interventions, further refinements might better support outbreak prevention [4]. As an example, GIS technology has not yet been used to detect and map plant species that are particularly favorable to desert locust population increases [4] (GIS mapping of a selected low-growing plant species in Texas; however, has been accomplished to determine ixodid (hard tick) distribution [78]). Transect counts of desert locusts and important plant species might be useful for making risk assessments [4,46] and associated intervention determinations.

Other relatively recent reports indicate that there are other facets of spatio-temporal aggregation patterns inherent to desert locust ecology that might be relevant to surveillance efforts. A field study conducted in Mauritania [77] confirmed observations that marching nymphal desert locust bands passed some plants and eventually roosted on patchily distributed trees near dusk, forming aggregations [79]. Maeno and Babah [77] found that gregarious bands of nymphs condensed on, and around, the largest tree in the local plant community; adults mostly roosted on large trees and substantial bushes [80,81]. Surveillance might also benefit from findings that indicate variability desert locust defensive responses to predators and likely to approaching terrestrial surveillance personnel (on foot and in vehicles). Whereas adult desert locusts usually take flight when temperatures exceed the minimum threshold for flight ($\approx$22 °C) [80,82], at cooler temperatures, locusts in low bushes tend to move into the middle of the bush upon which they are roosting in order to hide; individuals on taller plants ($\geq$2 m) tend to remain stationary [80].

5. Impediments

Lecoq [25] and Doré et al. [29] emphasized that impediments to proactive and preventive strategies require thorough analysis and documentation. Although some authors (e.g., [13,14]) have suggested that the sole obstacle has been funding shortfalls, the scope of impediments is considerably wider and more complex. The array of challenges confronting many afflicted countries has been fundamental to the development of plagues for $\geq$40 years. Challenges that intertwine with and compound one another have been described [3–5,8,10,11,16,83]: unpreparedness [5,11,18,20,25,62,67], insecurity [11,83], remote, rugged terrain and poor infrastructure [8,11], environmental concerns [11,18,19,40,67], political hindrances [3,5,8,16,20,62], dogmas and assumptions [11], and research impediments [11]. The negative impacts of many challenges can be mitigated by advances in technology and appropriate funding. One notably intractable exception is insecurity [11,83,84], which can cripple desert locust surveillance and control.

6. Changing Priorities

Major outbreak areas, and the ecological conditions and gregarization mechanism that initiate outbreaks, are known, and an early warning system of surveillance and reporting is in place [3,4,11,36]. Now that we have a substantially improved grasp of the desert locust's complex ecology, the next steps should focus to an increasing extent on the complexities of control systems governance [3,85]. The weakest link in desert locust control is unpreparedness due to problems with organization, coordination, logistics at international and national levels [3,4], and insecurity [83]. Unpreparedness is largely the result of inactive recessions when the complex web of interactions that enable control seizes up and donor funding declines [16,86]. Additionally, personnel turnover depletes trained cadres and institutional memory fades in the afflicted countries [16]. Hence, when proactive interventions hold desert locusts in the solitary phase for long periods, maintenance of surveillance and control capabilities declines, accompanied by loss of focus, equipment, insecticides, personnel, and expertise [16,86].

To muster coherent early responses to dynamic desert locust episodes, regional and international cooperation must be well coordinated. Increasing preparedness will involve strengthening national locust control units and making coordination among national, bilateral, and multilateral stakeholders a sustainable priority [3]. At the urging of international aid agencies, the FAO instituted the Emergency Prevention System (EMPRES) in 1996 to operate in parts of the desert locust recession areas [3]. EMPRES aimed to increase the capabilities of afflicted countries, regional desert locust control organizations, and the FAO [3,4,6,87]. The program has coordinated the development and application of technologies, and training, that improved upon proactive control [3,11]. Because the issue of financial sustainability of desert locust control systems is of critical importance, an institutional framework providing for graduated mobilization of funding and sources, depending on the severity of the desert locust episode, has been suggested [87,88]. An initial effort to develop such a fund involves the Commission for Controlling the Desert Locust in the Western Africa Region (CLCPRO), based in Algiers, in 2016 developed a regional fund to proactively control desert locust invasions, endowed with ≈6.5 million US dollars from member countries (Algeria, Burkina Faso, Chad, Libya, Mali, Mauritania, Morocco, Niger, Senegal, and Tunisia) [89,90]. A risk management plan, possibly including intervention thresholds, can be formulated prescribing coordinated stakeholder actions for different levels of desert locust activity [4]. Coordination must be strong at all times in order to maintain vigilance and early response capability during recessions [3,16]. Multi-agent modeling has been suggested as a way of better understanding stakeholder interactions and system responses [16].

7. Conclusions: New Upsurge

The most recent major desert locust upsurge became apparent during winter and spring 2020. The episode originated when heavy rains fell during the latter half of 2018 in the Rub al Khali, or Empty Quarter, of Saudi Arabia [90]. The resulting favorable breeding conditions supported three undetected generations [90], and in December, swarms spread elsewhere in Saudi Arabia, as well as to Yemen, Egypt, Eritrea, Sudan, and southwestern Iran, where breeding occurred (Figure 2) [91–93]. Control operations were conducted in Saudi Arabia, Iran, Eritrea, Egypt, and Sudan through spring 2019. Yemen, beset by armed conflict, and Iran mustered weak responses and swarms from Iran invaded the India-Pakistan border where three generations produced more swarms (Figure 2) [91]. Swarms from Yemen crossed the Red Sea to northern Somalia and Ethiopia (Figure 2), where breeding produced additional swarms [90]. Survey reports were not obtained from Somalia through spring 2019 [92–96], but it is likely that gregarious populations continued to arrive from Yemen and reproduce. During summer 2019, desert locust activity and control operations occurred in Ethiopia, Saudi Arabia, Iran, India, and Pakistan [96–98]. Surveillance remained weak in Somalia and Yemen, where gregarious populations were not controlled [97–99]. During fall and winter 2019–2020, desert locust activity resumed in Eritrea and Sudan, continued in Saudi Arabia, India, Pakistan, and, in December, Iran (Figure 2) [100–105]. Kenya was invaded (Figure 2) in December 2019, and infestations there intensified through February 2020, accompanied by control operations [104,105]. Furthermore, populations arriving from Saudi Arabia increased in Oman until, in the latter half of winter, intervention occurred [102–106]. Gregarious populations in Somalia and Yemen persisted while surveillance remained weak and control was limited [100–106]. During spring (March–June) 2020, infestations in Eritrea declined, while substantial gregarious populations and control operations continued in Ethiopia, Kenya, Oman, Saudi Arabia, Iran, Pakistan, and, by late spring, India [72,73,107]. Surveillance and control operations in Somalia and Yemen were spotty, but gregarious populations were likely to have been substantial [72,73,107]. During the episode, other countries were invaded and undertook control operations, particularly Iraq, Jordan, Kuwait and the United Arab Emirates, and infestations were comparatively light in Afghanistan, Bahrain, Djibouti, Libya, South Sudan, Tanzania, and Uganda (Table 3) [72,73,92–108]. During the first two years of the episode, June 2018 to June 2020, swarms were observed in $\geq$22 countries, and spray operations occurred in $\geq$19 countries, covering >2,792,840 ha (Table 3) [72,73,92–107].

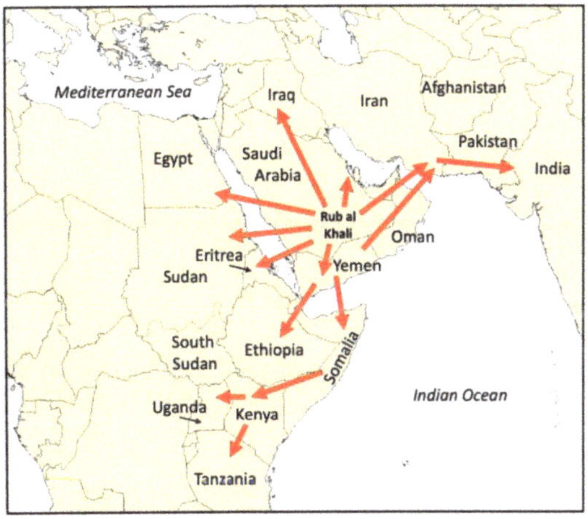

Figure 2. Generalized movements of major gregarious desert locust swarms, beginning June 2018 in the Rub al Khali of Saudi Arabia, moving out from the Rub al Khali by December 2018 and beyond Saudi Arabia through June 2020 [72,73,90–106].

We suggest that, had the initial outbreak in the Rub al Khali of Saudi Arabia been detected in summer 2018, swarms might not have reached Yemen, the Red Sea coast of Africa, and parts of Iran. While substantial control operations commenced December 2018 in Saudi Arabia, Egypt, Eritrea, and Sudan, the capabilities of Yemen and Somalia, both hobbled by insecurity, harbored and produced swarms that invaded other countries, and the initially weak response in Iran permitted swarms to move to Pakistan and India [91]. Because control was conducted in most of the other afflicted countries (Table 3), it is likely that once Saudi Arabia had begun to control its desert locusts, Somalia and Yemen continued to drive expansion and intensification. Effective surveillance and control in those two countries would have eliminated further spread in concert with the substantial control operations that occurred in Egypt, Eritrea, Ethiopia, and Sudan, and some other countries adjacent to them (e.g., Tanzania, Uganda). Similarly, effective early intervention in Iran would have spared Pakistan and India. The episode illustrates how a chain of worsening desert locust population events resulted because of unpreparedness and insecurity in a few countries [109].

Table 3. Countries invaded by desert locust swarms and areas sprayed with insecticides, December 2018–June 2020 [72,91–106].

Country	Area (ha) Sprayed
Afghanistan	20
Bahrain	34
Djibouti	0
Egypt	19,802
Eritrea	93,586
Ethiopia	288,205
India	427,233
Iran	887,657
Iraq	2610
Jordan	2900
Kenya	102,342
Kuwait	15,653
Libya	70
Oman	13,164
Pakistan	453,892
Saudi Arabia	402,881
Somalia	26,845
South Sudan	0
Tanzania	0
Uganda	607
United Arab Emirates	5904
Yemen	49,435
Total: 22	**Total: 2,792,840**

The episode also demonstrates that proactive interventions, which occurred in many of the afflicted countries, impeded significant spread beyond their borders [72,73,92–107]. Once gregarious populations arrived in India and Pakistan, for example, relatively few swarms escaped those two countries, and the same is true for Egypt, Eritrea, Ethiopia, and Sudan. Saudi Arabia, the initially unprepared source of the upsurge, could not avert movement into Bahrain, Iraq, Kuwait, Oman, United Arab Emirates, Yemen, and likely parts of the Red Sea coast of Africa. A timely proactive approach (or, hypothetically, an outbreak prevention approach) in the Rub al Khali, a relatively limited area, would have pre-empted the possibility of an upsurge.

Based upon the advantages of early intervention, stakeholders must recognize the difference between proaction and outbreak prevention to distinguish intermediate from ultimate goals. While proaction has been successfully implemented starting in the 1990s (Figure 1) [4,9,11,12], a new approach to outbreak prevention that does not rely on environmentally persistent, broad-spectrum pesticides has not yet materialized. Advances that might contribute toward a viable outbreak prevention strategy, however, continue to unfold. Although surveillance and control were weak in Somalia, entomopathogens were applied there on at least two occasions [72,73], representing a step toward alternative, increasingly environmentally benign control, and possibly toward development of persistent formulations for prophylactic outbreak prevention. We also suggest that proaction and outbreak prevention require more smoothly operating national and regional organizations dedicated to early intervention, technical and funding mechanisms that provide sustainable support during recessions and gregarious episodes, heightened sustained vigilance, alleviation of impediments, and, with particular regard to armed conflict, some luck, even when mitigation measures are available [84].

Author Contributions: Conceptualization, A.T.S.; methodology, A.T.S., M.A.O.B.E., M.L. and K.O.M.; software, A.T.S. and M.L.; validation, A.T.S., M.A.O.B.E., M.L. and K.O.M.; formal analysis, A.T.S. and M.L.; investigation, A.T.S., M.A.O.B.E., M.L. and K.O.M.; resources, A.T.S. and M.L.; data curation, A.T.S. and M.L.; writing–original draft preparation, A.T.S.; writing–review and editing, A.T.S., M.A.O.B.E., M.L. and K.O.M.; visualization, A.T.S. and M.L.; and supervision, A.T.S. All authors have read and agreed to the published version of the manuscript.

Funding: This research received no external funding.

Institutional Review Board Statement: Not applicable.

Informed Consent Statement: Not applicable.

Data Availability Statement: Data sharing not applicable.

Conflicts of Interest: The authors declare no conflict of interest.

Disclaimer: Mention of trade names or commercial products in this publication is solely for the purpose of providing specific information and does not imply recommendation or endorsement by the U.S. Department of Agriculture. USDA is an equal opportunity provider and employer.

References

1. Uvarov, B.P. A revision of genus *Locusta* (= *Pachytylus* Fieb) with a new theory as to the periodicity and migrations of locusts. *Bull. Entomol. Res.* **1921**, *12*, 135–163. [CrossRef]
2. International War on Locusts. *Nature* **1938**, *142*, 565. [CrossRef]
3. Lecoq, M. Desert locust management: From ecology to anthropology. *J. Orthopt. Res.* **2005**, *14*, 179–186. [CrossRef]
4. Sword, G.A.; Lecoq, M.; Simpson, S.J. Phase polyphenism and preventative locust management. *J. Insect Physiol.* **2010**, *56*, 949–957. [CrossRef] [PubMed]
5. Showler, A.T.; Potter, C.S. Synopsis of the desert locust, *Schistocerca gregaria* (Forskål), plague 1986–1989 and the concept of strategic control. *Am. Entomol.* **1991**, *37*, 106–110. [CrossRef]
6. FAO (Food and Agriculture Organization of the United Nations). *Annual Report 1999. Emergency Prevention System for Transboundary Animal and Plant Pests and Diseases. Desert Locust Central Region Programme*; FAO: Rome, Italy, 1999. Available online: http://www.fao.org/ag/locusts/oldsite/PDFs/meetings/EMPRES99e.pdf (accessed on 29 October 2020).
7. FAO (Food and Agriculture Organization of the United Nations). *Emergency Prevention System (EMPRES) for Transboundary Animal and Plant Pests and Diseases (Desert Locust Component), Central Region. Summary Report of the Evaluation Mission*; FAO: Rome, Italy, 2003. Available online: http://www.fao.org/3/a-bd101e.pdf (accessed on 29 October 2020).
8. Showler, A.T. Locust outbreak in Africa and Asia, 1992–1994. *Am. Entomol.* **1995**, *41*, 179–185. [CrossRef]
9. Showler, A.T. Proaction: Strategic framework for today's reality. In *New Strategies in Locust Control*; Krall, S., Peveling, R., Ba Diallo, D., Eds.; Birkhauser: Berlin, Germany, 1997; pp. 461–465. [CrossRef]
10. Showler, A.T. A summary of control strategies for the desert locust, *Schistocerca gregaria* (Forskål). *Agric. Ecosyst. Environ.* **2002**, *90*, 97–103. [CrossRef]
11. Showler, A.T. Desert locust control: The effectiveness of proactive interventions and the goal of outbreak prevention. *Am. Entomol.* **2018**, *65*, 180–191. [CrossRef]
12. Babah, M.A.O. Strategy for controlling the desert locust in Mauritania. In *New strategies in Locust Control*; Krall, S., Peveling, R., Ba Diallo, D., Eds.; Birkhauser: Berlin, Germany, 1997; pp. 487–491. [CrossRef]
13. Van Huis, A.; Cressman, K.; Magor, J.L. Preventing desert locust plagues: Optimizing management interventions. *Entomol. Exp. Appl.* **2007**, *122*, 191–214. [CrossRef]
14. Magor, J.I.; Lecoq, M.; Hunter, D.M. Preventive control and desert locust plagues. *Crop Prot.* **2008**, *27*, 1527–1533. [CrossRef]
15. Piou, C.; Bacar, M.H.J.; Babah, M.A.O.; Chiharane, J.; Ghaout, S.; Cisse, S.; Lecoq, M.; Halima, T.B. Mapping the spatiotemporal distributions of the desert locust in Mauritania and Morocco to improve preventive management. *Basic Appl. Ecol.* **2017**, *25*, 37–47. [CrossRef]
16. Gay, P.M.; Lecoq, M.; Piou, C. Improving preventive locust management: Insights from a multi-agent model. *Pest Manag. Sci.* **2018**, *74*, 46–58. [CrossRef]
17. Steedman, A. *Locust Handbook*; Overseas Development Natural Resources Institute: London, England, 1988.
18. United States Congress, Office of Technology Assessment. *A Plague of Locusts*; Special Report, OTA-F-450; Government Printing Office: Washington, DC, USA, 1990. Available online: https://ota.fas.org/reports/9001.pdf (accessed on 29 October 2020).
19. Showler, A.T. Desert locust control, public health, and environmental sustainability in North Africa. In *The North African Environment at Risk*; Swearingen, W.D., Bencherifa, A., Eds.; Westview Press: Boulder, CO, USA, 1995; pp. 217–239.
20. Belayneh, Y.T. Acridid pest management in the developing world: A challenge to the rural population, a dilemma to the international community. *J. Orthopt. Res.* **2005**, *14*, 187–195. Available online: https://www.jstor.org/stable/3657127 (accessed on 15 October 2020). [CrossRef]

21. Brader, L.; Djibo, H.; Faye, F.G.; Ghaout, S.; Lazar, M.; Luzietoso, P.N.; Babah, M.A.O. *Multilateral Evaluation of the 2003-05 Desert Locust Campaign*; Food and Agriculture Organization of the United Nations: Rome, Italy, 2006. Available online: http://www.fao.org/ag/locusts/common/ecg/1913/en/DesertLocustEvalReportE.pdf (accessed on 3 November 2020).
22. Krall, S.; Herok, C. Economics of desert locust control. In *New Strategies in Locust Control*; Krall, S., Peveling, R., Ba Diallo, D., Eds.; Birkhauser: Berlin, Germany, 1997; pp. 401–413. [CrossRef]
23. Lockwood, J.A. *Locust: The Devastation, Rise and Mysterious Disappearance of the Insect That Shaped the American Frontier*; Basic Books: New York, NY, USA, 2004.
24. Krall, S. Importance of locusts and grasshoppers for determining crop losses. In *New Trends in Locust Control: Ecotoxicology, Botanicals, Pathogens, Attractants, Hormones, Pheromones, Remote Sensing*; Krall, S., Wilps, H., Eds.; Deutsche Gesellchaft für Technische Zusammenarbeit (GTZ): Eschborn, Germany, 1994; pp. 7–22.
25. Lecoq, M. Recent progress in desert and migratory locust management in Africa: Are preventative actions possible? *J. Orthopt. Res.* **2001**, *10*, 277–291. [CrossRef]
26. Rowley, J.; Bennet, O. *Grasshoppers and Locusts: The Plague of the Sahel*; Panos dossier, the Panos Institute: London, England, 1993.
27. Joffe, S. *Economics and Policy Issues in Desert Locust Management: A Preliminary Analysis*; Report no. AGPP/DL/TSD/27; Food and Agriculture Organization of the United Nations: Rome, Italy, 1998; 27p. Available online: http://www.fao.org/ag/locusts/oldsite/PDFs/TS27p1.pdf (accessed on 3 November 2020).
28. FAO (Food and Agriculture Organization of the United Nations). *Biology and Behaviour. Desert Locust Guidelines*; FAO: Rome, Italy, 2001. Available online: http://www.fao.org/ag/locusts/common/ecg/347_en_DLG1e.pdf (accessed on 29 October 2020).
29. Doré, A.; Babah, M.A.O.; Barbier, M.; Lecoq, M. Rethinking organization and governance for surveillance and risk management of desert locust outbreaks. In Proceedings of the 9th International Conference of the Orthopterists' Society, Canmore, AB, Canada, 14–18 August 2005; p. 33. Available online: http://publications.cirad.fr/une_notice.php?dk=528208 (accessed on 29 October 2020).
30. Moussaoui, M. *Etude sur L'intérêt Economique de la Lutte Préventive Contre le Criquet Pèlerin*; FAO: Rome, Italy, 2010.
31. De Vreyer, P.; Guilbert, N.; Mesple-Sompsa, S. Impact of natural disasters on education outcomes: Evidence from the 1987–89 locust plague in Mali. *J. Afr. Econ.* **2014**, *24*, 57–100. [CrossRef]
32. FAO (Food and Agriculture Organization of the United Nations). *Global Information and Early Warning System on Food and Agriculture World Food Program: Special Report, FAO/WFP Crop and Food Supply Assessment Mission to Mauritania with Special Focus on Losses Due to the Desert Locust, 21 December 2004*; FAO: Rome, Italy, 2004. Available online: https://reliefweb.int/report/mauritania/faowfp-crop-and-food-supply-assessment-mission-mauritania-special-focus-losses-due (accessed on 5 November 2020).
33. FAO (Food and Agriculture Organization of the United Nations). *Global Information and Early Warning System on Food and Agriculture World Food Program: Special Report, FAO/WFP Crop and Food Supply Assessment Mission to Mali with Special Focus on Losses Due to the Desert Locust, 21 December 2004*; FAO: Rome, Italy, 2004. Available online: https://documents.wfp.org/stellent/groups/public/documents/ena/wfp043774.pdf?iframe (accessed on 5 November 2020).
34. FAO (Food and Agriculture Organization of the United Nations). *Global Information and Early Warning System on Food and Agriculture World Food Program: Special Report, FAO/WFP Crop and Food Supply Assessment Mission to Niger with Special Focus on Losses Due to the Desert Locust, 21 December 2004*; FAO: Rome, Italy, 2004. Available online: http://www.fao.org/3/J3969e/J3969e00.htm (accessed on 5 November 2020).
35. Lecoq, M.; Duranton, J.F.; Rachadi, T. Towards an integrated strategy for the control of the desert locust. In *New Strategies in Locust Control*; Krall, S., Peveling, R., Ba Diallo, D., Eds.; Birkhauser: Berlin, Germany, 1997; pp. 467–473. [CrossRef]
36. Lecoq, M. Desert locust threat to agricultural development and food security and FAO's international role in its control. *Arab J. Plant Prot.* **2003**, *21*, 188–193.
37. FAO (Food and Agriculture Organization of the United Nations). *The FAO Commission for controlling the Desert Locust in South-West Asia: A celebration of 50 Years*; FAO: Rome, Italy, 2014. Available online: http://www.fao.org/3/a-i4202e.pdf (accessed on 29 October 2020).
38. Latchininsky, A.V.; Sivanpillai, R. Locust habitat monitoring and risk assessment using remote sensing and GIS technologies. In *Integrated Management of Arthropod Pests and Insect Borne Diseases*; Ciancio, A., Mukerji, K.G., Eds.; Springer: Dordrecht, The Netherlands, 2010; pp. 163–188. [CrossRef]
39. FAO (Food and Agriculture Organization of the United Nations). *Desert Locust Outbreak in Yemen, June 2013-February 2014*; FAO: Rome, Italy, 2014. Available online: http://www.fao.org/ag/locusts/en/archives/1032/2136/YEM2013/index.html (accessed on 29 October 2020).
40. Wiktelius, S.; Edwards, S. Organochlorine insecticide residues in Africa fauna: 1971–1995. *Rev. Environ. Contamin. Toxicol.* **1997**, *151*, 1–37.
41. Roy, J. *Histoire d'un Siècle de Lutte Anti-Acridienne en Afrique. Contributions de la France*; L'Harmatan: Paris, France, 2001.
42. Skaf, R. A story of a disaster: Why locust plagues are still possible. *Disasters* **1988**, *12*, 122–126. [CrossRef]
43. Skaf, R.; Popov, G.B.; Roffey, J. The Desert Locust: An international challenge. *Philos. T. R. Soc. B* **1990**, *328*, 525–538.
44. Sharma, A. Locust control management: Moving from traditional to new technologies—An empirical analysis. *Entomol. Ornithol. Herpetol.* **2015**, *4*, 141. [CrossRef]
45. FAO (Food and Agriculture Organization of the United Nations). Desert Locust Watch. 2017. Available online: http://www.fao.org/ag/locusts/en/info/info/index.html (accessed on 3 November 2020).

46. Cisse, S.; Ghaout, S.; Mazih, A.; Babah, M.O.A.; Benahi, S.; Piou, C. Effect of vegetation on density thresholds of adult desert locust gregarization from survey data in Mauritania. *Entomol. Exper. Applic.* **2013**, *149*, 159–165. [CrossRef]
47. Bouïchi, A.; Simpson, S.J.; Roessingh, P. The influence of environmental microstructure on the behavioural phase state of distribution of the desert locust *Schistocerca gegaria. Physiol. Entomol.* **1996**, *21*, 247–256. [CrossRef]
48. McCafferty, A.R.; Simpson, S.J.; Islam, M.; Roessingh, P. A gregarizing factor present in the egg pod foam of the desert locust *Schistocerca gregaria. J. Exp. Biol.* **1998**, *201*, 347–363.
49. Simpson, S.J.; McCafferty, A.R.; Hägele, B. A behavioural analysis of phase change in the desert locust. *Biol. Rev.* **1999**, *74*, 461–480. [CrossRef]
50. Despland, E.; Simpson, S.J. The role of food distribution and nutritional quality in behavioral phase change in the desert locust. *Anim. Behav.* **2000**, *59*, 643–652. [CrossRef]
51. Despland, E.; Simpson, S.J. Small-scale vegetation patterns in the parental environment influence the phase state of hatchlings of the desert locust. *Physiol. Entomol.* **2000**, *25*, 74–81. [CrossRef]
52. Babah, M.A.O.; Sword, G.A. Linking locust gregarization to local resource distribution patterns across a large spatial scale. *Environ. Entomol.* **2004**, *33*, 1577–1583. [CrossRef]
53. Rodgers, S.M.; Matheson, T.; Anstey, M.; Simpson, S.J. Early induction of behavioural phase change in the desert locust. In Proceedings of the 9th International Conference of the Orthopterists' Society, Canmore, AB, Canada, 14–18 August 2005; p. 71. Available online: https://ybo.ewq.mybluehost.me/wp-content/uploads/2018/10/Metaleptea_9th_ICO_Canada.pdf (accessed on 3 November 2020).
54. Pener, M.P.; Simpson, S.J. Locust phase polyphenism: An update. *Adv. Insect Physiol.* **2009**, *36*, 1–286. [CrossRef]
55. Babah, M.A.O. Brief overview on the desert locust problem. *Tunis. Plant Prot. J.* **2011**, 6. Available online: https://www.jircas.go.jp/sites/default/files/publication/proceedings/2012-session-41_0.pdf (accessed on 29 October 2020).
56. *Improved Ways to Prevent the Desert Locust in Mauritania and the Sahel*; World Bank: Washington, DC, USA, 2010. Available online: https://www.worldbank.org/en/news/feature/2010/01/07/improved-ways-to-prevent-the-desert-locust-in-mauritania-and-the-sahel (accessed on 5 February 2021).
57. Byaruhanga, E.K. Developments in the strategy of Red Locust plague prevention in the fifty-year history of the International Red Locust Control Organisation for Central and Southern Africa. *Int. J. Trop. Insect Sci.* **1999**, *19*, 259–264. [CrossRef]
58. Hunter, D.M.; Cosenzo, E.L. The origin of plagues and recent outbreaks of the South American locust *Schistocerca cancellata* (Orthoptera: Acrididae) in Argentina. *Bull. Entomol. Res.* **1990**, *80*, 295–300. [CrossRef]
59. Babah, M.O.A. *Biogéographie du Criquet Pèlerin Schistocerca Gregaria Forskål, 1775: Identification, Caracterisation et Originalité d'un foyer Grégarigène en Mauritanie Centrale*; Hermann: Paris, France, 2010.
60. Tucker, C.J.; Hielkema, J.U.; Roffey, J. The potential of satellite remote sensing of ecological conditions for survey and forecasting desert locust activity. *Int. J. Remote Sens.* **1985**, *6*, 127–138. [CrossRef]
61. Ceccato, P. Operational early warning system using SPOT-VEGETATION and TERRA-MODIS to predict desert locust outbreaks. In Proceedings of the 2nd International SPOT-VEGETATION User's Conference; Veroustraete, F., Bartholomé, E., Eds.; European Commission Official Publication of the European Communities: Brussels, Belgium, 2005. [CrossRef]
62. Ceccato, P.; Cressman, K.; Giannini, A.; Trzaska, S. The desert locust upsurge in West Africa (2003–2005): Information on the desert locust early warning system and the prospects for seasonal climate forecasting. *Int. J. Pest Manag.* **2007**, *53*, 7–13. [CrossRef]
63. Renier, C.; Waldner, F.; Jacques, D.C.; Babah, M.O.A.; Cressman, K.; Defourny, A. A dynamic vegetation senescence indicator for near-real-time desert locust habitats monitoring with MODIS. *Remote Sens.* **2015**, *7*, 7545–7570. [CrossRef]
64. Waldner, F.; Babah, M.A.O.; Cressman, K.; Defourny, P. Operational monitoring of the desert locust habitat with Earth observation: An assessment. *Int. J. Geo-Inform.* **2015**, *4*, 2379–2400. [CrossRef]
65. Escorihuela, M.J.; Merlin, O.; Stefan, V.; Moyano, G.; Eweys, O.A.; Zribi, M.; Kamara, S.; Benahi, A.S.; Babah Ebbe, M.A.; Chihrane, J.; et al. SMOS based high resolution soil moisture estimates for desert locust preventive management. *Remote Sens. Appl. Soc. Environ.* **2018**, *11*, 140–150. [CrossRef]
66. Zhang, L.; Lecoq, M.; Latchininsky, A.; Hunter, D. Locust and grasshopper management. *Annu. Rev. Entomol.* **2019**, *64*, 15–34. [CrossRef] [PubMed]
67. Appleby, G.; Settle, W.; Showler, A.T. *Mid-Term Evaluation of the African Emergency Locust/Grasshopper Assistance (AELGA) Project*; Tropical Research and Development: Gainesville, FL, USA, 1989.
68. Cressman, K.; Van der Elstraeten, A.; Pedrick, C. *Elocust3: An Innovative Tool for Crop Pest Control*; Food and Agriculture Organization of the United Nations: Rome, Italy, 2018. Available online: www.fao.org/3/a-i6058e.pdf (accessed on 29 October 2020).
69. Cooper, J.F.; Coppen, G.D.A.; Dobson, H.M.; Rakotonandrasana, A.; Scherer, R. Sprayed barriers of diflubenzuron (ULV) as a control technique against marching hopper bands of a migratory locust *Locusta migratoria capito* (Sauss.) (Orthoptera: Acrididae) in southern Madagascar. *Crop Prot.* **1995**, *14*, 137–143. [CrossRef]
70. Rachadi, T.; Foucart, A. Barrier treatment of fipronil for controlling hopper bands of desert locust infesting a large area in Mauritania. *Int. J. Pest Manag.* **1999**, *45*, 263–273. [CrossRef]
71. Lomer, C.J.; Bateman, R.P.; Johnson, D.L.; Langwald, J.; Thomas, M. Biological control of locusts and grasshoppers. *Annu. Rev. Entomol.* **2001**, *46*, 667–702. [CrossRef]

72. FAO (Food and Agriculture Organization of the United Nations). *Desert Locust Bulletin, 4 May 2020*; No. 499; FAO: Rome, Italy, 2020. Available online: http://www.fao.org/ag/locusts/common/ecg/2551/en/DL499e.pdf (accessed on 3 November 2020).
73. FAO (Food and Agriculture Organization of the United Nations). *Desert Locust Bulletin, 4 June 2020*; No. 500; FAO: Rome, Italy, 2020. Available online: http://www.fao.org/ag/locusts/common/ecg/2555/en/DL500e.pdf (accessed on 3 November 2020).
74. Mahjoub, N. Le problème du criquet pèlerin et les perspectives de sa résolution. *Nat. Faune* **1988**, *4*, 16–20.
75. Woldewahid, G.; van der Werf, W.; van Huis, A.; Stein, A. Spatial distribution of populations of solitarious adult desert locust (*Schistocerca gregaria* Forsk.) on the coastal plain of Sudan. *Agric. Forest Entomol.* **2004**, *6*, 181–191. [CrossRef]
76. Lazar, M.; Piou, C.; Doumandji-Mitiche, B.; Lecoq, M. Importance of solitarious Desert locust population dynamics: Lessons from historical survey data in Algeria. *Entomol. Exp. Appl.* **2016**, *161*, 168–180. [CrossRef]
77. Maeno, K.O.; Ould Babah Ebbe, M.A. Aggregation site choice by gregarious nymphs of the desert locust, *Schistocerca gregaria*, in the Sahara Desert of Mauritania. *Insects* **2018**, *9*, 99. [CrossRef] [PubMed]
78. Showler, A.T.; Osbrink, W.L.A.; Abrigo, V.; Phillips, P.I. Relationships of salinity, relative humidity, mud flat fiddler crabs, ants, and sea ox-eye daisy with ixodid distribution and egg survival on the South Texas coastal plains. *Environ. Entomol.* **2019**, *48*, 733–746. [CrossRef] [PubMed]
79. Ellis, P.E.; Ashall, C. Field studies on diurnal behaviour, movement and aggregation in the desert locust (*Schistocerca gregaria* Forskål). *Anti-Locust Bull.* **1957**, *25*, 1–103.
80. Maeno, K.O.; Ould Ely, S.; Ould Mohamed, S.; Jaavar, M.E.H.; Nakamura, S.; Ould Babah Ebbe, M.A. Defence tactics cycle with diel microhabitat choice and body temperature in the desert locust, *Schistocerca gregaria*. *Ethology* **2019**, *125*, 250–261. [CrossRef]
81. Maeno, K.O.; Ould Ely, S.; Ould Mohamed, S.; Jaavar, M.E.H.; Ould Babah Ebbe, M.A. Adult desert locust swarms, *Schistocerca gregaria*, preferentially roost in the tallest plants at any given site in the Sahara Desert. *Agronomy* **2020**, *10*, 1923. [CrossRef]
82. Maeno, K.O.; Ould Ely, S.; Ould Mohamed, S.; Jaavar, M.E.H.; Nakamura, S.; Ould Babah Ebbe, M.A. Behavioral plasticity in anti-predator defense in the desert locust. *J. Arid. Environ.* **2018**, *158*, 47–50. [CrossRef]
83. Showler, A.T. The importance of armed conflict in the desert locust recession distribution, 1986–2002. *J. Orthopt. Res.* **2003**, *12*, 127–133. [CrossRef]
84. Showler, A.T.; Lecoq, M.; Babah, M.A.O.B. Incidence and ramifications of armed conflict in countries with major desert locust breeding areas. *Agronomy* **2021**, *11*, 114. [CrossRef]
85. Doré, A.; Barbier, M.; Lecoq, M.; Ould Babah, M.A. Prévention des invasions de criquets pèlerins: Analyse socio-technique d'un dispositif de gestion du risque. *Cah. Agric.* **2008**, *17*, 457–464. [CrossRef]
86. Lecoq, M. Lecoq, M. Le Criquet pèlerin. Enseignements de la dernière invasion et perspectives offertes par la biomodélisation. In *La Lutte Anti-Acridienne*; Essaid, A., Ed.; AUPELF-UREF, John Libbey Eurotext: Paris, France, 1991; pp. 71–98.
87. Martini, P.; Lecoq, M.; Soumaré, L.; Chara, B. *Proposition de Programme de Lutte Contre le Criquet Pèlerin dans la Partie Occidentale de Son Aire D'habitat*; Rapport 9963/F; FAO: Rome, Italy, 1998. Available online: http://www.fao.org/ag/locusts/oldsite/PDF/meetings/EMPRESwest98f.pdf (accessed on 20 October 2020).
88. Deshormes, A. *Institutional Study to Enhance the Roles and Responsibilities of the Desert Locust Control Commissions Established under Article XIV. Financial Governance Final Report*; Food and Agriculture Organization of the United Nations: Rome, Italy, 2011. Available online: http://www.fao.org/ag/locusts/common/ecg/2148/en/Financial_Governance_Report_E.pdf (accessed on 20 October 2020).
89. FAO (Food and Agriculture Organization of the United Nations). *Programme EMPRES en Région Occidentale*; FAO: Rome, Italy, 2016. Available online: http://www.clcpro-empres.org (accessed on 5 November 2020).
90. FAO (Food and Agriculture Organization of the United Nations). *Evaluation Final de la Phase II (2014–2017) du Programme EMPRES/Composante Criquet Pèlerin en Région Occidentale (EMPRES-RO)*; FAO: Rome, Italy, 2018. Available online: http://www.fao.org/3/ca4005fr/ca4005fr.pdf (accessed on 5 November 2020).
91. FAO (Food and Agriculture Organization of the United Nations). Current Upsurge (2019–2020). *Desert Locust Watch.* 2020. Available online: http://www.fao.org/ag/locusts/en/info/2094/index.html (accessed on 3 November 2020).
92. FAO (Food and Agriculture Organization of the United Nations). *Desert Locust Bulletin, 9 January 2019*; No. 483; FAO: Rome, Italy, 2019. Available online: http://www.fao.org/ag/locusts/common/ecg/2455/en/DL483e.pdf (accessed on 3 November 2020).
93. FAO (Food and Agriculture Organization of the United Nations). *Desert Locust Bulletin, 4 February 2019*; No. 484; FAO: Rome, Italy, 2019. Available online: http://www.fao.org/ag/locusts/common/ecg/2462/en/DL484e.pdf (accessed on 3 November 2020).
94. FAO (Food and Agriculture Organization of the United Nations). *Desert Locust Bulletin, 4 March 2019*; No. 485; FAO: Rome, Italy, 2019. Available online: http://www.fao.org/ag/locusts/common/ecg/2468/en/DL485e.pdf (accessed on 3 November 2020).
95. FAO (Food and Agriculture Organization of the United Nations). *Desert Locust Bulletin, 3 April 2019*; No. 486; FAO: Rome, Italy, 2019. Available online: http://www.fao.org/ag/locusts/common/ecg/2471/en/DL486e.pdf (accessed on 3 November 2020).
96. FAO (Food and Agriculture Organization of the United Nations). *Desert Locust Bulletin, 3 May 2019*; No. 487; FAO: Rome, Italy, 2019. Available online: http://www.fao.org/ag/locusts/common/ecg/2472/en/DL487e.pdf (accessed on 3 November 2020).
97. FAO (Food and Agriculture Organization of the United Nations). *Desert Locust Bulletin, 5 June 2019*; No. 488; FAO: Rome, Italy, 2019. Available online: http://www.fao.org/ag/locusts/common/ecg/2475/en/DL488e.pdf (accessed on 3 November 2020).
98. FAO (Food and Agriculture Organization of the United Nations). *Desert Locust Bulletin, 4 July 2019*; No. 489; FAO: Rome, Italy, 2019. Available online: http://www.fao.org/ag/locusts/common/ecg/2480/en/DL489e.pdf (accessed on 3 November 2020).

99. FAO (Food and Agriculture Organization of the United Nations). *Desert Locust Bulletin, 2 August 2019*; No. 490; FAO: Rome, Italy, 2019. Available online: http://www.fao.org/ag/locusts/common/ecg/2482/en/DL490e.pdf (accessed on 3 November 2020).
100. FAO (Food and Agriculture Organization of the United Nations). *Desert Locust Bulletin, 3 September 2019*; No. 491; FAO: Rome, Italy, 2019. Available online: http://www.fao.org/ag/locusts/common/ecg/2483/en/DL491e.pdf (accessed on 3 November 2020).
101. FAO (Food and Agriculture Organization of the United Nations). *Desert Locust Bulletin, 3 October 2019*; No. 492; FAO: Rome, Italy, 2019. Available online: http://www.fao.org/ag/locusts/common/ecg/2484/en/DL492e.pdf (accessed on 3 November 2020).
102. FAO (Food and Agriculture Organization of the United Nations). *Desert Locust Bulletin, 4 November 2019*; No. 493; FAO: Rome, Italy, 2019. Available online: http://www.fao.org/ag/locusts/common/ecg/2488/en/DL493e.pdf (accessed on 3 November 2020).
103. FAO (Food and Agriculture Organization of the United Nations). *Desert Locust Bulletin, 4 December 2019*; No. 494; FAO: Rome, Italy, 2019. Available online: http://www.fao.org/ag/locusts/common/ecg/2491/en/DL494e.pdf (accessed on 3 November 2020).
104. FAO (Food and Agriculture Organization of the United Nations). *Desert Locust Bulletin, 6 January 2020*; No. 495; FAO: Rome, Italy, 2020. Available online: http://www.fao.org/ag/locusts/common/ecg/2514/en/DL495e.pdf (accessed on 3 November 2020).
105. FAO (Food and Agriculture Organization of the United Nations). *Desert Locust Bulletin, 3 February 2020*; No. 496; FAO: Rome, Italy, 2020. Available online: http://www.fao.org/ag/locusts/common/ecg/2526/en/DL496e.pdf (accessed on 3 November 2020).
106. FAO (Food and Agriculture Organization of the United Nations). *Desert Locust Bulletin, 5 March 2020*; No. 497; FAO: Rome, Italy, 2020. Available online: http://www.fao.org/ag/locusts/common/ecg/2531/en/DL497e.pdf (accessed on 3 November 2020).
107. FAO (Food and Agriculture Organization of the United Nations). *Desert Locust Bulletin, 4 April 2020*; No. 498; FAO: Rome, Italy, 2020. Available online: http://www.fao.org/ag/locusts/common/ecg/2536/en/DL498e.pdf (accessed on 3 November 2020).
108. FAO (Food and Agriculture Organization of the United Nations). *Desert Locust Bulletin, 3 December 2018*; No. 482; FAO: Rome, Italy, 2018. Available online: http://www.fao.org/ag/locusts/common/ecg/2448/en/DL482e.pdf (accessed on 3 November 2020).
109. Meynard, C.N.; Lecoq, M.; Chapuis, M.P.; Piou, C. On the relative role of climate change and management in the current Desert Locust outbreak in East Africa. *Glob Chang. Biol.* **2020**, *26*, 3753–3755. [CrossRef]

agronomy

Article

Incidence and Ramifications of Armed Conflict in Countries with Major Desert Locust Breeding Areas

Allan T. Showler [1,*] and Michel Lecoq [2]

1 Knipling-Bushland U.S. Livestock Insects Research Laboratory, USDA-ARS, 2700 Fredericksburg Road, Kerrville, TX 78028, USA
2 CIRAD, UMR CBGP, F-34398 Montpellier, France; mlecoq34@gmail.com
* Correspondence: allan.showler@ars.usda.gov; Tel.: +1-(830)-792-0341

Abstract: Despite many areas of progress in recent years, desert locust surveillance and control is impaired by many obstacles, the most intractable of which is insecurity. Insecurity involves rebellions, insurgencies, civil and international war, banditry, terrorism, and minefields. Obstruction of desert locust operations in breeding areas by ongoing armed conflict and landmines constitutes "direct" insecurity. "Indirect" insecurity, although less obvious, is arguably more broadly deleterious by debilitating government function and diverting funds, personnel, and equipment from desert locust management. Indirect "active" insecurity is armed conflict and civil unrest that is occurring at the same time as a desert locust episode, but not in the breeding areas. Indirect "inactive" insecurity refers to the after-effects of insecurity, including weak funding because of prior inattention to capacity maintenance during times of direct and indirect active insecurity, disabled or militarily-appropriated vehicles and other resources, destruction of infrastructure, and deployment of mines. We provide examples of direct and indirect insecurity across 35 years, from 1986 through May 2020, in 13 African and Asian countries (Chad, Eritrea, Ethiopia, India, Mali, Mauritania, Niger, Pakistan, Saudi Arabia, Somalia, Sudan, Western Sahara, and Yemen) with desert locust breeding areas to illustrate the complexity, pervasiveness, and chronic occurrence of insecurity. The upsurge of 2020 is used to show how direct insecurity still contributes to the genesis and expansion of desert locust episodes. Possible mitigation of direct insecurity effects on some desert locust operations is discussed.

Keywords: control; impediments; insecurity; plague; *Schistocerca gregaria*; surveillance; war

Citation: Showler, A.T.; Lecoq, M. Incidence and Ramifications of Armed Conflict in Countries with Major Desert Locust Breeding Areas. *Agronomy* **2021**, *11*, 114. https://doi.org/10.3390/agronomy11010114

Received: 1 December 2020
Accepted: 5 January 2021
Published: 8 January 2021

1. Introduction

The desert locust, *Schistocerca gregaria* (Forskål) (Orthoptera: Acrididae), is arguably the oldest and most agriculturally threatening migratory pest globally [1]. Since 1989, following a desert locust plague, control has evolved toward proactive intervention [2,3] during outbreaks to avert the development of upsurges and plagues with notable success [4,5]. Further, control is evolving past proactive control toward outbreak prevention with the aim of holding desert locust populations in their nongregarious recession stage for indefinite periods of time [4,5]. Despite notable successes and advances in the science of desert locust management and available technologies [4,5], major constraints remain, including insufficient funding, political issues, unpreparedness, impediments to research, and rugged and remote terrain with weak infrastructure [4]. Another key constraint, insecurity, involves rebellions, banditry, civil and international war, terrorism, and minefields (Figure 1) [4–8]. Large areas in the desert locust distribution [7,9] cannot be surveyed for desert locust activity due to chronic insecurity, which constitutes a serious risk for proactive and preventative strategies [4,5,10]. Insecurity has been a major challenge to surveillance and control by blocking access [4,8] to desert locust breeding areas in Africa and Asia [9] (Figure 2). Desert locust control is unique in that it is conducted across vast expanses and it requires coordination within and between afflicted countries. Surveillance should chiefly focus on the main breeding, or "outbreak areas" [11,12]. Major desert locust breeding areas are mainly located around massifs across the Sahara Desert and Mauritania, the Red Sea plains on the east and west coasts, parts of the southern Arabian Peninsula, and the India–Pakistan border [4,7,9], all of which are frequently beset with insecurity [8]. Occurrences and effects of insecurity are described herein from 1986 through May 2020 [4,6]. Desert locusts can invade and reproduce in countries beyond those harboring key breeding areas [4–7,9], but we selected examples that have major breeding areas for illustrative purposes.

Figure 1. Landmines limit access to some areas frequented by desert locusts; photo credit Neil Rickards.

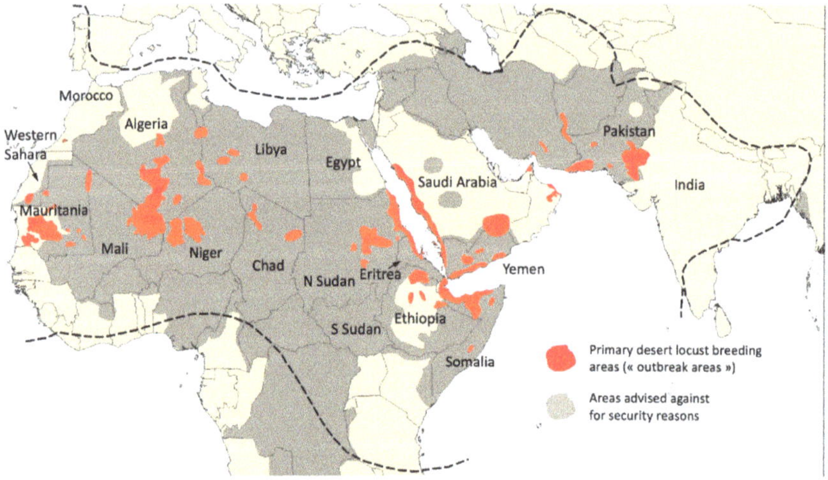

Figure 2. Proximity and overlap of major desert locust breeding areas with areas advised against and formally advised against by the French Ministry of Europe and Foreign Affairs for safety concerns [13] and other foreign ministries and departments of other "western" countries; the dotted lines indicate the limits of the desert locust invasion area [12]; 1986 through October 2020.

Insecurity can impede desert locust operations directly and indirectly. Direct insecurity obstructs ground and aerial surveillance and control operations in breeding areas by imperiling personnel and equipment. Further, swarms originating in non-contested areas can fly into conflicted zones where they might not be accessible for control. Unlike most other kinds of insect control campaigns, desert locust operations are, with relative frequency in some countries, impeded by direct insecurity.

Indirect insecurity, although less obvious than direct insecurity, is arguably more broadly deleterious by debilitating government function and diverting focus away from agricultural issues, and by diverting funding, personnel, and equipment (e.g., vehicles) from desert locust management. Indirect insecurity can be "active" and "inactive". Indirect active insecurity is armed conflict and civil unrest that is occurring at the same time as the desert locust episode, but not in the breeding areas. Because of the apparent constancy and diversity of conflicts in many countries with desert locust breeding areas, indirect active insecurity is widespread, persistent, and dynamic. Indirect inactive insecurity refers to the after-effects of insecurity regardless of where it had occurred. Impacts include weak funding due to prior inattention to capacity maintenance during times of direct and indirect active insecurity, disabled or government-appropriated vehicles and other resources, destruction of infrastructure, and deployment of mines. Indirect inactive insecurity is, because of the frequency of indirect active insecurity, also persistent. Dormant hostilities, such as those between Israel and some of its neighboring countries, are examples of indirect inactive insecurity. Further, desert locust upsurges and plagues are linked to loss of feed production; displacements of human populations; conflicts between farmers, nomads, and pastoralists over limited resources [14], all of which are exacerbated and compounded by the chronic, sweeping, and pernicious effects of indirect (active and inactive) insecurity. While we cannot measure the influence of indirect insecurity on desert locust management, we can report the incidence of insecurity to provide a sense of scope and magnitude, as well as variability between afflicted countries.

Because early intervention and reactive control require concentrated use of resources during time-limited windows of opportunity, depletion of resources contributes toward non-sustainability, which underlies the primary obstacle to early intervention (proactive and preventive): Unpreparedness [4,6,10,15–17]. Insecurity is often accompanied by increased reliance on bilateral and multilateral international aid agencies for development, routine operations, and non-conflict-related disasters and emergencies, such as combatting desert locust upsurges and plagues. Ramifications of indirect insecurity erode national desert locust control capabilities particularly during recession periods [16,17]. In addition, chronic demand for external support to suppress a recurrent emergency has resulted in "donor fatigue", a condition associated with dwindling funds, especially during recessions when there is no immediate emergency to mobilize and drive continued international aid agency support.

2. Insecurity in Countries with Major Desert Locust Breeding Areas

Algeria. *Direct Insecurity.* "Traditional" desert locust breeding areas are located in southern Algeria, particularly around the Ahaggar Mountains, and many swarms invaded north of the Atlas Mountains during the 1986–1989 plague and the 2003–2005 upsurge (Figure 3) [6,14]. Some areas in the 1980s and 1990s were insecure due to limited Tuareg unrest in the south and travel was restricted near Western Sahara. From 2007 through 2016, however, outbreaks were effectively controlled [4,18].

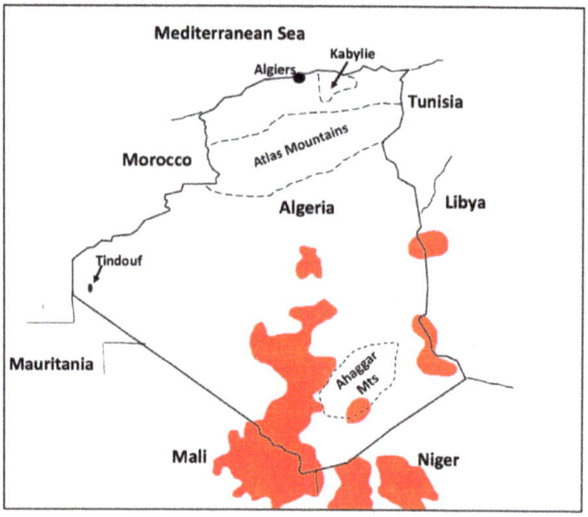

Figure 3. Locations and regions mentioned in the text in relation to desert locust activity and insecurity in Algeria. Primary breeding areas in red.

Indirect Insecurity. Islamic extremism began after a general election was curtailed by the government in 1992 when Islamic fundamentalists were expected to win [19]. During the next 10 years, ≈150,000 people were killed and the government sank into political paralysis [20,21]. Attendant civil unrest included assassinations, massacres, and rioting, hindering travel to and within Algeria [8]. In 2001, a Berber uprising occurred in the Kabylie region (Figure 3) [22,23], and in 2007 an *al-Qaeda* faction carried out assassinations, kidnappings, and bombings; the army responded with anti-insurgency offensives [22–25]. Although a 19-year-long national emergency was lifted in 2011, public protests and terrorist attacks by *al-Qaeda* and other insurgent groups continued through 2016 [23]. The north was most insecure [22], but *al-Qaeda's* southern branch accrued strength [26]; Libya's civil war produced a weapons influx which helped arm *al-Qaeda* in Algeria [22,27], increasing ambushes and use of concealed explosives against security forces [28]. Indirect active and inactive insecurity obstructed governance, limited funding, and restricted movement, hindering anti-locust capabilities [8].

Chad. *Direct Insecurity.* Desert locust breeding largely occurs in the northwestern mountainous Tibesti and Ennedi regions (Figure 4) [29,30]. In 1973, Libya attempted to annex Chad's Aouzou strip, part of the desert locust's Tibesti (Figure 4) breeding area [29], but Chad retook it militarily in 1986 and travel there was prohibited until the early 1990s [31–34]. During the 1986–1989 campaign, control operations were often halted because of conflict with Libya [8]. Civil war, mostly in the Tibesti region between the government, and the Movement for Democracy and Justice and the National Resistance Army [35], obstructed travel to the region and impaired desert locust operations through 2016 [4,18].

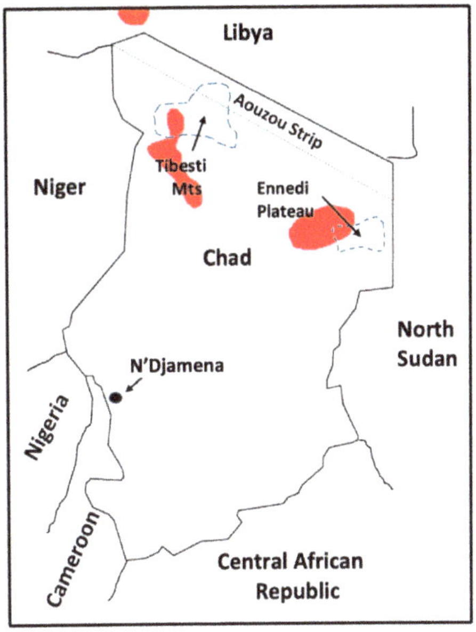

Figure 4. Locations and regions mentioned in the text in relation to desert locust activity and insecurity in Chad. Primary breeding areas in red.

Indirect Insecurity. The *Mouvement Patriotique du Salut* and other rebel groups replaced Chad's government in 1990 [27,36], and insurgency-based coup attempts became chronic [36,37]. During the 1990s, the border with the Central African Republic (Figure 4) was contested [38] between a group called the Federal Republic of Laouken and Chadian troops [8].

Border tension in eastern Chad deteriorated in 2005 when Chadian rebel groups used the region for bases (supported by Sudan), Sudanese militia groups also entered Chad and attacked civilians, instigating violence between eastern ethnic groups [36]. In April 2006, three rebel groupls, *Union des Forces pour la Démocratie et le Dévelopment*, *Rassemblement des Forces pour le Changement*, and the *Front Uni pour le Changement*, started an unresolved offensive against the national army with casualties on both sides [36], prompting the government to declare a national emergency in 2007 [37]. In early 2008, the same rebel groups entered N'Djamena (Figure 4), but failed to topple the government [36]. A rebel advance in May 2009 elicited government airstrikes against their sanctuaries across the border in Sudan [37]. Sudan countered by launching an airstrike on a Chadian refugee camp and by pursuing Sudanese rebels that had taken refuge in Chad; Chad retaliated with July airstrikes against Sudan [37]. In 2010, Chad and Sudan deployed a 3000-soldier combined cross-border force to quell problematic opposition groups [37]. Other rebel organizations have since emerged and conflict continues. We suggest that indirect insecurity, active and inactive, have had a substantial deleterious impact on Chadian governmental functions, including locust surveillance and control.

Eritrea. *Direct Insecurity.* The Red Sea coastal plain of Eritrea, particularly north of Massawa (Figure 5), has, along with the contiguous coastal plains of Sudan, likely been the most prolific breeding area for desert locusts. The 1986–1989 plague and the 1992–1994 upsurge both originated on the Red Sea coastal plains in Eritrea and Sudan [6,7]. A war for independence waged by the Eritrean People's Liberation Front (EPLF) against Ethiopia began in 1962 and continued for 30 years [39] (Figures 6 and 7), escalating to large tank battles. Fighting occurred everywhere, excluding the capital, Asmara (Figure 5) [39]. The war prevented desert locust control during the 1986–1989 plague [4,6,15], and during the

1992–1994 upsurge, minefields obstructed access to breeding areas [7]. Following independence in 1993, minefields still made off-road travel in the lowlands so dangerous [40] that some operations had to be conducted using aircraft [8]. The Sudan border was intermittently conflicted after Eritrea's independence [8] because of Eritrean support for Sudanese rebels [41] and a suspected Sudanese-backed insurgency within Eritrea [8,42]. In 1999, an Eritrean desert locust surveillance team was captured near the border, detained, and tortured [8]. Despite continued hostilities, 1997–2020 desert locust control was reasonably successful [4].

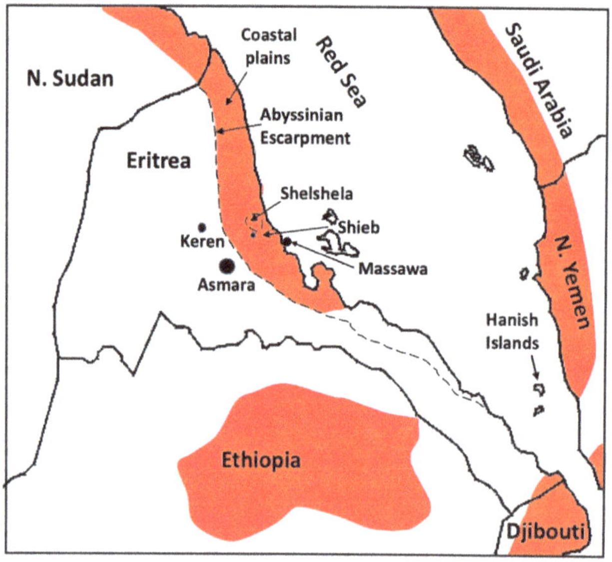

Figure 5. Locations and regions mentioned in the text in relation to desert locust activity and insecurity in Eritrea. Primary breeding areas in red.

Figure 6. Eritrean People's Liberation Front (EPLF) soldiers during the war for independence from Ethiopia; photo credit Gwynne Roberts (copyright Gwynne Roberts).

Figure 7. EPLF fighter, holding AK-47, who served 15 years on the front lines, with his wife who served on the front lines for 10 years; Eritrean women train and fight alongside men and they were reputed to be ferocious in battle; photo credit Allan Showler.

Indirect Insecurity. In addition to the direct danger of landmines, the 1992–1994 upsurge started on the coastal plain [7] because of unpreparedness following the long war for independence [7,8]. In 1996, Eritrean forces displaced Yemeni troops on the Red Sea Hanish Islands (Figure 5), disrupting communication and travel between the two countries until resolution in 1998 [8,43]. Additionally in 1996, Islamic (possibly Sudanese) insurgents killed Eritreans and Belgians along the Abyssinian escarpment "Filfil" road connecting Keren to the coastal plain near Shelshela, a prolific desert locust breeding area (Figure 5) [8].

Ethiopia launched airstrikes against Asmara in 1998 over a border dispute and Eritrea responded with an airstrike against Mekele in northern Ethiopia (Figure 5) [8]. War ensued, involving tens of thousands of casualties until Ethiopia invaded part of southwestern Eritrea in June 2000. A United Nations peacekeeping force arrived in 2002 but disbanded in 2008 despite continuing tension [44] and the border remained inaccessible and tense through 2020, even after the two countries reached a peace agreement in 2018. Continuous military mobilization since independence has prevented Eritrea from reaching optimal capacity for desert locust control.

Ethiopia. *Direct Insecurity.* From 1986, parts of the Ogaden Desert (Figure 8), where desert locusts breed, were unsafe because of rebellions and banditry; in 1992, for example, a surveillance helicopter was shot down there, killing three [8]. The senior author, on low-flying reconnaissance near Jijiga (Figure 8), 1994, was targeted by small arms. Although control operations have more recently been effective, dangers in some areas, particularly the Ogaden, remain.

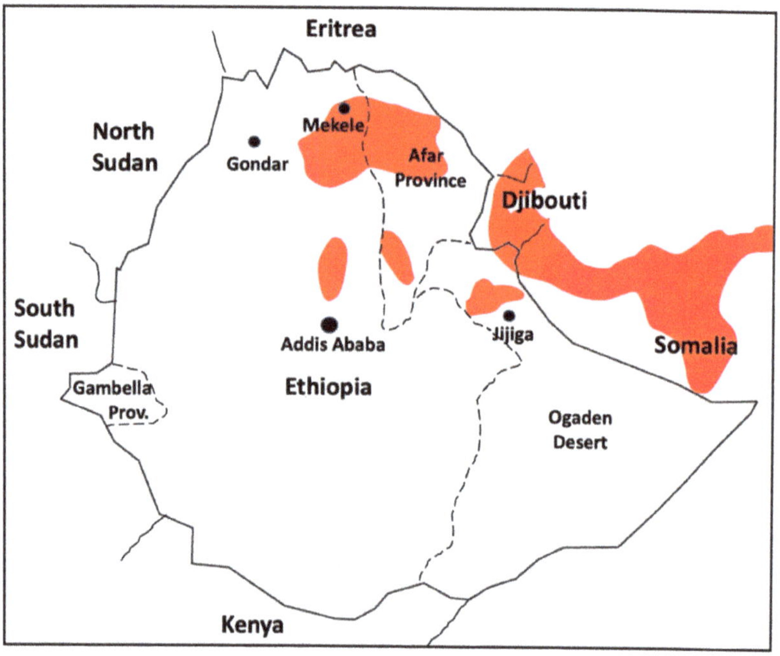

Figure 8. Locations and regions mentioned in the text in relation to desert locust activity and insecurity in Ethiopia. Primary breeding areas in red.

Indirect Insecurity. Aside from the 1962–1992 war of Eritrean secession, rebellions and civil unrest besieged Ethiopia from 1986 through 2020. A civil war during the 1980s involved the Ethiopian People's Revolutionary Democratic Front, which replaced the dictatorial Derg regime in 1991 [45]. The Sudan border was intermittently tense, and in 2000 Kenya reported that Ethiopian soldiers repeatedly crossed the border (Figure 8), killing Kenyan citizens and stealing cattle [8].

In 2004, ≈200 people died in ethnic clashes in Gambella Province (Figure 8), causing massive population displacement [46]. In 2005, disputed elections led to violent protests lasting months, in 2006 several bombs exploded in Addis Ababa, and Ethiopian troops were sent into Somalia for two years to combat Islamist militants [46]. In 2007, insurgents killed 74 workers at a Chinese oil facility in the southeast [46]. The Ogaden National Liberation Front (ONLF) captured several eastern towns in 2009 after heavy fighting [46], and clashes with the military continued into 2012, one raid killing 168 troops [46]. Further, European tourists were executed in the Afar region (Figure 8). In 2013, Ethiopia detained Somali *al-Shabaab* Islamist militants plotting to kidnap United Nations personnel [47], and, in 2015–2016, protests erupted in the northern city of Gondar (Figure 8) associated with the killing of >500 people and other human rights abuses by the government, prompting a state of emergency in late 2016 [46]. Occasionally, violent civil unrest in Ethiopia persisted through 2020. Although effects of indirect insecurity on Ethiopia's ability to manage desert locusts are highly likely, it is difficult to assess the extent of impairment.

India-Pakistan. *Indirect Insecurity.* The Thar Desert (Figure 9) on the southern border of India and Pakistan comprise the easternmost key desert locust breeding area [9,29]. Direct insecurity in the breeding areas, 1986–2020, has not been reported. During the desert locust campaigns of 1986–1989 and 1992–1994, involving breeding in the border deserts, control occurred before swarms escaped to other regions [6,7], but cooperation between India and Pakistan was problematic because of limited communications, travel restrictions to contested areas, and landmines [48]. Mutual hostility between the two countries escalated into a nuclear arms race [49,50]. From 1986 through 2020, conflict has mostly been focused on the Kashmir region (Figure 9). Terrorism and military skirmishes, including airstrikes, were conducted there, resulting in thousands of casualties [51–56]. India and Pakistan are likely among the countries (of the 13 presented) least impeded by indirect insecurity.

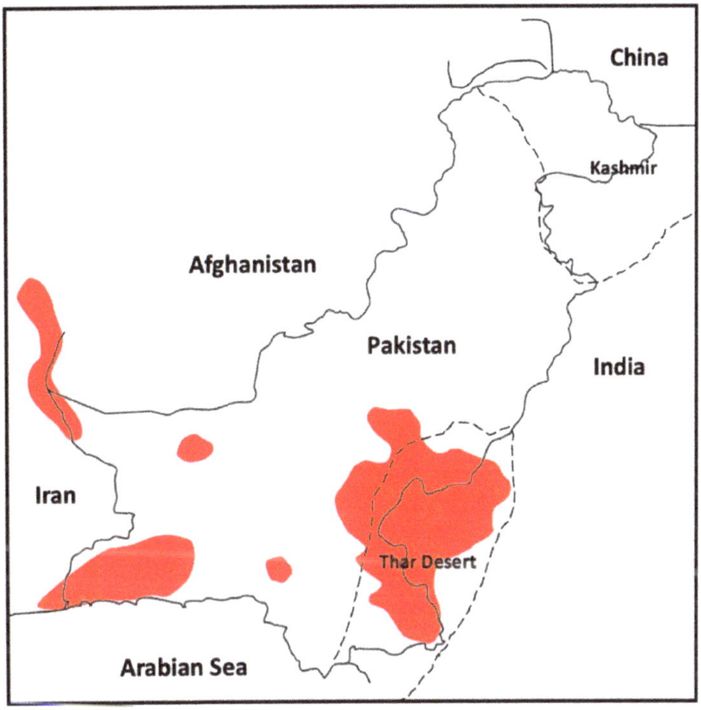

Figure 9. Locations and regions mentioned in the text in relation to desert locust activity and insecurity on the India-Pakistan border. Primary breeding areas in red.

Mali. *Direct Insecurity.* The Adrar des Iforas region of northern Mali (Figure 10) is an important desert locust breeding area [29], inhabited by the Tuareg people [57]. In the early 1990s, the Tuareg turned from occasional banditry [58] to rebellion [59]. Five Malian troops protecting a desert locust surveillance team were killed by Tuareg rebels in 1994 [8] and the insurrection restricted travel around breeding areas [7,8]. Desert locust operations in northern Mali were impeded relatively often through 2016 [4,18]. If gregarious desert locusts were to become active in Mali in the near future, surveillance and control operations might experience direct insecurity-related impediments.

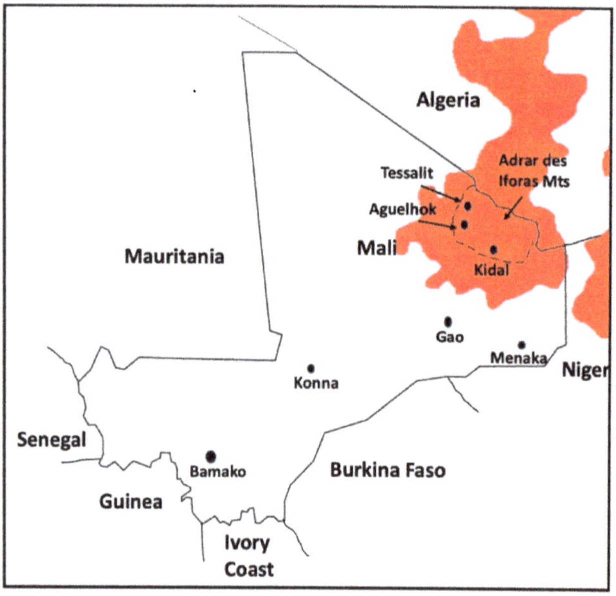

Figure 10. Locations and regions mentioned in the text in relation to desert locust activity and inScheme 2006. A Tuareg insurgency intensified when government army bases in Kidal, Menaka, and Tessalit (Figure 10) were attacked, but the conflict was resolved two months later [59,60]. In late 2011, another Tuareg rebellion began, strengthened by mercenaries from the Libyan civil war [59–61], and sporadic violent Tuareg unrest has continued through 2020.

Five Europeans were kidnapped by *al-Qaeda* in eastern and northern Mali in 2006, and a German tourist was killed [61]. A massacre of Malian troops and civilians at Aguelhok (Figure 10) occurred in early 2012, three of four military units in northern Mali defected to Malian National Liberation Army (MNLA) rebels, United States military airdrops supplied besieged Malian troops in Tessalit (Figure 10) [61], and Senegal sent troops to assist the Mali government [60]. A coup in March 2012, however, overthrew the government, collapsing military command, and the MNLA and Islamist *Ansar al Din* seized the north [61]. By April, the MNLA declared part of Mali a separate state, but it retreated after being ousted from the town of Gao (Figure 10) by a faction of *al-Qaeda* [61]. In early 2013, France attacked insurgents in the north after government forces were defeated in Konna and Bamako (Figure 10) [61]. While the impacts of indirect insecurity on desert locust operations have not been assessed, they are likely to be relatively substantial.

Mauritania. *Direct Insecurity.* The Mauritanian desert is mostly favorable to desert locust breeding [29], which occurred during the 1986–1989, 1992–1994, and the 2003–2005 desert locust episodes [6,7,14]. Although landmines had at times obstructed desert locust operations, they have largely been cleared or marked and have not interfered with desert locust control operations [4,62]. Intervention against desert locusts from 2007 to 2020 were reasonably effective [18].

Indirect Insecurity. In 1989–1991, racial tensions triggered ethnic violence that resulted in hundreds of deaths in Mauritania and in Senegal [63–65]. After those incidents, emphasis on conflict prevention and ethnic reconciliation has occurred with a number of constitutional amendments affirming the multiethnic character of the state [66]. *Al-Qaeda* killed four French tourists in 2007, launched attacks in Nouakchott, and conducted ransom kidnappings beginning in 2008 [65]. Border tensions involved Western Sahara and its Polisario separatist guerrillas [65]. Indirect insecurity probably has had a low to moderate effect on desert locust control capability.

Niger. *Direct Insecurity.* A key desert locust breeding area encompasses the Aïr Mountains and the Tamesna region (Figure 11), which are inhabited by Tuareg. Tuareg insurrection hindered desert locust surveillance and control through 1994 [7]. In 1991, an encampment of scientists, including one of the authors (ML) researching desert locusts in the Tamesna was attacked by armed men, resulting in one casualty [67]. During 2007–2016, insecurity did not impede locust operations [4,18].

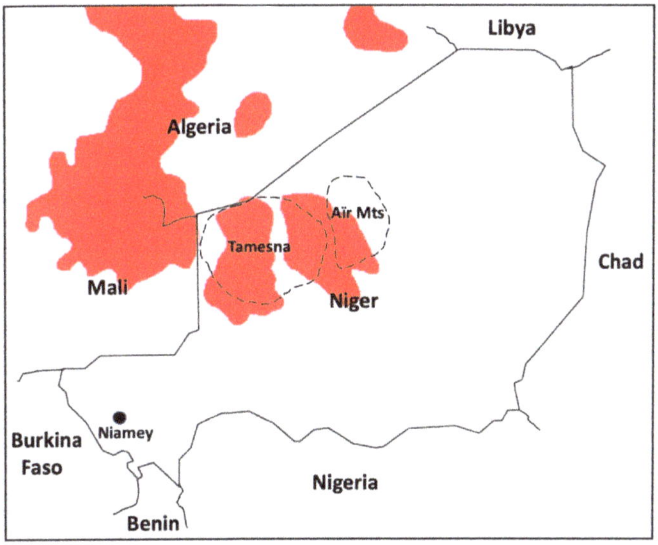

Figure 11. Locations and regions mentioned in the text in relation to desert locust activity and insecurity in Niger. Primary breeding areas in red.

Indirect Insecurity. While violent incidents were associated with Tuareg bandits and rebels, by the late 1990s insecurity was less tumultuous than in Mali because government responses were more accommodating [59], and some Tuareg groups laid down arms [68]. It appears likely that, were gregarious desert locust populations to develop or arrive in 2020, insecurity would not pose significant obstacles.

Saudi Arabia. *Indirect Insecurity.* Desert locust breeding in Saudi Arabia mainly occurs along the Tihama, the *Rub al-Khali* (Empty Quarter), and the north-central desert [29]. No direct insecurity regarding desert locust control has been reported. Despite Saudi Arabian involvement in external wars (i.e., Iraq, Kuwait, Yemen) and limited internal civil unrest and Islamic terrorism confined to cities, indirect insecurity has had a negligible impact on desert locust control.

Somalia. *Direct Insecurity.* Desert locust breeding mostly occurs in northern parts of Somalia, and these areas have been particularly problematic in 2020 [5,69]. National locust control capabilities were eliminated in Somalia, from the 1980s through 2020, because of armed conflict [70]; hence, desert locust surveillance during the late 1990s was conducted by one United Nations volunteer, a Somali counterpart in Hargeisa (Figure 12), and occasional FAO (Food and Agriculture Organization of the United Nations)-funded aerial missions by the Desert Locust Control Organization for Eastern Africa (DLCO-EA), based in Addis Ababa, Ethiopia (Figure 8) [8]. During the relatively limited 1997–1998 desert locust episode [15], swarms developed in Somaliland (Figure 12) and flew into Ethiopia's Ogaden region [4,8]. Direct insecurity has impeded desert locust operations in Somalia through 2020 [5], and Somalia is arguably the most problematic country with regard to insecurity in the desert locust's distribution.

Figure 12. Locations and regions mentioned in the text in relation to desert locust activity and insecurity in Somalia. Primary breeding areas in red.

Indirect Insecurity. After the Somali president was ousted and Mogadishu (Figure 12) was captured by rebel militias in 1991, thousands of civilians became casualties and Somalia lost its central government [71,72]. In 1992, ≈350,000 people died from disease, starvation, and civil war, prompting United States emergency aid airlifts [72–74]. In 1993, United States troops were deployed in Mogadishu to assist United Nations peacekeepers but heavy fighting resulted in 18 United States soldiers and >500 Somali fighters and civilians killed [72,75]. United States forces were withdrawn [74] and United Nations efforts to stabilize the government were aborted [75]. Puntaland and Jubaland (Figure 12) declared themselves as regional administrations in 1998 and factional fighting persisted after a transitional government was installed in 2000 [72,76]. A United Nations arms embargo on Somalia was in effect since the early 1990s, and in 2001–2002 Mogadishu was declared too dangerous for a United Nations office [72,77].

From 1996 through 1997, the *al-Qaeda*-linked Somali group, *al-Ittihad*, conducted attacks and bombings in Ethiopia. Starting in 1999, Ethiopia responded with incursions into southern Somalia to destroy *al-Ittihad* [78], capturing the regional capital of Garba Harre (Figure 12) [72]. In 2002, Ethiopian troops moved into Puntaland and, again, into southern Somalia [79–81]. Because the United States government suspected *al-Qaeda* activity in Somalia, military operations there increased [72]. An interim Somali government was formed in Kenya in 2003, and on moving to Somalia in 2005, internal divisions begat more violence [72]. Inter-militia fighting in Mogadishu during 2006 caused hundreds of civilian casualties and a militia-backed rival government, the Islamic Courts Union, seized most of southern Somalia and Mogadishu [72]. When Ethiopian troops re-invaded Somalia, Islamic government hard-liners declared *jihad* against Ethiopia, and subsequent fighting involved tanks and helicopters [72,82]. By 2007, the transitional government regained control, a United States airstrike killed the *al-Shabaab* insurgency's leader, and other airstrikes hit *al-Qaeda* targets in southern Somalia [72]. The government declared a state of emergency as fighting continued into 2008, including coordinated suicide bombings in semiautonomous regions [72].

By September 2011, >20 separate regional governments had developed across Somalia in addition to Puntaland and Somaliland (Figure 12), some warring against others [74]. Kenyan troops battled *al-Shabaab* in Somalia from late 2011 through 2016, and *al-Shabaab* claimed that it killed ≈100 African Union soldiers in El-Adda (Figure 12) [74]. In October 2016, Ethiopia pulled its 2000 troops (part of a 22,000-strong African Union force, Figure 13) from four towns in southern Somalia which *al-Shabaab* captured within hours [83]. Widespread insecurity in Somalia has continued through 2020.

Figure 13. African Union Mission in Somalia (AMISOM) soldiers in Jubaland, Somalia, to combat insurgent militias; photo credit Stuart Price.

Over decades of conflict, thousands of landmines were scattered across parts of Somalia [84]. The Somalia–Ethiopia border and the regions of Galgadud, Bakool, Bay, Hiraan, and the Afgooye Corridor, and parts of Mogadishu (Figure 12) were the most heavily mined [84]. In addition to the instances of insecurity described above, regions of Somalia were also unsafe because of banditry and piracy. Indirect insecurity, like direct insecurity, continue to be major obstacles to desert locust control.

Sudan. *Direct Insecurity.* Sudan's Red Sea coastal plain, particularly the region from the Tokar Delta (Figure 14) to the Eritrean border, is a prolific desert locust breeding area [29]. Breeding also tends to occur in the western desert [29]. Civil war and insurgencies, especially around the Tokar Delta and Kassala [85] (Figure 14), impeded surveillance and control during the 1986–1989 plague. The Tokar Delta area was held by National Defense Alliance (Beja Congress) rebels in the 1990s and swarms developed there during the 1992–1994 upsurge and the 1997–1998 desert locust outbreak [4,15]. The northern coastal plain around Halaib (Figure 14) has been contested from 1986 to the present, by Egypt [86], rendering it off limits to locust scouts from both countries. Sudan is vulnerable to desert locust outbreaks becoming upsurges because direct insecurity persists in some breeding areas.

Figure 14. Locations and regions mentioned in the text in relation to desert locust activity and insecurity in North and South Sudan. Primary breeding areas in red.

Indirect Insecurity. The government also fought other rebel armies, mostly in the south, following Sudan's 1956 independence [87,88] through 2020; the largest rebel force was the 40,000-strong Sudan People's Liberation Army (SPLA) [89]. During peak war years, 2003–2005, the government attacked villages using helicopters and fixed-wing Antonov bombers [90]. Ground forces followed, targeting the Fur, Massalit, and Zaghawa tribes especially in the Darfur provinces (Figure 14) [90]. The military was augmented by the *Janjaweed*, a militia comprised of former bandits, demobilized soldiers, young tribesmen, and unemployed youths [90]. The aim was ethnic cleansing accomplished through atrocities, killing hundreds of thousands [70,91]. Less than a year after Sudan split into the Republic of (North) Sudan and the Republic of South Sudan (Figure 14) in 2005, the two countries were fighting again [88], and limited conflict in the Darfur provinces erupted into another civil war. Further, in April 2012, war recommenced between North Sudan and South Sudan, with the SPLA crossing into the north and the North Sudan Armed Forces bombing southern villages [88]. In addition, North Sudan fought insurgencies in the South Kordofan, Blue Nile, and Abyei provinces (Figure 14) [88]. Conflict also occurred within South Sudan against the Juba-based government (Figure 14) [88]. In 2014 the president of North Sudan renamed the *Janjaweed* the Rapid Support Forces (RSF), comprised of $\approx$10,000 fighters, as his personal army to counter diminishing military loyalty [91]. The RSF continued inflicting atrocities, including aerial bombing of civilians in North and South Darfur, South Kordofan, and Blue Nile provinces (Figure 14) [91]. In 2016, in addition to its many other conflicts, the North Sudan government fought the Justice and Equality Movement, two different factions of the Sudan Liberation Movement, and the Liberation and Justice Movement [90].

Further, Sudan was cut off from international aid for harboring international terrorist cells, including *al-Qaeda* [8]; this was underscored by a 1997 United States missile strike on a "pharmaceutical factory" in Khartoum (Figure 14) suspected of manufacturing chemical weapons [8]. Despite the chronic and intense violence in Sudan that must certainly detract from optimal desert locust control capacity, and breeding areas compromised by direct insecurity, desert locust control operations in Sudan have been substantial and effective since 2003.

Western Sahara. *Direct Insecurity.* In 1988, a United States C-130 desert locust spray aircraft, flying from Senegal to Morocco, was downed by a shoulder-fired Polisario missile, killing the five-person crew [8]. Locust operations have been, at times, unfeasible in Western Sahara because of Polisario activity. During the 1986–1989, 1992–1994, and 2003–2005 control campaigns, swarms entering Western Sahara from Mauritania and Algeria could not be tracked until they crossed into Morocco or back into Algeria because of direct insecurity [4,6,7,18]. It is possible that direct insecurity will continue to be problematic for desert locust control for the foreseeable future.

Indirect Insecurity. Western Sahara (formerly Spanish Sahara), claimed by Morocco, has been vigorously contested by the Polisario Front since 1975 when Spain relinquished the territory [92,93]. The Polisario reported killing > 5000 Moroccan soldiers between 1982 and 1985 alone [53]. Many Polisario guerrillas took refuge in Algeria from Morocco's military [94], heightening tensions between Morocco and Algeria [8]. Additionally, the Polisario held 1360 Moroccan prisoners of war in Algeria as "bargaining chips", some of whom had been imprisoned for >20 years [94,95]. A United Nations-monitored ceasefire began in 1991, but Western Sahara's status has remained undecided (pending referendum), interspersed by ceasefire violations [53]. Insecurity also persisted when violent extremist groups emerged [96]. In 2010, Moroccan forces stormed a protest camp in Western Sahara, triggering violent demonstrations in the capital, Elayoun. In late 2011, three European aid workers were kidnapped from Polisario-administered refugee camps in Tindouf, Algeria (Figure 2), by the Movement for Unity and Jihad in West Africa (a faction of *al-Qaeda*) [96]. Polisario-Morocco tensions continued through 2016 and, although peace talks have been underway since 2018, tensions remain high and part of the country is held by the Polisario. Uncleared landmines constituted an additional source of persistent insecurity [97]. The continual conditions of insecurity, akin to Somalia, have stymied the development of a functional government that has crop protection capabilities. It is likely that, in Western Sahara, desert locust incursions and breeding will have to be controlled by external interventions (e.g., by Morocco, Mauritania, or cross-border multinational strike forces), assuming infested areas are accessible [6–8].

Yemen. *Direct Insecurity.* The two main Yemeni desert locust breeding areas are in the Tihama and Hadramawt regions (Figure 15) [29]. Insecurity in Yemen has resulted in a variety of obstacles to desert locust operations since 1986 [4]. The division of Yemen, in addition to chronic tribal conflicts, was partially responsible for exacerbating the 1986–1989 desert locust plague [8]. Tribal and clan frictions also impaired surveillance and control to a lesser extent during the 1992–1994 upsurge. Border conflicts with Saudi Arabia have impaired communication, travel, and desert locust surveillance on the border [8].

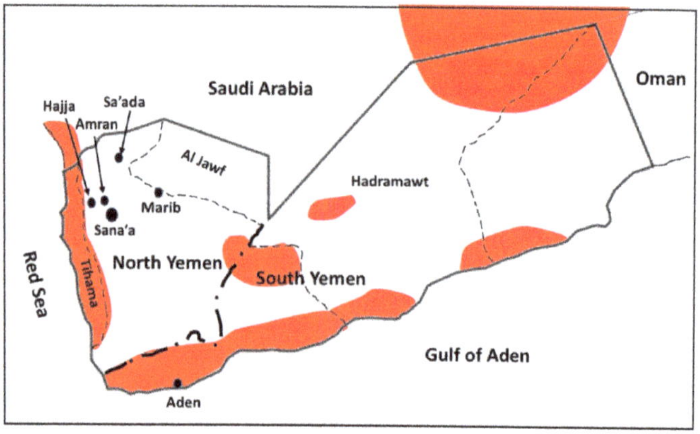

Figure 15. Locations and regions mentioned in the text in relation to desert locust activity and inScheme 1990. when traditionalist North Yemen and Marxist South Yemen (Figure 15) merged after years of skirmishing [98,99]. In mid-1994, a civil war erupted for several weeks, ending in defeat for separatist southern Yemen [99,100]. The war damaged infrastructure [100] and ministries were looted, including the ministry of agriculture, weakening Yemen's anti-locust capability [8]. North versus south enmity persisted [100,101] through 2020.

There are ≈60 million firearms in Yemen, nearly three per citizen [102]. Assault rifles, grenades, and light firearms were, into the 1990s, commonly sold from street stalls and at arms bazaars [102]. Long-standing animosities resulted in many inter-clan and anti-government clashes [98], and Yemeni clans often used foreign hostages to extort the government for funds and infrastructural improvements [103]. Tribal resistance [104–107], however, also intensified, involving surrounding and laying siege to Sana'a (Figure 15) in 2011 [108].

Chronic disputes over tribal homelands and national frontiers have afflicted Yemen since the 1930s [109]. As an example of border disputes, in 1998, Yemen alleged that Saudi Arabia violated its border 73 times over four weeks; transgressions included military deployments, shelling, and violations of air space and waters [8,110–113]. During 1999, a fundamentalist imam exhorted Muslims to kill all "infidels" in Yemen [8]. In 2000, *al-Qaeda* suicide bombers attacked the U.S.S. Cole in the Port of Aden (Figure 15) [8] and a French tanker [114]. The United States and Yemeni militaries struck at *al-Qaeda* by air and ground mainly in remote Marib and Hadramawt areas (Figure 15) [8,115]. In November 2002, a United States missile strike killed the leader of *al-Qaeda* in the Arabian Peninsula [100].

Mass protests in 2004 over myriad grievances led to arrests and military clashes with insurrectionists [100]. Emergent combatant groups in 2004 included the Houthis and the Believing Youth, and through 2016, six consecutive wars enveloped the Sa'ada, Hajja, Amran, and Al Jawf governorats (Figure 15), killing thousands [100,101].

The 2011 rise of Arab Spring resulted in President Saleh's removal in 2012, fracturing the army and the government [100]. Associated protests pitted civilians against law enforcement and military personnel, and many people were consequently attacked by activists and gangs [100,116]. Houthi rebels captured Sana'a (Figure 15) in September 2014 and the insurgency became a civil war [116]. The Houthis also battled *al-Qaeda* and Sunni Salafist movements [117].

In 2009, *al-Qaeda* called for an Islamic caliphate in Yemen [116] and seized territory in outlying provinces [118]. In March 2015, a Saudi-led military coalition (10 countries) intervened to defend the government against the Houthis and *al-Qaeda* and its affiliates [116,117]. By October, the coalition uprooted fighters of *Zaydi Shiite Ansar Allah* (Partisans of God) from most of southern Yemen, and the initial air campaign escalated into a ground conflict committing thousands of troops [118]. From March to October, ≈5000 people were killed, including 2355 civilians and >150 coalition soldiers [118]; a Saudi airstrike (Figure 16) in September, for example, killed 80–130 civilians [118], and the United States launched airstrikes against *al-Qaeda* leaders [118]. In late September, Islamic State of Iraq and Syria (ISIS) suicide bombers killed 30 at a Sana'a mosque and battled against Houthis [118]. The conflicts seriously damaged Yemen's inadequate, already war-torn, infrastructure [117]. The conflicts have continued through 2020. Indirect insecurity has crippled Yemeni capacity and efforts to monitor and control desert locusts.

Figure 16. Yemeni child walks though rubble from Saudi/Emirati airstrike on Sana'a; photo credit Felton Davis/Flickr.

3. Insecurity and the Recent Desert Locust Upsurge

The most recent desert locust upsurge originated in the remote *Rub al Khali*, or Empty Quarter, of Saudi Arabia during the late spring and early summer of 2018 [5,69,119,120]. The outbreak was not detected until late 2018 when swarms moved into Yemen, countries of the Horn of Africa, and Egypt [5,121]. While Egypt, Eritrea, Ethiopia, and Sudan mounted substantial and largely effective control operations [5,121–132], Somalia and Yemen, both crippled by direct and indirect [active and inactive] effects of insurrection, insurgency, international and civil war, lawless interclan fighting, terrorism, and, in the instance of Somalia, landmines, were unable to conduct effective surveillance and control. After Saudi Arabia began combatting gregarious desert locust populations [121–138], Somalia and Yemen produced many swarms; from Somalia they moved into Kenya, Tanzania, Uganda, and Ethiopia, and from Yemen they invaded Oman, Iran, Pakistan, and India [swarms continued to escape from Saudi Arabia into Yemen, Oman, Iraq, Jordan, and other neighboring countries [5,121–138]. In all, ≥22 countries in Africa and Asia were involved by May 2020, and ≈2.8 million hectares were sprayed with insecticides [5]. While it would have been best to proactively intervene against the initial outbreak in the *Rub al Khali*, possibly averting further intensification and dissemination of gregarious desert locust populations, efficacious control in Somalia and Yemen might also have had a pivotal impact on suppressing the upsurge [5]. Because of the effects of armed conflict, Somalia and Yemen became engines that continued to drive the upsurge further as Saudi Arabia, by May 2020, had treated ≈400,000 ha, mitigating its role, to an extent, as a source [5].

4. Insecurity and Possible Mitigation Measures

Inaccessibility of desert locust breeding areas due to insecurity is generally an intractable challenge requiring political solutions rather than technological advances. Conflicts can be acute, long-term, fluid, and sporadic as they erupt, subside, spread, morph, and multiply. Reliable forecasting in neighboring countries will likely enable prepositioning of resources to control swarms when crossing borders from inaccessible areas. In the past, multinational teams, such as the now-defunct Maghrebian Strike Force [MSF], conducted cross-border operations when certain desert locust afflicted countries failed to do so [7]. The MSF was formed by the *Commission de lutte contre le criquet pèlerin en Afrique du Nord-Ouest* (CLCPANO) using Islamic Development Bank, FAO, and member country (Algeria, Libya, Mauritania, Morocco, Tunisia) funds [139]. Active in the early and mid-1990s, the MSF mostly operated in Mauritania and Mali [8]. The MSF, however, was as vulnerable to insecurity as national units. DLCO-EA also does not operate in war zones and most of its member countries (Djibouti, Eritrea, Ethiopia, Kenya, Somalia, South Sudan, Sudan, Tanzania, Uganda) have been in arrears [8], and other regional locust control organizations were nonfunctional or abolished [8]. A new mechanism, the Commission for Controlling the Desert Locust in the Western Africa Region (CLCPRO), established in 2016, makes a regional fund available for proactively controlling desert locust invasions, has been endowed with ≈6.5 million dollars from its member countries (Algeria, Burkina Faso, Chad, Libya, Mali, Mauritania, Morocco, Niger, Senegal, and Tunisia) [5,140,141].

In situations where an outbreak developed in rebel-held territory adjoining a sympathetic country, covert cross-border cooperation has occurred [4,8]. Tensions between nations, however, might have increased if the collaboration was discovered, and further complicated if involved resources originated from international aid agencies.

In relatively recent years, the FAO's EMPRES program (Emergency Prevention System) in West and North Africa has proposed some approaches for dealing with insecurity. In the event of some forms of insecurity in a given area, the country should make efforts to ensure the safety of surveillance and control activities when possible, and, in consultation with the other countries of the region, establish reinforced surveillance in breeding areas of neighboring countries as a safeguard against possible desert locust swarm movement from the insecure area [142]. In addition, satellite imagery information on breeding conditions in the inaccessible areas is almost immediately available, making it possible to assess the probability of outbreaks and the value of strengthening surveillance in accessible breeding areas of neighboring countries [142]. As a contingency against actively developing gregarious desert locust episodes in insecure areas, emergency control plans for the existing national and regional systems can enable rapid response at lower cost compared to times when control systems and plans were not in place [142]. It has been suggested that desert locust survey teams should be comprised of indigenous personnel, familiar with the remote breeding areas, and with the consent and protection of local populations [142]. While the problem of insecurity has long been known, the recent desert locust episode demonstrates that this crucial aspect of desert locust management has not yet been sufficiently addressed.

5. Conclusions

While insecurity is mostly envisaged as a direct obstruction to surveillance and control operations, we suggest that insecurity is considerably more complex and crippling than direct obstruction. In 10 of the 13 countries (77%) we cited as examples, insecurity, whether direct or indirect (Tables 1 and 2), occurred. Insecurity in some instances has been chronic, and diverse, often involving multiple simultaneous conflicts. Indirect, active and inactive, insecurity does not obviously impair surveillance and control, but the adverse impacts of indirect insecurity, although unmeasured, are likely substantial, particularly in countries with sever insecurity problems, such as, at present, Somalia, Yemen, Sudan, and possibly Chad.

Table 1. Thirteen countries with major desert locust breeding areas, and where direct insecurity occurred during different time periods [4–8,14,15,121–138].

Countries with Breeding Areas	Direct Insecurity [a]				
	1986–1989 [b]	1992–1994 [b]	2003–2005 [b,c]	2007–2016 [b]	2018–2020 [b,d]
Chad	√				
Eritrea	√	√			
Ethiopia		√			
India, Pakistan					
Mali		√		√	
Mauritania	√	√			
Niger		√			
Saudi Arabia					
Somalia	√	√			√
Sudan	√	√			
Western Sahara	√				
Yemen	√	√		√	√

[a] Includes areas made inaccessible because of land mines. [b] 1986–1989 was a plague, 1992–1994 and 2003–2005 were upsurges, 2007–2016 had no major desert locust episodes and many outbreaks, and 2018–2020 was an ongoing upsurge at the time of this writing. [c] No reports of direct insecurity, 2003–2005; in July 2005, one area in Sudan was not accessible to ground surveillance due to "access difficulties" [143]. [d] This upsurge was ongoing at the time of this writing.

Table 2. Thirteen countries with major desert locust breeding areas, and where indirect insecurity occurred during different time periods.

Countries with Breeding Areas	Indirect Insecurity [a]				
	1986–1989 [d]	1992–1994 [d]	2003–2005 [d]	2007–2016 [d]	2018–2020 [d]
Chad	√	√	√	√	√
Eritrea	√	√	√		
Ethiopia	√	√	√	√	√
India, Pakistan [b]					
Mali	√	√	√	√	√
Mauritania	√	√			
Niger	√	√			
Saudi Arabia [c]					
Somalia	√	√	√	√	√
Sudan	√	√	√	√	√
Western Sahara	√	√	√	√	
Yemen	√	√	√	√	√

[a] Both active and inactive indirect insecurity. [b] India and Pakistan desert locust control capacities were likely not impaired by the Kashmir conflict [8]. [c] Despite Saudi Arabia's involvement in its ongoing war with Yemen, desert locust control capacity was likely not impaired. [d] 1986–1989 was a plague, 1992–1994 and 2003–2005 were upsurges, 2007–2016 had no major desert locust episodes and many outbreaks, and 2018–2020 was an ongoing upsurge at the time of this writing.

The challenge of insecurity on grand, varied, and complex scales presents a unique obstacle to desert locust control in breeding areas. Further, other desert locust afflicted countries are also beset by insecurity (e.g., Afghanistan, Iraq, Libya, Syria). While some desert locust scenarios might be managed within conflict zones through one creative means or another, operations will likely continue to be impeded and thwarted by insecurity.

Impediments posed by insecurity must be better taken into account for desert locust risk management planning, and realistic mitigation measures must be carefully designed and implemented.

Author Contributions: A.T.S. conceived the article and drafted the manuscript with substantial input from M.L. All authors have read and agreed to the published version of the manuscript.

Funding: This research received no external funding.

Conflicts of Interest: The authors declare no conflict of interest.

Disclaimer on Maps: The designations employed and the presentation of material in the maps do not imply the expression of any opinion on the parts of the U.S. Department of Agriculture and *Agronomy* concerning the legal or constitutional status of any country, territory or sea area, or concerning the delimitation of frontiers.

Mention of trade names or commercial products in this publication is solely for the purpose of providing specific information and does not imply recommendation or endorsement by the U.S. Department of Agriculture. USDA is an equal opportunity provider and employer.

References

1. Lecoq, M.; Zhang, L. *Encyclopedia of Pest Orthoptera of the World*; China Agricultural University Press: Beijing, China, 2019.
2. Babah, M.A.O. Strategy for controlling the desert locust in Mauritania. In *New Strategies in Locust Control*; Krall, S., Peveling, R., Ba Diallo, D., Eds.; Birkhauser: Berlin, Germany, 1997; pp. 487–491.
3. Showler, A.T. Proaction: Strategic framework for today's reality. In *New Strategies in Locust Control*; Krall, S., Peveling, R., Ba Diallo, D., Eds.; Birkhauser: Berlin, Germany, 1997; pp. 461–465.
4. Showler, A.T. Desert locust control: The effectiveness of proactive interventions and the goal of outbreak prevention. *Am. Entomol.* **2018**, *65*, 180–191. [CrossRef]
5. Showler, A.T.; Babah, M.A.O.; Lecoq, M.; Maeno, K. Early intervention against desert locusts: Current proactive approach and the possibility of sustainable outbreak prevention. *Agronomy* **2021**. (submitted).
6. Showler, A.T.; Potter, C.S. Synopsis of the desert locust, *Schistocerca gregaria* (Forskål), plague 1986–1989 and the concept of strategic control. *Am. Entomol.* **1991**, *37*, 106–110. [CrossRef]
7. Showler, A.T. Locust (*Orthoptera: Acrididae*) outbreak in Africa and Asia, 1992–1994. *Am. Entomol.* **1995**, *41*, 179–185. [CrossRef]
8. Showler, A.T. The importance of armed conflict to desert locust control, 1986–2002. *J. Orthopt. Res.* **2003**, *12*, 127–133. [CrossRef]
9. Steedman, A. *Locust Handbook*; Overseas Development Natural Resources Institute: London, UK, 1988.
10. Cossé, O.; Lazar, M.; Hassane, S. *Rapport de l'Evaluation à mi-Parcours du Programme EMPRES Composante Criquet Pèlerin en Région Occidentale*; Food and Agriculture Organization of the United Nations: Rome, Italy, 2009.
11. Lecoq, M. Vers une solution durable au problème du criquet pèlerin? *Science Chang. Planétaires/Sécheresse* **2004**, *15*, 217–224.
12. Sword, G.A.; Lecoq, M.; Simpson, S.J. Phase polymorphism and preventative locust management. *J. Insect Physiol.* **2010**, *56*, 949–957. [CrossRef] [PubMed]
13. France Diplomatie Home Page. Available online: https://www.diplomatie.gouv.fr/fr/ (accessed on 20 October 2020).
14. Belayneh, Y.T. Acridid pest management in the developing world: A challenge to the rural population, a dilemma to the international community. *J. Orthopt. Res.* **2005**, *14*, 187–195. [CrossRef]
15. Showler, A.T. A summary of control strategies for the desert locust, *Schistocerca gregaria* (Forskål). *Agric. Ecosyst. Environ.* **2002**, *90*, 97–103. [CrossRef]
16. Lecoq, M. Desert locust management: From ecology to anthropology. *J. Orthopt. Res.* **2005**, *14*, 179–186. [CrossRef]
17. Gay, P.M.; Lecoq, M.; Piou, C. Improving preventive locust management: Insights from a multi-agent model. *Pest Manag. Sci.* **2018**, *74*, 46–58. [CrossRef] [PubMed]
18. FAO (Food and Agriculture Organization of the United Nations). Desert Locust Watch. Available online: http://www.fao.org/ag/locusts/en/info/info/indiz/html (accessed on 23 October 2020).
19. Roberts, H. *The Battlefield: Algeria 1988–2002, Studies in a Broken Polity*; Verso Books: London, UK, 2003.
20. Algeria Hit by Two Massacres. Available online: http://news.bbc.co.uk/2/hi/middle_east/1963518.stm (accessed on 20 October 2020).
21. Mellah, S. The massacres in Algeria, 1992–2004. In Proceedings of the 32nd Session of the Permanent Peoples' Tribunal on Human Rights Violation in Algeria (1992–2004), Paris, France, 5–8 November 2004.
22. Strachan, M.L. *Conflict Analysis of Algeria*; Governance Social Development Humanitarian Conflict, Applied Knowledge Services: Washington, DC, USA, 2014.
23. Algeria Profile—Timeline. Available online: https://www.bbc.com/news/world-africa-14118856 (accessed on 24 October 2020).
24. Jesus, C.E. Terrorism financing: The particular case of al-Qaeda in the Islamic Maghreb (AQIM). *Afr. J. Prev. Combat. Terror.* **2011**, *2*, 39–59.

25. Boukhars, A. What's next for Mali and Algeria? Available online: http://carnegieendowment.org/2013/01/23/what-s-next-for-mali-and-algeria/f4sl (accessed on 6 January 2021).
26. Spencer, C. Strategic posture review: Algeria. Available online: http://www.chathamhouse.org/sites/default/files/public/Research/Middle%20East/0712wpr_spencer.pdf (accessed on 20 October 2020).
27. Ammour, L.A. *Regional Security Cooperation in the Maghreb and Sahel: Algeria's Pivotal Ambivalence*; Africa security brief no. 18; African Center for Strategic Studies: Washington, DC, USA, 2012.
28. Raleigh, C.; Dowd, C. Real-Time Analysis of African Political Violence. Conflict Trends No. 11. Available online: http://www.acleddata.com/wp-content/uploads/2013/02/ACLED-Conflict-Trends-Report-No.-11-February-2013.pdf (accessed on 20 October 2020).
29. Pedgley, D. *Desert Locust Forecasting Manual*; Centre for Overseas Pest Research: London, UK, 1981.
30. Kayalto, M.; Idrissi Hassani, M.; Lecoq, M.; Gay, P.F.; Piou, C. Cartographie des zones de reproduction et de grégarisation du criquet pèlerin au Tchad. *Cah. Agric.* **2020**, *29*, 14. [CrossRef]
31. Collins, R.O.; Burr, M.I.; Burr, M. *Africa's Thirty Years' War: Libya, Chad, and the Sudan 1963–1993*; Westview Press: Boulder, CO, USA, 1999.
32. More Clashes in Northern Chad. Available online: http://news.bbc.co.uk/2/hi/africa/1092482.stm (accessed on 20 October 2020).
33. Mine Wounds Chad Rebel Leader. Available online: http://news.bbc.co.uk/2/hi/africa/2225280.stm (accessed on 20 October 2020).
34. Chad profile—Timeline. Available online: https://www.bbc.com/news/world-africa-13164690 (accessed on 24 October 2020).
35. Chad Peace Deal Signed. Available online: http://news.bbc.co.uk/2/hi/africa/1747940.stm (accessed on 14 October 2020).
36. Berg, P. *The Dynamics of Conflict in the Tri-Border Region of the Sudan, Chad and the Central African Republic*; Friedrich Ebert Foundation: Berlin, Germany, 2008.
37. Ploch, L. *Instability and Humanitarian Conditions in Chad*; CRS report for Congress; Congressional Research Service: Washington, DC, USA, 2010.
38. Clash on Chad-CAR Border. Available online: http://news.bbc.co.uk/2/hi/africa/2177926.stm (accessed on 20 October 2020).
39. Papstein, R. *Eritrea: Revolution at Dusk*; The Red Sea Press: Trenton, NJ, USA, 1991.
40. Bereketeab, R. The complex roots of the second Eritrea-Ethiopia war: Re-examining the causes. *Afr. J. Int. Aff.* **2013**, *13*, 15–59.
41. Sudan and Eritrea in Reconciliation Talks. Available online: http://news.bbc.co.uk/2/hi/africa/839356.stm (accessed on 20 October 2020).
42. Sudan and Eritrea Begin Repairing Relations. Available online: http://news.bbc.co.uk/2/hi/africa/590910.stm (accessed on 20 October 2020).
43. Kwiatkowska, B. The Eritrea/Yemen arbitration: Landmark progress in the acquisition of territorial sovereignty and equitable maritime boundary delimitation. *Boundary Secur. Bull.* **2000**, *8*, 66–86. [CrossRef]
44. United Nations Mission in Ethiopia and Eritrea. Available online: https://peacekeeping.un.org/en/mission/past/unmee (accessed on 29 October 2020).
45. Young, J. *Beja: Local Conflict, Marginalization, and the Threat to Regional Security*; Institute for Security Studies: Pretoria, South Africa, 2011.
46. Ethiopia Profile—Timeline. Available online: https://www.bbc.com/news/world-africa-13351397 (accessed on 24 October 2020).
47. Ethiopia Says Foils Somali Rebel Plot to Seize U.N. Staff. Available online: https://news.trust.org/item/20130325200500-jfusq/ (accessed on 20 October 2020).
48. Recent Landmine Use by India and Pakistan. Available online: https://www.hrw.org/report/2002/05/22/recent-landmine-use-india-and-pakistan (accessed on 20 October 2020).
49. Ganguly, S. *Conflict Unending: India and Pakistan*; Columbia University Press: New York, NY, USA, 2002.
50. Jones, O.B. *Pakistan*; Yale University Press: New Haven, CT, USA, 2002.
51. India and Pakistan: Tense Neighbors. Available online: http://news.bbc.co.uk/2/hi/south_asia/102201.stm (accessed on 20 October 2020).
52. Kashmir Conflict: Tension on the India Pakistan Border. Available online: https://www.bbc.com/news/world-asia-37531900 (accessed on 24 October 2020).
53. Morocco Profile—Timeline. Available online: https://www.bbc.com/news/world-africa-14123260 (accessed on 24 October 2020).
54. Kashmir: Pakistan Call Emergency Meeting Amid 'Deteriorating Situation'. Available online: http://www.cnn.com/2016/09/29/homepage2/kashmir-pakistani-soldiers-killed-india-surgical-strikes (accessed on 23 October 2020).
55. Marszal, A. India Build Underground Bunkers on Kashmir Frontier as Border Clashes with Pakistan Escalate. Available online: http://www.telegraph.co.uk/news/2016/10/04/india-builds-underground-bunkers-on-kashmir-frontier-as-border-clashes-with-pakistan-escalate (accessed on 20 October 2020).
56. Le Miere, J. India-Pakistan War Timeline: Recent Deadly Attacks Are Just the Latest Between Nuclear Neighbors. Available online: http://www.ibtimes.com/india-pakistan-war-timeline-recent-deadly-attacks-are-just-latest-between-nuclear-neighbors (accessed on 20 October 2020).
57. Nicolaisen, J.; Nicolaisen, I. *The Pastoral Tuareg: Ecology, Culture, and Society*; Thames & Hudson: London, UK, 1997.

58. Malian Tuaregs Accused of Killings in Niger. Available online: http://news.bbc.co.uk/2/hi/africa/123742.stm (accessed on 23 October 2020).
59. Ba, O. *Tuareg Nationalism and Cyclical Pattern of Rebellions: How the Past and Present Explain Each Other*; Sahel Research Group, Working Paper 007; University of Florida: Gainesville, FL, USA, 2014.
60. Brahim, E.O. Conflict in Northern Mali: Internal Facts and Regional Impacts. Available online: https://studies.aljazeera.net/en/reports/2012/02/2012219113630111169.html (accessed on 20 October 2020).
61. Arieff, A. *Crisis in Mali*; R42664; Congressional Research Service: Washington, DC, USA, 2013.
62. Mauritania Declares Itself Landmine Free Nearly Two Decades After Mine Clearance Began. Available online: https://reliefweb.int/report/mauritania/mauritania-declares-itself-landmine-free-nearly-two-decades-after-mine-clearance (accessed on 8 November 2020).
63. Fievet, J.-L. Senegalese-Mauritanian Riots Spread. Available online: http://www.washingtonpost.com/archive/politics/198904/29/senegalese-mauritanian-riots-spread/54125c6fa-4815-bdfe-c4ab4db2482c/ (accessed on 8 November 2020).
64. Ethnic Riots Spread across Senegal border. Available online: http://www.chicagotribune.com/news/et-xpm-1989-05-01-8904080799-story.html (accessed on 8 November 2020).
65. Rao, S. *Conflict Analysis of Mauritania*; Governance, Social Development, Humanitarian, Conflict Applied Knowledge Services, University of Birmingham: Birmingham, UK, 2014.
66. Boukhars, A. The Drivers of Insecurity in Mauritania. Available online: https://carnegieendowment.org/2012/04/30/drivers-of-insecurity-in-mauritania-pub-47955 (accessed on 8 November 2020).
67. Chouvenc, H. *Pilote de Brousse*; Editions du Miroir: Montvicq, France, 1998.
68. Chronology for Tuareg in Niger. Minorities at Risk Project. Available online: https://www.refworld.org/docid/469f38c2104.html (accessed on 23 October 2020).
69. Current Upsurge (2019–2020). Available online: http://www.fao.org/ag/locusts/en/info/info/indiz/html (accessed on 23 October 2020).
70. Desert Locusts Ravage Breakaway Somalia Region. Available online: http://www.earthweek.com/2009/ew090612/ew090612c.html (accessed on 20 October 2020).
71. Peterson, S. *Me Against My Brother, at War in Somalia, Sudan and Rwanda*; Routledge Press: New York, NY, USA, 2000.
72. Hogg, A.L. Timeline: Somalia, 1991–2008. From Troubled to Dire. Available online: http://www.theatlantic.com/magazine/archive/2008/12/timeline-somalia-1991-2008/307190 (accessed on 23 October 2020).
73. Country profile—Somalia. Available online: https://www.bbc.com/news/world-africa-14094503 (accessed on 23 October 2020).
74. Somalia Civil War. Available online: https://www.globalsecurity.org/military/world/war/somalia.htm (accessed on 23 October 2020).
75. Bowden, M. *Black Hawk Down*; Signet: New York, NY, USA, 1999.
76. Factional Fighting Erupts in Somalia. Available online: http://news.bbc.co.uk/2/hi/africa/2146701.stm (accessed on 20 October 2020).
77. Somalia: General Overview of the Security Situation in the City of MOGADISHU (2000–2002). Available online: https://www.refworld.org/docid/3f7d4e1f2a.html (accessed on 23 October 2020).
78. Somalia's Role in Terror. Available online: http://news.bbc.co.uk/2/hi/africa/1723586.stm (accessed on 20 October 2020).
79. Ethiopian Troops Deploy in Somalia. Available online: http://news.bbc.co.uk/2/hi/africa/1747396.stm (accessed on 20 October 2020).
80. Ethiopian Troops in Somalia. Available online: http://news.bbc.co.uk/2/hi/africa/1989954.stm (accessed on 20 October 2020).
81. U.S. Watches Somali al-Qaeda Links. Available online: http://news.bbc.co.uk/2/hi/africa/1877878.stm (accessed on 24 October 2020).
82. Somali Rebels Shoot Down Ethiopian Helicopter as Clashes Worsen. Available online: https://www.pbs.org/newshour/nation/africa-jan-june07-somalia_03-30 (accessed on 20 October 2020).
83. Somalia's al-Shabaab in Comeback, as Ethiopia Pulls Troops. Available online: http://www.foxnews.com/world/2016/10/27/somalia-al-shabab-in-comeback-as-ethiopia-pulls-troops (accessed on 20 October 2020).
84. Landmine Danger Persists in Somalia. Available online: https://www.thenewhumanitarian.org/news/2013/02/01/landmine-danger-persists-somalia (accessed on 20 October 2020).
85. Kassala in Government Hands. Available online: http://news.bbc.co.uk/2/hi/africa/1014872.stm (accessed on 20 October 2020).
86. Mohyeldeen, S. *The Egypt-Sudan Border: A Story of Unfulfilled Promise*; Carnegie Endowment for International Peace: Washington, DC, USA, 2020.
87. Country Profile—Sudan. Available online: https://www.bbc.com/news/world-africa-14094995 (accessed on 23 October 2020).
88. Ottaway, M.; El-Sadany, M. *Sudan: From Conflict to Conflict*; The Carnegie Papers, Carnegie Endowment for International Peace: Washington, DC, USA, 2012.
89. Salopek, P. Shattered Sudan: Drilling for oil, hoping for peace. *Nat. Geograph.* **2003**, *203*, 30–59.
90. Rothbart, D.; Yousif, A. Ideology and Cultural Violence in Darfur. Available online: https://www.accord.org.za/conflict-trends/ideology-cultural-violence-darfur/ (accessed on 20 October 2020).
91. Adam, A.H. In Sudan, the Janjaweed Rides Again. Available online: https://www.nytimes.com.2014/07/17/opinion/in-sudan-the-janjaweed-rides-again.html (accessed on 6 January 2021).

92. Polisario Rejects UN Peace Plan. Available online: http://news.bbc.co.uk/2/hi/africa/1547452.stm (accessed on 20 October 2020).
93. Polisario Blasts Morocco's Sahara Claim. Available online: http://news.bbc.co.uk/2/hi/africa/1857421.stm (accessed on 20 October 2020).
94. Sahara Conflict Drags On. Available online: http://news.bbc.co.uk/2/hi/africa/1191810.stm (accessed on 20 October 2020).
95. Morocco Dismisses POW Move. Available online: http://news.bbc.co.uk/2/hi/africa/2052019.stm (accessed on 20 October 2020).
96. Arieff, A. *Western Sahara*; RS20962; Congressional Research Service: Washington, DC, USA, 2014. Available online: https://fas.org/sgp/crs/row/RS20962.pdf (accessed on 20 October 2020).
97. United Nations Security Council. *Report of the Secretary-General on the Situation Concerning Western Sahara*; S/2008/251; United Nations Security Council: New York, NY, USA, 2008; Available online: https://digitallibrary.un.org/record/624602?ln=fr (accessed on 20 October 2020).
98. Dresch, P. *A History of Modern Yemen*; Cambridge University Press: Cambridge, UK, 2001.
99. Country Profile: Yemen. Available online: https://www.bbc.com/news/world-middle-east-14704852 (accessed on 23 October 2020).
100. Lewis, A. Violence in Yemen: Thinking about violence in fragile states beyond the confines of conflict and terrorism. *Stab. Int. J. Secur. Dev.* **2013**, *2*, 13. [CrossRef]
101. Key Facts about the War in Yemen. Available online: httpps://www.aljazeera.com/news/2018/3/25/key-facts-about-the-war-in-yemen (accessed on 6 January 2021).
102. Yemen's Weapon Culture. Available online: http://news.bbc.co.uk/2/hi/middle_east/1775938.stm (accessed on 24 October 2020).
103. Kidnapping of Foreigners in Yemen, 1996–2001. Available online: https://al-bab.com/kidnapping-foreigners-yemen-1996-2001 (accessed on 6 January 2021).
104. Britons Unlawfully Killed in Yemen. Available online: http://news.bbc.co.uk/2/hi/uk_news/542899.stm (accessed on 20 October 2020).
105. Eight Die in Yemen Tribal Clash. Available online: http://news.bbc.co.uk/2/hi/middle_east/373554.stm (accessed on 23 October 2020).
106. Yemen: Arabia's Wild West. Available online: http://news.bbc.co.uk/2/hi/programmes/from_our_own_correspondent/253004.stm (accessed on 20 October 2020).
107. Yemen Troops Surround Two Villages. Available online: http://news.bbc.co.uk/2/hi/middle_east/521018.stm (accessed on 20 October 2020).
108. Schmitz, C. The Rise of Yemen's Houthi Rebels. Available online: https://www.bbc.com/news/world-middle-east-31645145 (accessed on 20 October 2020).
109. Smith, G.R.Y.; Auchterlonie, P. *Yemen*, 2nd ed.; ABC-CLIO: Santa Barbara, CA, USA, 1998.
110. Yemen Accuses Saudi Arabia of Over Seventy Border Violations. Available online: https://www.bbc.com/news/world-middle-east-37723472 (accessed on 20 October 2020).
111. Jehl, D. Border Clash Heightens Yemen-Saudi Tensions. Available online: https://www.nytimes.com/1998/07/26/world/border-clash-heightens-yemen-saudi-tensions.html (accessed on 25 October 2020).
112. Okruhlik, G.; Conge, P.J. The politics of border disputes: On the Arabian Peninsula. *Int. J.* **1999**, *54*, 230–248. [CrossRef]
113. Yemen, Saudi Arabia Sign Border Deal. Available online: http://news.bbc.co.uk/2/hi/middle_east/788153.stm (accessed on 20 October 2020).
114. New Front in War on Terror. Available online: http://news.bbc.co.uk/2/hi/middle_east/2407597.stm (accessed on 20 October 2020).
115. U.S. Defends Yemen Strike. Available online: http://news.bbc.co.uk/2/hi/americas/2439305.stm (accessed on 23 October 2020).
116. Mitreski, A. *Civil War in Yemen: A Complex Conflict with Multiple Futures*; Arab Center for Research and Policy Studies: Doha, Qatar, 2015.
117. Popp, R. *War in Yemen: Revolution and Saudi Intervention*; No. 175; Center for Security Studies: Zurich, Switzerland, 2015.
118. Sharp, J.M. *Yemen: Civil War and Regional Intervention*; R43960; Congressional Research Service: Washington, DC, USA, 2015.
119. Lecoq, M. Some considerations on the current desert locust situation in East Africa. *Metaleptea* **2020**, *40*, 25–26.
120. Meynard, C.N.; Lecoq, M.; Chapuis, M.P.; Piou, C. On the relative role of climate change and management in the current desert locust outbrteak in East Africa. *Glob. Chang. Biol.* **2020**, *26*, 3753–3755. [CrossRef] [PubMed]
121. FAO (Food and Agriculture Organization of the United Nations). *Desert Locust Bulletin*; No. 483; FAO: Rome, Italy, 2019; Available online: http://www.Fao.org/ag/locusts/en/archives/archive/1366/2003/index/html (accessed on 23 October 2020).
122. FAO (Food and Agriculture Organization of the United Nations). *Desert Locust Bulletin*; No. 484; FAO: Rome, Italy, 2019; Available online: http://www.Fao.org/ag/locusts/en/archives/archive/1366/2003/index/html (accessed on 23 October 2020).
123. FAO (Food and Agriculture Organization of the United Nations). *Desert Locust Bulletin*; No. 485; FAO: Rome, Italy, 2019; Available online: http://www.Fao.org/ag/locusts/en/archives/archive/1366/2003/index/html (accessed on 23 October 2020).
124. FAO (Food and Agriculture Organization of the United Nations). *Desert Locust Bulletin*; No. 486; FAO: Rome, Italy, 2019; Available online: http://www.Fao.org/ag/locusts/en/archives/archive/1366/2003/index/html (accessed on 23 October 2020).

125. FAO (Food and Agriculture Organization of the United Nations). *Desert Locust Bulletin*; No. 487; FAO: Rome, Italy, 2019; Available online: http://www.Fao.org/ag/locusts/en/archives/archive/1366/2003/index/html (accessed on 23 October 2020).
126. FAO (Food and Agriculture Organization of the United Nations). *Desert Locust Bulletin*; No. 488; FAO: Rome, Italy, 2019; Available online: http://www.Fao.org/ag/locusts/en/archives/archive/1366/2003/index/html (accessed on 23 October 2020).
127. FAO (Food and Agriculture Organization of the United Nations). *Desert Locust Bulletin*; No. 489; FAO: Rome, Italy, 2019; Available online: http://www.Fao.org/ag/locusts/en/archives/archive/1366/2003/index/html (accessed on 23 October 2020).
128. FAO (Food and Agriculture Organization of the United Nations). *Desert Locust Bulletin*; No. 490; FAO: Rome, Italy, 2019; Available online: http://www.Fao.org/ag/locusts/en/archives/archive/1366/2003/index/html (accessed on 23 October 2020).
129. FAO (Food and Agriculture Organization of the United Nations). *Desert Locust Bulletin*; No. 491; FAO: Rome, Italy, 2019; Available online: http://www.Fao.org/ag/locusts/en/archives/archive/1366/2003/index/html (accessed on 23 October 2020).
130. FAO (Food and Agriculture Organization of the United Nations). *Desert Locust Bulletin*; No. 492; FAO: Rome, Italy, 2019; Available online: http://www.Fao.org/ag/locusts/en/archives/archive/1366/2003/index/html (accessed on 23 October 2020).
131. FAO (Food and Agriculture Organization of the United Nations). *Desert Locust Bulletin*; No. 493; FAO: Rome, Italy, 2019; Available online: http://www.Fao.org/ag/locusts/en/archives/archive/1366/2003/index/html (accessed on 23 October 2020).
132. FAO (Food and Agriculture Organization of the United Nations). *Desert Locust Bulletin*; No. 494; FAO: Rome, Italy, 2019; Available online: http://www.Fao.org/ag/locusts/en/archives/archive/1366/2003/index/html (accessed on 23 October 2020).
133. FAO (Food and Agriculture Organization of the United Nations). *Desert Locust Bulletin*; No. 495; FAO: Rome, Italy, 2020; Available online: http://www.Fao.org/ag/locusts/en/archives/archive/1366/2003/index/html (accessed on 23 October 2020).
134. FAO (Food and Agriculture Organization of the United Nations). *Desert Locust Bulletin*; No. 496; FAO: Rome, Italy, 2020; Available online: http://www.Fao.org/ag/locusts/en/archives/archive/1366/2003/index/html (accessed on 23 October 2020).
135. FAO (Food and Agriculture Organization of the United Nations). *Desert Locust Bulletin*; No. 497; FAO: Rome, Italy, 2020; Available online: http://www.Fao.org/ag/locusts/en/archives/archive/1366/2003/index/html (accessed on 23 October 2020).
136. FAO (Food and Agriculture Organization of the United Nations). *Desert Locust Bulletin*; No. 498; FAO: Rome, Italy, 2020; Available online: http://www.Fao.org/ag/locusts/en/archives/archive/1366/2003/index/html (accessed on 23 October 2020).
137. FAO (Food and Agriculture Organization of the United Nations). *Desert Locust Bulletin*; No. 499; FAO: Rome, Italy, 2020; Available online: http://www.Fao.org/ag/locusts/en/archives/archive/1366/2003/index/html (accessed on 23 October 2020).
138. FAO (Food and Agriculture Organization of the United Nations). *Desert Locust Bulletin*; No. 500; FAO: Rome, Italy, 2020; Available online: http://www.Fao.org/ag/locusts/en/archives/archive/1366/2003/index/html (accessed on 23 October 2020).
139. CLCPANO (Commission de Lutte Contre le Criquet Pelerin en Afrique du Nord-Ouest). *Report of the 16th session of the Commission de Lutte Contre le Criquet Pelerin en Afrique du Nord-Ouest (CLCPANO)*; CLCPANA: Algiers, Algeria, 1991; Available online: http://www.fao.org/ag/locusts/fr/publicat/meeting/topic/regc/clcpano/1734/documents_1767.html (accessed on 20 October 2020).
140. FAO (Food and Agriculture Organization of the United Nations). *Programme EMPRES en Region. Occidentale*; FAO: Rome, Italy, 2016; Available online: http://www.clcpro-empres.org (accessed on 5 November 2020).
141. FAO (Food and Agriculture Organization of the United Nations). *Evaluation Final de la Phase II (2014–2017) du Programme EMPRES/Composante Criquet Pèlerin en Région Occidentale (EMPRES-RO)*; FAO: Rome, Italy, 2018; Available online: http://www.fao.org/3/ca4005fr/ca4005fr.pdf (accessed on 5 November 2020).
142. Martini, P.; Lecoq, M.; Soumare, L.; Chara, B. *Proposition de Programme de Lutte Contre le Criquet Pèlerin dans la Partie Occidentale de Son Aire d'habitat. Système de Prévention et de Réaction Rapide (EMPRES). Composante Acridienne (Criquet Pèlerin) en Région Occidentale*; Food and Agriculture Organization of the United Nations: Rome, Italy, 1998; Available online: http://www.fao.org/ag/locusts/oldsite/PDF/meetings/EMPRESwest98f.pdf (accessed on 20 October 2020).
143. FAO (Food and Agriculture Organization of the United Nations). Desert Locust Watch; Archives 2005. Available online: http://www.fao.org/ag/locusts/en/archives/archive/1366/2003/index.html (accessed on 23 October 2020).

agronomy

MDPI

Article

Ups and Downs of the Italian Locust (*Calliptamus italicus* L.) Populations in the Siberian Steppes: On the Horns of Dilemmas

Michael G. Sergeev [1,2]

[1] Department of General Biology and Ecology, Novosibirsk State University, 2 Pirogova Street, 630090 Novosibirsk, Russia; mgs@fen.nsu.ru

[2] Laboratory of Invertebrate Ecology, Institute of Systematics and Ecology of Animals, Siberian Branch, Russian Academy of Sciences, 11 Frunze Street, 630091 Novosibirsk, Russia

Abstract: The Italian locust is a common species and one of the most important pest in the steppes, semi-deserts, and deserts of Central and Southwestern Asia and also in the Mediterranean Region. The aim of this paper is to discuss some problems arising from studies of eco-geographical distribution and long-term dynamics of this species, especially in the southern part of West Siberia, and their consequences for management of pest species and rare forms. Peculiarities of the Italian locust populations' distribution and long-term dynamics are discussed for the north-eastern part of the species range (south-eastern West Siberian Plain, North-East and, partly, East Kazakhstan). Notable differences between local populations are revealed. Some issues arising from results of insecticide experimental treatments (mainly effectiveness of barrier applications with different compounds in variable environment) and ecological association between the Italian locust and rare Orthoptera are also discussed. Some suggestions as to requirements for improved monitoring, supervision, control, and forecasting of the Italian locust populations' distribution and dynamics are also provided.

Keywords: West Siberian Plain; outbreak; distribution; dynamics; population; population management; plant protection

Citation: Sergeev, M.G. Ups and Downs of the Italian Locust (*Calliptamus italicus* L.) Populations in the Siberian Steppes: On the Horns of Dilemmas. *Agronomy* **2021**, *11*, 746. https://doi.org/10.3390/agronomy11040746

Academic Editor: Salvatore Davino

Received: 2 April 2021
Accepted: 10 April 2021
Published: 12 April 2021

Publisher's Note: MDPI stays neutral with regard to jurisdictional claims in published maps and institutional affiliations.

1. Introduction

The Italian locust (*Calliptamus italicus* (Linnaeus, 1758)) is a common and widely distributed species and the most important acridid pest in the steppes and semi-deserts of Eurasia, from Ukraine to the southern part of West Siberia, East Kazakhstan, and North-West China. Many of its outbreaks were in these areas, in mountains and oases of Central Asia, and also in the Mediterranean region. In the steppes, during the 20th century, its outbreaks occurred in all decades, but the situation became especially dangerous in the end of the last century, when in 2000 more than 16 million ha were infested and more than 10 million ha were treated by different acridicides from organophosphates to insect growth regulators [1,2].

The Italian locust is an intermediate form between typical gregarious and solitarious acridid species [1,3,4]. During outbreaks, the dense bands and swarms are common; however, their dispersal rates are usually limited. Both hopper bands and swarms can migrate over comparatively small distances, usually up to several hundred meters (rarely up to several kilometers) for bands and up to 100–200 km (rarely up to 750–800 km) for swarms [5–7]. Morphologically, the gregarious form of adults may be distinguished only by relatively long wings [3,8]. Gregarious nymphs of the Italian locust are usually characterized by dominance of dark colors (brown, grey, and black) [5,9].

Ecological peculiarities and distribution patterns of *C. italicus* are described in many publications for some parts of its range. An analysis of many publications was given by B.P. Uvarov [3]. He emphasized an insufficient level of our knowledge on the Italian locust ecology for periods between outbreaks. During recent decades, results of many studies concerning this species were published as well; however, nearly all of them are limited by outbreak periods [1,4,10,11]. Recently, the main part of all data was summarized for the territory of the former USSR [9].

The Italian locust is a typical univoltine form with overwintering eggs. Its spatial population distribution during periods between outbreaks shows that, in the northern part of its range, the species occurs in very dry habitats, in the central one, it prefers relatively dry and diverse habitats of the steppe and semi-desert life zones, while in the southern part, its colonies are usually localized in meadow habitats of river valleys or in mountains [4,12]. The species prefers highly heterogeneous semi-arid landscapes (especially with dominance of sagebrushes—*Artemisia* spp.) in the dry steppes and the semi-deserts. During outbreaks, the Italian locust can penetrate and colonize different types of fields cultivated and abandoned, and also other types of transformed habitats, e.g., roadsides, lawns, overgrazed sandy plots, etc. However, the Italian locust usually prefers to consume dicotyledon plants, e.g., *Artemisia* Linnaeus, *Lactuca tatarica* (Linnaeus) C.A.Mey, *Kochia* Roth, *Medicago* Linnaeus, *Glyzyrrhiza* Linnaeus, etc. [3,9]. This means the species can seriously damage some cash crop fields, for instance, sunflowers, and, at least on the first stage of infestation, may eat and destroy different weeds across wheat fields.

The aim of this paper is to discuss some problems arising from studies of long-term dynamics of the Italian locust populations, especially in the southern part of West Siberia, and their consequences for population management of both possible pest and rare orthopteran species.

2. Materials and Methods

2.1. Study Territory

Original data were collected from 1979 until 2019 in the south-eastern part of West Siberian Plain and in the adjacent north-eastern and eastern parts of the Kazakh Uplands (Saryarqa). This region borders the Ishim River (left tributary of the Irtysh) to the west and south-west, the southern edge of the Kazakh Uplands to the south, the Altay Mts. to the south-east, and the Ob River to the east. Its northern boundary is approximately defined by the 56th parallel north. Average temperatures are relatively low (mean temperatures of the warmest month vary from 17 to 22 °C, the same for the coldest month—from −17 to −20 °C), and annual precipitation amounts vary from 280 to 520 mm [13]. Originally, this territory was covered with grasslands (from meadows to semi-deserts) and forests (mainly birch and pine) [13–15]. The main part of the plains was transformed in agricultural lands (fields and pastures), while the mountain steppes and semi-deserts are used for livestock grazing. Besides, there are some flood-plains with meadows and forest patches, sandy plots, solonchaks, and swamps. The Kazakh Uplands consist of small low ranges, numerous hills and small mountains, where some elements of altitudinal zonation may be observed.

2.2. Observation Sites

Nine fixed sites were selected to study long-term dynamics of the Italian locust populations in the central part of the so-called Kulunda steppe (between the Irtysh River and the central part of the Altaj Region). Almost all plots were covered by more or less typical steppe zonal vegetation. In some cases, vegetation cover was damaged by moderate grazing or haymaking. These sites were distributed from the northern steppes to the southern ones:

1. SE Aleksandrovskij settlement (Novosibirskaya Oblast (Novosibirsk Region), 53.67° N 78.25° E, northern steppe, studied in 2000–2008, 2015, 2018, 2019; in 2003 the local control, i.e., untreated, plot was moved about 100 m southwards, because the main part of the area was plowed) (Figure 1A).
2. S Burla settlement (Altai Krai (Altaj Region), 53.23° N 78.43° E, old abandoned field, 2000–2008).
3. SW Yarovoe town (Altai Krai (Altaj Region), 52.85° N 78.57° E, dry steppe (actually very old crested wheatgrass field), 1979, 1992, 1999–2008, 2015).

4. E Platovka settlement (Altai Krai (Altaj Region), 52.40° N 79.05° E, dry steppe, 1999–2008, 2015).
5. SE Severka settlement (Altai Krai (Altaj Region), 52.12° N 79.32° E, typical steppe, 1999–2008, 2015).
6. S Pokrovka settlement (Altai Krai (Altaj Region), 52.00° N 79.37° E, typical steppe, 1999–2008, 2015).
7. NW Mikhaylovskoye settlement (Altai Krai (Altaj Region), 51.82° N 79.62° E, typical steppe, 1999–2008, 2015; in 2003, we counted locusts on the adjacent steppe plot, 51.83° N 79.58° E, because we could not visit the stationary plot due to some technical problems).
8. W Bor-Forpost settlement (Altai Krai (Altaj Region), 51.87° N 80.03° E, typical steppe (actually very old crested wheatgrass field), 2000–2009, 2015).
9. E Ust-Volchikha settlement (Altai Krai (Altaj Region), 51.93° N 80.28° E, dry steppe, 1999–2009, 2015, 2018) (Figure 1B).

(**A**) (**B**)

Figure 1. Typical habitats of the Italian locust in the Kulunda steppe: (**A**)—northern steppe (near Aleksandrovskij settlement); (**B**)—dry steppe (near Ust-Volchikha settlement).

As a rule, locust abundance/density was counted in the first half and middle of July, while adults prevailed (see the Section 2.4). In this study, I analyzed data mainly for 2000–2008 when all nine sites were explored.

2.3. Experimental Area

In 2000, during the Italian locust outbreak, the dynamics of its local population was studied on the experimental area near Alexandrovskij settlement (see Section 2.2). This area included some plots with steppe vegetation, some agricultural fields, several small birch forests, a shallow depression with halophytes vegetation, meadows and willow bushes, and forest belts [16]. The steppe vegetation was quite diverse and heterogeneous (Figure 2). Several plant species, namely, *Phleum phleoides* (Linnaeus) H. Karst., *Agrostis capillaris* Linnaeus, *Koeleria macrantha* (Ledeb.) Schult., *Potentilla argentea* Linnaeus, *Galium verum* Linnaeus, *Veronica spuria* Linnaeus, *Artemisia glauca* Pall. ex Willd., *A. gr. frigida*, were common dominants. The local steppe habitats were used for very moderate livestock grazing and haying. The general distribution of *C. italicus* was studied over this area in June and July. These studies were continued in 2001 and 2002. Later we checked this population of the Italian locust once a summer, commonly in July.

Figure 2. Typical heterogeneous steppe habitat of the Italian locust with vegetation mosaic (grey—short sagebrushes, green—grasses or forbs) in the Kulunda steppe (near Aleksandrovskij settlement): red dot—exact position of an adult (31 July 2019).

In 2000 and 2001, some experimental treatments were organized. The steppe parts were split into several experimental and control plots. In 2000, experiments were conducted with a water-based formulation of fipronil (phenylpyrazole insecticide). The acridicide was applied to 50% of the main experimental plot, by alternating, 15-m-wide swaths. The dose rate in the "barriers" was 4 g of a.i. per ha (i.e., 2 g of a.i. per protected ha). In 2001 the main experiments included different types of treatments with a suspension formulation of teflubenzuron (insect grow regulator—IGR). Four experimental treatments were conducted: (1) blanket, (2) barrier (alternating swaths of 15 m wide), (3) alternating treated (15 m) and untreated (30 m) swaths, and (4) alternating treated (15 m) and untreated (45 m) swaths. The dose rate in the "barriers" was 30 g of a.i./ha, while the rate in the "blanket" area was 22.5 g of a.i./ha. In 2003, some parts of the experimental area (north-eastern and south-western) were plowed to cultivate perennial grasses (mainly *Bromus inermis* Leyss.) However, after several years, the steppe vegetation has been recovered almost completely.

2.4. Field Studies

General distribution patterns of the Italian locust over the south-eastern part of West Siberian Plain and the adjacent north-eastern and eastern parts of the Kazakh Uplands were revealed on the basis of quantitative and qualitative samples collected in natural and transformed ecosystems, usually in the middle of summer when adults were dominant.

Field sampling was organized along gradient (usually transzonal), regional and local transects [4,17]. Each transzonal transect crossed the set of life zones, e.g., from the southern taiga to the northern deserts, and consisted of the set of regional and local transects. Each regional transect laid out a geographic region (e.g., the Kulunda steppe), commonly along river valleys, and was composed of local transects. Each local transect crossed a river valley or a lake basin from a local lower flood-plain to a watershed plain (plakor or flat interfluve) or mountain slopes. The length of local transects was from hundreds of meters to several kilometers. Grass ecosystems and similar transformed habitats were studied along local transects. As a rule, each part (habitat) of the local transect (lower and upper flood-plains, lower and upper terraces, watershed plain, in some cases, their variants) was explored separately. Samples captured during a fixed period of time were made in every habitat studied [4,17,18]. Using this method, insects were captured with a standard net (40 cm diameter) over a period of 10–30 min. Results for each habitat were extrapolated to an hour. In some cases, a simple sweep netting was used too. Many habitats were also investigated to find species with very low abundance. This approach allowed us to obtain repeatable and comparable results over a number of years and habitats and independently of collectors (Figure 3).

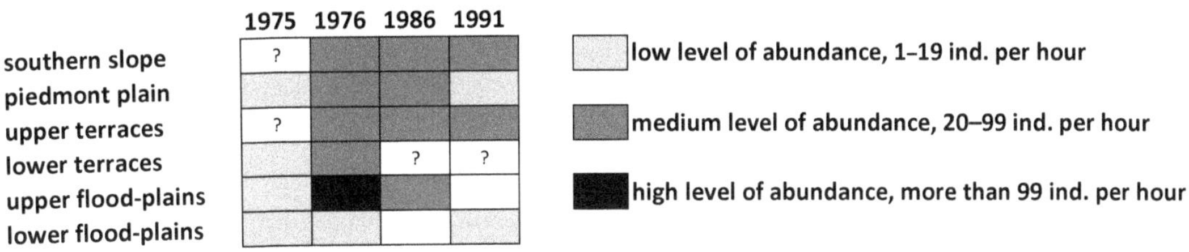

Figure 3. Population distribution of the Italian locust along the local transect crossing the Ayagöz River Valley (semi-desert, eastern part of the Kazakh Uplands, 47.92° N 80.25° E). Collectors: 1975—I. Stebaev, E. Moiseeva; 1976—A. Lopatkin; 1986 and 1991—M. Sergeev).

Some old data, mainly from the expeditions of Novosibirsk State University (1972–1977), were also used. We employed the Glonass/GPS receivers to determine geographical coordinates. We also used Google Earth Pro (©Google, 2020) to get the same parameters for habitats explored before 2000. The main part of studied specimens is in the collections of Novosibirsk State University and the Institute of Systematics and Ecology of Animals (Novosibirsk).

Peculiarities of long-term and seasonal dynamics and results of experimental treatments were revealed on the basis of the Italian locust density estimations. In each habitat studied, locusts were counted on arbitrarily placed plots 0.25×0.25 m^2 (in some cases—0.5×0.5 m^2 or 1×1 m^2) [4,16,19]. We determined densities on 25 plots, while locust densities were high, and on 50–200 plots (sometimes more), when densities were low. After that, the average density was estimated for every habitat studied.

2.5. Data Analysis

The general distribution patterns of the Italian locust along transects and habitats were analyzed by simple ratio scaling to avoid some problems with variations of abundance estimations [4,20] (see also Figure 3). Three scales were used: from very low up to 19 ind. per hour; from 20 to 99 ind. per hour; more than 99 ind. per hour.

Data on the long-term and seasonal dynamics and results of experimental treatments are commonly non-parametric. The long-term dynamics patterns of the Italian locust populations on the model plots (see Section 2.2) were compared by the Spearman rank order correlation analysis. Significance of difference between years for each model plot was estimated by the Kruskal–Wallis ANOVA and median tests. The Mann–Whitney U test was also calculated for every consecutive pair of years on each plot. The same tests were used for the analysis of seasonal dynamics peculiarities and results of treatments (see Section 2.3). Effectiveness of treatments was estimated by the modified Abbott's formula considering population dynamics on both test and control plots [21]. Data analyses were mainly conducted using PAST 4.02 [22] and Statistica 10 (© StatSoft. Inc., Tulsa, OK, USA; now—© TIBCO Software Inc., Palo Alto, CA, USA).

The tegmen length/posterior femur length ratios (E/F) were also counted for populations studied [8,19], especially for outbreak seasons. According to K.A. Vasil'ev [8], the typical solitarious specimens have relatively short tegmina (E/F for females < 1.418 and for males < 1.402), and the tegmina of the typical gregarious adults are visibly longer (E/F for females > 1.608; for males > 1.625).

Numerous publications describing the Italian locust spatio-temporal distribution in different parts of its range were also analyzed.

3. Results

3.1. Distribution Patterns of the Italian Locust over the South-Eastern Part of West Siberian Plain and the Adjacent North-Eastern and Eastern Parts of the Kazakh Uplands

A comparative analysis of three gradient transects crossing the eastern part of the Italian locust range from the forest-steppes of West Siberian Plain to Central Asia: (1) mainly along the 77–79th meridians east [4,9,12]; (2) along the 80th meridian east (Figure 4B), and (3) along the Irtysh River (Figure 4A)—allowed to reveal the general patterns of the species populations' distribution in the periods between outbreaks.

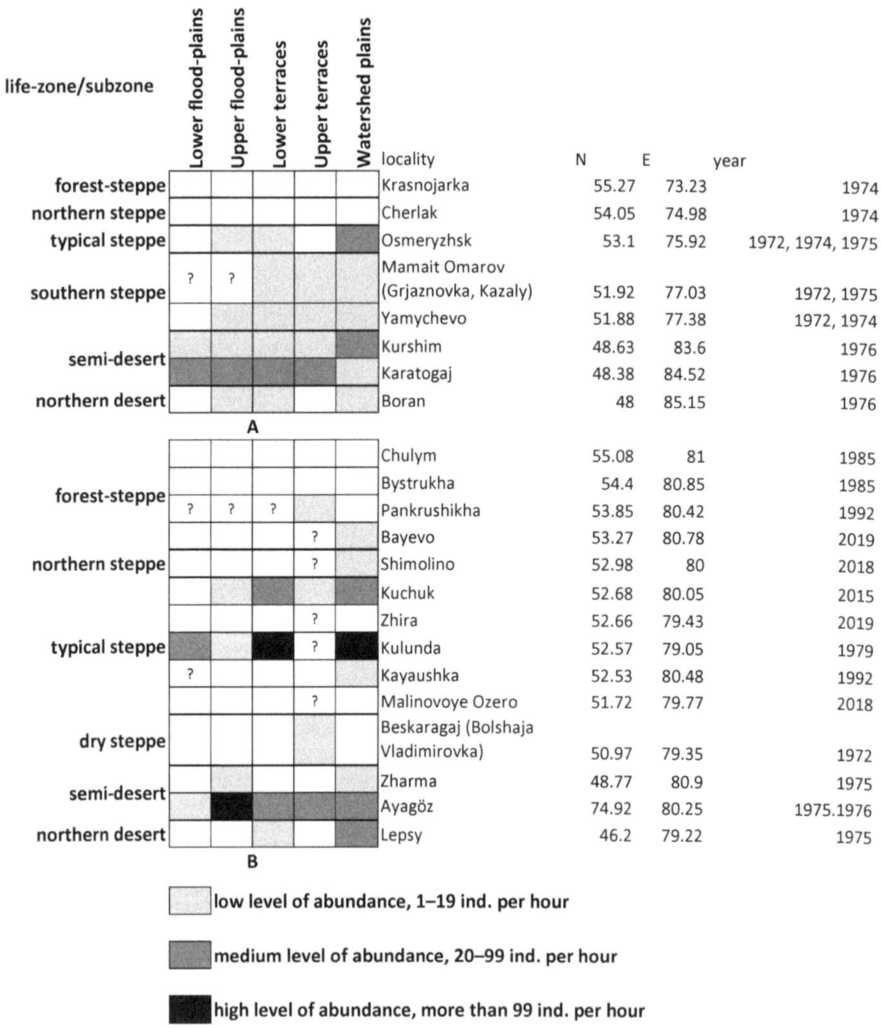

Figure 4. Population distribution of the Italian locust along two gradient (transzonal) transects crossing the south-eastern part of West Siberian Plain and the adjacent north-eastern and eastern parts of the Kazakh Uplands along the Irtysh River (**A**) and near the 80th meridian east (**B**). N—north; E—east.

In the southern forest-steppes and in the northern steppes of West Siberia [23], the Italian locust is distributed sporadically, especially during years without outbreaks. Its local colonies occur mainly in dry habitats of watershed plains, including overgrazed pastures and road-sides, while its abundance may be very low. Sometimes a surveyor may spend more than an hour to find one adult. The species is more common in the typical steppes. Its populations are also associated with the steppe habitats on watershed plains, but they are more or less common over terraces and dry parts of upper flood-plains. In the dry steppes of the Kulunda Plain [23], there is the local optimum of the Italian locust. Its populations are in almost all applicable habitats, from flood-plains to watershed plains [6,12,24,25]. In the southernmost steppes and in the northern parts of the semi-deserts, the species colonies occur either in watershed plains and southern slopes of hills or along upper flood-plains. The main optimal area of the Italian locust is in the southern parts of the semi-deserts. Its populations occur in all or almost all favorable habitats, and the abundance level is usually moderate (Figures 3 and 4). Further south, its populations are again sporadic and scarce. In the deserts, they are commonly associated with some parts of watershed plains, mountain slopes, and upper floodplains. The sketchy pattern of the Italian locust distribution along the Irtysh River in the 1920s described by G.Ya. Bey-Bienko [26] corresponds mainly to our data (Figure 5).

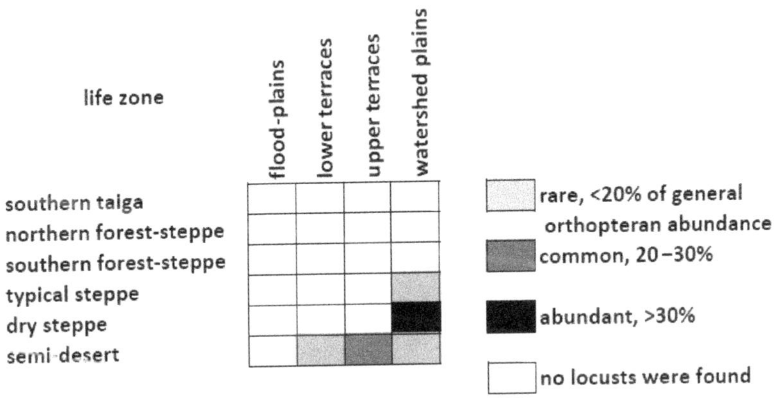

Figure 5. Population distribution of the Italian locust along the Irtysh River in the 1920s (after [26]).

3.2. *Long-Term Dynamics of the Italian Locust in the Central and Eastern Parts of Its Range: General Patterns*

General trends of the Italian locust long-term dynamics were mainly evaluated on the basis of published data [10,11,27–31]. Recently many of them were described for the main regions of the former USSR countries [9].

In the 19th century, the situation with the Italian locust and other grasshoppers in the Russian Empire was relatively calm. Many acridid outbreaks were registered in 1821–1830 and 1841–1850. However, K. Lindemann [27] described in detail a series of outbreaks of the Italian locust per se in the forest-steppes of European Russia. The situation changed during the next century. Some serious outbreaks occurred in 1921–1940. However, the Italian locust had remained the secondary pest.

For instance, in the comprehensive report on the acridid pests of the USSR for 1925–1933 [29], authors described the problems associated with the migratory locust, *Locusta migratoria* (Linnaeus), on 32 pages, with the Moroccan locust, *Dociostaurus maroccanus* (Thunberg)—on 33 pages, and even with the desert locust, *Schistocerca gregaria* (Forskål)—on four pages, while the part concerning the Italian locust took up slightly more than one page only. As a rule, the breeding areas of *C. italicus* were localized and limited by hundreds or thousands hectares, mainly in the semi-deserts [28]. For instance, during the outbreak in 1932–1933 in Kazakhstan, the general infested area was less than 190,000 ha (cf. with the area infested in 2000—more than 8 million ha [1]). However, its local bands and swarms were characterized by very high densities and active migrations, especially in the semi-deserts.

From 1940 until 1990 upsurges had been more or less common but remained highly localized. However, at the end of the 20th century, the outbreaks became extremely serious and common. For the North Caucasus—Lower Volga area, M.V. Stolyarov [10,11] described four outbreaks of the Italian locust (in 1954–1957, 1972–1974, 1982–1984, and 1992–1998). However, in the southern part of West Siberia, the outbreaks occurred in 1952–1956, 1967–1971, 1977–1982, and 1999–2002 [9]. Thus, the outbreaks of the Italian locust did not exactly coincide in the different part of its range. In many cases, outbreaks in West Siberia started earlier than in the North Caucasus—Lower Volga region, but, as a rule, were relatively weak and sporadic. However, the last and worst outbreak had begun in 1991 in the North Caucasus—Lower Volga region. After that, the outbreak spread over a huge area from Azov Sea to the eastern part of Kazakhstan (1996–1998). In 1998–2001 the main breeding areas moved northward and eastward. On the last stage (2002–2014), these areas became smaller and occupied the territory from south Ukraine to Caspian Sea and to the Ural Mts. [9]. The last decade is characterized by some sporadic outbreaks of the Italian locust in the central and eastern parts of its range, e.g., mainly in some regions of the North Caucasus [32–34].

Thus, the long-term population dynamics of the Italian locust can be very distinct in the different part of its range. Such pattern may be determined by regional climatic peculiarities, by variations in soils and vegetation covering, and by some changes in human activities.

3.3. Peculiarities of Long-Term Dynamics of the Italian Locust in the South-Eastern Part of West Siberian Plain and in the Adjacent Parts of the Kazakh Uplands

The south-eastern part of West Siberian Plain and the adjacent parts of the Kazakh Uplands are the typical territories where outbreaks of the Italian locust began and may begin [1,4,6,9,30,35,36]. First data on its local outbreaks appeared in the 1910–1930s [29,30,37,38]. However, these outbreaks were sporadic and restricted by one or a few years [29,30]. In the 1930s I.A. Chetyrkina [36] studied the Italian locust distribution in the semi-deserts and in the dry steppes of the eastern part of Kazakhstan. She revealed the main peculiarities of its population distribution when its colonies were found in almost all habitats except pine forests and some plots with halophytes. Actually, her observations showed the Italian locust preferred habitats with sagebrush dominance, including abandoned fields. These shifts in the Italian locust distributions could be associated with some transformation of the territory, because in the beginning of the 20th century, many areas had been plowed, and after that, many fields had become abandoned.

Later, in the middle of the 1950s, during the co-called Virgin Land campaign, vast steppe areas have been plowed and many remaining steppe habitats have become overgrazed. As a result, many habitats of *C. italicus* have been destroyed or damaged. Such regional and local transformations and/or also climate changes could and can result in some shifts in its population dynamics and spatial distribution (Figure 6). Extensive plowing often resulted in very significant decreasing of grasshopper abundance, especially over steppe regions [39].

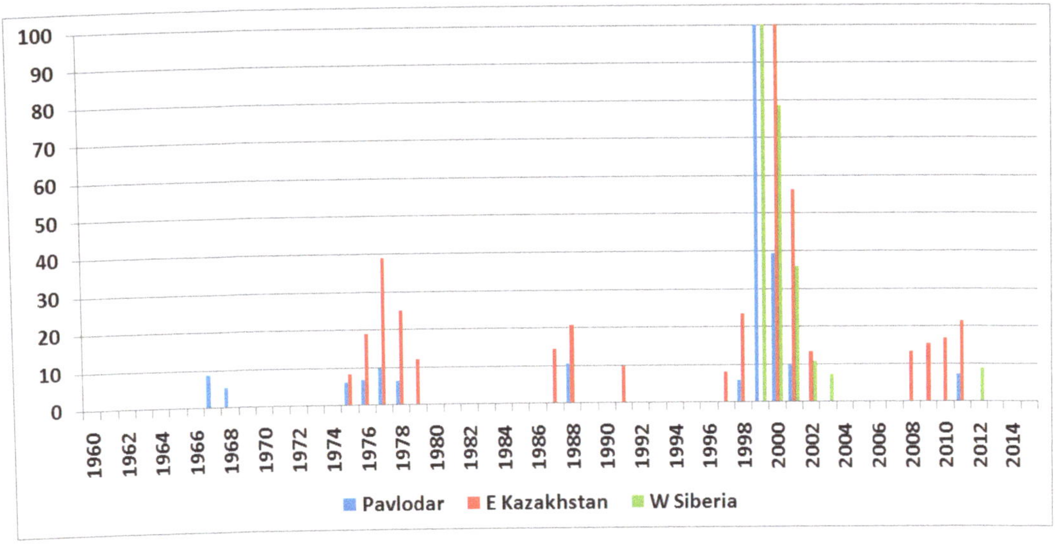

Figure 6. Long-term dynamics (1960–2015) of the Italian locust infestations in West Siberia (Altaj, Novosibirsk and Omsk Regions of Russia) and in the Pavlodar and East Kazakhstan Regions of Kazakhstan (% of the maximal registered infested areas [9]).

Later, in the 1960–1980s, some development of technologies preventing soil erosion caused again local but relatively weak and rare outbreaks of grasshoppers and locust, including *C. italicus* [6], but mainly in the semi-deserts of East Kazakhstan. The last and most devastating outbreak of the Italian locust started in 1998. The maximal infested areas were registered in 1999 in the Pavlodar Region and in the south-eastern parts of West Siberia (i.e., mainly in the steppes) and in 2000 in East Kazakhstan (i.e., in the semi-deserts). This outbreak may be qualified as a plague, because the large-scale infestation occupied huge areas and continued at least three years [1,9,16]. Besides, there were numerous bands and swarms in almost all parts of the territory, except local forests and the upper altitudinal belts of mountains. Some swarms actively migrated and could cross the state and regional borders. This extreme outbreak might be promoted either by economical transformations in the former USSR countries in the 1990s, while plant protection services were partly destroyed, trans-boundary cooperation in breeding areas was limited, and abandoned fields spread over huge territories, or by climatic changes of recent decades as well [1].

3.4. *Dynamics of the Italian Locust Populations in the Kulunda Steppe in 2000–2008*

In the Kulunda steppe, the last outbreak started in 1999. This area might be colonized by a specific mixture of local and vagrant populations [19,35], but our observations showed the low possibility of long migrations (cf. [36]). The Italian locust occurred in natural, semi-natural and anthropogenic habitats, including cultivated and abandoned fields. In 1999–2000, the average densities of the Italian locust were usually about 0.3–9.6 adults/m^2, but in some places, the density was significantly higher (~24 adults/m^2) [16,19]. Swarms of this species were common in 2000 and 2001. In 2000, they had occupied about 84 km (from 337 km) along a road from Ust-Volchikha to Aleksandrovskij (see Section 2.2), while in 2001 the swarms were observed over a distance of 60 km. Later, in 2002–2008, there were no swarms along this route. However, in 2015, very small bands (up to several square meters) were observed near Severka and Bor-Forpost.

Dynamics patterns were different in all observation sites (Table 1). One can formally split them into four groups based on the position of the maximum average population density. The first group includes the majority of studied populations (Aleksandrovskij,

Burla, Yarovoe, Platovka, and Mikhaylovskoye). They were at maximum levels during 2000. This group includes mainly the populations from the northern part of the Kulunda steppe, except Mikhaylovskoye. The other groups had the maximum densities of the Italian locust in 2001 (Severka), in 2002 (Bor-Forpost and Ust-Volchikha), and in 2000 and 2002 (Pokrovka), respectively.

This grouping is partly supported by the Spearman rank order correlation analysis (Rs from 0.900 up to 0.975 at the significance level $p < 0.05$) for 2000–2004 (Figure 7) [19]. The analysis of relatively long rows of data (from 2000–2005 to 2000–2008) allows to gradually combine almost all populations (except Bor-Forpost) but with less significant support (Rs > 0.7, $p < 0.05$) (Figure 7). Such similarity is determined by low levels of population densities from 2003 until 2008 in almost all sites. The population near Bor-Forpost only was characterized by more or less gradually increasing from 2003 until 2008 (actually this trend continued in 2009, with the average density 2.69 ± 0.54).

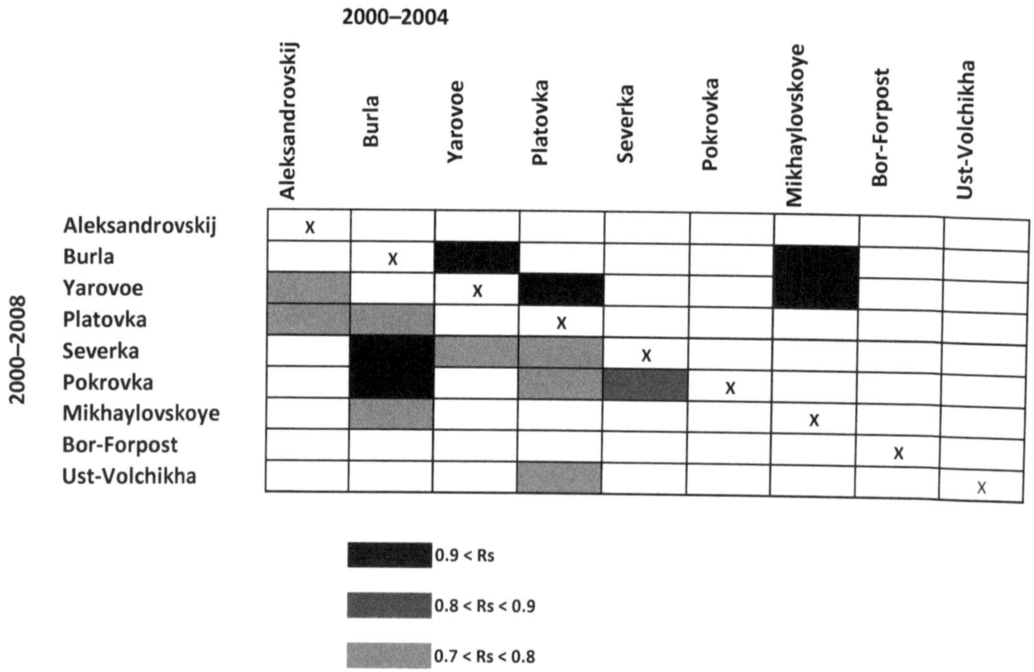

Figure 7. Associations between the long-term dynamics of the Italian locust populations in the Kulunda steppe for 2000–2004 and 2000–2008 (levels of the Spearman rank order correlation—Rs, $p < 0.05$).

The Kruskal–Wallis ANOVA and median tests show significant inter-annual difference for each site studied at $p < 0.0001$ for all tests. Thus, in all cases, the null hypotheses (the different samples are from the same distribution) should be rejected. The Mann–Whitney U test calculated for every consecutive pair of years for each population (Table 2) reveals significant differences between years with high densities and the adjacent ones for almost all populations [19]. In all cases, variations between consecutive years with densities lower than 3.0 ind. per m^2 are insignificant.

Table 1. Dynamics of the average densities (ind. per $m^2 \pm$ s.e.) of the Italian locust in the Kulunda steppe.

Year	Aleksandrovskij	Burla	Yarovoe	Platovka	Severka	Pokrovka	Mikhaylovskoye	Bor-For- post	Ust-Volchikha
1999	?	?	0.48 ± 0.27	0.80 ± 0.33	0.32 ± 0.22	~1	3.2 ± 0.61	?	0.32 ± 0.22
2000	24.32 ± 2.34	9.60 ± 0.85	6.40 ± 0.65	4.48 ± 0.48	2.88 ± 0.67	5.28 ± 0.86	6.56 ± 0.80	0.96 ± 0.35	0.48 ± 0.27
2001	2.77 ± 0.70	5.60 ± 0.73	4.16 ± 0.63	3.04 ± 0.53	7.36 ± 0.82	4.00 ± 0.65	0.96 ± 0.35	0.08 ± 0.08	1.60 ± 0.46
2002	0.16 ± 0.16	3.20 ± 1.31	0.16 ± 0.16	1.28 ± 0.89	1.92 ± 1.06	6.40 ± 1.85	0.32 ± 0.32	1.92 ± 0.74	1.92 ± 1.06
2003	+	0.16 ± 0.16	+	0.16 ± 0.16	0.32 ± 0.23	0.13 ± 0.13	0.16 ± 0.16	0.32 ± 0.23	0
2004	0.32 ± 0.32	+	+	0.96 ± 0.54	+	0	0	0.40± 0.27	0.32± 0.32
2005	0.32 ± 0.23	0	0.38 ± 0.22	0.53 ± 0.37	0.05 ± 0.05	0	0	0.85 ± 0.29	+
2006	+	0.64 ± 0.45	0.21 ± 0.15	0.43 ± 0.30	1.28 ± 0.62	0.21 ± 0.21	0	0.75 ± 0.028	0.16 ± 0.16
2007	0.32 ± 0.23	0.96 ± 0.54	0.64 ± 0.32	1.16 ± 0.69	1.92 ± 0.74	1.07 ± 0.46	0	1.15 ± 0.37	+
2008	0.48 ± 0.27	0.16 ± 0.16	0.51 ± 0.25	0.96 ± 0.54	0.64 ± 0.45	0.32 ± 0.32	0	1.54 ± 0.46	0

+—One or several specimens were found beyond counts; ?—no data for this year.

Table 2. Mann–Whitney U test ($U/p < 0.05$) for the status of Italian locust populations for every consecutive pair of years in the Kulunda steppe.

Year	Aleksandrovskij	Burla	Yarovoe	Platovka	Severka	Pokrovka	Mikhaylovskoye	Bor-Forpost	Ust-Volchikha
1999–2000	?	?	41.5 0.0000	85 0.00001	202.5 0.0328	125 0.0003	167.5 0.000	?	n.s.
2000–2001	469 0.0000	195.5 0.0024	196.5 0.0250	n.s.	126.5 0.0003	n.s.	166 0.0045	n.s.	n.s.
2001–2002	n.s.	165.5 0.0042	84.5 0.00001	153.5 0.0020	94 0.00002	n.s.	n.s.	n.s.	n.s.
2002–2003	n.s.	n.s.	n.s.	n.s.	n.s.	1012 0.0055	n.s.	n.s.	n.s.

?—no data for this pairs; n.s.—not significant difference.

The seasons with the high locust abundance can be considered as outbreak periods. Our data showed that, in the region, the outbreaks were characterized by the average densities of adults more than 3 ind. per m^2. However, intensity and duration of each local outbreak may be quite different. For instance, the maximal registered densities varied very significantly between the populations studied (Tables 1 and 2). As for duration, 5 from 9 populations were characterized by the explicit one-year maximum, while one population (near Platovka) showed the high levels both in 2000 and 2001, and the population near Pokrovka was distinguished by the relatively long outbreak from 2000 until 2002. Two other stages in long-term dynamics of the Italian locust populations can be revealed as well:

(1) A recession, when adults' abundance in a local population in the middle and the end of a summer varies between relatively low and moderate (according our estimation, in the Kulunda steppe, from 0.1 up to 2.0 ind. per m^2, perhaps in some cases—up to 3.2). This means that recommendations of some published manuals concerning monitoring and management of pest locusts and grasshoppers [40,41] to organize partial treatments in the next year, when the density of the Italian locust adults is more than 1 ind. per m^2, are based on too conservative estimations.

(2) A depression, when density is very low, commonly significantly less than 0.1 ind. per m^2. In this case, a population may be stable, but scarce locusts are distributed very sporadically. For instance, in 31 July 2019, I spent about two hours on our main model site in the northern part of the Kulunda steppe (near Aleksandrovskij) finding only 10 adults of the Italian locust and estimated their average density as about 25 ind. per ha (Figure 2).

Besides, there were no evident correlations between species abundance, its morphological characteristics (actually the E/F ratio), and its behavior (Table 3). In 2000, when the Italian locust was very abundant on our main model plot near Aleksandrovskij settlement and there were numerous and dense bands and small swarms, the average level of the E/F ratio corresponded to the transiens status [19]. In 2008, when its abundance was low and there were no bands or swarms, the E/F ratio for females corresponded to transiens again (moreover, one female morphologically looked like the gregarious one). Similar discords could be likewise observed on other sites (Table 3) [19]. Our data for 1999–2000 also show that presence of the gregarious form was limited (for all samples—about 4.9% of females and 7.1% of males) [19]. This also means that the rate of long (including transborder) migrations during the outbreaks was low.

Thus, the adjacent local populations of the Italian locust could differ significantly in their long-term dynamics patterns. In the south-eastern part of West Siberian Plain, these changes often look like some waves of spatial redistribution of maxima and minima of abundance [19,20]. The akin peculiarities of the Italian locust population dynamics were described for the North Caucasus as well [42].

Table 3. Changes of the average densities (ind. per $m^2 \pm$ s.e.) and average tegmen length/posterior femur length ratios (E/F ± s.e.) of the Italian locust on two plots with quite different long-term dynamics.

Year	Aleksandrovskij				Ust-Volchikha			
	Density	Species Behavior	E/F Females	E/F Males	Density	Species Behavior	E/F Females	E/F Males
1999	?	?	?	?	0.32 ± 0.22	No bands and swarms	**1.576 ± 0.004**	**1.609 ± 0.006**
2000	24.32 ± 2.34	Numerous bands, small swarms, active local flights	**1.514 ± 0.016**	**1.539 ± 0.013**	0.48 ± 0.27	Very small and scarce swarms	*1.368 ± 0.010*	*1.392 ± 0.012*
2001	2.77 ± 0.70	No bands and swarms	*1.395 ± 0.013*	**1.406 ± 0.018**	1.60 ± 0.46	Very small and scarce swarms	**1.473 ± 0.011**	**1.479 ± 0.009**
2008	0.48 ± 0.27	No bands and swarms	**1.484 ± 0.038**	*1.384 ± 0.024*	0	No bands and swarms	?	?

In bold—the average E/F ratios correspond to the intermediate (transiens) status of the local population; in italics—the solitarious status of the population; ?—no data for this year.

3.5. Seasonal Dynamics and Distribution of the Italian Locust Local Population in the Northern Steppe in 2000–2019

We started to study the dynamics of the Italian locust population in the northern part of the Kulunda steppe (near Aleksandrovskij settlement) in 2000, on 20 June. The steppe areas and the edges of agricultural fields were occupied by numerous hopper bands. These bands mainly consisted of the first stage hoppers (about 95%) [16]. The average density of locusts was 24.67 ± 4.42 ind. per m^2 on the steppe plot and 430.4 ± 161.6 ind. per m^2 on the field edge where some bands concentrated. The density in bands could be more than 1000 ind. per m^2, while band areas varied between 2 and 300 m^2 [16].

After a week, the bands included the hoppers of all five instars, while the second and third instars dominated. On 5 July, the main part of the population consisted of the fourth instar hoppers; however, it also included many nymphs of the third and fifth instars [16]. The hopper band areas varied between 4 and 3500 m^2. On 20 July, adults comprised almost half of the population. The second half included mainly the fifth instar hoppers and a few fourth stage hoppers [16]. Some bands occupied more than 5 ha. On the control plot, the average density was relatively stable [16]. Adults tried to fly, but their flights were limited to meters to dozens of meters. Besides, there were several huge colonies of rooks. Our estimations show that these birds can withdraw about 10–15% of the locust population [9].

Treatments with fipronil showed very high effectiveness within the 15 m of the treated swaths—more than 95% after several days, while on the untreated 15 m swaths only some weak effects (17–33% reduction) could be observed only after a week or two [16,43]. This pattern can be explained by extremely low rates of young hopper migrations till the end of June, because there were sufficient quantities of preferred foods. In July, insects (last instars and adults) started to move more actively [16,43].

In 2001, general abundance of the Italian locust decreased, and there were no bands. In the beginning of our studies, on 5 June, the average density of the Italian locust was only 3.41 ± 0.54 ind. per m^2, while the population mainly consisted of the first stage hoppers (about 87%). The early fourth instar hoppers were observed on 15 June, and the first adults—on 5 July. On 15 July, adults comprised about 64% of the population. Besides, there were also the fourth instar (about 21%) and fifth instar nymphs (about 15%). On the control plot, the average density was relatively stable and remained almost the same during the season (on 15 July—2.77 ± 0.70 ind. per m^2).

Experiments with teflubenzuron showed that this IGR was highly effective in blanket treatments and also within the treated 15 m swaths (Table 4). Effectiveness was most obvious after about a week which corresponds to the molting rates of hoppers—every 5–7 days [9]. The high levels of effectiveness were observed on treated plots. Untreated swaths were characterized by moderate effectiveness of treatments, usually about 80–84%.

Table 4. Effectiveness (%) of teflubenzuron treatments (6 June) on the experimental site in the northern part of the Kulunda steppe (near Aleksandrovskij settlement) in 2001.

Plot		7 June	12 June	13 June	15 June	16 June	23 June	26 June	5 July	15 July
Type	**Treated/Untreated**	**D1**	**D6**	**D7**	**D9**	**D10**	**D17**	**D20**	**D29**	**D39**
	Blanket	31.4	70.0	100	93.7	100	100	76.0	100	86.1
1:1	Treated	57.2	81.3	100	100	93.2	100	100	91.7	100
	Untreated	35.7	43.8	77.5	80.3	79.5	70.8	55.0	50.0	65.4
1:2	Treated	46.4	81.3	100	96.1	100	100	100	100	82.7
	Untreated	57.2	62.5	70.0	56.6	86.4	66.7	-	8.3	56.7
1:3	Treated	14.3	70.0	76.0	81.1	90.1	86.8	100	33.3	86.1
	Untreated	14.3	10.0	52.0	36.9	78.2	33.9	4.0	-	58.4

These experiments showed that teflubenzuron (like other IGRs) could be very suitable for the Italian locust management in the steppes of South Siberia. However, the efficacy of the barrier treatments with relatively wide untreated swaths was low. Hence, blanket application and barrier treatments with narrow equal alternating swaths may be optimal.

3.6. The Italian Locust and Rare Orthopteran Species

Among local species of Orthoptera, *Saga pedo* (Pallas), *Gampsocleis glabra* (Herbst), and *Onconotus servillei* (Fischer de Waldheim) are on the IUCN Red List of Threatened Species [44]. The first species is also in the Red Book of the Russian Federation. Besides, in the south-western part of West Siberian Plain, there are several rare species of Orthoptera. Recently, we analyzed shifts in rare acridid species distribution over the Baraba and Kulunda steppes [45]. At least two grasshoppers, namely, *Notostaurus albicornis* (Eversmann) and *Mesasippus arenosus* (Bey-Bienko), could be found in the habitats where the Italian locust may be very abundant. *Miramiola pusilla* (Miram) (Tettigoniidae) and *Aeropedellus baliolus* Mistshenko. (Acrididae) may be observed on these plots as well. This means, in the steppe habitats, anti-locust treatments can result in decreasing and even elimination of rare species populations.

Rare species were found on almost all sites studied or just nearby (see Section 2.2), except vicinities of Platovka and Pokrovka (Table 5). *G. glabra* was the most common rare species in the steppe habitats. *M. pusilla* was found on four sites, both *Saga pedo* and *A. baliolus*—on two, and *N. albicornis* occurred only in one place.

Table 5. Occurrence of rare Orthoptera on sites studied in the Kulunda steppe.

Year	Aleksandrovskij	Burla	Yarovoe	Platovka	Severka	Pokrovka	Mikhaylovskoye	Bor-Forpost	Ust-Volchikha
1992	?	?	AB, NA					?	?
1999	?	?	MP					?	
2000	GG, MP [b]		AB, NA						
2001	AB, GG [c], MP	GG	NA				GG, SP [a]		
2002	AB, MP [d]		AB						
2003									
2004	GG, MP [e]								
2005	GG	GG	AB					GG, SP	
2006	GG		AB, MP		MP [a]				
2007			MP						GG [a]
2008	GG [e], MP [e]		GG				GG	GG	
2015	GG		MP					MP	

AB—*Aeropedellus baliolus* Mistsh.; GG—*Gampsocleis glabra* (Hbst.); MP—*Miramiolla pusilla* (Mir.); NA—*Notostaurus albicornis* (Ev.); SP—*Saga pedo* (Pall.); [a]—the species was found on some adjacent plot; [b]—the species was also found after the fipronil treatment (30 days since—20 July); [c]—the species was also found after the teflubenzuron treatment (39 days since—15 July); [d]—the species was found only on the plot treated by the synthetic pyrethroid in 2001; [e]—the species was also found on the plots treated by the synthetic pyrethroid in 2001; ?—no data for this year.

In 2000 and 2001 anti-locust treatments in the northern part of the Kulunda steppe (near Aleksandrovskij settlement) did not result in elimination of the local populations ofrare Orthoptera. Three species, namely, *A. baliolus*, *G. glabra*, and *M. pusilla*, were found more or less often. Moreover, in 2000, the last species was observed on the treated swath after a month. In 2001, a similar situation was registered for *G. glabra*. Later, in 2002, 2004, and 2008, *M. pusilla* was also found on the plots treated by the synthetic pyrethroid in 2001. This means some localized treatments with acridicides (especially the barrier ones) do not destroy significantly populations of rare Orthoptera.

4. Discussion

Our studies of the Italian locust ecology and distribution as well as studies of our colleagues during last two decades resulted in revealing several dilemmas concerning approaches to monitor, manage, and forecast its population dynamics in time and space.

4.1. May We Use for the Italian Locust Approaches and Technologies Developed for the Desert and Migratory Locusts or Not?

Nowadays, the Italian locust is the only locust species mainly associated with temperate grasslands, namely, steppes and northern semi-deserts. The main breeding (outbreak) areas of the desert and migratory locusts as well as some other important pests are in the tropical and subtropical regions [2,3,9,46]. This is why many ecological and biological peculiarities of the Italian locust are quite different [1,2,9]. The species is univoltine and is, even during depressions (see Section 3.4), often widely distributed across different ecosystems, especially in the dry steppes and the semi-deserts. Its behavior during outbreaks is common for locusts, but areas occupied by bands and swarms and ranges of dispersal rates are relatively limited, while outbreaks are relatively short and commonly can last for one or several years. This means that, in steppes and semi-deserts, it is almost impossible to identify precisely preferred breeding areas of the Italian locust [47]. In the 1950s K.A. Vasil'ev [5] tried to delimit some such breeding areas in Central Kazakhstan, but in reality, he described several plots where gregarization already had started.

In the steppes, the species occurs usually in all applicable habitats, but their local populations are often scarce and characterized by low abundance, while, in the semi-deserts, its populations may be usually relatively abundant for many decades [4,9,12,36]. Besides, the Italian locust long-term dynamics may be described as quite specific. In the dynamics of the migratory and desert locust, several main periods, namely, plagues, outbreaks, and recessions (and sometimes—also peak plagues, upsurges, and deep recessions), may be often distinguished [48–50]. In the dynamics of the Italian locust, plagues are extremely rare. Only last huge outbreaks in the end of 20th century can be qualified as the plague. Actually, numerous outbreaks of *C. italicus* were and are usually limited in space and time, and their decline may result in either the recession or the depression characterized by stable populations with extremely low average densities and scarce individuals' distribution (see Section 3.4). If a population drops off, the recession may transform into the depression. On the contrary, if a population with very low abundance begins to increase (and may be crowding), the depression may transform into the recession.

One may suggest that, in this situation, we should monitor all habitats colonized by the Italian locust across huge territories of its range and, inside the range, try to check carefully populations with relatively high densities of adults (probably, more than 1 ind. per m^2 in the steppes and 5–7 ind. per m^2 in the semi-deserts). Therefore, many (or all) approaches developed to monitor and control the desert and migratory locusts are not applicable for the Italian locust, or they should be adapted for the situation.

4.2. Monitoring, Management, and Forecasting: Is the Regional Level of Exploration Enough or Not?

The environmental factors and the biological and ecological characteristics of each species can significantly vary in time and space so that their relations can be revealed only by consistent explorations of species populations across its total range [3]. This is important for both potential pest species and rare forms. Actually, in many cases, pest infestations can be small-scale events, localized in space and time and dependent on local conditions [51].

Local outbreaks of the Italian locust usually start after several years with warm and relatively dry periods in the end of spring and in the beginning of summer, while their regressions are mainly associated with cool and rainy summers [3,5,10,11]. However, rates of increasing and decreasing, gregarization and dissociation in local populations may be quite different [4,9,16] and look like depending on local environmental conditions.

As a result, the local outbreak in one population can develop faster (or slower) than the outbreaks in adjacent colonies. Some similar trends can be revealed for regression.

However, forecasting systems are usually based on the regional generalization and simplification of assessment data [32,33,40,41], while local colonies of *C. italicus* can be significantly different in their ecology (including dynamics) and diversity (including intrapopulation characteristics) [4,9,19,52]. This means that we should use often different management procedures for two neighboring local populations of this species; moreover, in many cases, we are not able to delimit a local population without special studies [52].

We should reveal seasonal and long-term dynamics of local populations (ideally, all such populations) and try to evaluate their temporal patterns relative to climatic and ecosystem changes and to peculiarities of each colony. We should reveal also spatial structures (patches and gaps, distances, population connections, barriers and corridors) and local dispersal systems (parameters of emigration and immigration).

These substantial localized variations in population dynamics mean large scale forecasting of Italian locust upsurges and declines is very difficult. Local forecasting and management may be possible based on increased understanding of local population dynamics, but having local knowledge in the many parts of its vast range and in quite different habitats would require substantial effort that at present is just not possible. However, if we eventually understand the exact patterns of spatial population structures, we will be able to understand, at least, some peculiarities of dynamic patterns in each local population.

4.3. Monitoring, Management, and Forecasting: When Should We Begin (or End) to Trouble or When Not?

National and international plant protection services usually use one or several parameters to determine when and where control operations should start. J.A. Lockwood [53] designated this approach as the administrative one. For example, in the USA and in Russia, these operations usually begin when population densities of grasshoppers reach more than 9–10 ind. per m^2 [1,40,41]. In the case of the Italian locust, according different authors [1,9,40,41], the critical values for hoppers are about 2–5 and for adults— about 1 ind. per m^2. However, these levels of density are actually too low [53,54]. For instance, in the dry steppes of Central Kazakhstan, the average adult density of solitarious populations of *C. italicus* could be more than 2–7 ind. per m^2 [5], while in the semi-deserts and the mountain steppes of East Kazakhstan, this density might reach 26 ind. per m^2 [36]. This means that the critical values for the Italian locust may vary remarkably from one population to another and from one year to another and can depend on numerous parameters (climatic fluctuations, vegetation and soil dynamics, human resources, etc.) Moreover, density effects can be also very notable [55]. They may be rather complicated and more or less unpredictable due to complex nature of ecosystem and population dynamics. Besides, according the catastrophe theory [56], small variations in weather may result in different population dynamics. There are no exact criteria to distinguish consecutive stages of long-term dynamics of locusts as well [50].

This means in many cases we do not have applicable criteria where, when, why and how to monitor and control early infestations [57]. This is especially important for the Italian locust, since its populations between outbreaks are sprawling and scarce. Some formal criteria (densities, band formation, morphometric indices, etc.) may mean almost nothing, because (1) critical values can vary significantly between different ecosystems and between different periods, (2) some traits of gregarization may be revealed during regressions when a population density is relatively low, (3) small changes can result in different dynamics [55,56], and (4) at least in some cases usual criteria do not correlate [19] (see also Section 3.4).

4.4. Treatments: When Do Barriers Work and When Not?

Barrier treatments are widely used for locust populations management in different regions. The so-called reduced area and agent insecticide treatments (RAATs) technology often result in evident decline of locust densities, preserve some level of biological diversity, and significantly decrease direct expenses [58,59]. However, the experimental barrier treatment with fipronil showed that its general efficacy was low [16] (see also Section 3.5). The hopper bands usually stayed on the steppe plots with well-developed vegetation, did not move or moved only over the very short distances (about several meters per day) and, thus, did not actively cross over the treated areas. In 2001 the experimental treatments with teflubenzuron showed the similar pattern. The maximal effect on the untreated swaths was observed in the middle of June (after 9–10 days since the treatment) (Table 4), and it was relatively significant only for the experiment with equal narrow swaths. When hopper bands begin to move more or less rapidly [5,60], barrier treatments, especially with low dosage, may be very effective. Besides, the developmental stages of hoppers are also important, because one can use low dosage of an insecticide against young hoppers [16,60,61], but such hoppers usually move slowly [5,60]. This means that we should take into consideration not only pest abundance, situations with biodiversity, and management costs but also general conditions of ecological systems and local population statuses.

4.5. Treatments: Can We Preserve Local Populations of Rare Orthoptera or Not?

Steppes and other grasslands are ecosystems where both pest acridids and rare Orthoptera are distributed [62–65]. This means there is a clash between approaches to manage pest populations (to monitor, forecast and control their outbreaks) and to preserve rare forms and their populations. In the steppes of the south-eastern part of West Siberian Plain, some rare species may occur in the same habitats and for the same time as the Italian locust, especially its bands [20,45,62]. Thus, acridicide treatments against the bands could result in damage or full elimination of local populations of rare Orthoptera, such as *S. pedo*, M. *pusilla* or *A. baliolus*. However, our observations show that the harmful effects of acridicides on rare steppe species of Orthoptera may be relatively weak (cf. Section 3.6), when treatments are localized in space, i.e., do not cover all area of a habitat, e.g., the RAATs type or even simple variants of barrier applications.

5. Conclusions

The Italian locust is characterized by populations, which are widely distributed across its range and through quite different types of grasslands. In the dry steppes and especially in the semi-deserts, it may colonize all or almost all applicable habitats and may be abundant, while in the forest-steppes and the northern and typical steppes, between outbreaks, its abundance is commonly low or extremely low, and often surveyors may observe very few specimens during several hours of observations. Its local populations often respond differently to climatic fluctuations and changes of human activity. In the steppes, hopper bands often move very slowly, or they stay in one place during several days or even weeks, while adults prefer very short local flights. Besides, the Italian locust often colonize the same grasslands as rare orthopteran species. This is why we cannot use many of the ideas, approaches, and technologies suggested for the desert and migratory locust to monitor, supervise, and control populations of *C. italicus*.

Recently we suggested to change considerably the strategy and tactics to avoid the Italian locust outbreaks [66]:

(1) Crucial parts of the range, habitats, and time lapses should be determined for the species across all species range, each region, and landscape.
(2) New approaches have to be elaborated to manage populations on the landscape scale and to treat agricultural fields and rangelands.
(3) Agricultural technologies per se ought to be modified significantly. For instance, in grasslands, overgrazing should be minimized, and restoration of abandoned fields that commonly harbor *C. italicus* should be accelerated.
(4) Information technologies, especially associated with big data storage and analysis, ecologo-geographic modeling, global information systems, and remote sensing, should become the main basis for monitoring, supervising and forecasting (see also [67–69]).

(5) New approaches should be very important for field assessments of the Italian locust populations as well. For instance, real time observations by using unmanned aerial vehicles are very hopeful.

Besides, one should consider that the Italian locust is the member of grasshoppers' group (Acridoidea), and these insects are common inhabitants of all or almost all grasslands. They often considered as one of the most important consumers in steppes, prairies, savannas, semi-deserts, and even deserts [70–72]. Moreover, grasshoppers may enhance net primary production over years [73]. In the dry steppes and the semi-deserts, even between outbreaks, the Italian locust may consume significant part of aboveground vegetation, to damage growing points of sagebrushes and to accelerate their growth [36]. During outbreaks the species can also change considerably fluxes of some chemical elements in the steppe ecosystems [74]. One may also hypothesize that the Italian locust can also interact with other species of Orthoptera in ecosystems, and some variations in its populations should result in changes of orthopteran assemblages and populations, including colonies of rare species. In any case, attempts to control the Italian locust populations affect colonies of other species, either positively or negatively. This means we should alter widely distributed common management approaches to the holistic ones arising not only from exploration of species populations or their assemblages but also from investigations of ecosystemsper se and their changes in space and time.

Funding: These studies were financially supported by the joint programme of the Russian Foundation for Basic Researches and the Government of Novosibirsk Region (18-416-540001 and 20-416-540004) and also by the Federal Fundamental Scientific Research Programme (FWGS-2021-0002).

Institutional Review Board Statement: Not applicable.

Informed Consent Statement: Not applicable.

Data Availability Statement: Not applicable.

Acknowledgments: I thank all my colleagues and students (especially O. Denisova, O. Efremova, O. Ermolaeva, K. Popova, and I. Vanjkova) for their valuable help during collecting field data. I am also indebted to all companions during our numerous field trips. I wish to express special thanks to the late L.L. Mistshenko and the late I.V. Stebaev for their advices and cooperation and to L.B. Pshenitsyna for indentification of plant species. Finally, to the late V.I. Strizhak, K.K. Musinov, N.I. Vastshenko, and A.N. Shedenko, I am indebted for helping us with organization of experimental treatments in 2000 and 2001. I also thank two anonymous reviewers for their valuable comments and proposals.

Conflicts of Interest: The author declares no conflict of interest.

References

1. Latchininsky, A.V.; Sergeev, M.G.; Childebaev, M.K.; Chernijakhovskij, M.E.; Lockwood, J.A.; Kambulin, V.E.; Gapparov, F.A. *The Grasshoppers of Kazakhstan, Middle Asia and Adjacent Territories*; Association for Applied Acridology International & University of Wyoming: Laramie, WY, USA, 2002; p. vii + 387. (In Russian)
2. Zhang, L.; Lecoq, M.; Latchininsky, A.; Hunter, D. Locust and grasshopper management. *Ann. Rev. Entomol.* **2019**, *64*, 15–34. [CrossRef]
3. Uvarov, B.P. *Grasshoppers and Locusts*; Centre for Overseas Pest Research: London, UK, 1977; Volume 2, p. 613.
4. Sergeev, M.G.; Denisova, O.V.; Vanjkova, I.A. How do spatial population structures affect acridid management? In *Grasshoppers and Grassland Health: Managing Grasshopper Outbreaks without Risking Environmental Disaster*; Lockwood, J.A., Latchininsky, A.V., Sergeev, M.G., Eds.; Kluwer Academic Publishers: Dordrecht, The Netherlands; Boston, MA, USA; London, UK, 2000; pp. 71–87. [CrossRef]
5. Vasil'ev, K.A. The Italian locust (*Calliptamus italicus* L.) in Central Kazakhstan. *Trudy Nauchno-Issledovatelskogo Inst. Zastshity Rastenij* **1962**, *7*, 123–190. (In Russian)
6. Bunin, L.D. The Italian Locust (*Calliptamus italicus* L.) in the Soil-Saving Agriculture Zone in East. Kazakhstan and Development of Treatments. Ph.D. Thesis, VIZR, Leningrad, Russia, 1979; p. 24. (In Russian)
7. Azhbenov, V.K. Outbreaks and migrations of locusts in Kazakhstan. *Steppe Bull.* **2000**, *6*, 16–20. (In Russian)
8. Vasil'ev, K.A. Phases of the Italian locust (*Calliptamus italicus* L.). *Proc. USSR Acad. Sci.* **1950**, *74*, 639–642. (In Russian)

9. Sergeev, M.G.; Childebaev, M.K.; Van'kova, I.A.; Gapparov, F.A.; Kambulin, V.E.; Kokanova, E.; Latchininsky, A.V.; Pshenitsyna, L.B.; Temreshev, I.I.; Chernijakhovskij, M.E.; et al. *The Italian Locust [Calliptamus italicus (Linnaeus, 1758)]: Morphology, Distribution, Ecology, Population Management, Draft Version*; FAO: Rome, Italy, 2016; p. 330. Available online: http://www.fao.org/locusts-cca/bioecology/italian-locust-cit/en/ (accessed on 21 December 2020). (In Russian)
10. Stolyarov, M.V. Massenvermehrungen von *Calliptamus italicus* L. in Südrussland im zwanzigsten Jahrhundert. *Articulata* **2000**, *15*, 99–108.
11. Stolyarov, M.V. Cyclicity and some characteristics of mass reproduction of *Calliptamus italicus* L. in Southern Russia. *Russ. J. Ecol.* **2000**, *31*, 43–48. [CrossRef]
12. Sergeev, M.G.; Van'kova, I.A. Zonal-landscape distribution of population groups of *Calliptamus italicus* L. (Insecta, Orthoptera, Acrididae). *Sib. J. Ecol.* **1996**, *3*, 219–225.
13. Isachenko, A.G. *Landscapes of the USSR*; Leningrad University Publ.: Leningrad, Russia, 1985; p. 320. (In Russian)
14. Smelansky, I.E.; Tishkov, A.A. The steppe biome in Russia: Ecosystem services, conservation status, and actual challenges. In *Eurasian Steppes. Ecological Problems and Livelihoods in A Changing World*; Werger, M.J.A., van Staalduinen, M.A., Eds.; Plant and Vegetation; Springer Science+Business Media, B.V.: Dordrecht, The Netherlands, 2012; Volume 6, pp. 45–101. [CrossRef]
15. Chibilev, A.A. *Steppe Eurasia: A Regional Review of Natural Diversity*; Institute of Steppe and RGS: Moscow, Russia; Orenburg, Russia, 2017; p. 324. (In Russian)
16. Sergeev, M.G.; Van'kova, I.A. The dynamics of a local population of the Italian locust (*Calliptamus italicus* L.) in an anthropogenic landscape. *Contemp. Probl. Ecol.* **2008**, *1*, 88–95. [CrossRef]
17. Sergeev, M.G. *Patterns of Orthoptera Distribution in North Asia*; Nauka Publ.: Novosibirsk, Russia, 1986; p. 237. (In Russian)
18. Gause, G.F. Studies on the ecology of the Orthoptera. *Ecology* **1930**, *11*, 307–325. [CrossRef]
19. Sergeev, M.G.; Van'kova, I.A. Dynamics of the Italian locust *Calliptamus italicus* L. population in the southeast of the West Siberian Plain. *Contemp. Probl. Ecol.* **2008**, *1*, 204–209. [CrossRef]
20. Sergeev, M.G. Distribution patterns of grasshoppers and their kin over the Eurasian Steppes. *Insects* **2021**, *12*, 77. [CrossRef] [PubMed]
21. Henderson, C.F.; Tilton, E.W. Tests with acaricides against the brown wheat mite. *J. Econ. Entomol.* **1955**, *48*, 157–161. [CrossRef]
22. Hammer, Ø.; Harper, D.A.T.; Ryan, P.D. PAST: Paleontological statistics software package for education and data analysis. *Palaeontol. Electron.* **2001**, *4*, 9.
23. Ili'ina, I.S.; Lapshina, E.I.; Lavrenko, N.N.; Meltser, L.I.; Romanova, E.A.; Bogojavlenskij, B.A.; Mahno, V.D. *Vegetation of West Siberian Plain*; Nauka Pub.: Novosibirsk, Russia, 1985; p. 251. (In Russian)
24. Sergeev, M.G. Metapopulations of locusts and grasshoppers: Spatial structures, their dynamics and early warning systems. In *New Strategies in Locust Control*; Krall, S., Peveling, R., Ba Diallo, D., Eds.; Birkhauser Verlag: Basel, Switzerland, 1997; pp. 75–80. [CrossRef]
25. Bugaev, G.S. Habitat distribution of grasshoppers in the zone of band pine forests in North-East Kazakhstan. *Vestn. Selskokhoziastvennoj Nauk. Kazakhstana* **1977**, *6*, 37–40. (In Russian)
26. Bey-Bienko, G.J. The zonal and ecological distribution of Acrididae in West Siberian and Zaisan Plains. *Tr. po Zastshite Rasteniy Ser. Entomol.* **1930**, *1*, 51–90. (In Russian)
27. Lindemann, K. *The Italian Locust in Voronezh Province*; V.I. Isaev Publ.: Voronezh, Russia, 1892; p. 52. (In Russian)
28. Filip'ev, I.N. Common pests. Grasshoppers. Acridoidea. *Tr. po Prikl. Entomol.* **1926**, *13*, 57–176. (In Russian)
29. Predtechensky, S.A.; Zhdanov, S.P.; Popova, A.A. Pest locusts in the USSR. *Tr. po Zastshite Rastenij. Ser. Entomol.* **1935**, *18*, 1–168. (In Russian)
30. Berezhkov, R.P. *The Grasshoppers of West Siberia*; Tomsk University Publ.: Tomsk, Russia, 1956; p. 174. (In Russian)
31. Borisenkov, E.P.; Pasetskij, V.M. *Thousand-Year Chronicle of Extraordinary Natural Phenomena*; Mysl Publ.: Moscow, Russia, 1988; p. 524. (In Russian)
32. Govorov, D.N.; Zhivykh, A.V. (Eds.) *The Report on Phytosanitary Conditions of Crops in the Russian Federation in 2018 and Pest Forecasts for 2019*; Rosselkhozcenter: Moscow, Russia, 2019; p. 900. (In Russian)
33. Govorov, D.N.; Zhivykh, A.V. (Eds.) *The Report on Phytosanitary Conditions of Crops in the Russian Federation in 2019 and Pest Forecasts for 2020*; Rosselkhozcenter: Moscow, Russia, 2020; p. 897. (In Russian)
34. Kushalieva, S.A.; Avtaeva, T.A. Peculiarities of the Italian locust (*Calliptamus italicus*) ecology in the semi-desert life zone of the Chechen Republic. *Her. Kursk State Agric. Acad.* **2019**, *6*, 73–78. (In Russian)
35. Kamynina, L.D. Locusts in the Altai Region. *Steppe Bull.* **2000**, *6*, 9–10. (In Russian)
36. Chetyrkina, I.A. The Italian locust *Calliptamus italicus* L. in East Kazakhstan. *Horae Soc. Entomol. Unionis Sov.* **1958**, *46*, 5–67. (In Russian)
37. Raevskij, V.G. Observation of the Italian locust (*Calliptamus italicus* L.) in the Slavgorod county in 1922. *Rep. Sib. Entomol. Bur.* **1924**, *2*, 53–58. (In Russian)
38. Troitskij, D. Pest grasshoppers of the Semipalatinsk Region in 1912. *Needs West. Sib. Agric.* **1914**, *1*, 23–49. (In Russian)
39. Tsyplenkov, E.P. *Pest. Grasshoppers in the USSR*; Kolos Publ.: Leningrad, Russia, 1970; p. 271. (In Russian)
40. Naumovich, O.N.; Stolyarov, M.V.; Dolzhenko, V.I.; Nikulin, A.A.; Alekhin, V.T. *Recommendation on Monitoring and Management of Pest Grasshoppers*; Kolos: Moscow, Russia, 2000; p. 21. (In Russian)

41. Dolzhenko, V.I.; Naumovich, O.N.; Nikulin, A.A. *Pest Grasshoppers*; Plant Protection and Quarantine: Moscow, Russia, 2003; p. 28. (In Russian)
42. Kovalenkov, V.G.; Kuznetsova, O.V. How the Italian locust outbreak may be stopped. *Plant. Prot. Quar.* **2011**, *9*, 14–17. (In Russian)
43. Sergeev, M.G.; Latchininsky, A.V.; Duranton, J.-F. Possibilities of fipronil (Adonis) application in Siberia. *Plant. Prot. Quar.* **2002**, *3*, 19–20. (In Russian)
44. IUCN. The IUCN Red List of Threatened Species. Version 2020-3. 2021. Available online: https://www.iucnredlist.org (accessed on 15 January 2021).
45. Popova, K.V.; Molodtsov, V.V.; Sergeev, M.G. Rare grasshoppers (Orthoptera, Acridoidea) of the Baraba and Kulunda steppes (South Siberia). *Acta Biol. Sib.* **2020**, *6*, 595–609. [CrossRef]
46. Le Gall, M.; Overson, R.; Cease, A. A global review on locusts (Orthoptera: Acrididae) and their interactions with livestock grazing practices. *Front. Ecol. Evol.* **2019**, *7*, 263. [CrossRef]
47. Uvarov, B.P. *Locusts and Grasshoppers*; Promizdat: Moscow, Russia; Leningrad, Russia, 1927; p. 306. (In Russian)
48. Hemming, C.F.; Popov, G.B.; Roffey, J.; Waloff, Z. Characteristics of desert locust plague upsurges. *Phil. Trans. R. Soc. Lond. B* **1979**, *287*, 375–386. [CrossRef]
49. Magor, J.I. Locust glossary. In *Desert Locust Control with Existing Techniques: An Evaluation of Strategies*; van Huis, A., Ed.; Wageningen Agricultural University: Wageningen, The Netherlands, 1994; pp. 119–127.
50. Magor, J.I.; Lecoq, M.; Hunter, D.M. Preventive control and desert locust plagues. *Crop. Prot.* **2008**, *27*, 1527–1533. [CrossRef]
51. Zimmerman, K.; Lockwood, J.A.; Latchininsky, A.V. A spatial, Markovian model of rangeland grasshopper (Orthoptera; Acrididae) population dynamics: Do long-term benefits justify suppression of infestations? *Environ. Entomol.* **2004**, *33*, 257–266. [CrossRef]
52. Kazakova, I.G.; Sergeev, M.G. The problem of delimitation of grasshopper populations. *Sib. J. Ecol.* **1997**, *4*, 315–321. (In Russian)
53. Lockwood, J.A. Rangeland grasshopper ecology. In *The Bionomics of Grasshoppers, Katydids and Their Kin*; Gangwere, S.K., Muralirangan, M.C., Muralirangan, M., Eds.; CAB International: Oxon, UK; NewYork, NY, USA, 1997; pp. 129–146.
54. Joern, A. What are the consequences of non-liner ecological interactions for grasshopper control strategies? In *Grasshoppers and Grassland Health: Managing Grasshopper Outbreaks Without Risking Environmental Disaster*; Lockwood, J.A., Latchininsky, A.V., Sergeev, M.G., Eds.; Kluwer Academic Publishers: Dordrecht, The Netherlands; Boston, MA, USA; London, UK, 2000; pp. 131–144. [CrossRef]
55. Uvarov, B.P. *Grasshoppers and Locusts*; Cambridge University Press: London, UK; New York, NY, USA; Ibadan, Nigeria, 1966; Volume 1, p. xi + 481.
56. Lockwood, J.A.; Lockwood, D.R. Catastrophe theory: A unified paradigm for rangeland ecosystem dynamics. *J. Range Manag.* **1993**, *46*, 282–288. [CrossRef]
57. van Huis, A. Can we prevent desert locust plagues? In *New Strategies in Locust Control*; Krall, S., Peveling, R., Ba Diallo, D., Eds.; Birkhauser Verlag: Basel, Switzerland, 1997; pp. 453–459. [CrossRef]
58. Lockwood, J.A.; Schell, S.P. Decreasing economic and environmental costs through Reduced Area and Agent Insecticide Treatments (RAATs) for the control of rangeland grasshoppers: Empirical results and their implications for pest management. *J. Orthoptera Res.* **1997**, *6*, 19–32. [CrossRef]
59. Latchininsky, A.V.; Lockwood, J.A.; Sergeev, M.G. Some experience to control grasshoppers in the North American prairies. *Plant. Prot. Quar.* **1999**, *8*, 12–14. (In Russian)
60. Wilps, H.; Levchenko, V.; Vernigor, A. Field trials with imidacloprid (Confidor) on the Italian locust *Calliptamus italicus* in Kazakhstan. *J. Appl. Entomol.* **2002**, *126*, 436–443. [CrossRef]
61. Sobolev, N.N.; Sergeev, M.G. Population dynamics of grasshoppers (Orthoptera, Acridoidea) in the agricultural ecosystems of North Kazakhstan. In *Human Influences on Insects' Assemblages*; Zolotarenko, G.S., Ed.; Nauka Publ.: Novosibirsk, Russia, 1985; pp. 96–104. (In Russian)
62. Sergeev, M.G. La sécheresse et les schémas de distribution des criquets en Asie centrale et septentrionale. *Sécheresse* **1996**, *7*, 129–132.
63. Samways, M.J.; Lockwood, J.A. Orthoptera conservation: Pests and paradoxes. *J. Insect Conserv.* **1998**, *2*, 143–149. [CrossRef]
64. Sergeev, M.G. Conservation of orthopteran biological diversity relative to landscape change in temperate Eurasia. *J. Insect Conserv.* **1998**, *2*, 247–252. [CrossRef]
65. Samways, M.J. Can locust control be compatible with conserving biodiversity? In *Grasshoppers and Grassland Health: Managing Grasshopper Outbreaks Without Risking Environmental Disaster*; Lockwood, J.A., Latchininsky, A.V., Sergeev, M.G., Eds.; Kluwer Academic Publishers: Dordrecht, The Netherlands; Boston, MA, USA; London, UK, 2000; pp. 173–179. [CrossRef]
66. Latchininsky, A.V.; Sergeev, M.G. The Italian locust—*Calliptamus italicus* (Linnaeus) (Orthoptera): A quarter of a century after. *Metaleptea* **2018**, *38*, 47–48.
67. Latchininsky, A.V. Locusts and remote sensing: A review. *J. Appl. Remote Sens.* **2013**, *7*, 075099. [CrossRef]
68. Tronin, A.A.; Gornyy, V.I.; Kiselev, A.V.; Kritsuk, S.G.; Latypov, I.S. Forecasting of locust mass breeding by using satellite data. *Curr. Probl. Remote Sens. Earth Space* **2014**, *11*, 37–50.
69. Klein, I.; Oppelt, N.; Kuenzer, C. Application of remote sensing data for locust research and management—A review. *Insects* **2021**, *12*, 233. [CrossRef] [PubMed]

70. Stebaev, I.V.; Gukasian, A.B. Orthoptera (Tettigonoidea and Acrididae) as stimulators of microbiological processes of decomposition and mineralization of phytomass in the meadow steppes of West Siberia. *Zool. Zh.* **1963**, *42*, 216–221. (In Russian)
71. Gandar, M.V. The dynamics and trophic ecology of grasshoppers (Acridoidea) in a South African savanna. *Oecologia* **1982**, *54*, 370–378. [CrossRef] [PubMed]
72. Hewitt, G.B.; Onsager, J.A. A method for forecasting potential losses from grasshopper feeding on northern mixed prairie forages. *J. Range Manag.* **1982**, *35*, 53–57. [CrossRef]
73. Belovsky, G.E. Do grasshoppers diminish grassland productivity? In *Grasshoppers and Grassland Health: Managing Grasshopper Outbreaks without Risking Environmental Disaster*; Lockwood, J.A., Latchininsky, A.V., Sergeev, M.G., Eds.; Kluwer Academic Publishers: Dordrecht, The Netherlands; Boston, MA, USA; London, UK, 2000; pp. 7–29. [CrossRef]
74. Pshenitsyna, L.B.; Sergeev, M.G.; Koutsenogij, K.P.; Kovalskaja, G.A.; Chankina, O.V. Biogeochemical role of the Italian locust *Calliptamusitalicus* L. (Orthoptera, Acrididae). In *XII Congress of Russian Entomological Society (Saint Petersburg, Russia, 19–24 August, 2002). Abstracts*; Russian Entomological Society: Saint Petersburg, Russia, 2002; pp. 299–300. (In Russian)

agronomy

Review

Locust and Grasshopper Outbreaks in the Near East: Review under Global Warming Context

Battal Çiplak

Department of Biology, Faculty of Science, Akdeniz University, 07058 Antalya, Turkey; ciplak@akdeniz.edu.tr

Abstract: Plagues of locust possibly date back to before humanity, as they evolved before humans. Following the Neolithic revolution and the permanent settlement of humans in Mesopotamia, locusts and grasshoppers have become a serious problem for people, as imprinted on archaeological remains. In the Near East, desert locust may be an important problem during invasion periods, in addition to various local species of locusts and grasshoppers. Past plagues caused serious disasters in the region, but there has been a pause since the 1960s, thanks to more effective monitoring and control. However, global warming and other anthropogenic activities change ecosystems, and these increase the potential for locust outbreaks, upsurges and plagues for the region. Outbreaks of some local species could also be a serious problem. Pest species of the locust and grasshopper of the Near East mainly belong to Caelifera and some to Ensifera. Global warming and extended agricultural activities can increase the potential for outbreaks of local species and create suitable conditions for desert locust invasions. This review is an attempt to (i) provide a historical background for locust invasions/outbreaks in the Near East, (ii) assess the potential for outbreaking of local species and (iii) define a perspective for future actions regarding global changes.

Keywords: locust swarms; Near East; Mesopotamia; swarming potential of resident species; *Schistocerca greraria; Dociostaurus maroccanus*

Citation: Çiplak, B. Locust and Grasshopper Outbreaks in the Near East: Review under Global Warming Context. *Agronomy* **2021**, *11*, 111. https://doi.org/10.3390/agronomy11010111

Received: 3 December 2020
Accepted: 5 January 2021
Published: 8 January 2021

Publisher's Note: MDPI stays neutral with regard to jurisdictional claims in published maps and institutional affiliations.

1. Introduction

Phylogenetically, the origin of the desert locust, *Schistocerca gregaria* Forskål, dates back to around eight million years ago [1], but that of humans as the genus *Homo* to date back to 2.5 million years and modern human *Homo sapiens* to about 200 thousand years [2]. As desert locust and *Homo sapiens* are both native to Africa, we can speculate that modern humans evolved under the selection pressures of the disasters caused by the desert locust plagues. Whether or not *Homo sapiens* has managed to adapt to this selection pressure is a provocative question, but the available data points to the side of "no". For instance, in recorded history, there are indications of a locust problem in Assyrians and New Testament of the Bible [3–5]. All these histories come from the Middle East and this study aims to present a review of the locust problem in a part of this region, starting from Sinai in the south, extending as far as to the Caucasus in the north, to the Aegean Sea in the west and Zagros Mountain range in the east (Figure 1). The southern part of the Arabian Peninsula is left out because this region is within the recession area of the desert locust and has already received considerable interests from locust experts (for a review see [6]). Comparing to North Africa and Southern part of the Arabian Peninsula, the locust and grasshopper problem in the Near East has other peculiarities, and thus requires a separate evaluation. In this region, the desert locust plagues have been relatively rare. The area is outside recession range of the species and is only contaminated during significant plague periods [7]. In addition, there have been no desert locust swarms in the region since 1960s. However, there are other outbreaking species, such as *Dociostaurus maroccanus* (Thunberg), *Locusta migratoria* Linnaeus and some species of *Calliptamus* Serville. Moreover, there have been occasionally and locally outbreaking species, such as *Heteracris pterosthica* (Fischer de Waldheim), *Notostaurus anatolicus* (Krauss) and *Arcyptera labiata* (Brulle), especially in Anatolia, and an assessment of their potential in the context of global warming seems of particular importance. Some species of Barbististini—*Isophya* Brunner von Wattenwyl, *Poecilimon* Fischer and *Phonochorion* Uvarov—will also be considered, although none of these long-horned Orthoptera could be considered a locust and there is no significant report on their outbreaks. The review will begin by summarizing the locust problem in the previous century, especially before the 1960s. In the second section, common and locally outbreaking species as well as their management will be considered, especially in Anatolia and Mesopotamia. Finally, an assessment of the possible impact of global warming and other anthropogenic activities on the locust and grasshopper problem will be discussed. The conclusion aims to provide a management perspective.

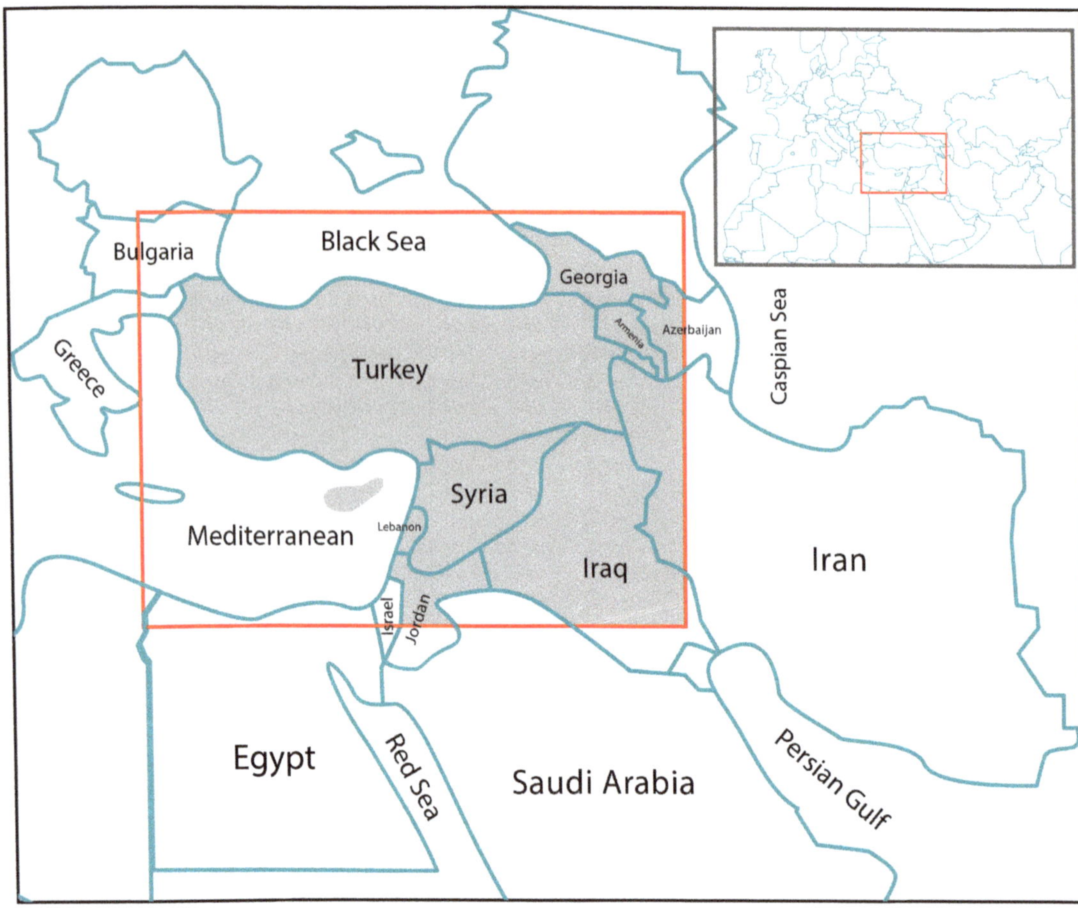

Figure 1. The geographic area considered in this review.

2. History of the Locust and Grasshopper Problem in the Near East—First Half of the 20th Century

Considering the Near East (Figure 1), publications on locust plagues in the 20th century can easily be assembled into two groups according to their dates. Publications included in the first group are rarely after 1960 [8–18] and if so, they review plague events of the past [19]. Rainey et al. [20] clearly defines ending of a period as no plagues reached to this region after 1962. A new period of the studies begins after 2000, although the publications by orthopterists or locust experts are scant among them (e.g., 6). The new period is mainly characterized by publications of historians [21–32], also dealing with locust invasions of the first half of the 20th century. They combine information from archives of governments and old newspapers and provide a significant amount of knowledge about the history of locust invasions in the Near East. For example, thanks to these publications, we discovered the acceptance of a special law for locust control by the Ottoman Parliament in 1912, which seems to be a unique event in this regard [28,29]. The law consists of nine articles. The first article defines responsibilities of local peoples in monitoring their district and informing the local governmental officers and regulate awarding of peoples who supply accurate information. The second article defines how to treat the areas contaminated with eggs and hoppers. The third article lists responsibilities of persons in a hierarchical way in destroying the eggs or hoppers. Later, six articles define the rules for offices and staffs of the government, the budget to be provided and its usage. This law was updated in subsequent years by Parliaments of Ottoman and Turkish Republic. Almost all of these publications report or discuss the plagues caused by two species—*S. gregaria* and *D. maroccanus*—indicating they were main locust pests in this area, and these will mainly be considered below. Damage to agriculture by other species such as *Calliptamus italicus* (Linnaeus) (or other species of the genus), *Gryllus campestris* Linnaeus, *Platycleis intermedia* (Serville), *Platycleis affinis* Fieber and *H. pterosthica* are negligible [14,16,29].

3. Desert Locust Plagues in the Near East

The range of the desert locust is variable depending on plague period. Regarding the present recession area [7], the desert locust range is between latitudes 10° N–35° N and longitudes 15° W–72° E, mainly associated with the Sahara Desert in North Africa, the Arabian Desert in the Arabian Peninsula and the Thar Desert in Pakistan/India. However, the extent of the range can double during periods of invasion. The Near East remains outside the recession area and constitutes a marginal part of the of invasion area [6,7,33]. The history of desert locust plagues during the last two centuries was summarized in Table 1 and the maximum northward extension of the invasions during the 20th century was shown in Figure 2. The extent of the past plagues in the Near East depended on some other peculiarities such as the severity of the plague, rain regime and the control efforts. Several desert locust plagues were reported for the period before the First World War (Table 1) [19,23,33], in particular for Mesopotamia and the Levant. Although data on the severity and extent of plagues are limited and mainly concern human aspects due to the destruction of food resources and control management by the Ottoman Government [12,22,31], they allows us to estimate that at least the Levantine and the Mesopotamian regions have been invaded on several occasions. According to reports, possibly the greatest plague occurred during First World War and its peak was in 1915 and ended in 1917. As controlling actions were primitive and ineffective, such as destroying the locusts by hand (Figure 3); its consequences were catastrophic. It resulted in a great famine, which was considered one of the greatest tragedies in the region, with the death toll attributed to it estimated at around 500,000 in 1918 [15,24,29–31]. Existing records, based on certain governmental documents, indicate that there was also a significant control effort (see [12,23,29,31] and references therein), but not sufficient to prevent a disaster. During these years, a local reproduction regularly took place, particularly in the Levant and Mesopotamia, leading to successive outbreaks and increasing the impact of the plague.

Table 1. History of *Schistocerca gregaria* plagues in the last two centuries (those in the 20th century are shown in Figure 2).

Year/Period	Invasion Area in Near East	Reference
1729	Aleppo	[22,32]
1865	Levant (from Sinai to Turkey), including Cyprus	[19]
1878	Sinai, Palestine, Syria (Levant)	[19]
1890	Sinai, Palestine, Syria (Levant)	[19]
1902	Sinai, Palestine, Syria (Levant)	[19]
1911–1915	Arabian Peninsula plus Mesopotamia (Syria, Lebanon, Iraq and Turkey) up to Southern Caucasian Plateau	[8,12,15,19,24,29,31]
1928–1930	Arabian Peninsula plus Mesopotamia (Syria, Lebanon, Iraq and Turkey) up to Southern Caucasian Plateau	[8,19,29]
1945	Arabian Peninsula plus Mesopotamia (Syria, Lebanon, Iraq and Turkey) up to Southern Caucasian Plateau	[8,19]
1952	Arabian Peninsula and a very small part of the Mesopotamian Turkey	[8,19]
1953	Arabian Peninsula and large parts of the East plus South East Turkey	[8,19]
1958	Arabian Peninsula and large parts of the East plus South Turkey	[19]
1959	Arabian Peninsula plus a small part of Mesopotamian Turkey	[19]
1960	Arabian Peninsula plus a small part of Mesopotamian Turkey	[19]
1962	Arabian Peninsula plus Mesopotamian Turkey	[19]

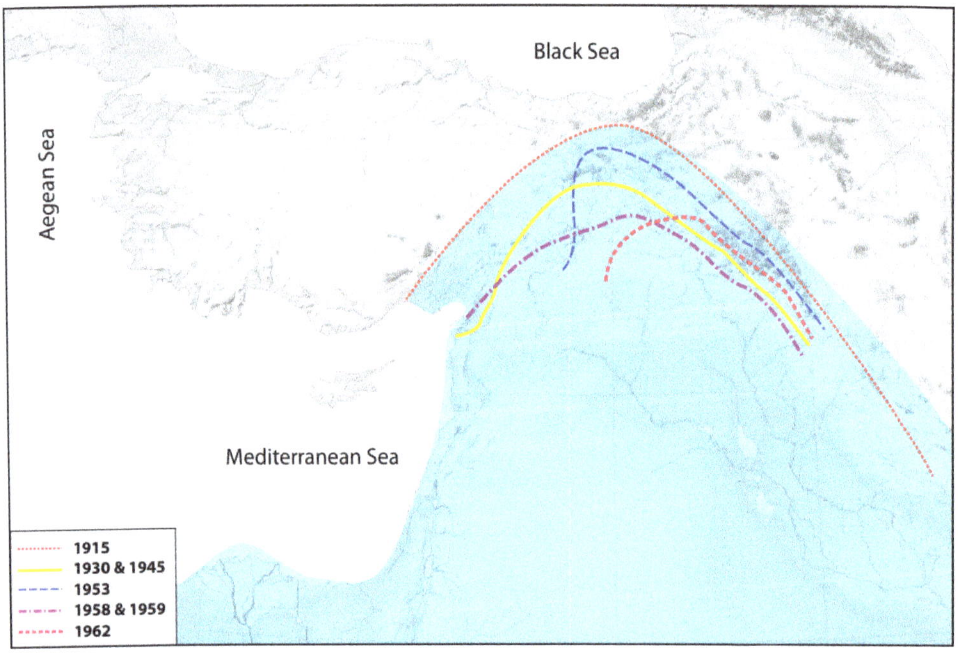

Figure 2. Northern invasion borders of *Schistocerca gregaria* plagues in the 20th century in the Near East (figure prepared using data by Balamir [19]).

Figure 3. Locust destruction by sticks at hand in Palestine (from the newspaper *Serveti Fünun*, 269, 1896, page 140).

Balamir [19] presented an extensive review of desert locust invasions in the Near East after the First World War, including extent, control activities and consequences of the invasions. Eight different plagues have been reported in the years 1930, 1945, 1952, 1953, 1958, 1959, 1960 and 1962 (Table 1). The size of these invasions and their northern boundaries were variable (Figure 2). The plagues extended to northern Syria and Iraq, but not to Turkey (or only into a small part), in 1952, 1959 and 1960. In 1962, the plagues invaded the Mesopotamian lowlands of Turkey, as well as, further south, the Levant, Jordan and Iraq. In the remaining years, plagues reached to mountain belts in the northern Mesopotamia. The two most extensive plagues were those of 1930 and 1958, and the invasion margin reached the Caucasus Plateau in north-eastern Turkey. Various preventive actions were taken. The use of sheet metal dams (made of zinc plates) for the management of hopper bands was the main control method. Hoppers were channeled into ditches and then crushed, burned or buried. For example, a total of 67,690,000 kg and 603,142,000 kg nymphs were destroyed in Turkey by this method during the 1930 and 1958 plagues, respectively [19]. Spraying insecticide (mainly the gamma-hexachlorocyclohexane preparations) by air became a dominant method after mid of the 20th century. Great success was achieved by this method and none of the 20 swarms of 1960 succeeded in laying eggs in Turkey [19]. Other control methods implemented in the past were ploughing egg-laying sites, collecting nymphs and adults by hand or spraying liquid soap or some other chemicals, such as arsenic [19,29,32,33].

There are no records of desert locust swarms in the Near East after the years mentioned above, which corresponds to the general trend in other parts of the invasion area [20]. This is possibly the reason why the countries in the region have halted governmental units for locust control or not established new ones [6]. Why did such a period of successive plagues, observed since at least the last quarter of the 19th century, end in the last half of the 20th century? In fact, after 1965, effective control activities, mainly in Africa and southern Arabia—efficient survey and timely response in desert locust breeding areas—appear to be the main reasons for this improvement and the absence of swarms in the Near East [7,20,34–39]. This is also applicable for the peripheral invasion areas, as using spraying aircrafts after 1950 prevented the spread of the swarms to the borders of Caucasus.

Although, in recent years, some swarms entered in the Near East, none of them were able to become harmful.

4. Moroccan Locust Plagues in the Near East

Considering the Near East, one of the most harmful locust species was the Moroccan locusts. This species seems to have recently lost its economic importance to the point that conservation measures have been suggested only for certain localities [40]. However, it was a serious problem for the Near East in the first half of the 20th century, especially important in West Anatolia and moderate highlands of Mesopotamia, including south-eastern Turkey and the northern regions of Iraq and Syria for around a hundred-year period. Several very harmful plagues of this species have been reported in the Near East during the period of 1847–1945 [12,18,25,29,31] (Table 2). Although there are no reliable reports for more archaic time, this locust species may have been a serious problem at all times in the Near East. West Anatolia, Mesopotamia and Southern Caucasus also were three important outbreak areas for this species (Figure 4). The outbreak periods were mostly different, although sometimes overlapping, from one region to another.

Seven plague periods have been reported in West Anatolia for a century, from the mid-19th century to the mid-20th century. Each plague lasted for 2 to 7 years. The plague extent varied from a few km^2 to the whole West Anatolia, from the Marmara Sea to the Mediterranean Sea, excluding narrow coastal areas. Some plagues also spread to the European part of Turkey. The three plagues of the second half of the 19th century—1847–1851, 1852–1864 (especially in 1861) and 1875–1881—invaded large parts of West Anatolia [25,29]. The plague in 1904–1905 was restricted to the eastern parts of the Manisa and Uşak provinces of Turkey and possibly played a significant role in the development of the next one, which lasted for about seven years from 1910 to 1917. It was possibly the largest in terms of extent and impact. The upsurge period started in 1909–1910 in Eşme and Kula, then reached its peak in 1916 and declined in 1917. All these stages were well monitored and documented [12,19]. As the Moroccan locust was a persistent problem for West Anatolia, Uvarov [18] was invited to study this species in the region; he considered the 1910–1917 plague as a reference with which to define the bio-ecology of the Moroccan locust. In addition, a considerable effort had been devoted to the prevention of future plagues [12]. Karabağ [16] reported that 125,000,000 kg nymphs and 12,500,000 kg egg pods were destroyed during the period of 1915–1917. Although it was largely controlled in West Anatolia, local swarms in the Marmara region, including parts of Thrace in 1919, were likely the aftershock of this plague period. For the aforementioned reasons, studies on this plague constitute a reference for understanding the upsurges characteristics of species in relation to ecology and for developing effective control strategies. These experiences have provided significant knowledge for the prevention of subsequent plagues. The later plagues, in 1930–1932 and 1939–1941, were also widespread and invaded large parts of West Anatolia up to Marmara Sea in the north, the Konya province-central Anatolia in the east and the north of Antalya in the south. However, both were controlled within one or two years and their damage was limited. Later, especially after 1962, only a few local swarms occurred and these were successfully controlled.

Mesopotamia (Syria, Iraq and Turkey) was another outbreak area for the Moroccan locust in the Near East (Figure 4). There are no data regarding plagues of this species for earlier periods; however, four plague periods are well known in first half of 20th century [12,13,29]. Those in 1919 and 1945 were one-year plagues affecting a restricted part of Mesopotamia. The two others, in 1931–1932 and 1939–1941, lasted at least two years each and invaded larger areas in the region. In addition, the plague of 1925 in Iraq was reported to cause serious damage [40]. The South Caucasus lowlands in the Basins Aras and Kura rivers appear to be a separate outbreak area, and the plague of 1920–1921 in Armenia and Azerbaijan caused significant damage to the region [29,40].

Table 2. History of *Dociostaurus moroccanus* plagues in the last two centuries (those in the first half the 20th century are shown in Figure 4).

Year/Period	Invasion Area	Reference
1830	West Anatolia, local swarms	[25,29]
1847–1851	Aegean Region and south Marmara	[25]
1852–1864 (peak in 1861)	Large parts of West Anatolia	[25,29]
1875–1881	Antalya, Inner part of Aegean Region up to Balıkesir Province	[29,31]
1904–1905	Manisa (inner parts) and Uşak provinces	[25,29,31]
1909	Local swarm in Eşme in Manisa Province	[18,29,31]
1910–1911	Swarms in Manisa province (Eşme, Kula, Demirci, Alaşehir)	[13,18,29]
1911–1912	West Anatolia (Denizli, Manisa, Aydın and İzmir provinces)	[13,18,29]
1915–1916, declined in 1917	Large parts of western Anatolia up to Antalya and Konya provinces	[13,18,29]
1919	Marmara Regions of Turkey	[13,18,29]
1919	A restricted part of Mesopotamian Turkey, Iraq and Syria	[29]
1920–1921	Armenia and Azerbaijan	[40]
1930–1932	Large parts of West Anatolia including Marmara Basin plus Konya province in Central Anatolia	
1931–32	Mesopotamian Turkey (from Siirt to Mersin), Iraq and Syria	[13,29]
1934–1938	Local swarms in several location especially in West Anatolia	[13,29]
1939–1941	Aegean and Southern Marmara regions, and local swarms in Mesopotamian Turkey, Iraq and Syria	[13,29]
1945	A restricted part of Mesopotamian Turkey and Syria	[29]

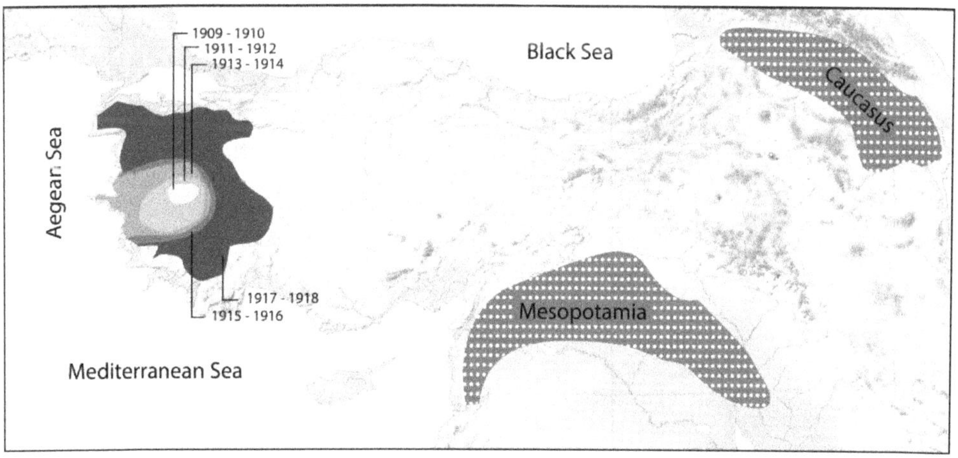

Figure 4. Outbreak areas of *Dociostaurus moroccanus* in the Near East: West Anatolia, Mesopotamia and Caucasus. Development of the most intense plague during 1909–1917 in West Anatolia is outlined by years. The borders for 1917–1918 also define the West Anatolian outbreak area. Figure is based on data mainly provided by Bücher [12] and Uvarov [18]).

The Moroccan locust plagues in the Near East exhibit some peculiarities. It seems that they are several independent outbreak areas and these can be classified as West Anatolia, Mesopotamia and the Caucasus (see Figure 4) [12,18,40]. In West Anatolia, the plains of the Inner Aegean Basin, especially those in Manisa, Uşak and the surrounding parts of the neighboring provinces of Denizli and Aydın, were the main outbreak areas.

Swarms that originated in these regions spread in subsequent years and extended to the surrounding parts of Aegean, Marmara, the Mediterranean and central Anatolia (Figure 4). The Mesopotamian outbreak area may include two places of origin: one in northern Syria and the other in northern Iraq. Uvarov [18] stated that the outbreak area of this species in West Anatolia exhibits special environmental conditions: (i) it is restricted to altitudes of 400–1000 m, (ii) it has a certain precipitation regime and (iii) it harbors a well-defined vegetation (the presence of *Plantago* spp. and *Hordeum* spp. is of particular importance). The determinative aspects of these environmental conditions have been reported from other parts of the species range [40]. This combination of factors restricts the species to parts of West Anatolia, but arising upsurges can invade the surrounding parts of the regions in Anatolia.

The Moroccan locust does not occur in Euro-Siberian vegetation along the Black Sea Basin up to the Great Caucasus. The average elevation of central Anatolia is around 1000 m and that of eastern Anatolia is around 1500 m [41], and these heights possibly prevent the species from colonizing these areas. Consistent with this viewpoint, no swarm or outbreak areas have been reported from central and eastern Anatolia, although the solitary form has been recorded frequently [9,42,43]. For similar reasons, the south-eastern Taurus Mountains and the Syrian Desert constitute the northern and southern borders of the Moroccan locust outbreak area in Mesopotamia, respectively. Thus, this outbreak area of the Moroccan locust in this region constitutes an arc starting from Mediterranean Syria and extending to Zagros. The differences between the West Anatolian and Mesopotamian outbreak areas are mainly altitudinal, with altitudes of 400–1000 m and 200–400 m, respectively [14,18], indicating that the latter outbreak area is less suitable. This is possibly the reason why the Moroccan locust plagues in this region were less severe than those in West Anatolia.

The moderate highlands of the inter-montane basins of the Aras and Kura rivers in southern Caucasus were the third important outbreak area in the Near East (Figure 4). Past reports indicate that this is a less significant outbreak area compared to the two others. It should be noted that these three outbreak areas are independent: no swarms in one area spread to another. The periodic overlap of some past plagues of these three distinct areas needs an explanation, but periodic climate changes lasting several years may be the main reason.

An issue worth highlighting is that there has been a pause in the Moroccan locust plagues since the mid-20th century. Although there have been a few local swarms ([44,45]; N. Babaroğlu—personal communication), none of them were invasive and they were controlled in their local outbreak area. This pause may be the result of a combination of several factors. The bio-ecology of the species is surely one of them, as indicated by Uvarov [18] and Latchininsky [40]. The reasons for the pause may be different for West Anatolia, Mesopotamia and the Caucasus. However, the topographic and ecological heterogeneity of the biogeographic Anatolia (see [41]) offers opportunities for the species to compensate for environmental changes. The most plausible reason for the pause seems to be the intense control of swarms in the outbreak areas, preventing upsurges in subsequent years. The swarm in 1996 in Eynif Plain in Akseki (Antalya, Turkey) was controlled by the application of 40,000 kg of insecticide [44,45]. Several similar local swarms occurred in the Aegean part of Turkey (A. Yüzbaş and N. Babaroğlu—personal communication), but none of them succeeded in invading other areas in the following years. This also indicates that control in the outbreak areas is the key reason for preventing long-lasting plagues.

5. Other Pest Orthopteran Species

The desert locust and the Moroccan locusts were the two most important agricultural pests in the Near East in 20th century. However, limiting to these two species seems insufficient for the region, especially for Anatolia. Although their characteristics are different from these two species, many other orthopteran pests have been reported in past studies [10,11,14,16,46–48]. We can cite some Acrididae species, such as *L. migratoria*, *Calliptamus* spp. (namely *C. italicus*, *C. barbarus* (Costa) and *C. tenuicercis* (Tarbinsky)), *H. pterosticha* and *Arcyptera labiata*; Ensifera species, such as *Melanogryllus desertus* (Pallas) and *P. intermedia*; and some species of *Poecilimon*, *Isophya* and *Phonochorion* (without certain specific identification). Of all these species, *L. migratoria* and *C. italicus* were the most important, corresponding to Uvarov's definition of locust [17] and requiring control measures [14,48].

The author himself observed many local swarms of orthopteran species during his field studies since 1990 across Anatolia. As mentioned above, the swarm of *D. maroccanus* in 1996 in the Eynif Plain (near Antalya) was observed from the merging of dense patches of hoppers into larger bands and then into migrating adults. Swarms of other species were also observed, but their swarming characteristics were different from typical locusts. These species present a great taxonomical variety, but mainly belong to Acrididae and Phaneropterinae. Among Acrididae, *L. migratoria*, *C. italicus*, *C. barbarus*, *C. tenuicercis* and *H. pterosticha* are the species reported in previous publications. *N. anatolicus* (Krauss), which coexists mainly with *Calliptamus* spp., is another species to be added to this list. *L. migratoria* is the model species on which Uvarov [17] based his phase theory. Local swarms of the species have been reported ([10]; N. Babaroğlu—personal communication) and control measures have been applied locally by the Turkish Ministry of Agriculture (by central and local plant protection institutions). *L. migratoria* is common in its solitary form and can pullulate in paddy fields, but there is no record of invasive plagues extended to other places. *H. pterosticha* occurs in humid plains irrigated by humans below 1000 m elevation. For example, its population showed a significant increase in 1961–1962 in Diyarbakır and was prevented by the application of chemical insecticide [11]. The author observed a similar case in the province of Malatya in Mesopotamian Turkey. The population density gradually increased in 2017–2020 in the plain part of the province. The increase in population density caused significant damage on crops and fruit trees. Apricot cultivation is the main agricultural activity in the region and these grasshoppers climb to apricot trees and consume both leaves and fruits, as complained about by farmers. This case should be monitored in the coming years. *Calliptamus* spp. and *N. anatolicus* can reach high densities, especially in arid agricultural fields in Central Anatolia, Mesopotamia and some plains of eastern Anatolia. The damage caused by *Calliptamus* spp., especially *C. italicus*, and insecticide application to manage their populations is known for a long time [48]. Although Tutkun ([48]; see also [29,31]) states that the species is the third most harmful locust species in the area, there has been no recent report of outbreaks or preventive control measures. However, monitoring seems to be a necessity.

The case of the phaneropterid species is more interesting. Highly dense populations of *Poecilimon* spp. and *Isophya* spp. have been reported in previous publications [47] or observed by the author himself. Only the two most striking cases are mentioned here. *Isophya hakkarica* Karabağ, whose range covers the south-eastern Taurus and the Anatolian Diagonal [49], reached to very high population densities in the Malatya, Elazığ and Adıyaman provinces of Turkey in 2016. Rural people asked the local government to control the pest as it caused extensive damage to pasture and destroyed animal feed. Residents observed that the increase in population density began in 2015 and peaked in 2016. Local governmental agencies provided partridges to be release into the wild to try to control the pest, but no insecticide was applied. A similar case has been observed with *Poecilimon celebi* Karabag in the Kastamonu province. During a field study in 2012, an extremely dense population was observed in the forest clearing areas with Euro-Siberian steppe vegetation (Figure 5). Locals complained about the destruction of fields and meadows where animal feed was produced. Many other similar cases with species belonging to *Poecilimon*, *Ispohya* and *Phonochorion* (all from Barbitistini) have been observed, especially in northern Anatolia along the Black Sea Basin and the highlands of the eastern Anatolia with a high grassland vegetation. The conclusion is that several Barbitistini species may reach high population densities, cause significant damage and require various control measures.

Figure 5. Collecting specimens of *Poecilimon celebi* in an outbreak area in North Anatolia.

The reports on the locust and grasshopper management from the Agriculture Ministry of Turkey since the second half of the 20th century, as well as the author's observations during last 30 years, allow us to make some general statements that could guide future plans and actions. Among the previously mentioned species, both *S. gregaria* and *D. moroccanus* show locust characteristics with phase polyphenism (shift from solitary to gregarious phase according to population density), as defined by Uvarov [18,50], and can invade new areas by migrating from the outbreak areas where the initial increase in density occus. *L. migratoria* and *Calliptamus* spp. (specifically, *C. italicus*, *C. barbarus* and *C. tenuicercis*) were also reported as locusts; however, in previous Near East studies or reports, no specific phase changes or long-distance migration leading to invasions of new areas were observed. Instead, they were reported to be harmful only by reaching high densities in their regular distribution range (N. Babaroğlu—personal communications). This statement is undoubtedly valid for the other acridid species: *A. labiata*, *H. pterosticha* and *N. anatolicus*.

The case of long-horned species requires a distinct assessment. All the long-horned or ensiferan species mentioned above have a limited range or are endemic to a specific location. In addition, they are short-winged and lack the ability to migrate. They can exhibit various pigmentation forms, such as green or black as in some *Isophya* and *Phonochorion*, or green and yellow/brown as in some species of *Poecilimon* [51], but these colors do not indicate phase polyphenism due to population density. Although they may pullulate and cause significant damage locally, their outbreaks are only local and population density increases within their regular range. This does not mean that they do not require monitoring and control measures. Several species of Barbitistini and, in addition, some other tettigoniids species can reach unexpectedly high population densities in their local range depending on annual environmental conditions, which may possibly require survey and control measures if a density threshold is exceeded.

6. Locust and Grasshopper Management in Turkey in Recent Years

Locust and grasshopper management activities in Turkey are organized under the authority of the Directorate of Plant Protection Central Research Institute (DPPCRI) a branch of the Ministry of Agriculture and Forestry (N. Babaroğlu, M. Çulcu and E. Akci from DPPCRI—personal communication). The main management activity is applying insecticide over outbreak areas. Releasing partridges is an uncommon practice; there is no report on the efficiency of this method. For the period of 2013–2020, the control of grasshopper outbreaks required, yearly, on average, an application of insecticides of over 55 km^2, in several locations, throughout 81 provinces of Turkey (Figure 6). During the period of 2013–2020, in some provinces, insecticide application has only been conducted in one year, while in some other provinces it has been conducted in two, three or more years. These data have some specific indications for the recent condition for outbreaks in Turkey. Although the grasshopper species are not generally identified and non-locust species are possible, most of the insecticide applications for the period of 2013–2020 are located in the Moroccan locust outbreak areas, i.e., in West Anatolia and Mesopotamian Turkey. This can be considered as an indication of the potential danger of the Moroccan locust for present and the future. Insecticide application in other areas is rare, just in restricted locations and only for one or two years, for the period 2013–2020. Rare application areas are mostly scattered along the one-third of Anatolia, in northern part. From previous publications and personal observation during the last 30 years, we assume that the species outbreaks in these districts are non-locust species mentioned above. In particular, the release of partridges in the provinces of Çorum and Kastamonu in northern Anatolia is consistent with this assumption. Comparison of these recent control activities with earlier records indicates that the potential of outbreak by resident species has increased and may increase further under global warming (see below). Overall, the data indicate that the outbreak potential of the non-locust species requires special attention.

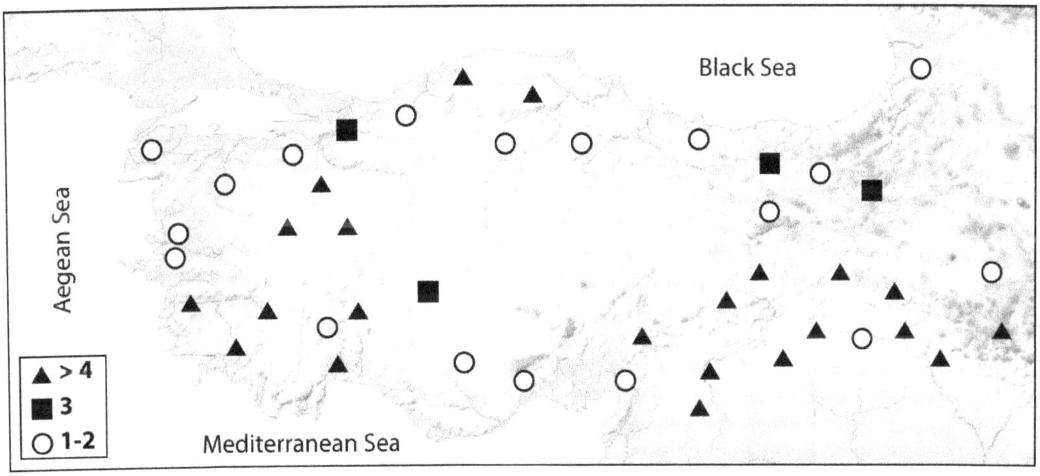

Figure 6. Insecticide application by the Directorate of Plant Protection Central Research Institute (DPPCRI) during the period of 2013–2020 throughout Turkey, indicated by the number of years (open circle application in ≤2 years, square in 3 years and triangle in ≥4 years).

7. Climate Change and Its Possible Impacts on Locust/Grasshopper Problem in the Near East

Global changes/warming directly or indirectly affect locust and grasshopper species [52–57]. The direct effects are those related to the auto-ecology (e.g., specific thermal thresholds for their development) and therefore to phenology, including changes in life history traits such as the hatching time, instar durations, adult emerging time, reproduction period, egg lying time and number of annual generations [53,54,58,59]. Such phenological changes depend on the resilience and adaptive capacity of the species [54,58–61]. The indirect effects of global warming on insect species are represented by changes in habitat as well as changes in food preferences. Changes in habitat characteristics can be particularly serious for phytophagous insects, as their presence is correlated with plant communities [53,54]. Locusts, generally phytophagous, may have some preference for a limited plant spectrum or may preferentially feed on some certain plant species. For example, Uvarov [18] reported that *Plantago* spp. and *Hordeum* spp. were determining plants for the presence of the Moroccan locust in western Anatolia. Thus, global warming will produce new conditions for locusts and grasshoppers and adaptation to the new conditions will require high resilience and/or adaptive plasticity, especially for the locally outbreaking species [41,54–57].

Depending on the adaptive capacity of a species, three main consequences may be expected: (i) change of ecological niche/habitat or phenology, (ii) dispersal to new areas with conditions similar to those of the original habitats (range shift) and (iii) extinction, if there is no capacity to adapt to the new conditions and/or to disperse into new areas [52–57]. A change in ecology/phenology of the pest orthopteran species is very likely. If their long-term adaptations provide the capacity for a broad ecological tolerance, they can overcome altered environmental conditions and continue to live in their regular range and outbreak if they meet favorable conditions. In the case of outbreaking species, one issue to be considered in the context of phenological changes is the number of generations per year. The lengthening of the warm period, a consequence of global warming [62,63], can lead to an increase in the number of generations. Shifting their ranges either diffusely over several generations or migrating long distances within a generation can be expected when the previous habitat becomes unavailable [41] or newly available areas appeared [57,61,63]. Regarding outbreaking and invasive locusts, the new conditions produced by global warming may also mean an increase in the available area to be invaded, in addition to their regular ranges [56,57]. Extinction or even range size reduction is less likely for common locusts but may be probable for locally outbreaking non-locust orthopteran. On the other hand, the severity or impacts of global warming/changes are expected to be different depending on geographic location [60,64]. The Mediterranean in general and the Near East in particular are among the places experiencing a high rate of warming [64] and the possible consequences for the species in this area are likely to be more serious.

The possible impact of global changes/warming on outbreaking species can be assessed according to four criteria: (i) phenological changes, including the number of annual generations, (ii) habitat/niche change, (iii) range shift and (iv) extinction. Although each species may be affected differently, a generalization may be possible for the locally outbreaking species, such as those of Barbitistini belonging to *Poecilimon*, *Isophya* and *Phonochrion*. Unfortunately, studies specifically examining the bio-ecology of the species belonging to these genera are scarce [51] and do not provide a basis with which to evaluate the possible impact of global warming for their future. Some speculations can be made from a phylogeographic point of view regarding the location of their range and habitat preference. These species are univoltine and this feature seems to be fixed evolutionarily; therefore, additional generations are unexpected. However, their hatching, nymph and adult periods may shift to earlier times of year. Outbreaking species of these genera prefer certain vegetation, mainly wet grasslands, and their ranges are mostly restricted to certain fragments in the heterogeneous topography of the Black Sea Basin and of the highlands of East Anatolia [41,65–67]. Furthermore, they are flightless, with a low dispersal capacity, and we believe that invasion of new areas by these species is unlikely. On the other hand, they can vertically shift their range and benefit from the buffer role of altitudinal heterogeneity to overcome the effect of global warming [68]. This is another reason why extinction is also unlikely. These species are likely to experience fluctuation in population density from year to year or at certain years and become harmful when the appropriate combination of ecological factors is met. Thus, they may require special monitoring and necessary control measures must be taken in the event of an increase in population density.

A generalization may also be possible for the locally outbreaking acridid species, i.e., *L. migratoria*, *H. pterosticha*, *N. anatolicus*, *C. italicus*, *C. barbarus*, *C. tenuicercis* and *A. labiata*. Different than all others, the last species is found in mountainous habitats and will not be considered. The other species are mainly associated with anthropogenic or agricultural habitats and prefer a relatively warm climate and moderate altitudes below 1000 m [43]. The first two species are found in humid and irrigated agricultural plains, while the last four in arid plains or arid habitats contiguous to humid plains. These species are known to be univoltine. However, our observations on *H. pterosticha* indicate that it can be multivoltine as nymphs and adults have been observed together at different periods from June to October. From the perspective of the above four criteria, the following statements can be made. The probability of extinction is less likely. Range shifting and an enlargement of the regular range is highly likely as warming provides new opportunities for these species. An additional generation seems less likely for most of the species, but this is probable for *L. migratoria* and *H. pterosticha*. *N. anatolicus*, and three species of *Calliptamus*, namely, *C. italicus*, *C. barbarous* and *C. tenuicercis*, occur in arid and mostly agricultural areas (48). Publications on locust generally report *C. italicus* as the main pest species in the genus; the author observed that these three species reach high population densities in the arid areas adjoining to irrigated plains. Modelling studies [56,57] estimate that *C. italicus* may benefit from global warming and enlarge its range or become a more serious pest in West Palearctic. Estimating the same consequences for *C. barbarous*, *C. tenuicercis*, and *N. anatolicus* will not be an exaggeration as their habitat preferences are quite similar. All these species may become more serious pests as temperatures and agricultural activities increase, and control measures may be an unavoidable necessity.

The two most important species showing real locust characteristics are *S. gregaria* and *D. moroccanus* and they should be considered separately. The first species, the desert locust, live in desertic habitats and the above model cannot be applied. Aridification due to global warming is an expectation for the whole region and especially Mesopotamia [62,64] and this can be considered as new opportunities for the desert locust. The Near East is not an outbreak area for this species, not even in its recession area, but it is on the margins of the invasion area. Larger breeding areas may become available for this locust, offering new opportunities to increase its populations in the case of new invasions. In particularly, enlargement of the dry desert-like areas in Central Anatolia requires special attention in the event of a new plague. Additionally, the inter-mountain plains that develop in Mesopotamia and eastern Anatolia also require caution. However, in any case, preventive or proactive control in the outbreak areas (far from Near East) is more important than the regional measures, which can only be considered during plague episodes.

The four criteria given above can be strictly applied to the Moroccan locust, the most economically important locust species in the region. Uvarov [18], also supported by other studies [40], constitutes a hallmark background for the bio-ecology of the species, to be used as a basis for the four criteria given above. This species is univoltine and a change in this characteristic due to new conditions seems unlikely. However, some climate changes, such as the lengthening of the warm season, can shorten various stages of the life cycle and cause earlier outbreaks. Shift of the distribution area is another parameter of our model. This species occurs in certain altitudes: 400–1000 m altitude in western Anatolia [18] and 200–400 m in Mesopotamia [13]. Thus, dispersing to higher altitudes along the outbreak areas that are suitable for other conditions is likely. However, species composition of vegetation constitutes another determinant of the outbreak. Thus, change in vegetation composition either in outbreak areas or in areas to be invaded further is also important. In addition, the precipitation regime is another significant factor in the ecology of this species, so fluctuation in the precipitation regime will be another determinant for the future outbreaks by the species. However, it should be noted that aridification in the region is an expectation [64] and this may interplay a role in an outbreak. Together, these conditions may further delimit the expansion and outbreaking of this species, as has been the case since the 1960s. As a result, it may become a less important pest, as observed in several other places [40]. Plagues of the species are possible in the case of upsurges during a few successive years. Thus, monitoring the species in local outbreak areas, as has been done during 2013–2020 by the DPPCRI, especially in the cases of random overlapping of the conditions mentioned above, remains the best practice to prevent possible plagues.

8. Conclusions

The above assay includes data gathered from studies in various scientific fields ranging from the humanities to phylogeography and such a rich combination indicates that locust plagues are natural events with a high impact on human life. As witnessed in the 20th century in the Levant, Mesopotamia and Anatolia, every society or state has to pay attention to it in order not to pay the price. In the light of past plagues, it can be stated that the two most important and damaging species were true locusts: *S. gregaria* and *D. moroccanus*. For *S. gregaria*, although the Near East does not include any outbreak area and remains just on the margins of the invasion area, societies living in this region have had to endure tremendous suffering and have paid hundred thousand lives. These two species prevailed in the region for nearly 100 years, between the mid-19th and 20th centuries, but large plagues came to a halt from the 1960s. *S. gregaria* is still a serious pest for North Africa and the Southern Arabian Peninsula, as well as in many other regions of its invasion area, but effective control in its outbreak areas [37,69] keeps the Near East apart of the disaster of this species. *D. moroccanus* was probably the most harmful species especially for West Anatolia and Mesopotamia, particularly in the first half of the 20th century. However, effective control in its outbreak areas/plains scattered in West Anatolia and Mesopotamia subsequently prevented large-scale plagues. As for *S. gregaria*, only a few local swarms were encountered after the 1960s. Although they lack the typical locust characteristics and do not swarm in extensive areas, other species pullulate locally and may cause significant damage in crop fields and pastures, especially in Anatolia. This is the case with *L. migratoria*, *C. italicus*, *C. barbarus*, *C. tenuicercis* and *H. pterosthica*, all from Acrididae, which should be closely monitored for early prevention. Outbreak of species belonging to the Ensifera, namely those of Barbitistini, is an event rarely reported. However, these species may cause significant damage to grasslands and require monitoring. Global changes, especially climate warming, can affect these species in various ways, changing the outbreak probability, as well as the course of invasions and migrations. We can assume that most of these species will benefit from these changes to invade new areas, enlarge their range or increase their population sizes, at the disadvantage of humans. Such global changes may present opportunities for the pest species of Acrididae, both in the irrigated and arid agricultural plains of Mesopotamia and Anatolia. Scientific attention on the part of local authorities is essential not to face such disasters as experienced during the First World War.

Funding: This research received no external funding.

Institutional Review Board Statement: Not applicable.

Informed Consent Statement: Not applicable.

Data Availability Statement: Not applicable.

Acknowledgments: This review would not have been possible without invitation, encouraging and valuable contributions by Michel Lecoq. Numan Babaroğlu, Mehmet Çulcu and Emre Akci from Directorate of Plant Protection Central Research Institute (Ankara) spent a great time and effort to provide data on grasshopper management for the period 2013–2020. Onur Uluar and Özgül Yahyaoğlu assisted during preparation of the review. Designer Helin Çıplak Palabıyık (helin@wodern.com) prepared the maps. Hasan H. Başıbüyük reviewed the manuscript for linguistic editing. We are grateful to all for their valuable contributions.

Conflicts of Interest: The author declares no conflict of interest.

References

1. Song, H.; Foquet, B.; Mariño-Pérez, R.; Woller, D.A. Phylogeny of locusts and grasshoppers reveals complex evolution of density-dependent phenotypic plasticity. *Sci. Rep.* **2017**, *7*, 6606. [CrossRef]
2. Freeman, S.; Herron, J.C. *Evolutionary Analysis*, 4th ed.; Pearson Education Inc.: San Francisco, CA, USA, 2009; pp. 753–798, ISBN 978-0132275842.
3. Buj Buj, A. International experimentation and control of the locust plague: Africa in the first half of the 20th Century. In *Nature et Environnement*; Chatelin, Y., Bonneuil, C., Eds.; Les Sciences Hors d'Occident Au XXe Siècle: Paris, France, 1995; Volume 3, pp. 93–108, ISBN 2-7099-1-297-X.
4. Lecoq, M. *Locusts*; CIRAD: Montpellier, France, 2001. Available online: https://www.researchgate.net/publication/233996579 (accessed on 1 October 2020).
5. Peloquin, C. Locust swarms and the spatial techno-politics of the French resistance in World War II. *Geoforum* **2013**, *49*, 103–113. [CrossRef]
6. Cressman, K. Monitoring desert locusts in the Middle East: An overview. *Yale F&S Bull.* **2001**, *103*, 123–140.
7. Symmons, P.M.; Cressman, K. *Desert Locust Guidelines: 1. Biology and Behaviour*; Food and Agriculture Organization of the United Nations: Rome, Italy, 2001.
8. Balamir, S. Yakın doğuda çöl çekirgesi durumu [Status of Desert locust in the Near East]. *Bitki Koruma Bülteni* **1952**, *3*, 14–26.
9. Balamir, S. Türkiye'de Fas çekirgesi (*Dociostaurus maroccanus*) durumu [Status of Moroccan locust (*Dociostaurus maroccanus*) in Turkey]. *Bitki Koruma Bülteni* **1952**, *1*, 11–15.
10. Balamir, S. *Zararlı Çekirgeler ve Mücadele Metotları [Pest Locusts and Control Methods]*; Yıldız Matbaacılık: Ankara, Turkey, 1956.
11. Balamir, S. Türkiye'de son yıllarda zararı görülen bir çekirge türü: *Thisoicetrinus pterostichus* [A pest grasshopper seen in recent years in Turkey: *Thisoicetrinus pterostichus*]. *Bitki Koruma Bülteni* **1962**, *2*, 3–10.
12. Bücher, H. Die Heuschreckenplage und ihre Bekämpfung. *Monogr. Angew. Entomol.* **1918**, *3*, 1–274.
13. Erkılıç, S.S. Anadolu'da Fas çekirgesi ve istilaları [Moroccan locust outbreaks in Anatolia]. *Ziraat Derg.* **1943**, *4*, 44–45.
14. Erkılıç, S.S. *Türkiye'nin Zararlı Çekirgelerini Birbirbinden ve Benzerlerinden Ayırma Anahtarı [The Key to Species to Distinguish Locust from Each Other and from Other Grasshoppers]*; Orhanbey Matbaası: Istanbul, Turkey, 1945.
15. Galazebrook, O.A. Remarkable details from American Consul on Palestine locust plague. *The New York Times*, 21 November 1915; p. 1.
16. Karabağ, T. *Ankara Vilayeti Dahilinde Mevcut Çekirgelerin Ekolojik, Coğrafi ve Sistematik Durumları Üzerine Araştırmalar [Research on the Ecological, Geographical and Systematic Status of Grasshoppers in Ankara Province]*; Ankara Üniversitesi Ziraat Fakültesi Yayınları: Ankara, Turkey, 1949.
17. Uvarov, B.P. A revision of genus *Locusta*, L. (=*Pachytylus* Fieb.) with a new theory as to the periodicity and migrations of locusts. *Bull. Entomol. Res.* **1921**, *12*, 135–163. [CrossRef]
18. Uvarov, B.P. Ecological studies on the Moroccan locust in Western Anatolia. *Bull. Entomol. Res.* **1932**, *23*, 273–287. [CrossRef]
19. Balamir, S. *Türkiye'de Çöl Çekirgesi (Schistocerca gregaria Forsk.) Salgınları [Desert Locust (Schistocerca gregaria Fors.) Outbreaks in Turkey]*; Şark Matbaası: Ankara, Turkey, 1973.
20. Rainey, R.G.; Betts, E.; Lumley, A. The decline of the desert locust plague in the 1960s: Control operations or natural causes? *Philos. Trans. R. Soc. Lond. B* **1979**, *287*, 315–344.
21. Asan, H. Diyarbakır vilayeti'nde meydana gelen çekirge afetleri (1886–1919) [Grasshopper disasters in Diyarbakır Province (1886–1919)]. *Int. J. Hist.* **2017**, *9*, 1–18. [CrossRef]
22. Ayalon, Y. *Natural Disasters in the Ottoman Empire: Plague, Famine, and Other Misfortunes*; Cambridge University Press: New York, NY, USA, 2014; ISBN 978-1-107-07297-8.
23. Candeğer, Ü. Locust problem in the Ottoman State in World War One. *Int. J. Hum. Sci.* **2016**, *13*, 2125–2134. [CrossRef]
24. Foster, Z.J. The 1915 locust attack in Syria and Palestine and its role in the famine during the First World War. *Middle East. Stud.* **2015**, *51*, 370–394. [CrossRef]
25. Gökmen, E. Batı anadolu'da çekirge felâketi (1850–1915) [The locust disaster in West Anatolia (1850–1915)]. *Belleten* **2010**, *74*, 127–180.
26. Gönüllü, A.R. Konya Vilayeti'nde meydana gelen çekirge istilaları (1866–1917) [Locust invasions in Konya Province (1866–1917)]. *Tarihin Peşinde* **2012**, *4*, 8–22.
27. Erler, M.Y. Kıbrıs'ta çekirge istilası (1845–1869) [Locust invasion in Cyprus (1845–1869)]. *Gazi Univ. Kast. Egit. Derg.* **2002**, *10*, 195–204.
28. Özbilge, N. *Çekirgeler, Devlet ve Kürtler: Erken Cumhuriyet Dönemine Yeniden Bakmak [Grasshoppers, the State and the Kurds: A Re-Look at the Early Republican Period]*; Tarih Vakfı Yayınları: Istanbul, Turkey, 2020; ISBN 978-975-333-373-3.
29. Özer, S. *Anadolu'da Görülen Çekirge Istilaları ve Halk Üzerindeki Etkisi (1914–1945) [Grasshopper Invasions in Anatolia and Its Effect on the People (1914–1945)]*; Türk Tarih Kurumu: Ankara, Turkey, 2016; ISBN 9789751632470.
30. Temel, M.; Baş, H. Cumhuriyetin ilk yıllarında izlenen bitki ve orman sağlığını koruma politikası [Policy of plant and forest health protection in the early years of the Republic]. *Muğla Üniversitesi Sosyal Bilimler Enstitüsü Dergisi (İLKE)* **2008**, *20*, 165–176.
31. Yıldırım, M.A. Birinci Dünya Savaşı yıllarında Osmanlı Devleti'nin beşinci düşmanı: Çekirgeler [Fifth enemy of the Ottoman Empire during the First World War: Locusts]. *Gaziantep Univ. J. Soc. Sci.* **2014**, *13*, 1017–1042.
32. Yıldırım, M.A. Ondokuzuncu yüzyıl sonlarından itibaren Halep Vilayetinde çekirge istilalarına dair notlar [Notes on locust invasions in Aleppo Province from the late 19th century]. *Tar. Okulu Derg.* **2014**, *7*, 535–560.

33. Cressman, K. Desert locust. In *Biological and Environmental Hazards, Risks, and Disasters*; Shroder, J.F., Sivanpillai, R., Eds.; Elsevier: Amsterdam, The Netherlands, 2015; pp. 87–105, ISBN 9780123948472.
34. Joffe, S. *Desert Locust Management: A Time for Change*; Discussion Papers; World Bank: Washington, DC, USA, 1995.
35. Lecoq, M. Recent progress in Desert and Migratory Locust management in Africa. Are preventive actions possible? *J. Orthoptera Res.* **2001**, *10*, 277–291. [CrossRef]
36. Lecoq, M. Desert locust management: From ecology to anthropology. *J. Orthoptera Res.* **2005**, *14*, 179–186. [CrossRef]
37. Sword, G.A.; Lecoq, M.; Simpson, S.J. Phase polyphenism and preventative locust management. *J. Insect Physiol.* **2010**, *56*, 949–957. [CrossRef] [PubMed]
38. Van Huis, A.; Cressman, K.; Magor, J.I. Preventing desert locust plagues: Optimizing management interventions. *Entomol. Exp. Appl.* **2007**, *122*, 191–214. [CrossRef]
39. Zhang, L.; Lecoq, M.; Latchininsky, A.; Hunter, D. Locust and grasshopper management. *Annu. Rev. Entomol.* **2019**, *64*, 15–34. [CrossRef] [PubMed]
40. Latchininsky, A.V. Moroccan locust Dociostaurus maroccanus (Thunberg, 1815): A faunistic rarity or an important economic pest? *J. Insect Conserv.* **1998**, *2*, 167–178. [CrossRef]
41. Çıplak, B. The analogy between glacial cycles and global warming for the glacial relicts in a refugium: A biogeographic perspective for conservation of Anatolian Orthoptera. In *Insect Ecology and Conservation*; Fattorini, S., Ed.; Research Signpost: Kerela, India, 2008; pp. 135–163, ISBN 978-81-308-0297-8.
42. Çıplak, B.; Demirsoy, A.; Bozcuk, A.N. Malatya Caelifera (Orthoptera, insecta) faunası (Caelifera (Orthoptera, Insecta) fauna of Malatya district). *Turk. J. Zool.* **1996**, *20*, 17–31.
43. Karabağ, T. *Orthoptera Fauna of Turkey: A Synonymic and Distributional Catalogue of Turkish Orthoptera*; Ankara Üniversitesi Fen Fakültesi Yayınları: Ankara, Turkey, 1958.
44. Çıplak, B. *Ibradı, Manavgat ve Akseki Civarında Görülen fas Cekirgesi (Dociostaurus (Dociostaurus) Moroccanus Thungerg Epidemisi) [Outbreak of Moroccan Locust (Dociostaurus (Dociostaurus) Moroccanus Thungerg) around Ibradi, Manavgat and Akseki]*; Department of Biology, Akdeniz University, Unpublished Report Submitted to Local Plant Protection Department of Antalya Province: Antalya, Turkey, 1996.
45. Özkan, A.; Gündüz, G.; Çıplak, B.; Fışkın, K. Kimyasal mücadele uygulanmış *Dociostaurus maroccanus* epidemik populasyonundan alınan örneklerde antioksidan enzim aktiviteleri [Antioxidant enzyme activities in samples taken from the chemically treated outbreaking population of *Dociostaurus maroccanus*]. *Turk. J. Biol.* **2000**, *24*, 141–149.
46. Gümüşsuyu, İ. *Orta Anadolu Bölgesinde Kültür Bitkilerine Zarar Yapan Kara Çekirge (Melanogryllus Desertus Pall.)'nin (Orthoptera, Gryllidae) Biyo-Ekolojisi Üzerine Araştırmalar [Research on the Bio-Ecology of Black Cricket (Melanogryllus desertus Pall.) (Orthoptera, Gryllidae) that Damage Cultivated Plants in the Central Anatolian Region]*; Yenigün Matbaası: Ankara, Turkey, 1973.
47. Özbek, H.; Yıldırım, E. Şenkaya (Erzurum) Yaylasında salgın yapan çayır çekirgeleri (Tettigonidae: Orthoptera) [Meadow grasshoppers (Tettigonidae: Orthoptera) causing damage in Şenkaya (Erzurum) Plateau]. *Ekoloji* **1994**, *11*, 12–15.
48. Tutkun, E. *Ankara İli Dahilinde Mevcut Caliptamus Türlerinin Morfolojisi ve Biyoekolojisi Üzerine Incelemeler [Studies on the Morphology and Bioecology of Caliptamus Species in Ankara Province]*; Yenigün Matbaası: Ankara, Turkey, 1974.
49. Çıplak, B.; Demirsoy, A.; Bozcuk, A.N. Distribution of Orthoptera in relation to the Anatolian Diagonal in Turkey. *Articuluta* **1993**, *8*, 1–20.
50. Uvarov, B.P. *Grasshoppers and Locusts. A Handbook of General Acridology-v. 2: Behaviour, Ecology, Biogeography Population Dynamics*; Centre for Overseas Pest Research: London, UK, 1977.
51. Çağlar, S.S.; Karacaoğlu, Ç.; Kuyucu, A.C.; Sağlam, İ.K. Humidity and seasonality drives body size patterns in males of the bush cricket *Isophya rizeensis* Sevgili, 2003 (Orthoptera: Tettigoniidae: Phaneropterinae). *Insect Sci.* **2014**, *21*, 213–226. [CrossRef] [PubMed]
52. Meynard, C.; Gay, P.-E.; Lecoq, M.; Foucart, A.; Piou, C.; Chapuis, M.P. Climate-driven geographic distribution of the desert locust during recession periods: Subspecies' niche differentiation and relative risks under scenarios of climate change. *Glob. Chang. Biol.* **2017**, *23*, 4739–4749. [CrossRef] [PubMed]
53. Cannon, R.J.C. The implications of predicted climate change for insect pests in the UK, with emphasis on non-indigenous species. *Glob. Chang. Biol.* **1998**, *4*, 785–796. [CrossRef]
54. Musolin, D.L. Insects in a warmer world: Ecological, physiological and life-history responses of true bugs (Heteroptera) to climate change. *Glob. Chang. Biol.* **2007**, *13*, 1565–1585. [CrossRef]
55. Cullen, D.A.; Cease, A.J.; Latchininsky, A.V.; Ayali, A.; Berry, K.; Buhl, J.; De Keyser, R.; Foquet, B.; Hadrich, J.C.; Matheson, T.; et al. From molecules to management: Mechanisms and consequences of locust phase polyphenism. *Adv. Insect Physiol.* **2017**, *53*, 167–285. [CrossRef]
56. Latchininsky, A.V. Climate change and locusts: What to expect? *Proc. Russ. State Hydrometeorol. Univ.* **2017**, *46*, 134–143.
57. Popova, E.N.; Semenov, S.M.; Popov, I.O. Assessment of possible expansion of the climatic range of Italian locust (*Calliptamus italicus* L.) in Russia in the 21st century at simulated climate changes. *Russ. Meteorol. Hydrol.* **2016**, *41*, 213–217. [CrossRef]
58. Kiritani, K. Predicting impacts of global warming on population dynamics and distribution of arthropods in Japan. *Popul. Ecol.* **2006**, *48*, 5–12. [CrossRef]
59. Buckley, L.B.; Huey, R.B. Temperature extremes: Geographic patterns, recent changes, and implications for organismal vulnerabilities. *Glob. Chang. Biol.* **2016**, *22*, 3829–3842. [CrossRef]

60. Deutsch, C.A.; Tewksbury, J.J.; Huey, R.B.; Sheldon, K.S.; Ghalambor, C.K.; Haak, D.C.; Martin, P.R. Impacts of climate warming on terrestrial ectotherms across latitude. *Proc. Natl. Acad. Sci. USA* **2008**, *105*, 6668–6672. [CrossRef]
61. Berg, M.P.; Tobykiers, E.; Driessen, G.; van der Heijden, M.; Kooi, B.W.; Kuenen, F.; Liefting, M.; Verhoef, H.A.; Ellers, J. Adapt or disperse: Understanding species persistence in a changing world. *Glob. Chang. Biol.* **2010**, *16*, 587–598. [CrossRef]
62. Lionello, P. *The Climate of the Mediterranean Region: From the Past to the Future*; Elsevier: Amsterdam, The Netherlands, 2012; ISBN 9780124160422.
63. Hannah, L. *Climate Change Biology*; Academic Press: Cambridge, MA, USA, 2014; ISBN 978-0-124-20218-4.
64. Lionello, P.; Scarascia, L. The relation between climate change in the Mediterranean region and global warming. *Reg. Environ. Chang.* **2018**, *18*, 1481–1493. [CrossRef]
65. Chobanov, D.P.; Grzywacz, B.; Iorgu, I.S.; Çıplak, B.; Ilieva, M.B.; Warchalowska-Silwa, E. Review of the Balkan *Isophya* (Orthoptera: Phaneropteridae) with particular emphasis on the *Isophya modesta* group and remarks on the systematics of the genus based on morphological and acoustic data. *Zootaxa* **2013**, *3658*, 1–81. [CrossRef]
66. Chobanov, D.P.; Kaya, S.; Grzywaczc, B.; Warchałowska-Śliwa, E.; Çıplak, B. The Anatolio-Balkan phylogeographic fault: A snapshot from the genus *Isophya* (Orthoptera, Tettigoniidae). *Zool. Scr.* **2017**, *46*, 165–179. [CrossRef]
67. Kaya, S.; Çıplak, B.; Chobanov, D.; Heller, K.-G. *Poecilimon bosphoricus* group (Orthoptera, Phaneropterinae): Iteration of morpho-taxonomy by song characteristics. *Zootaxa* **2012**, *3225*, 1–71. [CrossRef]
68. Kaya, S.; Gündüz, İ.; Çıplak, B. Estimating effects of global warming from past range changes for cold demanding refugial taxa: A case study on South-west Anatolian species *Poecilimon Birandi*. *Biologia* **2012**, *67*, 1152–1164. [CrossRef]
69. Magor, J.I.; Lecoq, M.; Hunter, D.M. Preventive control and Desert Locust plagues. *Crop Prot.* **2008**, *27*, 1527–1533. [CrossRef]

agronomy

Review

State of the Art Management of the Central American Locust *Schistocerca piceifrons piceifrons* (Walker, 1870)

Ludivina Barrientos-Lozano [1,*], Hojun Song [2], Aurora Y. Rocha-Sánchez [1] and Jorge Ariel Torres-Castillo [3]

1 Tecnológico Nacional de México-I. T. Ciudad Victoria. Blvd. Emilio Portes Gil No.1301, C.P. 87010 Cd. Victoria, Tamaulipas, Mexico; auro_3110@hotmail.com
2 Department of Entomology, Texas A&M University, College Station, TX 77843, USA; hsong@tamu.edu
3 Ciudad Victoria-Instituto de Ecología Aplicada, Universidad Autónoma de Tamaulipas-Campus, División del Golfo 356, Col. Libertad, C.P. 87019 Cd.Victoria, Tamaulipas, Mexico; jorgearieltorres@hotmail.com
* Correspondence: ludivinab@yahoo.com

Abstract: The Central American locust (CAL), *Schistocerca piceifrons piceifrons* (Walker, 1870), is a transboundary pest that is distributed from Mexico to Panama. It is a true locust species characterized by density-dependent phase polyphenism. The ancient record of the CAL is found in the Popol Vuh, the Mayan sacred book, demonstrating how it has affected humans for millennia. In Mexico, the CAL have been declared a national threat to agriculture since 1824. Serious locust plagues occurred in 1882–1883 when swarms of 20 km^2 in size invaded the Yucatán Peninsula and neighboring states in southern Mexico and, since then, management actions to suppress populations and economic damage have been implemented. A better understanding of the biology, ecology, and behavior of the CAL replaced a manual and mechanical collection of locust swarms, hopper bands, and egg pods with modern techniques such as the use of safer chemical products and environmentally friendly bioinsecticides. Presently, biomodels and GIS support the monitoring and forecasting of outbreaks. Currently, studies are conducted to investigate environmental factors that trigger locust gregarization, the evolution of phase polyphenism, and CAL bioactive compounds and nutritional contents, envisioning its potential use in biotechnological industries. Findings will be crucial to improve the management strategies of the CAL.

Keywords: locust plagues; ecology; field monitoring; outbreaks management

Citation: Barrientos-Lozano, L.; Song, H.; Rocha-Sánchez, A.Y.; Torres-Castillo, J.A. State of the Art Management of the Central American Locust *Schistocerca piceifrons piceifrons* (Walker, 1870). *Agronomy* **2021**, *11*, 1024. https://doi.org/10.3390/agronomy11061024

Academic Editors: Michel Lecoq and Arianne Cease

Received: 21 April 2021
Accepted: 18 May 2021
Published: 21 May 2021

Publisher's Note: MDPI stays neutral with regard to jurisdictional claims in published maps and institutional affiliations.

1. Introduction

The Central American locust (CAL), *Schistocerca piceifrons piceifrons* (Walker, 1870), is a menacing pest that has affected humanity for millennia [1]. It is believed that climate changes and recurrent locust plagues could have caused the decline of the Mayan civilization in Mesoamerica in the period 740–930 A.D. [1–3]. It is a transboundary migratory pest that is distributed throughout Mexico and Central America, affecting at least ten countries during outbreaks [4]. It is a true locust species characterized by density-dependent phase polyphenism [1,4–8]. At low density, individuals are solitary, avoid each other, and are harmless, and this phenotype is called the "solitary phase"; when environmental conditions promote population density increase, they transform into gregarious and conspicuous locusts that can form large migrating groups of nymphs (bands) or adults (swarms), and this phenotype is called the "gregarious phase". Solitarious locusts differ from gregarious locusts not only in behavior, but also in color, morphology, and physiology [4,7–9]. Locust bands march in compact groups in the same direction and in a cohesive way, while swarms may contain millions of individuals that migrate long distances in a few days, causing severe damage to agriculture and livestock. The phenomenon that triggers polyphenism is not "unknown", but it must be considered a very complex multifactorial process involving different mechanisms [10].

Throughout its distribution, permanent reproduction and gregarization areas for the CAL are located at 100 masl or below, which are characterized by optimal climatic conditions (27 °C average annual temperature, and at least 1000 mm annual precipitation) [4,5,9,11,12]. In modern Mexico, the first official report on the CAL plague dates back to 1882–1883 when swarms of 20 km^2 in size invaded the Yucatán Peninsula and other states of southern Mexico [1–4,7]. Along with its history of massive reproduction and invasions, different management actions have been used to suppress infestations and economic damage [2–4,9]. A better understanding of the biology, ecology, and behavior of the CAL has gradually replaced the manual and mechanical collection of locust swarms, hopper bands, and egg pods performed during the 19th century with more appropriate management techniques [4,9,11–16]. At present, the use of biomodels and Geographic Information Systems Technology (GIS) support CAL monitoring and the forecasting of outbreaks [17–19]. During the most recent outbreak in 2018–2020, the National Service of Agri-food Health, Safety and Quality (SENASICA) introduced the potential use of unmanned aerial vehicles (UAVs) or drones for the monitoring of locust populations in addition to land monitoring [20,21]. Currently, several research lines elucidate on the evolution of phase polyphenism, the environmental factors that trigger gregarization and swarm formation, the bioactive compounds and nutritional contents of the CAL, and the improvement of an early warning system (EWS) [8,15,19,21–26]. Hopefully this work will provide novel information and knowledge that helps to improve the management and control strategies of the CAL.

This work aims to provide a general view on the current knowledge of the CAL, management strategies, present research, and future work.

2. Life History of the Central American Locust

It is well established that in continental Mexico and Central America, the CAL exhibit two generations per year [4,11–13,23,27–29]. Interestingly, only one generation per year is reported to occur in Socorro Island, Archipelago Revillagigedo by the Mexican Pacific Coast, 480 km southwest of Baja California, Mexico where a locust invasion was reported in 2006 [5,30]. In this review, we will refer to the life history of the CAL populations that occur in continental Mexico and Central America. In these regions, the second generation overwinters as sexually immature adults from November to April–May, undergoing a reproductive diapause. This physiological state is a response to the onset of the autumn-winter dry season that brings about a decrease in temperature and shorter days [4,9,23,27,28]. Adults remain sexually immature until the onset of the rainy season around mid-April. As climatic conditions improve, and when environmental conditions become suitable for reproduction, adults of the second generation become sexually mature and start mating and laying eggs from May through June. Sexually mature adults turn yellow and mate en masse on bare sandy-clay soils, where each female lays one to four egg pods at a depth of approximately 3 to 10 cm. Each egg pod may contain 40–75 eggs based on field observations [29]. These eggs will hatch from mid-May through June and give rise to the first generation. First-generation nymphs are present from May-June through July, and adults from mid-July to the end of September. The first generation develops very rapidly, given the favorable climatic conditions (27 °C average temperature, 50–80% relative humidity, and 13L: 11D period, as well as abundant and lush vegetation for feeding); it may take 55–60 days to progress from egg through to sexually mature adult. Mating and oviposition occur from mid-September through October. Second-generation nymphs may occur from mid-September through mid-December. By the end of November, 70–80% of the second-generation nymphs will have reached the adult stage and will remain sexually immature. There is an overlap between second-generation adults and nymphs of the first generation, and also between adults of the first generation and second-generation nymphs. The second generation has a longer duration, 155–180 days since the adult enters diapause in response to a drier season, lower temperature, and shorter day length that occurs in the region from November through April [4,7,9,11,12,17,20,23,27,28]. Under these unfavorable environmental conditions, the CAL may not resume development, though relative humidity occasionally may be high at 70–85% [4,9,27–29]. If the population density is high (>25–30 adults/m^2) and environmental conditions become suitable, four to six weeks after reaching the adult stage, scattered populations may join to form swarms ready for migration. The first swarms in the region are observed and located from the second half of December onwards. The period, from December through April, is appropriate to undertake control measures since adult populations of the CAL remain sexually immature and in large aggregations (up to 5000 adults/m^2) [9]. When the first rains begin in April–May, the adults then enter the process of sexual maturation and changes in their behavior and color continue to occur. The solitary young adult is brown with light-brown tones, and has a great capacity for flight over short distances of 15–20 m when disturbed; while the gregarious young adult is brown with dark brown tones and moves in a cohesive manner in large groups or swarms. As they reach sexual maturity, this coloration disappears and turns yellow, observed first in males and later in females, and is more conspicuous in gregarious than in solitary populations [4,9,27,28]. During this period, large numbers of locusts are observed flying against the wind, which helps to increase their body temperature and strengthen their flight muscles [4,9,11].

3. Habitat and Ecology

The CAL is a polyphagous species throughout its distribution area. It has been recorded to feed on up to 400 plant species, feeding on both cropped and grazing areas, including dry vegetation [4,9,27,28]. Feeding ecology of the CAL is not fully understood, but a recent study on the effect of plant volatiles and feeding preference of the CAL showed that the CAL in the Yucatán Peninsula was highly attracted to the odor of *Pisonia aculeata* L., which it uses as a refuge but not for feeding [22]. As for the feeding preference, the study revealed that the CAL showed marked feeding preference for two native plant species, *Leucaena glauca* Benth and *Waltheria americana* L. [22]. However, it is not clear if this observation can be applied to other regions where the CAL is present. In Mexico and Central America, the breeding areas for the CAL are cultivated and natural grasslands that have replaced the low and thorny forests. For instance, in the coastal plain of the Gulf of Mexico and the Yucatán Peninsula, 80% of the areas of reproduction and gregarization occur in these agroecosystems [19]. The habitat of solitarious populations is characterized by grassland and vegetation patches that do not exceed 2.5 m in height [4]. Its distribution depends upon the vegetation pattern and, if the vegetation is homogeneous, locust populations may remain scattered. However, anthropogenic activities, such as burning or abandonment of cultivated areas, and environmental factors (convergent winds, droughts, flooding, and food shortage) may lead to gregarization and swarm formation [4,19,28,29]. In recent years, changes in land use through the conversion of forested areas into agricultural areas have aggravated the locust problem, since the area suitable for reproduction and gregarization has increased. For example, in the Yucatán Peninsula, which represents the most important breeding area in Mexico, the agricultural area increased from 394,236 ha in 1981 to 579,643 ha in 2014; in a 30 year period, 47% of the forested area was converted to livestock activities [31]. Recent studies found that the burning of grassland and agricultural areas that are ideal for feeding, reproduction, and oviposition of the CAL are associated with the emissions of methane (CH4) and carbon dioxide (CO2), and these greenhouse gases increase soil temperature up to 3 °C which favors locust breeding and gregarization, as occurs in northern Mexico, north of the Huasteca Potosina and southern Tamaulipas, where locust populations have increased [19,32].

4. Permanent Breeding Areas

Solitarious populations of the CAL reproduce and develop in permanent breeding areas located in the lowlands that provide suitable climatic conditions for reproduction, uncovered land or with little vegetation cover, and close to cultivated areas. The invasion area extends south of the Tropic of Cancer (23°26′), at no more than 2000 masl and 700–2500 mm annual precipitation [28,33,34]. The onset of the spring rains is critical to induce sexual maturation, mating, and oviposition of second-generation adults [4,9,16,28,33,34]. Four permanent breeding areas have been recognized [4,11,13] (Figure 1): (1) the Yucatán Peninsula in Mexico; (2) the Valley of Rio Aguan in Northern Honduras; (3) the Gulf of Fonseca in El Salvador, Honduras, and Nicaragua; and (4) the Province of Guanacaste in Costa Rica. Recently, four additional breeding areas have been identified in Mexico [34,35] (Figure 1): (5) the State of Veracruz, municipalities of Medellín, Boca del Río, Alvarado, Tlalixcoyan, and Tierra Blanca, which is second in importance after the Yucatán Peninsula; (6) Chahuites-Tepanatepec, located on the limits of Oaxaca and Chiapas, where the CAL becomes gregarious, causing invasions into the Isthmus of Tehuantepec and other localities in the state of Oaxaca; (7) the border of San Luis Potosí-Tamaulipas states, in the Valleys of the Sierra Nahola, the smallest of the four additional permanent breeding areas, which currently has given rise to serious invasions into the Huasteca region (East of San Luis Potosi, South Tamaulipas, and North of Veracruz); and (8) Tabasco, located on the banks of the Usumacinta River on the border with Guatemala, where it is considered that environmental conditions can lead to the development of CAL populations.

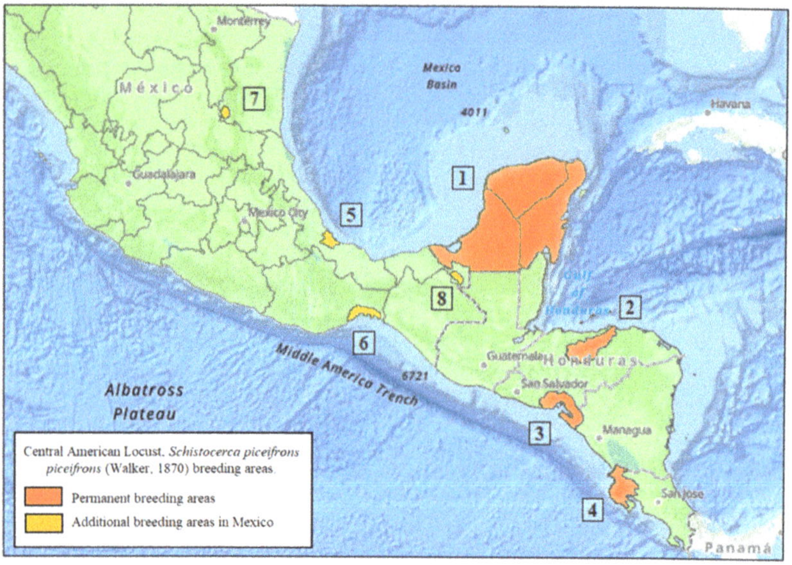

Figure 1. Central American locust, *Schistocerca piceifrons piceifrons* (Walker, 1870), permanent breeding areas: (1) Yucatán Península, México, (2) Valley of Rio Aguan, Northern Honduras, (3) Gulf of Fonseca in El Salvador, Honduras, and Nicaragua, (4) Province of Guanacaste, Costa Rica, (5) State of Veracruz, México, (6) Chahuites-Tepanatepec, on the limits of Oaxaca and Chiapas, Mexico, (7) Border of San Luis Potosí-Tamaulipas states, México, and (8) Tabasco, on the banks of the Usumacinta River, border with Guatemala.

Seasonal migration between habitats does not seem to be part of the life-cycle of the solitarious phase, although conclusive studies have not been made. Plagues occur when swarms form and migrate outside of the permanent breeding sites. The ecological factors that cause outbreaks are poorly understood; nonetheless, the extent of a suitable habitat created by anthropogenic activities plays a major role [4,9,11–13,16,19]. Several works have associated CAL outbreaks with the event of El Niño Southern Oscillation (ENSO) (a drought period followed by abundant rain) and La Niña [19,36]. Recent studies found that solitarious populations of the CAL in the Yucatán Peninsula are high in the rainy season and correlated with the abundance of the grass *Panicum maximum* Jacq. and precipitation [37]. Indeed, precipitation is one of the most influential factors in determining locust activities, such as mating and oviposition periods. Humidity is required in the soil to ensure the hatching of the eggs and to satisfy the water requirements of nymphal development [4,9,11–13,16,17,23,38]. Regarding the CAL plagues and invasions, the coastal lowlands of Mexico have been invaded as far north as the Tropic of Cancer by swarms originating in the Yucatán Peninsula or the Pacific Coast of Guatemala [13,39]. In 1923, a CAL plague invaded the state of Veracruz [13,40]. Swarms occupied the hot coastal plain (Tierra Caliente) where they bred. However, adults moved to the west, up the valleys towards the cooler highlands where they bred, but hoppers perished because of the cool weather. Prevailing winds and the occurrence of rain all contributed to determining the direction of swarm movements. It has been shown that, in the Yucatán Peninsula, the migration direction of the first-generation swarms is northeast and of the second-generation swarms is southwest [37].

5. Band and Swarm Formation

Solitarious nymphs of the CAL are green or tan in color, avoid encounters with other individuals, and cause no economic damage. As the local population density increases due to favorable environmental conditions, they become gregarious and attracted to each other. Soon they develop a pink to orange background color with intense black patterns (Figure 2) [4–6,9,41].

Figure 2. Central American locust, *Schistocerca piceifrons piceifrons* (Walker, 1870), solitarious (**a**,**b**) and gregarious (**c**) nymphs showing different color patterns.

Prolonged high-density conditions may result in an exponential population growth, which can lead to the formation of bands consisting of millions of individuals that march in the same direction in a cohesive manner [4,5,9,11–13,41]. Displacement of locust bands depends upon developmental stages and environmental factors; the first and second nymphal instars of the gregarious phase may remain on the host plant for a few days, but older nymphal instars (third-sixth) form bands and may displace 15–50 m per day [4,11,28]. It is reported that the black pigmentation allows the insects to be more efficient at absorbing solar radiation, raising their body temperature and metabolic activity; this is important during the winter (diapause stage) [4,11]. Other adaptive significance of the black patterns, such as aposematic coloration, which was shown in the desert locust [42], has not yet been demonstrated in the CAL. Gregarization is also associated with drastic changes in the insect's habitat or unusual climatic changes, such as convergent air currents. If the population density is high (>5 nymphs/m^2) and harvesting, burning of crop remnants, or soil preparation activities are performed, the nymphs gather and form compact groups (bands) on patches of vegetation nearby the cultivated land [9,29,41]. Upon reaching the adult stage, these small groups, which can vary in size from a few m^2 to several hectares, join with others and form swarms. This behavior may be observed at the end of crop harvest in the spring-summer cycle. At this stage, there is an opportunity to undertake control measures since it is easier to deal with small areas already located than large areas of land when the population is dispersed [4,9,16,27–29]. Further research is necessary to understand the gregarization mechanism and factors involved in the displacement of locust bands and swarms.

6. Phase Polyphenism, Gregarization, and Migration

The CAL is a true locust species that shows an extreme form of density-dependent phenotypic plasticity [4–6,9,16,27–29,41]. While the proximate mechanisms of phase transformation in the CAL are less understood than its congener *S. gregaria*, recent laboratory research has shown that the rearing density has a dramatic impact in resulting phenotypes including color, morphology, and behavior. Similar to other locust species, the solitarious CAL avoid each other, are inconspicuous, and harmless, while the gregarious CAL are attracted to each other, are conspicuously colored, and form large aggregations of nymphs (bands; Figure 3) or adults (swarms; Figure 4) that migrate long distances causing severe damage to agriculture and humankind [4–6,11–13,27–29,41].

Figure 3. Central American locust, *Schistocerca piceifrons piceifrons* (Walker, 1870), 1st (**a**) and 6th (**b**) instar bands.

Figure 4. Central American locust, *Schistocerca piceifrons piceifrons* (Walker, 1870), swarms.

It is unclear at this point what the relative importance of tactile, olfactory, and visual stimuli is in detecting changes in density in the CAL, but the preliminary laboratory data seem to suggest that the proximate mechanisms of phase transformation may be slightly different from what is known in the desert locust [43]. For example, the relative rates of gregarization and solitarization are different from the desert locust (Foquet et al., unpublished). This is not surprising given the fact that the CAL has evolved phase polyphenism independently from the desert locust [44]. The molecular basis of phase polyphenism in the CAL is currently being studied using transcriptomics and RNA interference. Phenotypic plasticity is often considered to be an adaptation to heterogeneous environment [45], but whether the phase polyphenism in the CAL is adaptive or not has not yet been formally studied. Currently, it is not known whether the conspicuous coloration of the gregarious nymphs is aposematic, because it is frequently observed in the field that vertebrate predators readily consume gregarious nymphs. However, it is possible that these insects may prefer to feed on toxin plants at the early stage of gregarization, which will provide a certain level of chemical protection. This is an untested hypothesis, but a previous study on the desert locust demonstrated that at the early stage of gregarization, the nymphs would preferentially feed on toxic plants, which makes their yellow and black coloration aposematic [42]. Other adaptive features of the nymphal coloration, such as thermoregulation or heightened immunity [45], have not yet been studied in the CAL [Song, pers. comm.]. Likewise, the adaptive nature of cryptic coloration and solitary behaviors associated with the solitarious phase have not been tested, but it would be quite easy to imagine that they would be adaptive.

The solitarious nymphs in the field setting are often green with small black dots and markings on the pronotum, abdomen, and hind legs, while the gregarious nymphs have a pink or peach background color with an extensive black pattern on the head, pronotum, abdomen, pads, and legs (Figures 2 and 3) [4,5]. Newly fledged adults are brown to pinkish in color, with abundant light black marks on the tegmina. Sexually immature adults in the solitarious phase are brown with light-brown tones and have a great capacity for flight. Sexually immature adults in the gregarious phase exhibit a darker pink color with darker marks on the tegmina and turn bright yellow on sexual maturity, the latter being observed first in males and later in females (Figure 5) [4,5,9,11,12,27–29,41].

Figure 5. Central American locust, *Schistocerca piceifrons piceifrons* (Walker, 1870), sexually immature adults in the solitarious (**a**,**b**) and gregarious phase (**c**); sexually mature adult in gregarious phase showing bright yellow color (**d**).

Sexual maturity is closely related to precipitation and relative humidity, with second-generation adults beginning this process with the onset of rains in April–May; during the next three to four months, thousands of yellow adult couples will mate and lay eggs simultaneously in bare, rich-clay soils [4,5,9,11,12,27–29,41]. Under laboratory conditions, a similar pattern has been observed, with small differences [4,9,11,12,27–29]: rearing density affected the hopper coloration greatly. Those reared in crowds invariably had a heavy black pattern on a pink or peach ground color; those reared in isolation had little or no black pattern and the ground color was green, brown, pink, or straw. The type of food also affected the coloration of the isolated hoppers, but had no effect on the crowded specimens. The color of fledgling adults was not correlated with nymphal coloration; all had a rather somber, brown appearance. Sexually mature males kept in crowds became predominantly yellow, but crowded females developed only a little yellow coloration, and isolated adults developed none at all upon maturity. In captivity, density also affected the length of the pre-oviposition period, which averaged 36 days from fledging to the production of the first egg pod by females kept in crowds and 61 days for those reared in isolation. Crowded adult females averaged six egg pods each, with 70 eggs per pod; while isolated females averaged eight pods, with 62 eggs per pod. Egg pods were produced at 6- to 7-day intervals, regardless of density. The incubation period was 18–19 days at 32 °C and 25 days at 28 °C. It is documented that adults of the CAL strengthen their flight muscles four to six weeks after reaching the adult stage, and if population density is high (>10–15/m^2) and environmental factors suitable, scattered populations join forming swarms that start migrating [4,9,16]. Swarm migration occurs mainly during the day, but occasionally swarms have been seen flying at night when the temperature is high (25 °C ± 2 °C) and on light moon nights. Second-generation swarms are observed and located in the region from the second half of December onwards [4,9,27–29]. During this stage, it is advisable to undertake all possible control actions since adults have not mated or oviposited and remain at very high population densities (up to 5000 adults/m^2) in large aggregations (Figure 4).

Large numbers of locusts are frequently observed flying against the wind to increase their body temperature [4,9,28]. In regard to environmental factors, such as temperature, precipitation, day length, distribution of vegetation patterns, and land use, it is well established that all play a major role in influencing and inducing gregarization and phase transformation [4,9,13,16,18,19,23,27,28]. Recent work showed that in the Yucatán peninsula, which represents the most important breeding area in Mexico, environmental factors associated with the population density of the CAL are, in order of importance, the relative species density of vegetation, the isotherm and isohyets, maximum precipitation, temperature, and land use; the population density of the CAL is influenced primarily by the abundance of the grass *P. maximum* and precipitation [23].

7. Central American Locust Outbreaks

The CAL, like its congeners, has caused fear and destruction throughout its distribution area for centuries. In Mexico, the most ancient records come from pre-Hispanic times [1–3,14] and the locust swarms are believed to be one of the causes that led to the decline of the Maya-Quiché cultures in the Yucatán Peninsula and the Toltec in the Anahuac Valley. The Mayan sacred book, the Popol-Vuh, mentions that the ancient people migrated out of hunger brought about by the agricultural losses inflicted by the locust plagues. During the colonial time, written records on locust plagues date back to the seventeenth century, when the first invasion in the territory currently occupied by the Yucatán Peninsula, Tabasco, Chiapas, Oaxaca, and Veracruz, was reported [38,39,46,47]. In Mexico, the CAL was declared a national pest in 1824 and provisions were established to fight off bands and swarms [47]. In 1924–1926, there was a great invasion and, since then, the presence of locust plagues has occurred without interruption in one or another part of its distribution area, especially in the Yucatán Peninsula and Central America [4,9,14,19,21]. The National Plant Protection Service in Mexico created a permanent locust control campaign in 1949 [48,49]. The OIRSA (Organismo Internacional Regional de Sanidad Agropecuaria) was created in 1947 as the CICLA (Comité Internacional Contra la Langosta), and in 1955 the CAL was declared a transboundary pest that required the coordinated efforts of Central American countries to deal with it. At present, the OIRSA represents the Central American anti-locust organization and continues to be in charge of coordinating the regional control efforts, funding anti-CAL campaigns, training, and providing information on the current locust situation [21,49,50]. From 2016 to date, there have been significant population increases of the CAL in the OIRSA Region. For example, in Nicaragua in 2016, locust swarms were controlled by the Institute for Agricultural Protection and Health (IPSA). There were outbreaks in the Yucatán Peninsula, Mexico (2014–2020), El Salvador (2016 and 2018) and, at present, after a long recession period, outbreaks were registered in Belize. Significant populations of the CAL have been reported in Guatemala, suggesting that there are factors that favor the development of this pest. On 2 July 2020, OIRSA issued a regional alert for the prevention of the CAL swarms [50].

8. Central American Locust Management Strategies

Considerable progress has been made on the management of the CAL during recession and outbreak periods [9,16–19,36,51–62]. Management techniques have evolved from the manual and mechanical collection of egg pods, hopper bands, and adult swarms [1–4,14] to the use of prediction models that aim to understand how population dynamics of the CAL relate to environmental factors and local conditions. For instance, a study [36] on the relationship between sea surface temperatures, the El Niño Southern Oscillation (ENSO), and the potential of the CAL plagues found a 72% coincidence between years of massive CAL attacks and ENSO years in the North Pacific of Costa Rica; an increase of environmental temperature and irregular distribution of precipitation, as occurs during the ENSO years, may result in vigorous copulation and oviposition in the CAL. A population increase is the first step for phase change and eventual plagues. A subsequent work assessed the risk of next-generation nymph-populations, considering an initial adult population and each instar mortality rate [51]. In another study, a Thermal Time Locust Development Clock (TTLDC) was built considering average values of temperature, day length, diapause cool hours, and calendar dates; this model helped to predict the CAL timing of phenological stages. This model implies a starting date (Bio-fix) established experimentally after quality data on population dynamics and the life cycle of the CAL. When the model was compared with field observations, the differences were relatively small. The reliability of predictions depends upon the accuracy of the starting date and deviations in weather conditions from the average year [17]. More recently, an early warning system (EWS) for monitoring and management of the CAL habitat in permanent breeding areas in the Huasteca region (Mexican Coastal Plain) and the Yucatán Peninsula was implemented by SENASICA-UASLP (Servicio Nacional de Sanidad, Inocuidad y Calidad Agroalimentaria-Dirección General de Sanidad Vegetal-Universidad Autónoma de San Luis Potosí). This EWS applies multi-criteria models and NOAA-AVHRR satellite images [15,19,32]. The model is based on two variables of the meteorological mesoscale: the presence of the ENSO and drought monitoring; it also takes the CAL behavior into account. Accordingly, the process of band or swarm formation begins when solitarious populations face environmentally unfavorable conditions (grass or crop remnant burning, convergent winds, droughts, or food shortage) and converge in places where vegetation is available. The increase in population density of individuals per unit area leads the insect to gregarize and prepare to migrate in search of food and oviposition sites. Hence, climatic effects are among the fundamental cause of behavioral change, and the host distribution pattern reflects spatially the areas of gregarization and oviposition (greater stress), and the areas of migration in search of food (higher biomass production in agricultural areas). Currently, this EWS proposes that not only the ENSO, but also the La Niña, affects locust breeding and gregarization. Furthermore, there is evidence that as a consequence of the increase of greenhouse gas emissions such as CH_4 and CO_2 in gregarious areas, in the last four years (2017–2020), the drought monitor showed an increase of more than 3 °C in soil temperature in areas proximate to the Gulf of Mexico. These greenhouse gas emissions (CH_4 and CO_2) are produced by anthropogenic livestock activity, which creates more favorable environmental conditions for the reproduction of the CAL [32]. An additional dynamic simulation model [18] showed that the growth and development of the CAL in the Yucatán Peninsula was associated with daily rainfall, temperature, and physical soil properties, such as texture and depth. In regard to the growth of non-cultivated grass in breeding zones and oviposition rates, the model estimated both variables as a function of soil moisture. The latest work [23] focused on a deeper analysis of the environmental factors that affect population dynamics of the CAL and it was found that plant species richness (PSR) and relative species density (RSD) in the Yucatán Peninsula were higher during the rainy season than in the dry season, with RSD the most important variable associated with locust density, followed by isotherm and isohyets, maximum precipitation and temperature, and land use. Locust density was positively correlated with the abundance of the grass *P. maximum*. Therefore, surveys for early detection and control of the CAL on the Yucatán Peninsula may focus on areas with the grass *P. maximum* to predict risk areas and target survey efforts. In the most recent outbreak (2014–2020) in the Yucatán Peninsula, the possibility of using drones or unmanned aerial vehicles (UAVs) for locust surveying was introduced [49,50,52]. Hopefully, the Mexican National Plant Protection Service and the OIRSA will have trained technicians in the near future to incorporate this tool in the surveying of locust populations and into the CAL campaigns. All these efforts have provided novel information and additional techniques for the CAL surveying and management. However, future research integrating multiple disciplines is needed to obtain finer data on the ecological factors and their correlation with the population dynamics of the CAL. Moreover, all information generated must be incorporated into the CAL management strategy since, at the moment, the use of the available techniques and tools is limited.

9. Control Measures

Control campaigns in Mexico and Central America aim to prevent swarm formation or migration by undertaking timely control of locust bands and swarms through the continuous surveying and monitoring of breeding areas in the Mexican Coastal Plain, southern Mexico, and six Central American countries [4,7,11–13,19–21,27,34,35,47]. This preventive control strategy is occasionally halted due to a lack of economic resources, shortage of personnel, or climatic conditions. In addition, deforestation, grassland burning, and livestock activity have intensified in recent years, increasing the area with suitable conditions for reproduction and growth of the CAL [22,23,41,49]. Up to the present, chemical control is perhaps the main alternative during outbreaks. Previous to 1998, most control operations against the CAL were carried out using methyl parathion. However, beginning in 1999–2000, chemical control was diversified with the validation and introduction of additional products such as phenylpyrazole, pyrethroids, insect growth regulators, and, above all, bioinsecticides [16,28,50,52–62]. From 2000–2010, a successful biological control program was established. This program employed searching for local strains of entomopathogenic fungi, screening for the most pathogenic and virulent strains, development and formulation of a biological insecticide supported in laboratory and field trials, and the transference of technology to the National Plant Protection Service to produce the bioinsecticide [16,28,53–62]. *Metarhizium anisopliae acridum*, a native isolate, is produced at present by the National Plant Protection Service in two state-run laboratories in Guanajuato and Yucatán, and a privately-owned business in Puebla [16,28,53–62]. Field trials in the Huasteca region and the Yucatán Peninsula (4×10^{12} conidia per hectare) in a mineral oil formulation provided up to 90% mortality in swarms following applications of the pathogen. At present, this biological insecticide is used to suppress locust bands to prevent swarm formation and, during the winter season, it is used to suppress locust swarms over grassland areas. In 2010, the fungus was applied to over 4000 ha of locust at a cost of approximately USD 10 per hectare [60]. Neem extracts (*Azadirachta indica* Juss) are recommended by OIRSA as a viable alternative to suppress nymph and adult populations of the CAL that invade forest or urban areas. It is reported that at a dose of 1 L/ha applied in 1–2 L of vegetable oil, neem extracts cause 67% mortality after 48 h and 100% after 7 days [34,50].

Safer chemical products, bioinsecticides, and neem extracts, as well as the introduction of new technologies, have been integrated into the locust preventive management strategy, making the Integrated Pest Management approach, adopted in 2005, possible [61,62].

10. Current Research

There are several ongoing research lines aiming to have a better understanding of major behavioral, ecological, environmental, and evolutionary factors that favor CAL reproduction, growth, gregarization, migration, and phase change. The main goal is to provide a multidisciplinary approach that strengthens and deepens current knowledge on the CAL. This may help improve preventive strategies and the suppression of locust plagues during outbreaks, as well as generate alternative management practices. Laboratory and field studies are performed to understand the molecular mechanisms that sustain the behavioral and physical changes in the CAL and how the environment influences phase polyphenism [8]. Research using remote sensing technology is being conducted to gain insight into the environmental factors that determine the CAL distribution and trigger concentration, reproduction, and gregarization [19,32,46]. This project encompasses the Huasteca Region and the Yucatán Peninsula. To date, an early warning system (EWS) that provides timely information on possible concentration and reproduction areas has been developed. Currently, it has been found that the CAL is distributed in areas of high CH_4 and CO_2 emissions north of the Huasteca Potosina and South Tamaulipas [19,32]. Livestock activity and the burning of grasslands are among the major variables that are modifying the CAL habitat. Additional research is being conducted to study the nutritional contents and bioactive compound levels of the CAL [24–26], aiming to incorporate this information into its management. The presence of phenolic compounds, alkaloids, tannins, saponins, flavonoids, and quantity of antioxidants against the DPPH (2, 2-diphenyl-1-picrylhydrazyl) and ABTS (2, 2′-azino-bis, 3-ethylbenzothiazoline-6-sulfonic acid) free radicals was reported. Proximate analysis showed that *S. p. piceifrons* has a high protein content (80.26%), low-fat content (6.21%), and fiber content (12.56%) similar to other Orthoptera species. The chitin and chitosan contents of the CAL were 11.88 and 9.11%, respectively, and the recovery percentage of chitosan from chitin was 76.71%. Among the Orthoptera, the protein content of the CAL is among the highest while its contents of chitin and chitosan are similar to those of other insect species, e.g., *Bombyx mori* Linnaeus (Lepidoptera: Bombycidae). Results suggest that this locust species is a potential source of bioactive compounds of biotechnological interest for use by pharmaceutical and food industries. A more recent work [26] assessed the digestibility of proteins of the CAL and the content of antioxidants from the peptides resulting from proteolysis. It was found that digestion with trypsin allowed the recovery of 0.68 and 0.055 mM TE/mg of protein, with a digestion time of 1 and 3 h, respectively. This work documents the potential of the CAL as a source of protein for the food industry and provides additional evidence on the potential use of this natural resource. Studies on the inhibition of intestinal proteases by plant species such as the fruit of hardy parlor palms (*Chamaedorea radicalis* Mart.) and its effect on the intestinal proteolytic activity of adult Orthoptera, the CAL and *Pterophylla beltrani* Bolívar and Bolívar, provided encouraging results [63]. The intestinal proteolytic activity of both orthopteran species was related to serine and cysteine proteases. Total proteolytic activity was 98,930.56 ± 7544.46 PAU/g (proteolytic activity units per gram) of intestinal tissue in the CAL and 52,847.22 ± 8509.11 PAU/g in *P. beltrani*. Fruit of hardy parlor palms is considered a strong candidate in future studies for identification and characterization of the inhibitors and their potential use in biological control.

These innovative research topics will enhance our understanding of the CAL plagues and may contribute substantially to the development of more effective and sustainable management mechanisms.

11. Future Work

In the last 20 years, a large amount of information has been generated to improve the management of the CAL. The integration of a bioinsecticide and the diversification of chemical products into the CAL campaign are, perhaps, among the most outstanding, for two reasons: (a) the development or introduction of these products allowed the opportunity of an integrated pest management approach for solitarious populations that might be of risk and improved management of locust outbreaks; and (b) the environmental impact caused by locust operations was considerably reduced. However, it is necessary to have a better understanding of the CAL population dynamics in relation to environmental factors, physical and physiological traits that promote gregarization, and swarm formation, its nutritional ecology, and new technologies that will contribute to a more efficient and effective management of this pest. Taking these remarks into account, some aspects for future research are outlined below.

To date, the proximate mechanisms of phase change are relatively unknown. However, we are in the beginning stage of understanding locust phase polyphenism in the CAL better. A long-term laboratory colony was established, which allows the possibility of manipulative experiments that can yield detailed understanding of the phase change in the CAL. Currently, the effect of rearing density on behavior, color, and morphology has been studied. Specifically, the behavioral polyphenism has been quantified using the same behavioral assay used in the desert locust. Density-specific transcriptomes from the CAL were generated and the number of differentially expressed genes between the density conditions were identified. The whole genome sequencing with chromosome-level assembly is currently underway for the CAL. These new developments will allow us to have a better understanding of density-dependent phase polyphenism in the CAL.

The environmental factors that affect the population dynamics of the CAL have not been sufficiently studied. It is necessary to carry out more in depth and systematic studies in critical areas where the CAL lives and reproduces. Environmental information generated (climatic variables: temperature, precipitation, relative humidity, photoperiod, soil types, vegetation types and its distribution patterns; and cultural practices such as the burning of grasslands and crop residues, and radiation, among others) must be integrated into biomodels that consider the largest possible number of environmental variables and their relationship with the presence of this pest. This will provide reliable information for locust surveying, monitoring, and the prediction of locust outbreaks to allow for timely and effective decisions.

The use of new technologies, such as remote sensing and unmanned aerial vehicles (UAVs), for the monitoring of sensitive areas is equally important. As is the training of field technicians in the use and management of these and other new technologies for surveying the CAL habitat and populations that might generate risky conditions.

Studies on the nutritional ecology vs. the population dynamics of the CAL will provide useful information on its distribution and habitat preferences; and this information will be a useful tool in its management.

The potential effect of climate change and land use or cover change on the CAL biology and outbreak frequency needs to be addressed. This is a promising research field that may provide insight into the development of gregarious populations, density-dependent phase polyphenism, and migration [64].

Funding: This research received no external funding.

Institutional Review Board Statement: Not applicable.

Informed Consent Statement: Not applicable.

Data Availability Statement: Not applicable.

Acknowledgments: To Tecnológico Nacional de México-I. T. Cd. Victoria for providing facilities to undertake this work. Special thanks to Michel Lecoq for the invitation to write this review.

Conflicts of Interest: The authors declare no conflict of interest.

References

1. Flores-Granados, F. Las Plagas de Langosta en el Área Maya: Ambiente e Historia de Una Calamidad en la Época Prehispánica. *Península* **2011**, *6*, 27–46. Available online: https://www.redalyc.org/pdf/3583/358333208002.pdf (accessed on 10 January 2021).
2. García-Quintanilla, A. La Langosta, los Mayas y el Colonialismo en Yucatán, México, 1883. *Relaciones* **2012**, *129*, 215–249. Available online: http://www.scielo.org.mx/pdf/rz/v33n129/v33n129a7.pdf (accessed on 10 January 2021).
3. García-Quintanilla, A. *Saak'* y el Retorno del Fin del Mundo: La Plaga de Langosta en las Profecías del Katun 13 Ahau. *Ancient Mesoamerica* **2005**, *16*, 327–344. Available online: https://www.jstor.org/stable/26309188?seq=1 (accessed on 10 January 2021). [CrossRef]
4. Barrientos-Lozano, L.; Astacio-Cabrera, O.; Poot-Pech, M.; Cruz-Martínez, O. *Manual Técnico Sobre la Langosta Voladora (Schistocerca Piceifrons Walker 1870) y Otros Acridoideos de Centro América y Sureste de México*, 1st ed.; Fao-Oirsa: San Salvador, El Salvador, 1992; pp. 1–162.
5. Song, H.; Weissman, D.B.; Barrientos-Lozano, L.; Cano-Santana, Z. The Locust Island. *Am. Entomol.* **2009**, *52*, 168–181. [CrossRef]
6. Song, H. Density-dependent phase polyphenism in nonmodel locusts: A minireview. *Psyche* **2011**. [CrossRef]
7. Senasica-DGSV. *Langosta Centroamericana [Schistocerca piceifrons piceifrons (Walker, 1870)] (Orthoptera: Acrididae), Ficha Técnica*; National Service of Agrifood Health, Safety and Quality-General Directorate of Plant Health-National Center for Phytosanitary Reference-Phytosanitary Specialist Group: Ciudad de México, Mexico, 2016; pp. 1–18. Available online: https://www.gob.mx/cms/uploads/attachment/file/157817/Ficha_t_cnica_langosta.pdf (accessed on 11 January 2021).
8. Cullen, D.A.; Cease, A.J.; Latchininsky, A.V.; Ayali, A.; Berry, K.; Buhl, J.; de Keyser, R.; Foquet, B.; Hadrich, J.C.; Matheson, T.; et al. From molecules to management: Mechanisms and consequences of locust phase polyphenism. *Adv. Insect Phys.* **2017**, *53*, 169–285. [CrossRef]
9. Ávila-Valdez, J.; Barrientos-Lozano, L.; García-Salazar, P. Aspectos Bio-Ecológicos de la Langosta Centroamericana en el Sur de Tamaulipas. In *Simposio Control Biológico y Manejo de la Langosta Centroamericana Schistocerca piceifrons piceifrons (Walker)*; Instituto Tecnológico de Cuidad Victoria: Ciudad Victoria, Tamaulipas, Mexico, 2007; pp. 1–21.
10. Pflüger, H.J.; Bräunig, P. One hundred years of phase polymorphism research in locusts. *J. Comp. Physiol.* **2021**, *207*, 321–326. [CrossRef]
11. Astacio-Cabrera, O. *La Langosta Voladora o Chapulín Schistocerca piceifrons piceifrons (Walker) en Centroamérica*, 1st ed.; Organismo Internacional Regional de Sanidad Agropecuaria (OIRSA): San Salvador, El Salvador, 1990; pp. 1–42.
12. Hunter-Jones, P. Life history of the Central American Locust, *Schistocerca* sp. (*Orthoptera: Acrididae*), in the laboratory. *Ann. Entomol. Soc. Am.* **1967**, *60*, 468–477. [CrossRef]
13. Harvey, A.W. *Schistocerca piceifrons* (Walker) (*Orthoptera: Acrididae*), the swarming locust of tropical America: A review. *Bull. Entomol. Res.* **1983**, *73*, 171–184. [CrossRef]
14. Ortiz-Yam, I.; Zuleta, M.C. Asuntos de vecinos: Langosta, defensa agrícola y la construcción de la sanidad vegetal en México y Centroamérica, siglo XX. *Historia Méxicana* **2019**, *70*, 313–373. [CrossRef]
15. Conteras-Servín, C.; Magaña-Ortíz, C. Ficha Técnica de la Langosta Centroamericana, Schistocerca piceifrons piceifrons (Walker). In *La Plaga de la Langosta (Schistocerca piceifrons piceifrons Walker) Una Visión Multidisciplinaria Desde la Perspectiva del Desastre Fitosanitario en México*, 1st ed.; Galindo Mendoza, G., Contreras Servín, C., Ibarra Zapata, E., Eds.; Universidad Autónoma de San Luis Potosí: San Luis Potosí, Mexico, 2013; Volume 1, pp. 17–36.
16. Barrientos-Lozano, L. Taller Sobre Control. In *Biológico y Manejo de Langosta Centroamericana (Schistocerca Piceifrons Piceifrons, Walker)*, 1st ed.; Servicio Nacional de Sanidad, Inocuidad y Calidad Agroalimentaria-Instituto Tecnológico de Ciudad Victoria: Ciudad Victoria, Tamaulipas, Mexico, 2007; pp. 1–197.
17. Rodríguez-Absi, J.; Almaguer-Sierra, P.; Barrientos-Lozano, L.; Rodríguez-Fuentes, H. Thermal time clock for estimating phenological development of *Schistocerca piceifrons piceifrons* Walker (*Orthoptera: Acrididae*) in northeastern Mexico. *J. Orthoptera Res.* **2009**, *18*, 65–73. [CrossRef]
18. Hernández-Zul, M.I.; Quijano-Carranza, J.A.; Yañez-López, R.; Ocampo-Velazquez, R.V.; Torres-Pacheco, I.; Guevara-Gonzalez, R.G.; Castro-Ramírez, E. Dynamic simulation model of Central American Locust *Schistocerca piceifrons* (*Orthoptera: Acrididae*). *Fla. Entomol.* **2013**, *96*, 1274–1283. [CrossRef]
19. Galindo-Mendoza, M.G.; Ibarra-Zapata, E.; Hipólito-Cruz, G.; Aldama-Aguilera, C. Sistema de Alerta Temprana Para el Control de la Langosta Centroamericana (*Schistocerca piceifrons piceifrons* (Walker) en la Península De Yucatán Aplicando Modelos Multicriteria e Imágenes de Satélite NOAA-AVHRR. In *La Plaga de la Langosta, Schistocerca piceifrons piceifrons (Walker), una Visión Multidisciplinaria Desde la Perspectiva del Desastre Fitosanitario en México*, 1st ed.; Universidad Autónoma de San Luis Potosí: San Luis Potosí, Mexico, 2013; pp. 173–215.
20. Senasica-DGSV. *Manual Operativo de la Campaña Contra la Langosta Centroamericana [Schistocerca piceifrons piceifrons (Walker, 1870)] (Orthoptera: Acrididae)*, 1st ed.; Servicio Nacional de Sanidad, Inocuidad y Calidad Agroalimentaria–Dirección General de Sanidad Vegetal: Ciudad de México, Mexico, 2019; pp. 1–80. Available online: https://www.gob.mx/cms/uploads/attachment/file/483901/Manual_Langosta_Junio_2019.pdf (accessed on 15 January 2021).
21. Organismo Internacional Regional de Sanidad Agropecuaria (OIRSA). *Plan de Acción de Manejo de la Langosta Centroamericana*, 1st ed.; Secretaría General del Sistema de la Integración Centroamericana: San Salvador, El Salvador, 2020; pp. 1–80. Available

online: https://www.oirsa.org/contenido/2020/(17%20marzo2020)%20Plan%20de%20accio%CC%81n%20y%20atencio%CC%81n%20de%20brotes%20corregido%202.pdf (accessed on 15 January 2021).
22. Poot-Pech, M.A.; Ruiz-Sánchez, E.; Ballina-Gómez, H.S.; Gamboa-Angulo, M.M.; Reyes-Ramírez, A. Olfactory response and host plant feeding of the Central American Locust *Schistocerca piceifrons piceifrons* Walker, to common plants in a gregarious zone. *Neotrop. Entomol.* **2016**, *45*, 382–388. [CrossRef]
23. Poot-Pech, M.A.; Ruiz-Sánchez, E.; Gamboa-Angulo, M.; Ballina-Gómez, H.S.; Reyes-Ramírez, A. Population fluctuation of *Schistocerca piceifrons piceifrons* (*Orthoptera: Acrididae*) in the Yucatán peninsula and its relation with the environmental conditions. *Rev. Biol. Trop.* **2018**, *66*, 403–414. [CrossRef]
24. Pérez-Ramírez, R.; Torres-Castillo, J.A.; Barrientos-Lozano, L.; Almaguer-Sierra, P.; Torres-Acosta, R.I. *Schistocerca piceifrons piceifrons* (*Orthoptera: Acrididae*) as a source of compounds of biotechnological and nutritional interest. *J. Insect Sci.* **2019**, *19*, 1–9. [CrossRef] [PubMed]
25. Viesca-González, F.C.; Romero-Contreras, A.T. La entomofagia en México. Algunos aspectos culturales. *El Periplo Sustentable* **2009**, *16*, 57–83. Available online: https://www.redalyc.org/articulo.oa?id=193414421003 (accessed on 16 January 2021). [CrossRef]
26. Garza-Sánchez, J. Evaluación del Potencial Alimenticio de la Langosta Centroamericana, *Schistocerca piceifrons piceifrons,* (Walker, 1870), (*Orthoptera: Acrididae*). Master's Thesis, Tecnológico Nacional de México-Instituto Tecnológico de Cd. Victoria, Victoria, Tamaulipas, Mexico, 21 October 2020.
27. Barrientos-Lozano, L. Population Dynamics, Biology and Ecology of the Central American Locust (*Schistocerca piceifrons piceifrons,* Walker) in Southern Mexico. In Proceedings of the Metaleptea—Eight International Meeting of the Orthopterists' Society, Montpellier, France, 19–22 August 2001; International Conference of Orthopteroid Insects. p. 75.
28. Barrientos-Lozano, L. Ecología, Manejo y Control de la Langosta Voladora, Schistocerca piceifrons piceifrons (Walker). In Proceedings of the Primer Curso Internacional: Ecología, Manejo y Control de la Langosta Voladora, Schistocerca piceifrons piceifrons (Walker), Altamira, Tamaulipas, México, 5–7 November 2001; Barrientos-Lozano, L., Ed.; Dinámica Impresa S.A.: Cuidad Victoria, Tamaulipas, Mexico, 2001; pp. 1–232.
29. Ávila-Valdez, J.; Barrientos-Lozano, L.; García-Salazar, P. Biología y comportamiento de la langosta centroamericana, Schistocerca piceifrons piceifrons (Walker, 1870), en el Sur de Tamaulipas. In Proceedings of the 2do Curso Internacional: Manejo Integrado de la Langosta Centroamericana y Acridoideos Plaga en América Latina, Cuidad Victoria, Tamaulipas, Mexico, 27–29 June 2005; Barrientos-Lozano, L., Almaguer-Sierra, P., Eds.; Dinámica Impresa S.A.: Cuidad Victoria, Tamaulipas, Mexico, 2005; pp. 31–35.
30. Cano-Santana, Z. *Ecología e Historia Natural de Schistocerca americana socorro y S. piceifrons piceifrons en Isla Socorro;* Informe Final Proyecto BS007; Universidad Nacional Autónoma de México: Ciudad de México, México, 2006; pp. 1–138. Available online: http://www.conabio.gob.mx/institucion/proyectos/resultados/InfBS007.pdf (accessed on 1 February 2021).
31. Poot-Pech, M.A. La Langosta Voladora Schistocerca piceifrons (Orthoptera: Acrididae), Hacia un Manejo Sustentable. In *El Patrimonio, su Importancia y Conservación: Conociendo el Patrimonio*, 1st ed.; TECCIS: Ciudad de Campeche, Campeche, Mexico, 2016; pp. 58–66. Available online: https://www.researchgate.net/profile/Marvel-Valencia-Gutierrez-publication/329761423GUAYA_Talisia_olivaeformis_R_AL_RESCATE_DEUN_PATRIMONIO_NATURAL/links/5c1985e7a6fdccfc70587065/GUAYA-Talisia-olivaeformis-R-AL-RESCATE-DEUN-PATRIMONIO-NATURAL.pdf (accessed on 4 February 2021).
32. Galindo-Mendoza, M.G.; Contreras Servín, C. Alerta Climática-Fitosanitaria Para: Langosta Centroamericana, Schistocerca piceifrons piceifrons (Walker, 1870), Zona Huasteca, San Luis Potosí. In *Sistemas de Alerta Temprana en Actividades Agropecuarias con Tecnología Espacial;* SENASICA-UASL: San Luis Potosí, Mexico, 2021.
33. Barrientos-Lozano, L. *Ortópteros Plaga de México y Centro América: Guía de Campo*, 1st ed.; Instituto Tecnológico de Cd. Victoria, COSNET, SEP-CONACYT: Cuidad Victoria, Tamaulipas, Mexico, 2003; pp. 1–114.
34. Garza-Urbina, E. *La langosta Schistocerca piceifrons piceifrons y su Manejo en la Planicie Huasteca;* Folleto Técnico No. 12; Instituto Nacional de Investigaciones Forestales, Agrícolas y Pecuarias (INIFAP), CIRNE: San Luís Potosí, Mexico, 2005; pp. 1–23. Available online: http://www.inifapcirne.gob.mx/Biblioteca/Publicaciones/149.pdf (accessed on 4 February 2021).
35. Senasica-DGSV. *Memoria del Curso Regional Sobre la Langosta (Schistocerca piceifrons piceifrons);* Servicio Nacional de Sanidad, Inocuidad y Calidad Agroalimentaria-Dirección General de Sanidad Vegetal, CIRNE, INIFAP, SAGARPA: San Luis Potosí, Mexico, 2003; pp. 1–93.
36. Retana, B.J.A. Relación Entre Algunos Aspectos Climatológicos y el Desarrollo de la Langosta Centroamericana *Schistocerca piceifrons piceifrons* en el Pacífico Norte de Costa Rica Durante la Fase Cálida del Fenómeno El Niño-Oscilación Sur (ENOS). *Tóp. Meteorol. Oceanogr.* **2000**, *7*, 73–87. Available online: http://www.bio-nica.info/biblioteca/retana2000acrididae.pdf (accessed on 4 February 2021).
37. Poot-Pech, M.A.; Ruiz-Sánchez, E.; Ballina-Gómez, H. Indicators Plants in Solitary Phase and Migration Behavior of *Schistocerca Piceifrons* in Yucatán, México. *Metaleptea* **2017**, *37*, 21.
38. Rainey, R. *Meteorology and the Migration of Desert Locusts: Applications of Synoptic Meteorology in Locust Control;* Technical Note No. 54; World Meteorological Organization (WMO): Geneva, Switzerland, 1988; pp. 1–115. Available online: https://library.wmo.int/doc_num.php?explnum_id=1746 (accessed on 4 February 2021).
39. Márquez-Delgado, A. *La Lucha Contra la Langosta en México*, 1st ed.; Colegio de Ingenieros Agrónomos: Ciudad de México, Mexico, 1963; pp. 1–220. Available online: https://agris.fao.org/agris-search/search.do?recordID=US201300278309 (accessed on 4 February 2021).

40. Hoffmann, C.C.; Dampf, A.; Varela, G. *Informe de la Comisión Científica Exploradora de la Plaga de la Langosta en el Estado de Veracruz*, 1st ed.; Talleres Gráficos de México: Ciudad de México, Mexico, 1925; Volume 3, pp. 1–140.
41. Ávila-Valdez, J.; Barrientos-Lozano, L.; García-Salazar, P. Aspectos de la Biología y hábitos de la langosta centroamericana, *Schistocerca piceifrons piceifrons* (Walker) (Orthoptera: Acrididae) bajo las condiciones del Sur de Tamaulipas. *Entomol. Mex.* **2007**, *3*, 177–181.
42. Sword, G.A.; Simpson, S.J.; El Hadi, O.M.; Wilps, H. Density-dependent aposematism in the desert locust. *Proc. Roy. Soc. B Biol. Sci.* **2000**, *267*, 63–68. [CrossRef]
43. Pener, M.P.; Simpson, S.J. Locust phase polyphenism: An update. *Adv. Insect Physiol.* **2009**, *36*, 1–286. [CrossRef]
44. Song, H.; Foquet, B.; Mariño-Pérez, R.; Woller, D.A. Phylogeny of locusts and grasshoppers reveals complex evolution of density-dependent phenotypic plasticity. *Sci. Rep.* **2017**, *7*, 6606. [CrossRef]
45. Simpson, S.J.; Sword, G.A. Phase Polyphenism in Locusts: Mechanisms, Population Consequences, Adaptive Significance and Evolution. In *Phenotypic Plasticity of Insects Mechanisms and Consequences*; Whitman, D.W., Ananthakrishnan, T.N., Eds.; Science Publishers: Enfield, NH, USA, 2009; pp. 147–189.
46. Contreras-Servín, C. Conexión Climática del Fenómeno del Niño con la Plaga de la Langosta en Centroamérica (*Schistocerca piceifrons piceifrons*, Walker) Localizada en el Estado de Yucatán y la Huasteca Potosina. *Entomol. Mex.* **2009**, *8*, 34–351. Available online: http://www.entomologia.socmexent.org/revista/2009/BHN/347-351.pdf (accessed on 5 February 2021).
47. Senasica-DGSV. *Central American Locust*; Servicio Nacional de Sanidad, Inocuidad y Calidad Agroalimentaria-Dirección General de Sanidad Vegetal: Ciudad de México, México, 2018; pp. 1–2. Available online: https://issuu.com/senasica/docs/info_senasica_no_2 (accessed on 8 February 2021).
48. Hernández-Velázquez, V.M.; Berlanga-Padilla, A.M.; Barrientos-Lozano, L. Vegetable and mineral oil formulations of *Metarhizium anisopliae* var. *acridum* to control the Central American Locust *Schistocerca piceifrons piceifrons* (Walker, 1870) (*Orthoptera: Acrididae*). *J. Orthop. Res.* **2000**, *9*, 223–227. [CrossRef]
49. Song, H.; Poot-Pech, M. Taller regional de manejo de la langosta centroamericana, *Schistocerca Piceifrons*. *Metaleptea* **2017**, *37*, 16–17.
50. Organismo Internacional Regional de Sanidad Agropecuaria (OIRSA). *Situación Actual y Recomendaciones para el Control de la Langosta Centroamericana en los Países de la Región OIRSA*, 1st ed.; Secretaría General del Sistema de la Integración Centroamericana: San Salvador, El Salvador, 2020; pp. 1–14. Available online: https://www.sica.int/documentos/situacion-actual-y-recomendaciones-para-el-control-de-la-langosta-centroamericana-en-los-paises-de-la-region-oirsa_1_122911.html (accessed on 9 February 2021).
51. Galván-Martínez, E.; Almaguer-Sierra, P.; Barrientos-Lozano, L. Dinámica poblacional de la langosta centroamericana, *Schistocerca piceifrons piceifrons* (Walker 1870), usando un modelo de simulación. *Entomol. Mex.* **2008**, *7*, 285–289. Available online: http://www.socmexent.org/entomologia/revista/2008/BHN/285-289.pdf (accessed on 10 February 2021).
52. Senasica-DGSV. Langosta Centroamericana. Servicio Nacional de Sanidad, Inocuidad y Calidad Agroalimentaria-Dirección General de Sanidad Vegetal. 2020. Available online: https://www.gob.mx/senasica/documentos/langosta-centroamericana-1109199 (accessed on 10 February 2021).
53. Barrientos-Lozano, L.; Hernández-Velázquez, V.M.; Milner, R.J.; Hunter, D.M. Advances in biological control of locusts and grasshoppers in México. *J. Orthop. Res.* **2002**, *11*, 77–82. [CrossRef]
54. Hernández-Velázquez, V.M.; Berlanga-Padilla, A.M.; Garza-González, E. Detección de *Metarhizium flavoviride* sobre *Schistocerca piceifrons piceifrons* (*Orthoptera: Acrididae*) en la Isla Socorro, Archipiélago de Revillagigedo, México. *Vedalia* **1997**, *4*, 45–46.
55. Barrientos-Lozano, L.; Milner, R.J. Comparison of the Pathogenicity of Three Isolates of *Metarhizium anisopliae* var. *acridum*, Effect of Dose and Temperature. In Proceedings of the XXI International Congress of Entomology, Foz de Iguaçu, Brazil, 20–26 August 2000; Sociedade Entomológica do Brasil: Londrina, PR, Brazil, 2000; pp. 513–514.
56. Hernández-Velázquez, V.M.; Hunter, D.M.; Barrientos-Lozano, L.; Lezama-Gutiérrez, R.; Reyes-Villanueva, F. Susceptibility of *Schistocerca piceifrons* (*Orthoptera: Acrididae*) to *Metarhizium anisopliae* var. *acridum* (Deuteromycotina: Hyphomycetes): Laboratory and field trials. *J. Orthop. Res.* **2003**, *12*, 89–92. [CrossRef]
57. Milner, R.J.; Barrientos-Lozano, L.; Driver, F.; Hunter, D.M. A comparative study of two Mexican isolates with an Australian isolate of *Metarhizium anisopliae* var. *acridum* strain characterization, temperature profile and virulence for wingless grasshopper, *Phaulacridium vittatum*. *BioControl* **2003**, *48*, 335–348. [CrossRef]
58. Barrientos-Lozano, L.; Hunter, D.M.; Ávila-Valdéz, J.; García-Salazar, P.; Hernández-Velázquez, V.M. Control Biológico de la Langosta Voladora (*Schistocerca piceifrons* Walker) con *Metarhizum anisopliae* var. *acridum* en el Noreste de México. In *Proceedings of the XXVII Congreso Nacional de Control Biológico, Los Mochis, Sinaloa, México, 8–13 November 2004*; Sociedad Mexicana de Control Biológico: Sinaloa, Mexico, 2004; pp. 137–141. Available online: https://www.cabi.org/wp-content/uploads/Barrientos-Lozano-2004-Biological-control-of-Schistocerca.pdf (accessed on 12 February 2021).
59. Barrientos-Lozano, L.; Hunter, D.M.; Ávila-Valdéz, J.; García-Salazar, P.; Horta-Vega, J.V. Control Biológico de la Langosta centroamericana *Schistocerca piceifrons piceifrons* Walker (Orthoptera: Acrididae) en el Noreste de México. *Vedalia* **2005**, *12*, 119–128.
60. Williams, T.; Arredondo-Bernal, H.C.; Rodríguez-del-Bosque, L.A. Biological Pest Control in Mexico. *Annu. Rev. Entomol.* **2013**, *58*, 119–140. [CrossRef] [PubMed]
61. Barrientos-Lozano, L. *Manejo Integrado de la Langosta Centroamericana (Schistocerca piceifrons piceifrons, Walker) y Acridoideos Plaga en América Latina*, 1st ed.; Tecnológico Nacional de México-Instituto Tecnológico de Cuidad Victoria: Cuidad Victoria, Tamaulipas, Mexico, 2005; pp. 1–303.

62. Barrientos-Lozano, L. Memoria Simposio Control. In *Biológico y Manejo de la Langosta Centroamericana Schistocerca piceifrons piceifrons (Walker)*; Servicio Nacional de Sanidad, Inocuidad y Calidad Agroalimentaria-Instituto Tecnológico de Cuidad Victoria: Cuidad Victoria, Tamaulipas, Mexico, 2007; pp. 1–112.
63. Montelongo-Ruíz, G.; Olazarán-Santibáñez, F.E.; Reyes-Zepeda, F.; Torres-Castillo, J.A.; De la Rosa-Manzano, E.; Martínez-Ávalos, J.G.; Arellano-Méndez, L.U.; Mora-Olivo, A. Inhibition of intestinal proteolytic activity of *Schistocerca piceifrons piceifrons* (Walker) and *Pterophylla beltrani* (Bolívar and Bolívar) using extracts from *Chamaedorea radicalis* Mart. Fruit. *Southwest Entomol.* **2020**, *45*, 137–146. [CrossRef]
64. Le Gall, M.; Overson, R.; Cease, A. A global review on locusts (*Orthoptera: Acrididae*) and their interactions with livestock grazing practices. *Front. Ecol. Evol.* **2019**, *7*, 1–24. [CrossRef]

agronomy

MDPI

Review

Alternative Strategies for Controlling the Brown Locust, *Locustana pardalina* (Walker)

Roger Price

Insect Ecology Division, ARC-Plant Health and Protection, Private Bag X134, Queenswood, Pretoria 0121, South Africa; pricer@arc.agric.za

Abstract: Regular and often intense outbreaks of the brown locust, *Locustana pardalina* (Walker), in the semi-arid Nama Karoo region of South Africa present a formidable pest control problem. Outbreak patterns over a 64-year period (1941–2005) were reviewed indicating a very high frequency of outbreak years with regular 'plague' periods being experienced, while a more detailed analysis of the numbers of locust targets controlled during a 22-year period (1983–2005) described the intensity and scale of the outbreaks. The operational constraints associated with the traditional ground-based control strategy employed against the thousands of individual roosting brown locust hopper band and swarm targets in the Karoo are discussed. A brief review of laboratory and field trials of alternative methods of controlling the brown locust, such as insecticide baits, barrier treatments and the Green Muscle® myco-insecticide, as an alternative to broad-spectrum pyrethroid insecticides are described. In addition, alternative control strategies to the current 'Commando' system of ground-based control operations are discussed. The recommendation is for a modernised and technology-equipped integrated brown locust management strategy (IPM), combining ground and aerial tactics that will have the flexibility and the capacity to deal effectively with outbreaks. The integrated management strategy should focus on ground-based control of hopper bands and fledgling swarms in the Upper and eastern Karoo, while outbreaks over most of the Central Karoo and arid Bushmanland areas should be left to fledge and coalesce into large-sized swarms that could then be targeted by spray aircraft as they migrate along their known swarm flight paths. The introduction of electronic reporting and GIS mapping technologies for brown locust campaign management is essential.

Keywords: brown locust; *Locustana pardalina*; Karoo; outbreak patterns; control strategy; integrated pest management

Citation: Price, R. Alternative Strategies for Controlling the Brown Locust, *Locustana pardalina* (Walker). *Agronomy* **2021**, *11*, 2212. https://doi.org/10.3390/agronomy11112212

Academic Editors: Michel Lecoq and Arianne Cease

Received: 27 July 2021
Accepted: 31 August 2021
Published: 31 October 2021

1. Introduction

The aim of this review article is to summarise the outbreak history of the brown locust, *Locustana pardalina* (Walker) along with the control tactics currently employed and alternative control methods evaluated, and to then discuss possible alternative strategies for more effective management of outbreaks. The brown locust has a recognised outbreak area that covers approximately 250,000 km^2 of the semi-arid Nama Karoo biome region of South Africa and southern Namibia [1–4], from where plagues have developed that have overrun the entire southern African sub-continent up to the Zambezi River [3]. The Nama Karoo occupies the central plateau of the western half of South Africa at an altitude of 500–2000 m, so frost is common in winter. The rainfall mainly falls in late summer and autumn, with a rainfall gradient ranging from approximately 100 mm in the western arid areas to 400 mm per annum in the eastern Karoo, although rainfall is typically erratic and extended droughts are common. The brown locust displays a wide range of phase polymorphism and sexual dimorphism [5], with the differences measured between the extreme solitaria and gregaria phases being more pronounced than with any other locust species [6,7]. The field biology and population dynamics of the brown locust have been extensively studied [2,8–13]. An important aspect of the brown locust lifecycle for survival in the semi-arid Karoo is the egg stage, which is highly resistant to drought involving a complex of diapause and quiescent stages [14]. Solitaria phase females only lay diapause eggs, with the obligate diapause being broken after 9–45 days under dry soil conditions, while gregaria swarming females only lay non-diapause eggs that develop continuously under moist soil conditions. Transient phase females lay egg pods containing various proportions of diapause and non-diapause eggs depending upon factors including day length, phase, age and nutrition of the female [14]. Under extended dry conditions in the Karoo, the non-diapause eggs, as well as the diapause eggs that have exited the diapause state, stop further development and enter a state of quiescence just before the embryonic stage of katatrepsis takes place [14]. Quiescent eggs are known to be very drought tolerant and small numbers of egg pods have been known to survive for up to 2–3 years in the field [9,15], Price, *pers. obs.*, but the contribution of any eggs surviving for more than 12 months was dismissed as being irrelevant to the population dynamics of the brown locust [15]. Eggs are typically laid in dry soil and hatch after 10–14 days following summer rainfall, with 20–25 mm of rain required for widespread hatching [10]. There are five hopper instars, with solitaria hoppers developing within 21–30 days depending upon temperature, while gregaria phase hoppers are much larger with the hopper bands typically taking 42–45 days to develop in summer [10,16]. Gregarious phase females in the field mature within 2–3 weeks after fledgling and will then lay 3–4 egg pods containing a mean of 45 eggs at weekly intervals [17], but under laboratory conditions the females mature more quickly and can lay up to 10–15 egg pods each [8]. Adult locusts live for 2–3 months under summer conditions, but live a lot longer during the cooler autumn season [8]. The multivoltine lifecycle along with a high fecundity of the brown locust allows a rapid rate of population increase with two or three generations normally possible per year from September to May [10], with four generations recorded during some of the more climatically favourable years [3], Price, *pers. obs.*

The brown locust is predominantly graminivorous and hoppers and adults prefer to feed on a range of Karoo grasses, but under dry field conditions the hoppers and adults can be frequently observed nibbling on the leaves and bark of Karoo bushes and anything else that is edible [10,11,16]. Eruptions of the brown locust pose a direct pest threat to the sheep grazing rangeland within the Karoo, to crops planted under irrigation within the Karoo, and more especially to the main maize and wheat cereal cultivation areas of the country that are within range of swarms escaping from the eastern Karoo. The brown locust therefore is a formidable pest problem and has been the target of chemical control campaigns waged by the South African Department of Agriculture since 1906 [18,19].

The historic periodicity of brown locust plague periods from 1797 to 1915 were reported by Lounsbury [20] as lasting an average of 13 years with a recession period of approximately 11 years between the plagues. After the introduction of Government-funded chemical control in 1906, the periodicity of outbreak cycles was considered to have shortened to 7–11 years of swarming activity followed by a similar period of recession [3,5,12,21,22]. However, gregarious brown locust populations requiring chemical control could be found somewhere or other in the Karoo in approximately 90% of years throughout the entire twentieth century, although the intensity and extent of outbreaks varied greatly [19]. The outbreak and invasion areas of the brown locust in southern Africa as described by different authors are depicted for reference in Figure 1.

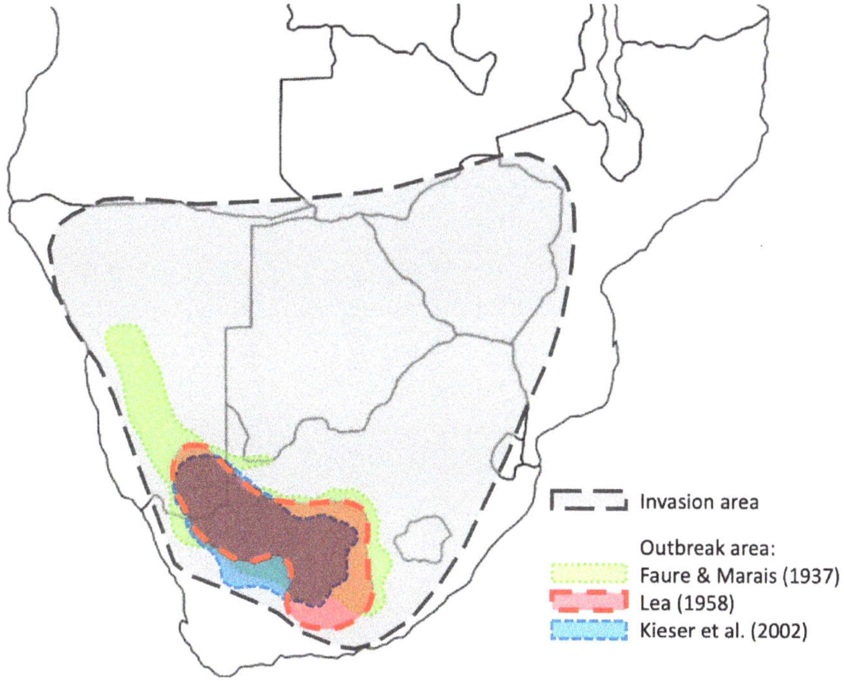

Figure 1. Outbreak area of the brown locust in southern Africa, as defined by Faure and Marais [1], Lea [3], Kieser [4]. Invasion area as defined by Lea [3].

2. History of Brown Locust Outbreaks and Control

2.1. Outbreak Cycles of the Brown Locust

Incipient outbreaks develop following the successful breeding and aggregation of solitary and transient phase adults leading to rapid gregarization of the next generation of hatching hoppers. Under favourable rainfall conditions over one or more summer seasons, such incipient outbreaks often develop simultaneously over a vast area of the Nama Karoo [5]. The scattered outbreaks rapidly develop into gregaria-phase population eruptions, which then have been traditionally combated at source within the Karoo [3], in a suppressive control strategy that can be described as upsurge elimination [23] before migrating swarms can escape and threaten cereal crop production areas outside the Karoo. Since the mid-1940s, when synthetic insecticides first became available, this control strategy has largely restricted outbreaks to the Karoo with only brief invasions of surrounding countries being recorded [19]. Long-term plague cycles have been prevented and food security within the southern African region has not been seriously threatened.

The distribution and abundance of brown locust populations for the period 1906–1969 were described by Lea [12,21,22], with the relative abundance per year based on figures of control expenditure. However, financial comparisons become inaccurate over time and a more comparative analysis of brown locust outbreaks was considered to be the number of Magisterial Districts that reported locust control action per year in South Africa, Namibia and Botswana [19], although this number gave no indication of the actual intensity of the outbreaks within each of the districts. An updated graph of the number of Magisterial Districts where control action was undertaken between 1941 and 2005 is shown in Figure 2.

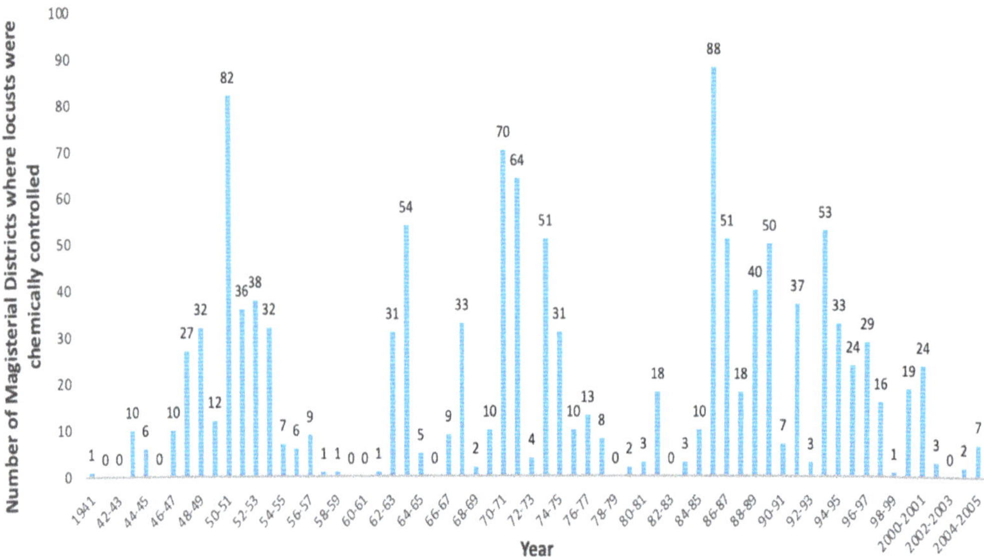

Figure 2. Number of Magisterial Districts that reported brown locust campaigns per locust season in South Africa, Namibia and Botswana between 1941 and 2005. Data obtained from Lea [3,22] for the period 1941–1954, from Lea [13,22] for the period 1954–1969, from annual reports of the South African National Dept. Agric. between 1970 and 1984, and from weekly locust control reports from the South African Dept. Agric. for the period 1984–2005. Data for Namibia and Botswana from Lea [22] and from Dept. Agric. reports from 1985 to 2005.

Figure 2, shows the high outbreak frequency and occurrence of major eruptions or 'plague' periods over a wide area, namely during the periods 1950–1951, 1963–1964, 1970–1975, 1985–19787, 1988–1990, 1993–1997 and 1999–2001. The actual intensity of the brown locust control campaigns can be demonstrated from more recent records of the number of individual locust targets chemically controlled during each 'locust outbreak season' (September–June) over the 22 year period in South Africa between 1983–1984 and 2004–2005 (Table 1).

Intense outbreak seasons can be clearly seen (Table 1), with 'plague' eruptions, as defined by the Southern African Regional Commission for the Conservation and Utilization of the Soil (SARCCUS) sub-committee for the control of migratory pests [24] as seasons with >5000 adults swarms being controlled, being evident in 1985–1986, 1988–1989, 1993–1994, 1995–1996, 1996–1997 and 1999–2000. The continual chemical control campaigns waged against the brown locust have evidently not dampened the ability of this locust species to produce serious eruptions on a regular basis [22], with enough of the gregarious populations surviving to lay eggs before winter to ensure the ongoing eruption into the next season [19]. It is evident that the preventative control action, in the sense of stopping the plagues from developing, has utterly failed [22]. The ever increasing economic costs of the almost annual control campaigns, as well as the environmental contamination that is inevitably caused by the application of broad-spectrum insecticides in the ecologically unique Nama Karoo biome, is a serious cause for concern and introspection.

Table 1. The number of individual locust targets (roosting hopper bands and adult swarms) controlled during each outbreak season (September to June) in South Africa between 1983–1984 and 2004–2005 (data from the weekly reports of the locust control officers of the South African National Department of Agriculture).

Outbreak Season	No. Hopper Bands Controlled	No. Adults Swarms Controlled
1983–1984	61	34
1984–1985	633	65
1985–1986	175,500	38,600
1986–1987	68,902	14
1987–1988	5618	1123
1988–1989	85,935	1642
1989–1990	36,553	1392
1990–1991	1142	357
1991–1992	18,131	1603
1992–1993	72	0
1993–1994	34,581	9565
1994–1995	20,895	663
1995–1996	24,489	6577
1996–1997	75,890	8081
1997–1998	1018	80
1998–1999	2	0
1999–2000	40,115	9021
2000–2001	28,642	1135
2001–2002	1905	137
2002–2003	0	0
2003–2004	128	154
2004–2005	1167	20

2.2. Control Strategy against the Brown Locust in South Africa

Due to the devastation caused to agriculture in South Africa at the turn of the 20th century by incessant locust plagues, the South African Government first took over the coordinated responsibility for locust control in 1906 using sodium arsenite solutions ('arsenite of soda' dissolved in warm water with sugar added) applied using spray pumps [18]. Sodium arsenite aqueous and dusting formulations were used for brown locust control for the next 30 years, even though the toxicity of arsenic compounds to the spray operators and livestock was already well known with safe handling and operator protection regulations being issued [25]. Plague locusts were declared as National Pests in South Africa in 1911 in one of the first legislations promulgated by the new Government of the Union of South Africa [20]. The national pest status of locusts and the legal responsibility of landowners to report the presence of locusts on their land and for the Minister of Agriculture to be responsible for combatting locust outbreaks was ratified under Article 6 of the South African Migratory Pests Act (Act No. 36 of 1983). The principles guiding locust control in South Africa were updated and consolidated in a new policy for the management of the locust problem in South Africa [26]. The National Department of Agriculture, Land Reform and Rural Development: Directorate of Disaster Management, who are responsible for locust control, currently maintains two locust control depots in the Nama Karoo, namely at De Aar in the Upper Karoo region and at Upington in the arid Bushmanland region, managed by a senior locust control officer at each depot who controls the stores of application equipment and pesticide stocks.

The pro-active control of incipient sub-swarming brown locust concentrations has historically been considered as being uneconomic and impractical [5,12,22] and brown locust control has always relied on an emergency response to swarming populations. Due to the seasonal nature and unpredictable location of brown locust outbreaks, the operational management of outbreaks has relied for decades on what is locally known as the "Commando system", which activates during outbreaks with an army of temporarily employed locust officers and spray machine operators tracking down and controlling individual hopper band or roosting swarm targets. The Directorate maintains a network of District Locust Control Officers in each of the Magisterial Districts within the brown locust outbreak region, who are usually farmers with previous experience with locust control campaigns and who are re-appointed during outbreaks in order to coordinating the control campaigns in their District by directing control teams and managing pesticide and equipment resources. These officers report any control actions directly to the Senior Officers at the two locust control depots. Farmers and town residents from the local communities within each of the districts are then appointed on a temporary basis as 'supervisors/drivers', who use their own private pick-up vehicles for undertaking the locust control operations and who then claim a mileage allowance rate. The supervisor/drivers then appoint a small team of pest control operators (usually local farm assistants) to manage the spray machines. The district locust officers, supervisors and assistants also claim day-wage allowances when on active duty [27]. The supervisors and assistants receive protective clothing and training on the effective and safe operation of the spray equipment.

Brown locust control currently relies on the ground-based spot application of broad-spectrum, fast-acting synthetic pyrethroid insecticides, such as deltamethrin (Decis®) UL 6g a.i./ℓ, and more recently esfenvalerate (Sumi-Alpha®) UL 8g a.i./ℓ), applied at an area dosage rate of 2.5 ℓ/ha as low-volume drift sprays from motorised knapsack sprayers (Solo Port 423®) or vehicle mounted (Power Solo®, Francisco 'Bakkie' pump® and Micron Ulvamast®) equipment [27]. The vehicle-mounted sprayers are used to control the larger-sized targets. In some areas, esfenvalerate (Sumi-Alpha DP®) dusting powder is applied at rates of 5–20 kg/ha using manual backpack dusters or motorized powder dusters (Osbornbirds®). Individual hopper band and fledgling swarm targets are located by the mobile control teams and are spot-treated early in the morning while still densely aggregated on their overnight roosts and while basking in the morning sun. Hopper control is focussed on the later instar (L4–L5) gregarious hopper bands which are red and black in colour, known locally in Afrikaans as 'rooibaadjies' (red jackets). Roosting band targets can be readily spotted with the naked eye or through binoculars over a distance of up to 1–2 km in the early morning sunshine as red-coloured masses roosting on the low Karoo bushes (usually 0.3–1 m high). Adult swarms are usually tracked by the locust officers and farmers during the afternoon until they roost on the Karoo bushes in the evening and are then controlled from vehicles during the night as this provides more time to complete the control operations against these often large targets [19].

Brown locust hopper band targets are usually small and discrete, measuring between 5 m^2 and 0.5 ha in extent when roosting over the Karoo bushes, with >90% of hopper band targets being <0.25 ha in extent [28]. The Department of Agriculture classifies small band targets as <120 m^2, medium targets as 121–2500 m^2 and large targets as >2500 m^2, while roosting adult swarms average 10–14 ha in extent, but can sometimes cover 2–3 km^2. The aerial spraying of locust targets using spray helicopters and fixed-wing aircraft is generally unsuitable for use against the typically small-sized brown locust targets, but aerial control is employed during emergencies if the ground teams are being overwhelmed by the numbers of locust targets or if they occur in terrain which is difficult to access, or if fast-flying swarms are starting to exit the Karoo and to potentially threaten crops. The Department of Agriculture maintains a spray aircraft and spray helicopter on a stand-by contract, but emergency aerial support is also contracted from commercial crop spraying companies when required. The aerial control of flying swarms (air-to-air spraying) is not undertaken in South Africa.

The history of the methods and insecticide products used for brown locust control are described in Table 2. Apart from the insecticides used operationally for locust control in the Karoo, a wide range of insecticide products have also been screened as possible acridicides since 1975 against the brown locust [19,28]. Please note that mention of trade names of spray equipment and commercial insecticide products in this review publication is for the purpose of providing specific information only.

Table 2. History of the methods and insecticide products used for brown locust control operations in South Africa (from the South African National Department of Agriculture and [19,28]).

Before 1906	Mechanical and cultural control methods (trampling, beating, burning pastures, digging up of egg beds, collecting adult locusts in sacks, spraying soap solutions)
1906–1934	Application of sodium arsenite (supplied free of charge to farmers and applied as aqueous or dusting formulations using hand-operated pumps)
1934–1944	Sodium arsenite baits (moistened bran bait applied by hand to roosting hopper bands)
1945–1986	Benzene hexachloride (BHC) (applied first as a bait agent and as aqueous spray, but mainly as a dust formulation (mainly 7% gamma isomer, but also as Lindane dust with 99% gamma BHC formulation), applied as a dusting powder from hand-operated or motorised dusting machines at area application rates of 15–20 kg/ha)
1975–1994	Organophosphate insecticides (diazinon and fenitrothion applied as an ultra-low-volume (ULV) sprays from a range of motorised mist-blower and stacked-disc sprayers). A standard 400 g/ℓ fenitrothion formulation was applied at a volume rate of 2.5 ℓ/ha, giving an area dose rate of 1 kg a.i./ha
1990–to date	Synthetic pyrethroid insecticides (deltamethrin and esfenvalerate applied as ULV spray at a volume rate of 2.5 ℓ/ha from motorised back-pack or vehicle-mounted sprayers). Esfenvalerate UL 8g a.i./ℓ formulation is now mainly employed in the Karoo
Future	New products already registered or trial work completed (e.g., alpha-cypermethrin, fipronil, *Metarhizium* myco-insecticide)

2.3. Outbreak Early Warning Systems

The complicated relationship between the patchy rainfall distribution typical in the semi-arid Karoo areas and the development of incipient outbreaks has been studied for decades [2,3,10,12], but no conclusive patterns were found to enable an effective outbreak early warning system to be developed. Both Smit and Lea clearly thought that drought conditions of dry winters and dry early-summers, followed by widespread rainfall in late-summer, supported large-scale population increases, whilst fewer outbreaks were recorded following wet winters and wet early-summer periods. By analysing the correlations between rainfall data and swarming activity for the development of an outbreak simulation model [29], the data supported the view that dry winters followed by good early rains in the Karoo were conducive to years of locust eruptions, while wet winters were conducive to years of reduced locust numbers. The hypothesis was that the dry conditions over several months, with no rain on any one day of more than 6 mm, ensured that both diapause and quiescence in the overwintering egg populations was broken and allowed the eggs to respond to rainfall events in early summer (October–December). However, this simulation model relied on estimates of egg populations to generate outbreaks and the reliability of such detailed field data will always be problematic. However, other studies could find no statistical evidence of any connection between brown locust eruptions and previous austral winter rainfall [30], but instead found a high correlation between rainfall over the previous 12 months prior to the locust season and in particular with rainfall during December. The hypothesis was that the high-frequency outbreak cycles were related to the El Nino/Southern Oscillation (ENSO) climatic patterns and that high-frequency locust activity was clearly evident during La Niña events (wet cycles in southern Africa) and with low locust activity during El Nino dry events. A dominant outbreak frequency of 17.3 years was proposed, with these extended cycles strongly related to the sea-surface temperatures and ENSO events [30].

The monitoring of farms with a known history of producing locust outbreaks was also employed for many years in the past as an early warning system of the build-up of solitary adult populations and the development of impending outbreaks [10,12]. The foot survey counts accurately predicted the development of incipient swarming populations based on critical population levels for phase transformation [12], but the monitoring was very time consuming and became redundant once widespread outbreaks developed.

2.4. *Environmental Impact of Synthetic Pyrethroid Insecticides*

Modern synthetic pyrethroid insecticides are safer to apply and pose less environmental risk than the arsenic, organochlorine and organophosphate compounds used for locust control in the past [31,32]. The tactics of the spot application of insecticide to densely aggregated, roosting brown locust targets also minimizes the area treated, compared with the aerial blanket-spraying undertaken against some other locust and grasshopper species.

As discussed previously, the roosting brown locust hopper band and swarm targets are usually relatively small and discrete, so only a limited area is actually spot-sprayed during control operations. Over a 20-year period, an estimated average of 75,000–90,000 ha were sprayed each year against the dense roosting targets during locust control operations in the Karoo, which is less than 0.25% of the total brown locust outbreak area [27]. Spot spraying with synthetic pyrethroids in the Karoo also has a relatively short-term environmental impact as most non-target arthropods recolonize the small treated areas within a few weeks in summer [33], while the chance that exactly the same areas are sprayed every year is also low [28]. Nevertheless, the intensity of the spraying campaigns and the negative impact that the repeated application of broad-spectrum insecticides may have on the rich diversity of endemic invertebrates and reptiles found in the Nama-Karoo biome is of concern to landholders and conservationists. How to reduce the insecticide load and minimize the environmental impact in the Karoo and yet at the same time control this serious agricultural pest has become a controversial issue. There is thus a need for more target specific and environmentally benign methods of controlling the brown locust as an alternative to the current spraying of broad-spectrum insecticides.

3. Alternative Control Methods

Various alternative methods of controlling the brown locust were evaluated against gregarious hopper populations in the Agricultural Research Council—Plant Health and Protection (ARC-PHP) laboratories in Pretoria and in small-scale field trials in the Nama Karoo between 1994 and 2003 [34].

3.1. *Natural Enemies*

A wide range of natural enemies are known to prey upon the various life stages of the brown locust [9,21,35,36] and their impact can sometimes be locally high [9]. The predation of egg pods by larvae of the woolly bee fly, *Systoechus* sp. (Bombyliidae) and the digging up of egg beds by small mammals and birds sometimes caused up to 12% mortality of egg pods [34]. The voracious ants, *Camponotus fulvopilosus* and *Anoploepis custodiens,* were also sometimes important predators of hatching hoppers, while various species of robber flies, ground beetles, hunting wasps, scorpions, spiders, lizards, mammals (e.g., jackals, mongooses, feral dogs) and a wide range of bird species are opportunistic predators of all life stages of the brown locust [34]. Of particular interest was the impact of the sarcophagid fly, *Wohlfarhtia pachytyli* (Walker), which is a generalist scavenger as well as a facultative parasite of moulting L5 brown locust hoppers and fledglings [37]. The fly larviposits on moulting locusts when they are immobilised and unable to defend themselves from fly attack. In contrast, no parasitism of active hoppers or adults was observed [37,38]. The fly was common throughout the Karoo and highest fly populations were always associated with brown locust outbreaks. Locusts parasitised by fly maggots were regularly found at roosting sites where hopper bands had moulted, or where fledgling had occurred. However, intensive sampling showed that *W. pachytyli* failed to cause more than 0.1% mortality of hoppers and 6% mortality of fledgling swarms [38]. These figures suggest that the potential of *W. pachytyli* as a biological control agent may have been overestimated in the past [38]. Despite the wide range of opportunistic predators and parasites reported, their impact on brown locust populations was usually very low. On no occasion were natural enemies observed in sufficient numbers to dampen the intensity of brown locust outbreaks [34].

3.2. Mechanical Control

Before the first insecticides became available at the turn of the 20th century, farmers had to resort to beating, trampling, trenching and burning to protect their crops and pastures from the ravages of locusts [18,39]. Turning back the clock, the destruction of locust egg beds and the harvesting of locusts were re-examined as control methods. Brown locust egg beds proved very difficult to locate in the remote Karoo, but once found, the excavation of locust eggs with picks and garden forks gave excellent control. However, the disturbance of the friable soils in the Karoo would cause severe erosion problems and is therefore not advocated. Harvesting of live locusts using nets or vacuum machines was not practical as locusts either scattered upon disturbance, or retreated into the interior of the spiny Karoo bushes to avoid capture. However, the harvesting of locust cadavers lying on the soil surface following insecticide spraying, once they had dried out and insecticide residues had broken down, was possible. With their high protein and fat content, the processing of locust cadavers into animal feed may become economically viable in future.

3.3. Insecticide Baits

Bran bait containing 2–3% sodium arsenite was extensively used for the control of brown locust and red locust hopper bands in South Africa during the 1930s [40]. The baiting technique was re-evaluated as a more target-specific, low cost method of locust control using two modern insecticides, the non-ester pyrethroid, silafluofen (Neophan®), and the phenyl-pyrazol, fipronil (Regent®). Bioassays showed that these insecticides, when dissolved in water and mixed into wheat bran as the edible carrier, had a potent stomach action and were effective in minute quantities against the brown locust. Bran baits containing 0.2% Neophan 80EC and 0.02% Regent 200SC were prepared in the field and were broadcast by gloved hand onto the soil surface around bushes occupied by hopper bands as overnight roosting sites [34,41]. The densely aggregated hoppers descended from their roosts to bask at the base of the bushes and fed avidly on the bait. Hoppers became visibly affected within 2–3 h and usually died within 24 h.

Batches of 5 and 10 kg of both baits gave excellent control (>95%) of small and medium sized hopper bands roosting over an area of 10–100 m^2 and 101–250 m^2, respectively. However, baiting was only effective if undertaken shortly after sunrise, before hoppers descended from their roosts and scattered from the baited area. Baiting large bands or diffuse targets, or baiting later in the day once hoppers became active, was not effective [41]. Baiting was more practical under autumn conditions in the Karoo when there was more time during the cool mornings to locate and bait bands before they descended from their roosts. Baiting was inexpensive and was easy to prepare and apply, requiring basic equipment and limited training. Baiting was also a more target-specific method of controlling locusts as there was no off-target drift of insecticide and it only killed the hoppers and some other insects that ate it. Baiting could thus be effectively used for small-scale locust control by rural communities, or in conservation areas where more target-specific locust control is required [34]. However, the logistics of the bulk transport, preparation and application of locust baits under large-scale operational conditions appear daunting.

3.4. Insecticide Barrier Treatments

Barrier treatments, based on the application of the organochlorine insecticide, dieldrin, applied to narrow strips of vegetation were extensively used in North Africa to control marching bands of the desert locust [42]. Following the withdrawl of dieldrin during the 1970s due to its high persistence and bio-accumulation in the environment [43], the most promising candidates to have emerged as a replacement for dieldrin in barrier treatments have been various insect growth regulator (IGR) compounds and the phenyl pyrazole insecticide, fipronil. These compounds provide moderate persistence in the environment without the bio-accumulation problems of dieldrin [23]. In recent years, fipronil (Adonis UL®) barriers have been successfully applied from aircraft under operational conditions against hopper bands of the Malagasy locust, the Australian plague locust and the desert locust. However, insecticide barrier treatments have never been used operationally against the brown locust.

The high efficacy of fipronil against the brown locust was established in laboratory bioassays and in small-scale field trials of ULV cover sprays. Field trials of fipronil (Adonis® 5UL) barriers, applied to 21 m-wide strips of Karoo vegetation at a volume rate of 2.5 ℓ/ha (giving an area dose rate of 12.5 g a.i./ha) from motorized 'Solo 432' knapsack sprayers, were used to intercept marching gregarious brown locust hopper bands. Barriers of Adonis® applied at 12.5 g a.i/ha proved very effective against L2–L3 bands and against small-size L4–L5 bands, giving >90% control within 48 h [34]. However, a lower dose rate of 7.5 g a.i./ha in similar barrier strips proved largely ineffective against L3–L5 marching bands. Even at the high 12.5 g a.i./ha dose rate, the barriers sometimes failed to adequately control large and mobile L5 hopper bands that maintained cohesion and had sufficient momentum to march quickly through the relatively narrow barrier strips before the majority of hoppers acquired a lethal dose. For maximum efficacy it was vital that hoppers consumed fipronil-treated vegetation and barriers proved less effective where the vegetation density was sparse or where the vegetation was less palatable to hoppers [34].

Despite these factors, Adonis® barriers were considered to have potential for the passive control of brown locust hopper bands in the more remote western areas of the Karoo, especially during the early stages of an outbreak when young-instar hopper bands become mobile. The high mobility of the hopper bands, which can march at least 1–3 km per day in summer, would suggest that even a low-density network of insecticide barriers, set 5–10 km apart, would soon be intercepted by marching bands. However, the fipronil product has a comparatively high environmental impact against non-target organisms, with the product assigned a high-risk rating by the FAO against terrestrial non-target organisms and a medium risk against mammals, reptiles and aquatic arthropods [44]. This is a similar environmental risk rating as malathion, which is extensively applied as an area treatment against the desert locust. The FAO Pesticide Referee Group recommends a dose rate of 4.2 g/ha Adonis® for wide barrier treatments applied from aircraft against other locust and grasshopper species [44]. Any barrier treatments would have to be judiciously applied to restrict the environmental impact of Adonis®against non-target organisms in the Nama Karoo environment.

3.5. Insect Growth Regulators (IGRs)

The benzoylphenyl urea ICR products only affect the juvenile stages of insects and are thus relatively target specific with a limited environmental impact [45]. Various IGRs have been used operationally as barrier treatments or cover sprays against a number of locust and pest grasshopper species [23]. Laboratory bioassays with the IGRs, flufenoxuron and teflubenzuron, applied to maize leaf discs and fed to L5 brown locust hoppers at dose rates of 3–15 μg/g per hopper, gave variable mortality of 30–70% after a single ingestion [34]. However, when diflubenzuron (Dimilin OF6®), was sprayed onto maize plants at volume rates of 1–3 ℓ/ha and subsequently fed to L2 hoppers, 100% mortality of hoppers was produced within 11 days at all application rates as long as the hoppers were continuously exposed to treated vegetation. However, irregular exposure to Dimilin® during the inter-moult period produced unsatisfactory mortality, as the product is evidently non-accumulative and is readily excreted [46]. The fact that brown locust hoppers have to feed regularly on IGR-treated vegetation during the entire inter-moult period, combined with the high mobility of hopper bands, would probably make IGR barriers unsuitable for brown locust control operations. To increase the likelihood of hoppers ingesting treated vegetation, large areas would have to be block-treated from aircraft, or multiple barriers would have to be applied in parallel across vast tracts of the Karoo. These application strategies are clearly uneconomic and are not environmentally acceptable.

3.6. *Metarhizium Myco-Insecticide*

Myco-insecticides, comprising the aerial conidia of certain isolates of the entomopath ogenic fungus, *Metarhizium acridum* ((Driver and Milner) J.F. Bisch., Rehner and Humber), formulated in oil and applied through conventional spray equipment, have proved effective against a range of locust and pest grasshopper species [47,48]. Myco-insecticides have a low environmental impact and are recommended for locust control in environmentally sensitive areas as replacements for synthetic insecticides [49]. Commercial myco-insecticide products have been shown to be effectively in large-scale operational use against the Australian plague locust, as well as against locust outbreaks in China [50]. Such products have also been used on occasion against the red locust and desert locust.

Under suitable application conditions the *M. acridum* (LUBILOSA isolate IMI 330189) myco-insecticide, applied at a standard dose of 100 g conidia/ha from hand-held ULV spinning disc apparatus, or from microlight aircraft fitted with micronair AU 4000 atomisers, regularly produced >90% mortality of L5 brown locust hoppers sprayed in the field and then maintained in cages [51,52]. However, speed of kill was slow, with median lethal times of 10–13 days for the ground and aerial application trials, respectively. In most cases, acceptable >90% mortality was not achieved for at least three weeks after application, although affected hoppers did show reduced feeding behaviour [53]. Despite the slow speed of kill in the small-scale trials, the myco-insecticide agent was considered a significant advance in brown locust control and was commercially registered as Green Muscle® in South Africa in 1998 under the South African Agricultural Remedies Act (Act 36 of 1947). However, subsequent studies showed that various factors negatively affected the performance of Green Muscle® against the brown locust. The hot and dry Karoo environment, combined with the high ultra-violet radiation typical of the latitude and altitude of the Karoo, is detrimental for the survival and transmission of fungal conidia. The most important constraint, however, proved to be the unusually active thermoregulation behaviour of brown locust hoppers in the field that enabled them to elevate and maintain their body temperature 2–3 °C above normal [54]. This 'behavioural fever' elevated the body temperature of the locust above the threshold for optimum growth of the fungal pathogen and effectively suppressed the development of mycosis under certain field conditions for up to 70 days [55]. The great variability of performance of the myco-insecticide was not evident in the cage samples taken during the earlier trial work, as the caged locusts were unable to thermo-regulate to the same extent to defer the onset of mycosis. This unfortunately gave a false impression of the true speed of kill in the field.

The lack of a rapid knock-down action and the slow speed of kill currently makes Green Muscle® unsuitable for spot spraying operations in the Karoo. The thousands of individual hopper bands treated during control campaigns, and the high mobility of marching bands, would make the easy recognition of treated and untreated targets by locust officers impossible. Its use in conservation areas is also currently not practical, as treated bands would soon march out of these relatively small and fragmented areas. An alternative application strategy to manage the slow speed of kill needs to be developed and tested, before Green Muscle® can be considered for brown locust control.

3.7. Pathogenic Micro-Organisms

Apart from entomopathogenic fungi, a range of other pathogenic micro-organisms have been isolated from locusts and grasshoppers, including bacteria, entomopoxvirus, microsporidia, protists and nematodes [56]. Biological control of locusts and grasshoppers using the augmentative release of pathogens has met with mixed success [57] and has been considered as being impractical because of the technical problems and expense involved in the mass production and application of pathogens.

Various strains of the spore forming bacteria, *Bacillus thuringiensis* Berliner (Bt), have been commercialised world-wide as microbial insecticides against various Lepidoptera, Coleoptera and Diptera pests. However, none of the commercially available Bt strains has proved sufficiently pathogenic against locusts because of the acid pH of the Orthoptera gut which prevents the Bt endotoxin crystals from dissolving [58]. In bioassays against the brown locust, four Bt strains that showed an enhanced acid-tolerance were fed to L3 and L4 brown locust hoppers as freeze-dried Bt spores mixed into bran bait at a dose rate of 1mg spores/g dry bran. However, none of the Bt baits produced any significant mortality compared with the untreated controls [34].

Another pathogenic micro-organism evaluated against the brown locust was an entomopoxvirus (EPV) isolated from a West African grasshopper, *Odaleus senegalensis* (De Geer), which had shown potential for the control of some rangeland grasshopper pests in the western USA [59]. However, bioassays of the virus suspension applied to maize leaf discs and fed to different brown locust hopper instars at a range of dosage rates did not produce any observed infection or mortality compared with untreated controls [34]. Entomopoxviruses are known to be relatively host specific [60] and the virulence of the imported EPV against acridids other than its direct grasshopper host may be low.

4. Alternative Control Strategies

The regular and often intense outbreaks of the brown locust present a formidable pest control problem. Locust control operations have evidently failed as they have been unable to stop the regular plague cycles from developing [22]. The typical large-scale incipient upsurges over a wide area of the Karoo are impossible to prevent and once the populations develop into gregaria phase eruptions then the only option is to undertake a large-scale and expensive chemical control campaign. However, chemical control operations are rarely able to subdue the large-scale eruptions on their own and such extensive control campaigns are usually assisted by the onset of unfavourable dry or cold climatic conditions that restricts further locust breeding and supresses the eruptions. Fipronil (Adonis®) barrier treatments and insecticide baits were considered to have sufficient potential to be incorporated into an integrated management strategy against brown locust hopper bands in certain areas of the Karoo. Although there is concern regarding the environmental impact of fipronil, the barriers could have potential if judiciously applied to intercept marching hopper bands in the more remote areas of the Karoo, while insecticide baits could have application in specific areas where a low environmental impact was required. However, no alternative control method was considered sufficiently flexible and effective enough under all field conditions to entirely replace the application of synthetic pyrethroid insecticides in the Karoo. Fast-acting ULV insecticides will hence continue to provide the backbone of the brown locust control programme for the foreseeable future.

The current 'Commando' system of ground-based locust control has been in operation for decades in the Karoo, partly because it has provided a relatively cost-effective strategy against the unpredictable intensity and distribution of brown locust outbreaks across the Karoo, while also ensuring that Government finances are channelled to pay local inhabitants to fight the locusts as temporary employees when needed. However, the tracking down and spraying of thousands of individual hopper bands and roosting swarms has long been considered as an inefficient use of manpower and resources. Records dating back to the mid-1960s and again confirmed in the 1990s showed that the targeting the hopper bands accounted for up to 90% of the locust control budget, but only accounted for 10%

of the actual number of locusts controlled, whereas the targeted spraying of roosting adult swarms accounted for an estimated 90% of the locust population [61,62]. In recent years, the traditional Commando system has also become more difficult to sustain due to the spiralling costs of transport and insecticides, as well as the costs of hiring the large temporary labour force. The ongoing depopulation of farms in the more remote and arid areas of the Karoo means that there is a very low density of resident farmers left to report the locusts. In addition, the changing attitudes of some landholders towards reporting locusts due to their conservation views on the spraying of pesticides, or more especially regarding the security concerns on the farms, has therefore limited access to farms for locust surveys and has negatively impacted the efficacy of the locust reporting and control network. Outbreaks on the large-sized farms in the remote Central Karoo and Bushmanland areas therefore often go undetected and the local control capacity can be suddenly overwhelmed by swarm escapes.

Alternative strategies of combatting outbreaks in the remote areas of the Karoo are required as an alternative to the traditional active location and spot spraying of individual locust targets. There is therefore a need to discuss possible alternative strategies based on sound economic and logistical principals.

4.1. Abandoning Brown Locust Control Entirely

Historical records showed that the uncontrolled brown locust plagues in the past (before 1940) all eventually collapsed due to a combination of factors including the build-up of natural enemies and diseases, the onset of unfavourable climatic conditions, or as a result of the exodus of swarms into areas that were unsuitable for further breeding, e.g., the mountain Kingdom of Lesotho. Earlier plague cycles were then always followed by a long recession period since the swarming populations had all vacated the optimal breeding areas in the Karoo [12,22]. By not controlling locust outbreaks, it can be argued that the outbreaks will always come to an end naturally and the money saved on control operations could be used instead to compensate for crop losses. However, the uncontrolled brown locust plague cycles between 1890 and 1940, as well as the various invasions of red locust and African migratory locust plagues into South Africa during this time, posed a devastating threat to agricultural production throughout southern Africa, causing the threat of starvation amongst rural communities. The South African and other African Governments will not tolerate this threat to food security again.

There is still an economic and environmental debate on whether locusts and grasshoppers should be chemically controlled to protect rangeland grazing [32], especially in the Karoo where the cost of locust control per hectare is usually far more than the grazing value of the land [63]. However, other studies estimated the amount of grass consumed by the brown locust hoppers and correlated this with the consumption by sheep, suggesting that it was economically viable to control the locusts in the Karoo to save the available grazing for the sheep flocks [64]. To counter these findings, it can be argued that most of the locust control operations are undertaken against late instar hopper bands and fledgling swarms when the damage to the local grazing has already been done. There is no doubt, however, regarding the need to control locust and grasshopper outbreaks within the rangeland areas to protect vulnerable staple food crops such as maize, wheat and millet grown outside the rangeland areas [32,65]. Likewise in the Karoo, the brown locust problem will have to be continually managed and contained within the outbreak area so that damage to food crops outside the Karoo is prevented as much as possible.

4.2. Update the Current "Commando System"

Within the Upper Karoo region (e.g., area containing the Districts of Prieska, Douglas, Strydenburg, Britstown, De Aar, Philipstown, Hopetown, Hanover, Richmond, Victoria West), as well as the adjacent Districts to the east and south-east, there is a higher population of resident farmers and farm workers on the relatively smaller-size farms (2000–5000 ha), as well as a good communication network. In these areas, the traditional Commando system of locust control can still work well once the manpower and operational resources are in place and is generally considered to be a viable and cost-effective strategy for brown locust control. However, in the more remote western areas of the arid Bushmanland and the Central Karoo, the Commando system has become ineffective due to the demographic and economic factors discussed earlier. If the Commando system is still to be applied throughout the Karoo it will have to be modernised and substantially resourced to improve the capacity to locate locust targets by implementing an effective locust monitoring and outbreak early warning system. This implies a substantial investment in regular monitoring of known locust outbreak 'hot-spot' areas, as well as resources for the implementation of technologies for effective target location and mapping systems to support the Commando teams. It is also highly recommended that the control teams should focus on the spraying of the fledgling swarms as they coalesce, as well as the larger-sized hopper bands as they aggregate, rather than spending most of their time and resources tracking down and spraying individual small-sized hopper band targets.

There are various digital field tools and GIS mapping systems used in locust control forecasting and field control operations in other parts of the world, such as the FAO's recently introduced 'eLocust3' system used in the desert locust campaigns. Survey teams can log the local survey observation data, as well as information on local climatic and vegetation conditions, onto a digital tablet that then transmits the data via satellite to a national locust command centre. All the data is precisely GPS geo-referenced and the survey or control data is then automatically downloaded onto the FAO's 'RAMSESv4' database software application and GIS mapping platform, which can combine multiple observation inputs along with dedicated layers of GIS spatial information to generate visual maps of the target areas. The GIS maps are then a powerful tool for early warning forecasting, or for directing field control operations.

Climate monitoring and vegetation greening index satellites are also employed for wide-scale desert locust forecasting in order to monitor rainfall events, soil moisture conditions and vegetation greening in remote desert regions. Such technology is readily available in South Africa and should be adopted by the brown locust management strategy to assist with outbreak forecasting. Likewise, there are sophisticated climatic models used in crop yield forecasting that could be readily adopted to model and predict rainfall conditions in the Karoo over the medium-term. The use of unmanned aerial vehicles (drones) is also showing potential for locust survey work in desert locust campaigns and can be used to locate green vegetation zones in desert areas where locusts are likely to aggregate and oviposit. The survey and location of isolated ecotone areas of tall grass and open bare patches of soil typically suitable for brown locust adult aggregation and for nearby oviposition, defined as the locust 'concentration belts' [16], could certainly benefit the locust survey teams in the Karoo by reducing the need for the extensive foot surveys.

4.3. Abandon Hopper Control and Target Adult Swarms Only

Using modern spray aircraft to target young swarms as they mill around and aggregate into large-size targets is standard control strategy in various countries, such as Australia and Madagascar. The similar convergence and coalescence of brown locust swarms into large swarm targets has also been regularly observed in the Karoo [61], Price, *pers. obs.* Apart from the high work rate of spray aircraft and the large areas that can be quickly spayed, adult locusts are also known to be more susceptible targets to kill with insecticides than late instar hoppers, which has important consequences for reduced insecticide application and environmental impact. Aircraft would only be brought into operation once a threshold number or size of swarms had developed and once cereal crops were threatened. The co-ordinated air-to-air spraying of flying swarms using low dosages of synthetic pyrethroid insecticides, accurately applied from aircraft fitted with GPS systems can effectively control even large-scale plagues. The active spotting and tracking of swarms could be done along strategically placed observation stations.

A limited number of small to medium capacity spray aircraft, such as the Piper Pawnee, Gippsland GA200C and the Air Tractor series, could intercept large-size swarms as they aggregated and migrated along their traditional seasonal flyways out of the Central and Upper Karoo. The active spotting and tracking of swarms could be done along strategically placed observation stations. Aircraft operating from the larger towns in the Upper Karoo, such as De Aar, Britstown and Hopetown, could attack swarms once they entered a 100–150 km wide buffer zone stretching along the Orange River. Outbreaks in the Great Karoo could be controlled with aircraft stationed at Beaufort West or other suitable locations. Small-scale outbreaks may even be managed with microlight aircraft equipped with appropriate spray gear for ULV application, as has been demonstrated in earlier trial work carried out by the ARC-PHP.

5. Conclusions

The current Commando system of managing the brown locust is only considered as being viable and effective in the Upper and eastern Karoo regions where there is a sufficient density of resident farmers and farm workers who can report outbreaks and where good communications exist to be able to rapidly respond and obtain access to the locust targets. However, the Commando system is not currently effective over a vast area of the Central and Great Karoo and Bushmanland. The scattered outbreaks throughout these areas should be left to fledge into adult swarms, which would then soon aggregate into large-sized swarms that would make viable targets for spray aircraft. The benefits (pros) and constraints (cons) of the alternative brown locust control strategies discussed are summarized in Table 3.

Table 3. Comparison of current and alternative strategies to manage the brown locust.

Control Strategies, Current and Alternatives	Benefits (pros)	Constraints (cons)
0-'Commando' system	Current strategy. Long history, community-based system, local knowledge and communication networks, per diem payments to local people for operations, good control of small outbreaks and able to dampen large upsurges. Spray roosting swarms at night so more time to control large-size targets.	Not effective in some areas, weak reporting from remote farms, reduced access to all farms, mainly target L5 bands so not cost effective, not able to stop large upsurges or prevent plague cycles. No central coordination of operations or situation over-view, no early warning or mapping of outbreaks, sub-optimal use of strategic resources.
1-Abandoning brown locust control entirely	With natural enemies and diseases, unfavourable climatic conditions, exodus of swarms into unsuitable areas, outbreaks will always end naturally. Ecologically sustainable. Money saved on control operations could be used to compensate for crop losses. No environmental damage from pesticides.	Large-scale outbreaks cause severe damage to Karoo grazing, widespread damage to commercial cereal crops outside the Karoo, threat of starvation amongst smallholder farmers and rural communities. Swarm invasions into neighbouring countries, political pressure on South Africa to contain locust outbreaks.
2-Update the current "Commando system"	Adoption of modern GIS technology for target reporting and campaign management, more focus on control of aggregating swarms, more effective deployment of resources, better planning and direction of operations, more tactical use of spray aircraft when required. A viable and cost-effective strategy for brown locust control in some areas, such as the Upper and eastern Karoo. Will work well once the manpower and operational resources are in place.	Outdated and not currently effective over a vast area of the Central and Great Karoo and Bushmanland. Substantial investment necessary in new technologies. Will still have weak reporting of targets in remote areas, risk of sub-optimal adoption and use of technologies. Fewer locust control teams required—loss of income for communities. Availability, coordination and high costs of spray aircraft.

Table 3. *Cont.*

Control Strategies, Current and Alternatives	Benefits (pros)	Constraints (cons)
3-Abandon hopper control and target adult swarms only	Adult swarms naturally coalesce to form large targets. Adult locusts are known to be more susceptible targets to kill with insecticides than late instar hoppers. Farmers and locust officers spot and report the swarms, limited number of small to medium capacity spray aircraft required. Sophisticated commercial aerial spray companies in South Africa. Reduced insecticide application and environmental impact.	During large outbreaks adult populations develop simultaneously over wide areas, swarm spotting capacity and spray aircraft resources overwhelmed, massed swarm escapes from the Upper and eastern Karoo. More locust damage to Karoo grazing and increased threat to cereal crops outside Karoo. Political pressure from commercial farmers and rural communities, swarm invasion of neighbouring countries.
4-Necessity of an IPM approach Modernised and well-resourced Commando system in the Upper Karoo and eastern Karoo, spraying fledgling swarms and large hopper bands. + Stop using ground-based 'Commando' system in remote areas of central and western Karoo where ineffective and switch mainly to aircraft control of aggregating swarms.	Modernize and support Commando system in Upper and eastern Karoo. Stop wasting finances and resources with current ineffective system in remote areas. Locust officers now employed to spot and track aggregating swarms and to direct spray aircraft to targets. Use modern technology to map outbreaks and to direct ground operations and aircraft campaign. More effective allocation of manpower and resources. More effective kill of locusts (dead locusts per litre of insecticide).	Farmer political pressure that remote areas of Karoo are being 'neglected', loss of potential income for locust officers and spray teams. Manpower for spotting and tracking swarm targets in remote areas. Availability and coordination of aircraft for swarm control. High costs of hire aircraft capacity. New ULV formulations required for aircraft. Short window to spray roosting swarm targets in the morning.

When reviewing possible options for controlling the brown locust, the most effective strategy probably lies with an integrated locust management strategy, with a modernised and well-resourced Commando system in the Upper Karoo and eastern Karoo that focusses on the spraying of coalescing fledgling swarms and the larger hopper band targets, supported by the judicious use of aircraft for the aerial control of swarms developing in the more remote regions of the Karoo. Such an integrated strategy combining ground and aerial tactics would have the flexibility and the capacity to deal effectively with all locust emergencies. It is also vital to invest more resources in technology for the early-warning forecasting of brown locust outbreaks in relation to climatic conditions, along with timely surveys of known outbreak hot-spot farms early in the season. The use of digital technology data gathering and mapping systems would also enhance the ability to manage the outbreaks as they developed and would allow resources to be deployed more judiciously.

Funding: Received from the ARC-Plant Health and Protection.

Institutional Review Board Statement: Not applicable.

Informed Consent Statement: None applicable.

Data Availability Statement: Data sharing not applicable.

Acknowledgments: To my late colleagues and mentors, Dick Brown and Peter Napier-Bax, for their camaraderie on our numerous locust field trips throughout the Karoo. Thanks to the many hospitable Karoo farmers who allowed us to work on their land and to the locust officers who enthusiastically guided us to the locust targets. Thanks to the ARC-Plant Health and Protection for the financial and administrative support, as well as the various Agro-Chemical companies and to CABI for financial support for specific projects within the study. Thanks to the South African Department of Agriculture for providing the weekly locust control data and to my colleague, Margaret Kieser, for analysing and computing the locust data.

Conflicts of Interest: The author declares no conflict of interest.

References

1. Faure, J.C.; Marais, S.J.S. The control of Locustana pardalina in its outbreak centres. In Proceedings of the 4th International Locust Conference, Cairo, Egypt, 22 April 1936. 5p.
2. Du Plessis, C. *The Influence of Weather Conditions on the Incipient Swarming of the Brown Locust, Locustana Pardalina (Walker)*; Science Bulletin, Department of Agriculture and Forestry, Union of South Africa Government Printer: Pretoria, South Africa, 1938; Volume 186, pp. 1–51.
3. Lea, A. Recent outbreaks of the brown locust with special reference to the influence of rainfall. *J. Entmol. Soc. S. Afr.* **1958**, *21*, 162–213.
4. Kieser, M.E.; Thackrah, A.; Rosenberg, J. Changes in the outbreak region of the brown locust in Southern Africa. *Grootfontein Agric.* **2002**, *4*, 20–23. Available online: http://gadi.agric.za/articles/Kieser_M/kieser_vol4_2002_locust.php (accessed on 9 August 2021).
5. Du Plessis, C. *The Incipient Outbreaks of the Brown Locust in 1937–1938 with Special Reference to Biometrical Analysis*; Science Bulletin, Department of Agriculture and Forestry, Union of South Africa Government Printer: Pretoria, South Africa, 1939; Volume 209, pp. 1–69.
6. Faure, J.C. The phases of locusts in South Africa. *Bull. Entmol. Res.* **1932**, *23*, 293–405. [CrossRef]
7. Uvarov, B.P. *Grasshoppers and Locusts*; Cambridge University Press: Cambridge, UK, 1966; Volume 2.
8. Faure, J.C. The life history of the brown locust. *J. Dept. Agric. Union S. Afr.* **1923**, *4*, 205–224.
9. Potgieter, J.T. *A Contribution to the Biology of the Brown Swarm Locust and Its Natural Enemies*; Science Bulletin, Department of Agriculture and Forestry, Union of South Africa Government Printer: Pretoria, South Africa, 1929; Volume 82, pp. 1–48.
10. Smit, C.J.B. *Field Observations on the Brown Locust in an Outbreak Centre*; Science Bulletin, Department of Agriculture and Forestry, Union of South Africa Government Printer: Pretoria, South Africa, 1939; Volume 190, pp. 1–143.
11. Smit, C.J.B. The behaviour of the brown locust in its solitary phase. In *Technical Communication*; Department of Agricultural and Technical Services: Pretoria, South Africa, 1960; Volume 1, pp. 1–132.
12. Lea, A. Natural regulation and artificial control of brown locust numbers. *J. Entomol. Soc. S. Afr.* **1968**, *31*, 89–112.
13. Lea, A. The distribution and abundance of brown locusts between 1954 and 1965. *J. Entomol. Soc. S. Afr.* **1969**, *32*, 367–398.
14. Matthee, J.J. *The Structure and Physiology of the Egg of Locustana pardalina (Walker)*; Science Bulletin, Department of Agriculture and Forestry, Union of South Africa Government Printer: Pretoria, South Africa, 1951; Volume 316, pp. 1–83.
15. Botha, D.H. The viability of brown locust eggs, *Locustana pardalina* (Walker). *S. Afr. J. Agric. Sci.* **1967**, *10*, 445–460.
16. De Wet, W.J.; Webb, D. *Field observations on the behaviour of hoppers of the brown locust in the swarming phase*; Science Bulletin, Department of Agriculture and Forestry, Union of South Africa. Government Printer: Pretoria, South Africa, 1951; Volume 337, pp. 1–38.
17. Price, R.E. The life cycle of the brown locust, with reference to egg viability. In Proceedings of the Locust Symposium, MacGregor Museum, Kimberley, South Africa, 11–13 May 1987; Mckensie, B., Longridge, M., Eds.; South African Institute of Ecologists Bulletin, Special Issue: Rondebosch, South Africa, 1988; pp. 27–40.
18. Lounsbury, C.P. *The locust invasion. Agricultural Journal of the Cape of Good Hope*; Department of Agriculture, Cape of Good Hope: Cape Town, South Africa, 1906; pp. 596–600.
19. Price, R.E.; Brown, H.D. A century of locust control in South Africa. In Proceedings of the Workshop on Research Priorities for Migrant Pests of Agriculture in Southern Africa, PPRI, Pretoria, South Africa, 24–26 March 1999; Cheke, R.A., Rosenburg, L.J., Kieser, M.E., Eds.; Natural Resources Institute: Chatham, UK, 2000; pp. 37–50.
20. Lounsbury, C.P. Some phases of the locust problem. *S. Afr. J. Sci.* **1915**, *11*, 33–45.
21. Lea, A. Some major factors in the population dynamics of the brown locust, *Locustana pardalina* (Walker). *Monogr. Biol.* **1964**, *14*, 269–283.
22. Lea, A. The plague dynamics of the brown locust, Locustana pardalina (Walk.). In Proceedings of the International Study Conference on Current and Future Problems of Acridology, London, UK, 6–16 July 1970; Hemming, C.F., Taylor, T.H.C., Eds.; Centre for Overseas Pest Research: London, UK, 1970; pp. 289–297.
23. Lecoq, M. Recent progress in Desert and Migratory locust management in Africa. Are preventative actions possible? *J. Orthoptera Res.* **2001**, *10*, 271–276. [CrossRef]
24. Brown, H.D.; Bashir, E.A. Criteria for classification of brown locust outbreaks. In Proceedings of the 4th Meeting of the Migratory Pests Sub-Committee of the Southern African Regional Commission for the Conservation and Utilization of the Soil (SARCCUS), Windhoek, Namibia, 4–6 June 1990.
25. Mally, C.W. Arsenite of soda as a locust poison. *J. Dept. Agric. Union. S. Afr.* **1923**, *13*, 1–15.
26. Policy for Managing the Locust Problem in South Africa. National Department of Agriculture. September 1988. Available online: https://www.nda.agric.za/docs/policy/policy98.htm (accessed on 9 August 2021).
27. Peters, S.M.; Lindeman, H.; van der Westhuizen, M.C. Locust control data analysis: Environmental, manpower and financial analysis of locust control in South Africa. In Proceedings of the Workshop on Research Priorities for Migrant Pests of Agriculture in Southern Africa, PPRI, Pretoria, South Africa, 24–26 March 1999; Cheke, R.A., Rosenburg, L.J., Kieser, M.E., Eds.; Natural Resources Institute. Chatham, UK, 2000; pp. 75 87.

28. Brown, H.D. Current pesticide application: Effectiveness and persistence. In Proceedings of the Locust Symposium, MacGregor Museum, Kimberley, South Africa, 11–13 May 1987; Mckensie, B., Longridge, M., Eds.; South African Institute of Ecologists Bulletin, Special Issue: Rondebosch, South Africa, 1988; pp. 101–117.
29. Nailand, P.; Hanrahan, S.A. Modelling brown locust, *Locustana pardalina* (Walker), outbreaks in the Karoo. *S. Afr. J. Sci.* **1993**, *89*, 420–424.
30. Todd, M.C.; Washington, R.; Cheke, R.A.; Kniverton, D. Brown locust outbreaks and climate variability in southern Africa. *J. App. Ecol.* **2002**, *39*, 31–42. [CrossRef]
31. Riegert, P.W.; Ewen, A.B.; Lockwood, J.A. A history of chemical control of grasshoppers and locusts 1940–1990. In *The Bionomics of Grasshoppers, Katydids and Their Kin*; Gangwere, S.K., Muralirangan, M.C., Muralirangan, M., Eds.; CAB International: Wallingford, UK, 1997; pp. 385–405.
32. Peveling, R. Environmental conservation and locust control—Possible conflicts and solutions. *J. Orthoptera Res.* **2001**, *10*, 171–187. [CrossRef]
33. Stewart, D.A.B. Non-target grasshoppers as indicators of the side effects of chemical locust control in the Karoo, South Africa. *J. Insect Conserv.* **1998**, *2*, 263–276. [CrossRef]
34. Price, R.E. Alternative Methods of Controlling the Brown Locust, *Locustana pardalina* (Walker). Ph.D. Thesis, University of the Free State, Bloemfontein, South Africa, 2003; pp. 1–152. Available online: http://hdl.handle.net/11660/6227 (accessed on 9 August 2021).
35. Botha, D.H.; Lea, A. Do brown locust periods come in cycles? *Farming S. Afr.* **1970**, *387*, 1–5.
36. Erasmus, B.H. Natural predators and parasites of the brown locust—an ecological appraisal. In Proceedings of the Locust Symposium, MacGregor Museum, Kimberley, South Africa, 11–13 May 1987; Mckensie, B., Longridge, M., Eds.; South African Institute of Ecologists Bulletin, Special Issue: Rondebosch, South Africa, 1988; pp. 118–137.
37. Van Schalkwijk, H.A.D. The status of *Wohlfahrtia euvitatta* Vill. (Diptera, Sarcophagidae) as a parasite of the brown locust. *J. Entmol. Soc. Sth. Afr.* **1939**, *2*, 18–35.
38. Price, R.E.; Brown, H.D. The status of the locust fly, *Wohlfahrtia pachytyli* (Diptera: Sarcophagidae), in the Karoo and the impact of locust control operations on its abundance. *Afr. Entmol.* **2006**, *14*, 35–43.
39. Anonymous. *First Report of the Committee of Control of the South African Central Locust Bureau*; Fuller, C., Ed.; Republic of South Africa, Locust Control and Research, Department of Agriculture, Cape of Good Hope: Cape Town, South Africa, 1907; Volume 1.
40. Du Plessis, C.; Botha, D.H. Preliminary field experiments on the attractiveness of certain chemicals and bait carriers to hoppers of the brown locust. *J. Entomol. Soc. S. Afr.* **1939**, *2*, 74–92.
41. Price, R.E.; Brown, H.D. Locust control by means of selective baiting. In *New Strategies for Locust Control*; Krall, S., Peveling, R., Ba Diallo, D., Eds.; Birkhäuser: Basel, Switzerland, 1997; pp. 209–217.
42. Bennett, L.V.; Symmons, P.M. A review of the effectiveness of certain control techniques against the Desert locust. *Anti. Locust Bull.* **1972**, *50*, 1–15.
43. Brader, L. Control of grasshopper and migratory locusts. In Proceedings of the Crop Protection Conference (Pests and Diseases), Brighton, UK, 21–24 November 1988; pp. 283–288.
44. Matthews, G.A. The Pesticide Referee Group of FAO and its contribution to locust control. *J. Orthoptera Res.* **2005**, *14*, 203–206. [CrossRef]
45. Everts, J.W. *Environmental Effects of Chemical Locust and Grasshopper Control. A Pilot Study Project ECLO/SEN/003/NET*; FAO: Rome, Italy, 1990; 227p.
46. Neuman, R.; Guyer, W. Biochemical and toxicological differences in the mode of action of the benzoylureas. *Pest. Sci.* **1987**, *20*, 147–156. [CrossRef]
47. Bateman, R.P. The development of a mycoinsecticide for the control of locusts and grasshoppers. *Outlook Agric.* **1997**, *26*, 13–18. [CrossRef]
48. Lomer, C.J.; Prior, C.; Kooyman, C. Development of *Metarhizium* spp. for the control of grasshoppers and locusts. *Mem. Entomol. Soc. Canada* **1997**, *171*, 265–286. [CrossRef]
49. Food and Agriculture Organisation of the United Nations. Evaluation of Field Trial Data on the Efficacy and Selectivity of Insecticides on Locusts and Grasshoppers. In Proceedings of the Report to FAO by the Pesticide Referee Group, 8th Meeting, Rome, Italy, 11–14 October 1999. Available online: http://www.fao.org/ag/locusts/common/ecg/800_en_PRG08e.pdf (accessed on 9 August 2021).
50. Love, G.; Riwoe, D. *Economic Costs and Benefits of Locust Control in Eastern Australia. ABARE eReport 05.14 Prepared for the Australian Plague Locust Commission*; Australian Bureau of Agricultural and Resource Economics: Canberra, Australia, 2005.
51. Bateman, R.P.; Price, R.E.; Müller, E.J.; Brown, H.D. Controlling brown locust hopper bands in South Africa with a myco-insecticide spray. In Proceedings of the Crop Protection Conference (Pests and Diseases), Brighton, UK, 21–24 November 1994; Volume 2, pp. 609–616.
52. Price, R.E.; Bateman, R.P.; Brown, H.D.; Butler, E.T.; Müller, E.J. Aerial spray trials against brown locust, *Locustana pardalina* (Walker), nymphs in South Africa using an oil- formulation of *Metarhizium flavoviride*. *Crop Prot.* **1997**, *16*, 345–351. [CrossRef]
53. Müller, E.J. Control of the brown locust, *Locustana pardalina* (Walker) (Orthoptera: Acrididae), using a myco-insecticide: Addressing the issues of speed of kill, dose rate, mortality and reduction in feeding. *Afr. Entmol.* **2001**, *8*, 217–221.

54. Blanford, S.; Thomas, M.B. Thermal behaviour of two acridid species: Effects of habitat and season on body temperature and the potential impact on biocontrol with pathogens. *Environ. Entmol.* **2000**, *29*, 1060–1069. [CrossRef]
55. Arthurs, S.P.; Thomas, M.B. Effects of a mycoinsecticide on feeding and fecundity of the brown locust, *Locustana pardalina*. *Biocontrol. Sci. Technol.* **2000**, *10*, 321–329. [CrossRef]
56. Streett, D.A.; McGuire, M.R. Pathogenic diseases of grasshoppers. In *Biology of Grasshoppers*; Chapman, R.F., Joern, A., Eds.; John Wiley & Sons: New York, NY, USA, 1990; pp. 483–516.
57. Streett, D.A.; Woods, S.A.; Erlandson, M.A. Entomopoxviruses associated with grasshoppers and locusts: Biology and biological control potential. In *Microbial Control of Grasshoppers and Locusts*; Goettel, M.S., Johnson, D.L., Eds.; Memoirs of the Entomological Society of Canada: Ottawa, ON, Canada, 1997; Volume 171, pp. 115–130.
58. Zelazny, B.; Goettel, M.S.; Keller, B. The potential of bacteria for the microbial control of grasshoppers and locusts. In *Microbial Control of Grasshoppers and Locusts*; Goettel, M.S., Johnson, D.L., Eds.; Memoirs of the Entomological Society of Canada: Ottawa, ON, Canada, 1997; Volume 171, pp. 147–156.
59. Streett, D.A. Future prospect for microbial control of grasshoppers. In *Integrated Pest Management on Rangeland*; Capinera, J.L., Ed.; Westview Press: Boulder, CO, USA, 1987; pp. 205–218.
60. Prior, C.; Greathead, D.J. Biological Control of Locusts: The Potential for the Exploitation of Pathogens. *FAO Plant Prot. Bull.* **1989**, *37*, 37–48.
61. Cilliers, J.; van Someren-Greve, J.; Lea, A. A case of converging swarms of the brown locust and implications for control strategy. *S. Afr. J. Agric. Sci.* **1964**, *7*, 867–874.
62. Steenkamp, D.; Botha, T.; van der Westhuizen, L. Cost benefit analysis of locust control in South Africa. In *Insects in African Entomology and Environment, Proceedings of the Joint Conference Entomological Society of South Africa (11th Congress) and the Africa Association Insect Scientists (12th Congress), Stellenbosch, South Africa, 30 June–4 July 1997*; Entomological Society of Southern Africa: Pretoria, South Africa, 1997; p. 81.
63. Esler, K.J.; Milton, S.J.; Dean, W.R.J. *Karoo Veld: Ecology and Management*; Briza Publications: Pretoria, South Africa, 2010; 214p.
64. Coetsee, J.C. Weidingsimpak van Die Bruinsprinkaan, *Locustana pardalina* (Walker) in Die Karoo-Uitbraakgebiet. Master's Thesis, University of Free State, Bloemfontein, South Africa, 1994; 132p.
65. Krall, S.; Herok, C. Economics of desert locust control. In *New Strategies for Locust Control*; Krall, S., Peveling, R., Ba Diallo, D., Eds.; Birkhäuser: Basel, Switzerland, 1997; pp. 401–414.

agronomy

MDPI

Review

A Review of the Biology, Ecology, and Management of the South American Locust, *Schistocerca cancellata* (Serville, 1838), and Future Prospects

Eduardo V. Trumper [1,*], Arianne J. Cease [2], María Marta Cigliano [3,4], Fernando Copa Bazán [5], Carlos E. Lange [4], Héctor E. Medina [6], Rick P. Overson [2], Clara Therville [2], Martina E. Pocco [3,4], Cyril Piou [7], Gustavo Zagaglia [8] and David Hunter [9]

[1] Estación Experimental Agropecuaria Manfredi, Instituto Nacional de Tecnología Agropecuaria (INTA), Manfredi X5988, Argentina
[2] School of Sustainability, Arizona State University, Tempe, AZ 85287, USA; acease@asu.edu (A.J.C.); roverso@asu.edu (R.P.O.); clara.therville@gmail.com (C.T.)
[3] División Entomología, Museo de La Plata, FCNyM-UNLP, La Plata 1900, Argentina; cigliano@fcnym.unlp.edu.ar (M.M.C.); martinapocco@fcnym.unlp.edu.ar (M.E.P.)
[4] Centro de Estudios Parasitológicos y de Vectores, UNLP-CONICET-CICPBA, La Plata 1900, Argentina; carlosl@cepave.edu.ar
[5] Instituto de Investigaciones Agrícolas El Vallecito, Universidad Autónoma Gabriel René Moreno, Km 7 1/2 Carretera al Norte, Santa Cruz de la Sierra, Santa Cruz, Bolivia; anfecopa@gmail.com
[6] Servicio Nacional de Sanidad y Calidad Agroalimentaria (SENASA), Av. Paseo Colón 367, Buenos Aires C1063ACD, Argentina; hmedina@senasa.gob.ar
[7] Unité Mixte de Recherche Centre de Biologie pour la Gestion des Populations (UMR CBGP), Centre de Coopération Internationale en Recherche Agronomique pour le Développement (CIRAD), 755 Avenue du Campus Agropolis, CS 30016, F-34398 Montpellier, France; cyril.piou@cirad.fr
[8] Servicio Nacional de Sanidad y Calidad Agroalimentaria (SENASA), España 366 Piso 3, Salta A4400, Argentina; gzagaglia@senasa.gob.ar
[9] Locust and Grasshopper Control, 125 William Webb Drive, McKellar, Canberra, ACT 2617, Australia; davidmhunter100@gmail.com
* Correspondence: trumper.eduardo@inta.gob.ar

Citation: Trumper, E.V.; Cease, A.J.; Cigliano, M.M.; Copa Bazán, F.; Lange, C.E.; Medina, H.E.; Overson, R.P.; Therville, C.; Pocco, M.E.; Piou, C.; et al. A Review of the Biology, Ecology, and Management of the South American Locust, *Schistocerca cancellata* (Serville, 1838), and Future Prospects. *Agronomy* **2022**, *12*, 135. https://doi.org/10.3390/agronomy12010135

Academic Editor: Stefano Bedini

Received: 22 October 2021
Accepted: 24 December 2021
Published: 6 January 2022

Abstract: In the first half of the twentieth century, the South American Locust (SAL), *Schistocerca cancellata* (Serville, 1838), was a major pest of agriculture in Argentina, Bolivia, Paraguay, Uruguay, and Brazil. From 1954–2014, a preventive management program appeared to limit SAL populations, with only small- to moderate-scale treatments required, limited to outbreak areas in northwest Argentina. However, the lack of major locust outbreaks led to a gradual reduction in resources, and in 2015, the sudden appearance of swarms marked the beginning of a substantial upsurge, with many swarms reported initially in Argentina in 2015, followed by expansion into neighboring countries over the next few years. The upsurge required a rapid allocation of resources for management of SAL and a detailed examination of the improvements needed for the successful management of this species. This paper provides a review of SAL biology, management history, and perspectives on navigating a plague period after a 60-year recession.

Keywords: locust plagues; population dynamics; management

1. Introduction

Locusts are among the most devastating pests globally and leave clear imprints in memories and historical records. The substantial impact of the South American Locust (SAL), *Schistocerca cancellata* (Serville, 1838), on agriculture and cattle production shaped the early stages of the government agricultural service in Argentina. Going back at least to the first half of the nineteenth century and during the first half of the twentieth century, there were frequent plague periods where SAL invaded other countries, including Bolivia,

Paraguay, Uruguay, and Brazil [1]. The available historic records show that recession periods were the exception rather than the rule [1], but between 1954 and 2014, coinciding with the implementation of a preventive management program, there were no large-scale upsurges, with only localized outbreaks from time to time within Argentina [2,3]. However, there was an upsurge in the number of swarms in Argentina in 2015, and by early 2017, swarms reached Bolivia and Paraguay [3]. This rapid expansion of the locust populations meant that the infrastructure for managing locusts had to be quickly rebuilt.

Locusts differ from most agricultural pests. They are sporadic, have the capacity to rapidly expand across borders, and then contract to localized pockets that act as reservoirs for future outbreaks [4]. The key to successful management is not only knowledge of locust biology and design of sound management strategies but also a robust organization and maintenance of stakeholder engagement. In the case of SAL, a preventive strategy was established by the Argentinean government during the mid-1950s. Organizations for permanent surveillance and control of locusts were established as part of Departments of Plant Health and Acridology of a number of provinces in northwest Argentina [1]. The preventive management system successfully kept SAL populations quite low and generally limited to Catamarca and La Rioja, where preventive management actions were systematically conducted by field teams. The preventive management went largely unnoticed so that the general public thought that the pest had been eradicated, and with no obvious outbreaks, political support and funding gradually declined [2,3]. Thus, the recent resurgence of SAL, six decades after the last major plague, meant public agencies as well as private stakeholders faced the challenge of recovering capacities to deal with an almost forgotten problem.

Across six decades, many things changed, with the most relevant and apparent being land use and the emergence of a more complex matrix of stakeholders. Additionally, there was a change in the way complex problems were dealt with, with a clear worldwide claim for and trend of more participatory approaches for diagnosis, research, and innovation. Naturally, the key questions are (1) "why" and "how" did SAL re-emergence happen, and (2) how to best manage the system into the future. The answer to the first is likely a complex combination of factors including changes in the natural system (land use and convergence of favorable meteorological conditions likely attributable to climate change), and a reduction in monitoring and control operations. To begin to answer these questions, we need to revisit what we know about the problem from different perspectives, considering the new context in terms of the scientific, technological, agricultural, and social matrix.

The present article reviews SAL biology, including a description of typical habitats, temporal changes, and migration, as well as how management changed during the twentieth century. With the sudden and unexpected resurgence of SAL in 2015 [3], a whole new process began, characterized by a steep learning curve in the face of a dramatic upsurge in locust numbers. This process involved the engagement of new stakeholders and cooperation on different fronts that had not been experienced previously: research and development, governance of the emergency, and management of the pest. The review also identifies old and new knowledge gaps, highlighting research and development needs, and thus provides a contribution to redesigning a robust management strategy.

2. Biology and Ecology of *S. cancellata*

Locusts are grasshoppers in the family Acrididae that have a phenotypic plasticity known as phase polyphenism in that they behave as solitarious individuals at low densities but shift to a completely different behavior and morphology when at high density [5]. At low density, locusts are *solitarious:* they avoid each other and are generally cryptic green or brown. Crowded locusts are *gregarious:* they tend to have darker or aposematic coloration and their attraction to each other leads to marching bands of nymphs and migrating swarms of adults. The genus *Schistocerca* (Acrididae, Cyrtacanthacridinae) includes about 50 species, and at least four are considered highly swarming locusts: the desert locust (*S. gregaria*); the Central American locust (*S. piceifrons*); *S. interrita*, which had several major outbreaks in

Perú; and the South American Locust (*S. cancellata*) [6]. *Schistocerca cancellata* is similar to the three others in coloration and body size and can be differentiated by the male cercus, with the lower lobe larger than the upper lobe; by the subgenital plate of males with a notch almost parallel sides; and by the presence of a waxy secretion on the abdomen on mature adults [7,8].

2.1. Phenotypic Plasticity Traits in S. cancellata

2.1.1. Nymphs

Morphology. In the laboratory, the nymphal body size is affected by rearing density, and there is a sex-dependent pattern. Crowded final instar female nymphs are smaller than those reared isolated, while crowded male nymphs are larger, resulting in a less pronounced body size difference between crowded males and females [9] (Table 1). In the field, sexual dimorphism has not been thoroughly studied, though recent observations in Argentina (Piou et al., unpublished data) suggest that female nymphs are larger than males. In both sexes, isolated nymphs have higher density hair on the outer surface of the hind femur, suggesting that this species might use the same mechanism as the desert locust (stimulation of these hairs from bumping into other locusts [10] to detect changes in density) [9].

Coloration. Solitarious nymphs are generally green, with variations from light brown to light green, and small black dots over part or all of the body (e.g., hind femur) (Table 1, Figure 1). At high density, gregarious nymphs exhibit the contrasting pattern of striking yellow, red, or reddish orange and black though some bright yellow nymphs are seen [9] (Figure 1). Between the two extreme colorations typical of the solitarious and gregarious phases, transitional colorations are commonly observed, corresponding to the "transiens" phase [9]. In the laboratory, recently hatched nymphs are generally light green with or without darker areas, and with nymphs kept in isolation, the greenish or light brown coloration is maintained [9]. With nymphs kept at high density, the black pattern is evident within a few hours of hatching. The first two instars are rather pale with the orange-red color of the head and the bright black and yellow becoming more evident after the third instar. However, nymphs crowded for several hours and then isolated initially developed coloration typical of crowding but lost their coloration after the first molt.

Behavior. In its natural habitat, after hatching, solitarious SAL nymphs disperse in the vegetation and develop in isolation, hiding in the plants protected by their cryptic coloration [11]. After hatching, gregarious nymphs are much more active and form dense groups that eventually constitute bands of marching nymphs [8]. Gregarious nymphs display 'hectic' behavior, with higher individual irritability and more ingestion of food [11]. When food is available and the temperature is high enough, nymphs feed almost continuously [12]. The gregarious bands display circadian cycles of activities: (1) going down their roosting plants and basking in the early morning; (2) basking in groups in the ground until body temperature is high enough to initiate marching; (3) marching and eventually jumping by late morning; (4) when the ground temperature is too high, hoppers climb on plants, jump, or hide in the shade from midday to mid-afternoon; (5) they resume marching in the late afternoon; and (6) they climb up trees or bushes at sunset to find high roosting plants to spend the night [12]. The marching bands can cover several hundreds of meters per day and display a large front of high density in a similar way as desert locust [12]. In the laboratory, Pocco et al. [9] demonstrated that crowded final instar nymphs are more active and more attracted to conspecifics than those reared in isolation, regardless of sex.

Figure 1. *Schistocerca cancellata*, (**A**,**B**) final instar nymphs and (**C**,**D**) adults. (**A**,**C**) solitarious phase (isolated locusts in laboratory); (**B**,**D**) gregarious phase (**B**) crowded nymph in laboratory; (**D**) adult female in nature). (Photographs: M. Pocco).

2.1.2. Adults

Coloration. SAL adults (Figure 1) are in general brownish, with a contrasting pattern of stripes in the pronotum, two brown dorsal bands delimited by dark brown lines, and a wide median longitudinal light-brown stripe extending from the fastigium along the tegmina. In the lateral lobes of the pronotum, there is a pattern of light brown, brown, and whitish stripes. The tegmina exhibit a pattern of dark maculae, and the hind wings are hyaline to light yellowish, with brownish veins in the remigium area. On the hind femur, there is a white stripe in the outer face, below the pinnae, and the hind tibia is purple on its dorsal face, with white spines and black tips. The coloration of SAL adults is quite similar in solitarious and gregarious individuals, although slight differences can be seen in the pattern of stripes in the pronotum. In gregarious mature adults, the contrasting coloration of stripes in pronotum is faintly evident, turning homogenously pale yellow, keeping the narrow dark brown stripes. In natural conditions, immature gregarious adults have a reddish coloration, turning to a general pale-yellow color as they mature [11].

Morphology. SAL is a large acridid: female adults range from 39 to 66 mm in length and males range from 28 to 49 mm [13]. In the laboratory, crowded male adults are significantly larger than isolated males (mean value of body length for crowded males: 58.1 mm and isolated males: 52.7 mm). The size of the female adults does not differ as much between the two density conditions (mean value of body length for crowded females: 67.2 mm and isolated females: 66.0 mm), although the head is significantly wider and the pronotum slightly larger in crowded females than in isolated ones [9].

The classic measures of phase change seen in *Schistocerca gregaria* [14] are the morphometric ratios F/C (where F is the length of the hind femur and C is the maximum width of the head) and E/F (where E is the length of the fore wing over the length of the hind femur (F)). Pocco et al. [9] found that the F/C ratio is significantly lower in crowded than in isolated adults, indicating that the head width is wider in crowded individuals, which could be related to the voracious feeding capability of the gregarious locusts [15]. However, there was no significant difference in the E/F ratio between isolated and crowded adults.

2.2. Habitats and Feeding Preferences

As with most locusts [4,16,17], SAL outbreaks tend to originate in arid regions, which are marked by low levels of human presence, so that early outbreaks are mostly detected by government surveys. Typical habitat in the central permanent gregarization zone of Catamarca and La Rioja (Figure 2: Outbreak Area) is largely desert and semi-desert, where mountain ranges alternate with basins containing inland drainage, saltpans, extensive silt deposits, and sand dunes [18]. The dominant vegetation is wooded steppe of *Prosopis* and *Larrea* bushes, with tufts of *Sporobolus* and *Panicum* grasses or with *Atriplex* or *Suaeda* where the soil is salty [19]. The mosaic of vegetation types associated with these soil types was considered by Köhler [11,18] to be of particular importance to locusts and was termed "Acridoflora". "Acridoflora" consists mainly of annuals that spring up following rain: (Poaceae: *Bouteloua aristidoides* (Kunth), *Aristida murina* Cav., *Trichloris* sp.; Amaranthaceae: *Gomphrena martiana* Gillies; Nyctaginaceae: *Boerhavia spicata* Choisy; Portulacaceae: *Portulaca* sp.; *Heliotropium* sp.). Indeed, SAL is highly polyphagous and eats plants from many different families. The reported list for Argentina includes nearly every cultivated plant from pasture grass and cereal crops such as maize and sorghum to soybeans, peanuts, and citrus [17]. In Bolivia, the identified host range of SAL includes *Amaranthus retuflenxus, Sida cordifolia, Cassia tora, Nicandra physaloides, Panicum maximum, Alternanthera* sp., *Zea mays, Croton argenteos* L., *Cynodon dactylon, Schinus* sp., *Sorghum sudanense, Braquiaria plantaginea, Echinochloa colonum*, and *Digitaria sachariflora* [20,21].

Studies on field populations have revealed that SAL marching bands are carbohydrate (not protein) hungry and that access to carbohydrates is likely important for supporting long-distance migration. Teams of researchers and practitioners set artificial diets in front of oncoming bands at eight unique sites across Argentina, Bolivia, and Paraguay and found that individuals stopped and ate most often at dishes high in carbohydrates [Cease et al. unpublished data]. In the Paraguay study [22], juveniles collected from a marching band and fed artificial diets differing in a protein–carbohydrate (p:c) ratio in the laboratory had increased survival as diets became more carbohydrate biased. When fed single diets of plants that they were seen eating in the field, locusts only gained weight on the plant with the lowest p:c ratio. The plants were from the genera *Paspalum* sp., *Celtis* sp., *Mikania* sp., *Grabowskia* sp., *Prosopis* sp., and *Digitaria* sp. and one from the subfamily Celastraceae. Most of these local plants had a p:c higher than optimal, suggesting that marching locusts must actively seek out carbohydrates, or their growth and survival would be carbohydrate limited.

In contrast to the general pattern of herbivores being nitrogen or protein limited [23], SAL's carbohydrate demand and preference are similar to what has been shown for field populations of other locust species [16]. Land management practices, such as heavy livestock grazing, that deplete soil nitrogen promote outbreaks of these species by lowering plant p:c ratios [24–26]. Lower p:c diets (high carbohydrate) are likely required to support migration. Locusts rely on lipids to fuel long-distance adult flight [5,27], and while the energetic costs of locust marching have not been measured, terrestrial locomotion increases metabolic rates by 2–12-fold above resting for other insects [28,29]. In the Talal et al. [22] study, for both artificial and plant diets, SAL body lipid content increased with decreasing dietary p:c. similar to that shown for migratory locusts (*Locusta migratoria*) eating low p:c meals [30,31]. Both migratory locusts and SAL elevate CO_2 production following a low p:c meal. Talal et al. [22] showed that this is likely due to de novo lipid synthesis, as indicated by a respiratory exchange ratio above 1 (ratio of CO_2 produced in metabolism to O_2 used). Further studies on *Oedaleus* spp. in Senegal and China corroborate the hypothesis that high carbohydrate diets support locust migration [32,33].

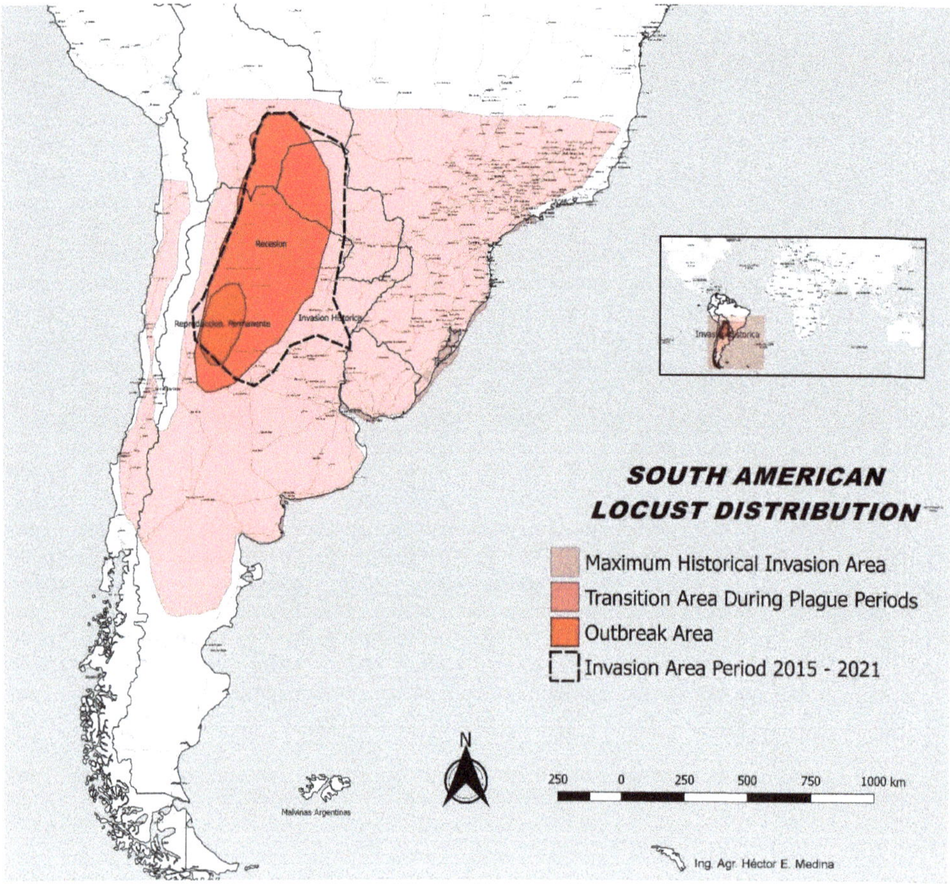

Figure 2. South American locust distribution map. Adapted from de Wysiecki and Lange, 2005.

2.3. *Population Dynamics*

2.3.1. Life Cycle Parameters

In natural conditions, gregarious *S. cancellata* has five nymphal instars and the total period of nymphal development is about 36–57 days [17]. Under laboratory conditions, six nymphal instars are recorded in both isolated and crowded conditions [9]. At constant 30 °C, nymphal development is 47.9 days for isolated nymphs but only 35.6 days for crowded nymphs (Table 1). As in other locusts, the number of eye stripes for both isolated and crowded nymphs is the same as the number of instars from hatching to adult [9].

Typically, there are two generations (spring and summer) per year. Adults from the second generation spend the dry season in a reproductive diapause until rains fall in spring [2,8]. During the summer non-diapause period, adults live for at least 6 weeks with maturation and laying occurring from the third week onwards when rain has fallen. During the diapause period, overwintering adults can live for up to 6–8 months from the beginning of the diapause period in autumn until the rains of the following spring [2,17]. Gregarious females lay eggs in dense groups on bare ground, laying up to six egg-pods over their lifetime, usually with 60–120 eggs per pod [17]. Egg development is continuous, and the incubation period in Argentina varies from 50–61 days for eggs laid in early spring when it is cool to 15–20 days for eggs laid in late spring or summer [2,19,34]. In Bolivia, laboratory studies found that the period from egg-laying to adult emergence of SAL in the gregarious phase at 25 °C (normal conditions of Santa Cruz, Bolivia) is 85–95 days, with the adults living an average of a further 114 days [20].

Table 1. Summary of the main differences between solitarious and gregarious phases for the nymphal and adult stages of *Schistocerca cancellata*.

Schistocerca cancellata					
	Solitarious Phase		Gregarious Phase		
Traits	**Nymphs**	**Adults**	**Nymphs**	**Adults**	**References**
Size	Final instar: males smaller; females larger	Males smaller; females similar size, narrower head	Final instar: males larger; females smaller	Males larger; females similar size, wider head	[9]
Coloration	Green, varying from light brown to light green,	General brownish, contrasting pattern of stripes in pronotum	Pattern of striking yellow, red, or reddish orange & black	General brownish; pattern of stripes in pronotum	[9,11]
	with small black dots			faintly evident In nature: reddish (immature) pale yellow (mature)	
Behavior	Sedentary, disperse in the vegetation and develop in isolation	Sedentary, disperse in the vegetation and develop in isolation	Active; dense groups (marching bands)	Active; dense groups (swarms)	[9,11]
Femur hairs (%)	Higher		Lower		[9]
F/C ratio		Higher		Lower	[9]
Life cycle: **N° instars**	6 (in lab.)		6 (in lab.); 5 (in nature)		[9,17]
Mean duration of stage	47.9 days	87 days	35.6 days	58 days	[9,34]

In the laboratory, at constant 30 °C, the mean duration of the adult stage is about 87 days for isolated individuals compared with the significantly shorter 58 days for crowded adults [9] (Table 1). In crowded conditions, adults take about two to three weeks after molting before mating. Sanchez et al. [34] recorded 73 eggs per pod (mean value) and a mean fecundity of 161 eggs/female in crowded locusts. Based on observations made in the laboratory, the number of hatchlings per pod was higher in the crowded than in the isolated locusts (mean 85.1 for crowded; 51.8 for isolated). In the laboratory, egg incubation takes about 15–20 days.

Similar to many other locust species [4], SAL demonstrates dramatic swings in both its abundance and its occupied geographic range [3]. The population dynamics of this species are characterized as having both recession and outbreak areas sensu Uvarov [35]. In most years, there are low to moderate populations in a localized area in northwest Argentina (Figure 2) and periodic treatment of the bands and swarms that appeared in this area [2,18] helped to prevent plagues for more than 60 years [1–3]. However, SAL shows a pattern of population fluctuation that fits into Berryman's [36] sustained irruption type of population dynamics, where a period of unusually favorable conditions [36,37] can lead to rapid population increases. Once populations reach high densities, even less than ideal conditions are sufficient to maintain the population, resulting in a stable equilibrium at high densities that can result in plagues lasting many years. In the past, dense bands and swarms spread to cover more than 1.5 million km^2 in Argentina, Bolivia, Paraguay, Brazil, and Uruguay [1], and the current upsurge has expanded to cover a significant proportion of this maximum area [3,20] (Figure 2). The stable equilibrium at high densities has profound effects on locust management efforts: if locust populations reach high levels, even substantial control efforts are often unable to reduce a population and end a plague unless aided by unfavorable climatic conditions [38].

Different factors influencing or determining insect outbreaks have been hypothesized including physiological [39] or top-down mechanisms [40]. Unfortunately, limited research on SAL has been carried out on the mechanisms underlying outbreaks. However, joining some known features of SAL and environmental conditions concomitant to recession and outbreak periods, a plausible hypothesis can be stated as follows.

2.3.2. The Third Generation Hypothesis

As mentioned earlier, under normal meteorological conditions, SAL has two generations [1,9]. As with the red locust *Nomadacris septemfasciata* (Audinet-Serville, 1883) in Africa and the spur throated locust *Austracris guttulosa* (Walker, 1870) in Australia [41], SAL has an adult diapause during the winter dry season. Oviposition only occurs between September and March and, then, only if rain has fallen to produce green vegetation [2]. During the June to August winter period, rainfall is very low in northwest Argentina, with an average of <7 mm per month in Catamarca and La Rioja and <15 mm per month in adjacent provinces. Rainfall during summer (December–February) is much higher, averaging > 60–90 mm/month in Catamarca and La Rioja and >100 mm/month elsewhere. Adult diapause ends around August, but adults remain immature until the first rains of the spring, which is often in October or even November. Oviposition begins a week or so after the rains, leading to the normal two (spring and summer) generations during the late spring/summer rainy period.

Importantly, rain sometimes falls during the June–September period, providing green vegetation for post-diapause adults to begin maturation and laying in September. Using a developmental model based on Barrera and Turk [42], Hunter and Cosenzo [2] found that after winter/early spring rain, three generations per year were possible: (1) early spring to early summer, (2) early summer to late summer, and (3) late summer+. In this scenario, the second-generation adults mature in late summer, just before the late March initiation of diapause, enabling a third (autumn) generation. However, only a few locations in Catamarca and La Rioja provinces were found to be hot enough to allow for these three generations [2]. Rainfall is low in Catamarca and La Rioja during winter, and for the 80 years after accurate records began for both provinces, there was significant rain (>25 mm) during any of the months between June and September for 20% of the years for Catamarca and 16% for La Rioja. Interestingly, in 1943, during the winter before the gregarization leading to the last plague of the twentieth century (1944–1954), rain fell during July in Catamarca (27.6 mm) and La Rioja (31.9 mm). Running of the model revealed that the bands of young nymphs seen during January 1944 [1] would have been nymphs of the second generation. These nymphs would have reached the adult stage by March and laid their eggs, allowing a third generation to produce the substantial number of swarms seen during winter 1944. To confirm the hypothesis of three generations and its possible importance in leading to the current upsurge, further studies should be conducted on how the higher temperatures and alterations in rainfall resulting from climate change might lead to an expanded area favorable for a third generation beyond the current Catamarca and La Rioja outbreak area. In particular, an analysis of the conditions throughout northern Argentina, Bolivia, and Paraguay that led to the resurgence of swarms in 2015 is needed.

2.4. *Natural Enemies and Biological Control*

SAL has many natural enemies or antagonists including predators, parasitoids, and pathogens. Numerous invertebrates, from nematodes to insects and spiders, and vertebrates, particularly insectivorous or opportunistic birds, are known to consume either embryonic or postembryonic stages of SAL [43]. Daguerre [44,45] found that *Sarcophaga* spp. parasitoid flies inflicted the highest levels of mortality in Argentina. In Bolivia, López and Copa Bazán [20] found that natural enemies included lizards, hawks, cuckoos, spiders, and different insects such as reduvids, carabids, ants, wasps, predatory mites, and the entomopathogenic fungus *Beauveria* sp. [46]. Although such predators or parasitoids may reduce locust populations or contribute to earlier collapses [47], their actual impact has not been studied for SAL. As for other locusts [48], heavy, widespread chemical campaigns against locusts may also have had serious negative effects on natural enemies but is so far understudied.

Pathogens are the only natural enemies that have so far received some attention as possible biocontrol agents of SAL. In fact, SAL in North Argentina was among the first targets (along with Australia) of using natural enemies for control: Bruner introduced a fungus from South Africa in 1898, and the microbiologist d'Hérelle [49] applied the bacterium *Coccobacillus acridiorum*, which he had isolated from diseased locusts in Mexico. Although some good results were reported, there was little success overall [50]. Under laboratory conditions, SAL proved to be susceptible to the amoeba *Malameba locustae* and the *Entomopox* virus of *Melanoplus sanguinipes* (MsEPV) [51]. However, both pathogens are unlikely to be useful. *Malameba locustae* cysts lose viability under storage, and the MsEPV is not a native natural enemy. Species of Eugregarines (Apicomplexa) are also known to occur in SAL populations [43,52] but normally do not seem to harm the host. Among the fungi, there are old records of *Sporotrichum paranenese, Fusarium* sp., and *Aspergillus parasiticus* [43] and new isolates of *Beauveria bassiana* and *Cordyceps locustiphila* that are under study as possible biocontrol agents [53,54]. In China, a species of *Aspergillus* has shown high virulence against locusts [55]. The naturalized microsporidium *Paranosema locustae*, which appears to continue causing epizootics and expanding its host range in grasshopper communities of the Pampas and Patagonia [56], turned out to be not infective for crowded gregarious nymphs of SAL in the laboratory as opposed to relatively uncrowded gregarious nymphs of earlier studies [57]. It remains to be determined if *P. locustae* may prevent phase transformation when it is administered to solitary nymphs of *S. cancellata* as observed in the Migratory locust *Locusta migratoria* [58,59].

3. Management of SAL: An Historical Perspective

3.1. Early Management Efforts against SAL Plagues

The first records of locusts in Argentina date from the 1500s, with swarms damaging crops of cassava (*Manihot esculenta*) [1]. While there were reports of locusts in the following centuries, including plagues in Córdoba and Santa Fe provinces during 1833–1840 and 1844–1849 [50], detailed records of the extent of locust infestations began in the late 1800s with the continuous record of the size of locust infestations from 1897–1968 as summarized by Gastón [1]. These detailed records were a response to an increasing realization that SAL caused a great deal of damage to Argentinian agriculture, even though early damage estimates were vague and indirect. The problem led to the gradual development of specific pest control policies, characterized by a strong interventionist profile. These policies underscore SAL's seriousness and importance for the national economy [60] and led the Argentine government to create an agency to implement the first locust control campaigns. This mission was assigned in 1891 to the National Commission for Locust Extinction (CNEL) [61], which having "extinction" in its name, was based on the idea that the problem could be solved for good by a short-term campaign. In 1898, the control of SAL was declared mandatory by law (https://www.argentina.gob.ar/normativa/nacional/ley-3708-284864 accessed on 17 December 2021) and the first agricultural pest in which control was declared compulsory in Argentina. SAL was a problem of such an impact, and the difficulties experienced were so deep that the government decided in 1912 to transform this low-level and temporary office into a permanent department, named the General Directorate of Agricultural Defense, which then turned into the National Direction for Agricultural Defense and Plant Health [61]. The NDADPH was first appointed with the task of controlling SAL attacks on crops, pastures, and trees in public places, but their role widened to include all kinds of pests as agriculture advanced throughout the country.

In those first years, there was little understanding of the true nature of the population dynamics of SAL, and this contributed to a lack of effectiveness in its management. The locust problem was widespread such that, between the first detailed records in 1897 until the late 1930s, more than half of the years had SAL invasions in >20% of the whole territory of Argentina, with a maximum invasion of 57% of Argentina in 1932 [1]. There were only five years when <5% of Argentina was invaded, leading to the idea that swarms were a permanent condition of this species [18,45]. All recorded expeditions were organized during periods of large-scale locust invasions, 1908, 1917, and 1934–1936 [18,62], so studies of the gregarious swarming phase predominated. While the occurrence of a solitarious phase consistent with Uvarov's phase theory [63] began to be considered by the late 1930s, its presence was initially not generally recognized [18,62].

There were similar early misunderstandings concerning the source of invasions. In 1880 and 1892, there were observations leading to the hypothesis that invasions came from the Great Chaco in Argentina and Bolivia [63]. The belief that the "permanent region" was in northern Argentina and especially southeast Bolivia was built on the expeditions that Enrique Lynch Arribálzaga and Carlos A. Lizer y Trelles carried out in Jujuy, the northernmost province of Argentina during 1908 and 1917, respectively, coinciding with periods of very expanded invasions [1,62]. However, even though these source area ideas were erroneous, what was recognized very early was the importance of migration. D'Hérelle [49] reported widespread swarm movements: from the north towards the south in winter months and from the center of Argentina towards the north at the end of the summer months, effectively a migratory circuit later found in other locusts such as the desert locust [64–67] and Australian plague locust [68]. He mentioned that migrations were at night during the summer but only seen during the day in winter. The importance of long-distance migrations would later be further elaborated for swarm movements in the 1930s to 1950s [17] and again during the recent 2015–2021 upsurge.

Before the first systematic surveys, it was believed that SAL could reproduce anywhere and therefore every region was suspected to be a potential source of new invasions. Initially, it was also thought that SAL persisted in winter refuges, based on the frequent observations that swarms came southward to agricultural areas from northern regions. To try to locate these refuges, the Central Commission for Locust Research was established in 1933 and conducted expeditions lasting 3–4 months that explored the north and the west. These expeditions were led by three entomologists and served not only for describing dispersion paths, behavior, meteorological factors affecting swarm dispersion, feeding preferences, and natural enemies but also to test different inorganic insecticides and application equipment [69]. These expeditions were unable to find overwintering swarms, leading to the rejection of the winter refuge hypothesis [18].

During this period, there were several hypotheses concerning the wide fluctuation in the extent of the invaded area including correlations with sunspot numbers [63]. This idea may have derived from the proposition by Archibald [70] that periodical invasions of *Locusta migratoria migratoria* (L.) into the temperate zone are apparently regulated by certain meteorological conditions, which in turn seem correlated with variations in sunspots. There were similar hypotheses for the Rocky Mountain Locust ((*Melanoplus spretus* (Walsh, 1866)) in North America [71]; for *Locusta migratoria migratorioides* (Reiche & Fairmaire), still referred to as *Locusta migratoria manilensis* (Meyen) in eastern China [72,73]; and for the desert locust, *S. gregaria* [73].

The initial misunderstandings about the population dynamics of SAL combined with the widespread nature of swarming populations and an inability to locate source areas meant that management efforts during the first three decades of the twentieth century were "primitive", to use the terminology of Lizer and Trelles [62]. The locust commissions were initially established with the appointed task of exploring territories affected by swarms or bands and of recording habitat characteristics and crops affected. Control of the swarms was generally limited to treating adults settled on trees or fences. Most control measures were aimed at containing nymphal bands, a "defensive" strategy [1] aimed at preventing bands from invading crops. The strategy was to interrupt band marching with perimetral metal barriers fixed to the ground. "Funnels" or collectors used the organized and uniform movement of the gregarized bands to direct the bands to deep trenches, where the millions of nymphs that accumulated were torched with flame throwers. Other methods included plowing to destroy the eggs, and allegedly, the adults as they lay. Hand or mechanical picking was promoted with economic compensation per bag of locusts collected. There are anecdotal records of turkeys and chickens released in the vineyards and orchards as a complementary control measure [60]. Additionally, setting fire to pastures and fallow fields was sometimes practiced but abandoned because of the low efficacy and risk of spread of uncontrolled fires [74].

During one of the long-lasting plagues in the 1920s, mechanical collection of locusts reached industrial proportions. In eastern Argentina, five hundred metal fences were installed, totaling 50 km of barriers to herd the nymphs towards corrals where locusts were sun-dried and then sold as fertilizer. The production of locust powder reached 95 million tonnes, but this practice was abandoned due to high costs [60].

3.2. The Offensive Management Period

The Locust Control Service was established in 1945 and was characterized by the heavy use of insecticides applied with blowers mounted on trucks, backpacks, airplanes, and helicopters either hired or specifically assigned to the aerial Aviation Department of the Agricultural Ministry [1]. This move was enabled by improvements in control techniques including more effective chemical pesticides (e.g., DDT) and aerial campaigns as has started in Africa [75]. While the environmental impacts had yet to be uncovered, these technical advancements allowed for the rapid treatment of large areas through offensive operations [1]. The Locust Control Service treated locusts as the upsurge increased after locusts invaded from Paraguay and Brazil in early 1945, and the infested area then expanded to 35° South by 1946–1947 [43]. In 1948, the large scale of the plague led to the formation of international cooperation between countries affected by SAL: Brazil, Uruguay, Paraguay, Bolivia, and Argentina. By 1952–1954, there was an extensive control campaign involving 12,000 tons of pesticide [17,43], and by early 1955, few locusts remained. While it was uncertain whether this decline was solely a result of the intensive treatment programs or in combination with unfavorable environmental conditions, by 1956, no gregarious locusts were detected, and this great plague period came to an end [1,3,43].

3.3. Advances That Formed the Basis for Preventive Management

The 1944–1954 plague was monitored closely not only regarding the efficiency of methods of control but also concerning the biology and population dynamics of SAL [18]. These observations led to substantial breakthroughs and to a new strategy for SAL: preventive management. The aim of preventive management was to have teams equipped for surveillance and control operations so that any gregarizing groups of nymphs or adults found early in an outbreak could be treated as a way of preventing an upsurge.

The first breakthrough was applying the phase theory of Uvarov [63] to SAL, indicating that SAL exhibited phenotypic plasticity by switching between solitarious (non-outbreak) and gregarious (outbreak) phases. The commonly encountered gregarious swarms and the scattered solitarious locusts found in northwestern drylands had been thought to be a different species [18,62]. Detailed observations in the late 1930s to early 1940s showed that they were different density-dependent phenotypes of the same species [18]. Rearing experiments demonstrated phase transformation when locusts were crowded [76]. Subsequent careful field observations revealed that the scattered locusts present during the population decline of the late 1930s started to change their behavior as numbers increased during the 1943–1944 season. Together, these advancements demonstrated that the isolated forms seen in northwest Argentina and the gregarious swarming forms of SAL were the same species [18].

The second breakthrough was that phase change (from solitarious to gregarious SAL) occurred in specific arid areas of La Rioja and Catamarca provinces [18]. This led to the realization that plagues occurred in the past because of a failure to control these newly gregarized nymphs and the resulting swarms that followed, leading to enormous costs from damage to agricultural production. The identification of localized outbreak areas meant that early targeted treatment might prevent plagues, as had been proposed by Uvarov [35] for other locusts with outbreak areas. Thus, regular surveillance for signs of gregarizing locusts and treatment of any bands or swarms found within the 100,000 km^2 outbreak area formed the basis of the preventive management program for SAL [1,18]. SAL invasions reaching Bolivia, Paraguay, Brazil, and Uruguay were correlated with swarms originating in the Argentinian outbreak area including the presence of SAL in Bolivia in 1947, 1949 [77,78] and in the recent upsurge during 2017 and 2020, where they were reported in the Bolivian Chaco region, Santa Cruz, Tarija, and Chuquisaca [79,80]. In Brazil, recorded invasions occurred in its southeast region in 1906, 1932–1933, and 1946–1948 [81], while in Uruguay, there were nine invasions that caused significant damage between 1890 and 1948 [82].

The third breakthrough was recognition that SAL had a migratory circuit. Locusts originating in the La Rioja and Catamarca regions initially as small bands and then adult groups transitioned into large migrating swarms that headed towards northern Argentina with converging winds [18]. These northward dispersions were called "concentration flights" [62], with a proposed hypothesis that summer/autumn wind currents and topography contributed to making swarms converge during this northward dispersion [11,18]. The swarms reached northern Argentina and even Bolivia and Paraguay at times. During the winter, the swarms returned south, and in the first waves of swarms, adults were reddish and tended to fly long distances following the prevailing northerly winds. As the southerly migratory movement continued, the locusts mature irregularly in pulses and disperse into ever smaller masses and begin the spring and summer breeding period.

The final breakthrough to support preventative management was uncovering the relationship between precipitation patterns and multiple SAL generations in one year. It had long been realized that SAL adults remained immature over winter, and it was thought that they matured and laid with warm weather of spring, with asynchronous and sporadic laying and hatching continuing from October to March as part of a single generation per year. However, Köhler [11,18] observed that the transition from solitarious to transiens to gregarious occurred over a period of two generations in a season. The existence of a second generation was further confirmed in February 1945, when new swarms invaded northern Argentina from Paraguay [1]. Due to a drought in the area, there was no maturation and laying despite the hot weather that was thought to lead to maturation and laying, but there was then heavy rain, which was followed by widespread oviposition and an "extraordinary quantity of nymphs" [1]. This meant not only that there was more than one generation per year [1,18] but also that rainfall was important in allowing maturation and laying, a finding reinforced by Hunter and Cosenzo [2]. Therefore, during rainy periods, adults matured rapidly and laid, allowing for more than one generation in a season, and since a female could lay 100 eggs or more, a substantial population increase was possible.

3.4. The Preventive Management Program

Uvarov [35] had suggested preventive management be implemented for locusts with outbreak areas. Such a program had begun to be implemented in the outbreak areas of the red locust, *N. septemfaciata* in Africa [83], and so, a similar preventive program was envisaged for the outbreak areas of SAL [1]. In 1954, the Service to Police and Prevent Locusts was established, which consisted of permanent scouting commissions with the aim of locating and controlling locust infestations [1,43]. At first, these commissions were deployed in an area of 300,000 Km^2, covering Catamarca, La Rioja, and adjacent provinces of northwest Argentina. Surveillance followed pre-established routes and occurred throughout the spring and summer generations. Teams were equipped for control operations so that any gregarizing groups of nymphs or adults found could be treated [1]. Treatments were carried out usually by ground equipment such as backpack sprayers or sprayers attached to Unimogs [2] though some years had slightly larger outbreaks requiring the use of aircraft [1,2,60]. With the information gathered in the first 14 years of implementation of the preventive management strategy, it became clear that there was an area of approximately 100,000 Km^2 where the build-up of SAL population was permanent, particularly during the summer [1]. The intensity of the scouting and control process was adjusted to concentrate more on this 100,000 Km^2 area, mainly in Catamarca and La Rioja [1,2].

Following the implementation of the preventive management program in the 1950s and 1960s, outbreaks of SAL were small, with treatments on the order of a few thousand hectares or less, reaching 10,000–30,000 ha at times [1,2], much less than the hundreds of thousands of hectares requiring treatment during plagues. The resulting long recession was a complete contrast to the first half of the twentieth century when locusts were in recession for only 10 of the 55 years from 1900 and 1954 [1,2]. However, the very success of the program in containing outbreaks before they could expand to agricultural areas meant that there was little impetus for studies to increase the understanding of SAL population dynamics. Research was limited to studies by Barrera and Turk [42] on aspects of SAL biology including development rates and by Waloff and Pedgley [19], who reviewed the biology of SAL in comparison with the South African desert locust, including mention of an Argentina–Bolivia/Paraguay migratory circuit for SAL. However, the successful preventive management strategy in Argentina caught the attention of locust workers in Australia and visits to Argentina led to a detailed analysis of the plague and contrasting preventive management recession periods. Hunter and Cosenzo [2] found that, when rain fell between June and September in one or both of Catamarca/La Rioja, three generations were possible in a season, resulting in a larger than normal treatment program, which turned out to be critical in preventing the population to further upsurge to plague proportions.

In a similar way, the very success of the preventive management program led to a gradual decline in finances and resources due to the "vicious cycle" of successful prevention, leading to loss of the memory of how damaging plagues had been and gradual erosion of management effectiveness [84]. During the 1960s, there were field teams for locust surveillance and control with substantial amounts of ground equipment and access to helicopters and airplanes for treatments of larger infestations [1,60]. In 1970, Daguerre [85] warned about not enough effort being put into controlling reproduction foci before SAL reaches the reproductive stage. By the late 1980s, budget constraints led to a reduction in staff and equipment such that Hunter and Cosenzo [2] reported that, during the 1987–88 season, there were few surveys in spring and no band treatments in summer, leading to an unusually large treatment campaign against adults that included the use of aircraft. An international workshop held in February 2020 [86] confirmed the importance of a gradual reduction in financial and political support as the recession progressed, which reduced the effectiveness of the preventive management system. Participants identified the lack of a budget and a strong and continuous state policy as major explanations of the current crisis: surveillance activities became increasingly restricted in location and timing so that, when conditions favorable for an outbreak occurred, the initial outbreak populations were not located and controlled. These reductions in resources are similar to the 'vicious governance cycle' described for the desert locust [87–89]: long recessions induce a loss of interest, coordination, and institutional memory such that the human, social, material, and financial infrastructure accumulated during and immediately following plague periods, slowly decreases in the face of other political priorities.

The result was a dramatic reduction in the amount of locust survey and control [3,90–92]. Surveys were largely limited to Catamarca and La Rioja, with a dramatic reduction in access to the permanent gregarization area by closure of access trails and legal difficulties of SENASA teams being allowed to enter private property (HM, unpublished). With locusts rarely being present in numbers to cause economic damage [3], there was little institutional resistance to these increasing restrictions, thus limiting the ability of field officers to effectively locate and control localized gregarious infestations [86], a key requirement of any preventive management program. Compounding these limitations was a gradual loss of institutional knowledge and experience [86], including the loss of the detailed records of the extent of bands and swarms and their treatment, so that the only records remaining were those retained in Australia by the Hunter and Cosenzo [2] study. With such decreases in knowledge and capacity to respond, it is not surprising that there was a sudden appearance of swarms in 2015 that marked the onset of the current upsurge.

4. Management under A New Plague Period

4.1. The Resurgence of SAL

The new upsurge of SAL began with the sudden appearance of in north-central Argentina during July 2015 [90,91]. In the years prior to 2015, the surveillance and control actions focused on a small area of Argentina [3], and so at present, there is no certainty about the geographical source of the original outbreak that triggered the subsequent regional emergency.

Following the appearance of swarms during July 2015, a program of widespread surveys was introduced, which found a number of swarms during spring 2015 [3]. Widespread control of nymph bands was instituted during the 2015–2016 season [90]. There was a small nymphal treatment program in Argentina during late 2016, and following sightings of swarms migrating at night in northern Salta province and at Tarija in Bolivia [HM, unpublished], many swarms were seen in Bolivia during January 2017 [91]. These northward migrations had not been documented during the 60-year period of preventive management [3].

Overall, the 2015–2021 plague was characterized by periods of breeding in Argentina alternating with breeding in Bolivia/Paraguay [3]. As a result, instead of bands and swarms being mainly limited to the outbreak area, they were present in a much larger area (Figure 2: "Transition Area During Plague Periods") [90,91] though not as large an area as sometimes occurred in the past (Figure 2: "Maximum Historical Invasion Area") [19]. During the years of migrations north (2016–2017, 2018–2019, and 2019–2020), there were generally two generations per year: a spring generation in Argentina, followed by a summer generation in Bolivia/Paraguay [3,90–92].

While there was the widespread treatment of nymphs in Argentina whenever they were present, treatment of swarms proved much more difficult: during 2015 and 2019, there was little or no treatment of swarms, and during 2017, treatments were conducted but were limited by social and governance factors [88,89]. Chemical control in horticultural areas was very difficult because the long recession period without locusts meant that there were few insecticides registered for use on such crops, and even for insecticides that were registered, strict regulations further limited their use, leading to substantial economic impacts on horticultural production [90]. When high-value fruit tree production was threatened [90,91], swarms were controlled with the decisive participation of local crisis committees that included the producers affected by the pest [86].

The new plague forced very fast learning and decision-making processes adapted to conditions that were very different from those existing in the previous plague period during 1944–1954 [60]. After 60 years, the situation changed significantly: there were many new stakeholders [93], evolving interests and priorities of citizens, farmers, and local decision-makers [86], huge changes in land use [94] particularly the extension of agricultural land [95], and greatly increased environmental concerns associated with large scale treatment programs using chemical pesticides [86]. The management program required an extremely rapid increase in tools and resources, so it was not surprising, therefore, that it took some time to facilitate the governance and oversight capable of coping with the resurgence, including the coordination of efforts and construction of links to ensure control was effective [86].

4.2. Re-Establishing a Sustained SAL Preventive Management Program

As stated by a participant to the 2020 workshop on SAL governance [86]: "the key to work with locusts in the long term is to go back to preventive management (...) so we don't have another crisis ... We need preventive management, not reactive and palliative". An effective preventive strategy should be the way to manage locusts [38], but it requires first getting out of the crisis and building back a preventive system that includes international best practices.

As part of adapting SAL management to the twenty-first century, some important improvements have been made: there have been substantial improvements in the overall organization of SAL management in that collaborations between State agencies in charge of locust management, the provinces, and producers have been strongly reinforced since 2015 [84]. This improvement has mostly relied on personal commitment, "good interpersonal relationships" that emerged in the context of emergency protocols implementation and crisis committees. To maintain this commitment and improvements beyond individuals and during recession, interactions and collaboration arenas need to be institutionalized and maintained as part of a preventive management program.

To ensure the successful implementation of a new preventive management system, there also needs to be improved efficiency of data collection and analysis. In 2020, a locust alert system that notified farmers, beekeepers, researchers, and members of government about the advance of the pest was implemented. Currently, work is underway to implement this system used in Argentina throughout the region affected by SAL as it was recognized that data collection needs to be harmonized between countries to ensure that outbreaks are detected quickly and reported, so they can be controlled.

Critical to efficient preparation and planning of treatment programs is to have a better forecasting system based on an improved understanding of factors leading to outbreaks [67]. Both field and laboratory studies are needed and should utilize the latest technologies including remote sensing to accurately identify suitable locust breeding habitats and habitat conditions [96], wind analyses for migration patterns, and rates of development under different environmental conditions. Such data lead to more accurate development models that forecast the extent and timing of outbreaks so that surveys can be concentrated in areas where outbreaks are more likely. The influence of climate change on outbreaks of different locust species in the world has been debated [97], and for SAL, there needs to be an analysis of altered rainfall and temperature patterns on the location and timing of gregarization processes leading to outbreaks. However, probably, a more parsimonious explanation is that effective preventive strategies put in place for different locust species in the world in the sixties [1,83], play key roles in long recession periods. The gregarization process in SAL has been compared with that of *S. gregaria* in the laboratory [9], but there needs to be field studies such as those already conducted for *S. gregaria* [67,98–102]. A detailed analysis of ways to improve methods of treatment are required, not only an investigation of the latest chemical treatments appropriate for governments and landholders but also an integration of biological control, as has been already accomplished in Australia [103], Mexico [104], China [55], Tanzania [105], and Somalia [106] as part of complete implementation of Integrated Pest Management programs. Increasing constraints on the widespread use of chemical pesticides mean a biological alternative is essential to ensure that locusts are treated wherever they are, including in environmentally sensitive areas and where restrictions make the use of chemical pesticides difficult. To facilitate treatments in sensitive areas, the environmental sector should be more involved, including better coordination between the ministries of environment and agriculture [92].

5. Conclusions

The very success of SAL preventive management program led to less survey and control, but even more importantly, there was a virtual absence of research into the latest developments in locust biology and management. An update on locust biology is critical in view of climate change effects on rainfall and temperature, possibly altering the location and intensity of initial SAL outbreaks, which might make regions outside the traditional provinces of Catamarca and La Rioja [1,2] important in the initiation of upsurges. Management of SAL has begun to implement some of the world's best practice, but many aspects need to be investigated and then implemented in ways relevant to the political and managerial system prevalent in the affected countries. The re-establishment of effective preventive management will rely on harmonizing coordination between Argentina, Bolivia, and Paraguay, but this "slow pace of regional coordination" encounters "obstacles coming from institutional instabilities" [86], including changing administrations. Only by "putting the three countries together" [86], through a regional plan, will an effective preventive management system be ensured and further be locust upsurges avoided.

Author Contributions: Conceptualization E.V.T., A.J.C., M.M.C., D.H.; methodology, all authors; formal analysis, all authors; writing—original draft E.V.T., M.M.C., C.E.L., M.E.P., C.P., D.H.; writing—review and editing, all authors; supervision, E.V.T., funding, A.J.C. All authors have read and agreed to the published version of the manuscript.

Funding: This research was supported in part by the Foundation for Food and Agricultural Research, grants no. 593561 to A.J.C.

Data Availability Statement: Not applicable.

Acknowledgments: Special thanks to Michel Lecoq for the invitation to write this review and suggestions to improve the graphical abstract.

Conflicts of Interest: The authors declare no conflict of interest.

References

1. Gastón, J. Síntesis histórica de las invasiones de langosta en la Argentina. *Sec. Agric. Gan. Misc.* **1969**, *434*, 1–32.
2. Hunter, D.; Cosenzo, E. The origin of plagues and recent outbreaks of the South American locust, *Schistocerca cancellata* (Orthoptera: Acrididae) in Argentina. *Bull. Èntomol. Res.* **1990**, *80*, 295–300. [CrossRef]
3. Medina, H.E.; Cease, A.J.; Trumper, E. The resurgence of the South American locust (*Schistocerca cancellata*). *Metaleptea* **2017**, *37*, 17–21.
4. Zhang, L.; Lecoq, A.; Latchininsky, A.V.; Hunter, D.M. Locust and Grasshopper Management. *Annual Rev. Entomol.* **2019**, *64*, 15–34. [CrossRef]
5. Cullen, D.A.; Cease, A.J.; Latchininsky, A.V.; Ayali, A.; Berry, K.; Buhl, J.; Rogers, S.M. From molecules to manangement: Mechanisms and consequences of locust phase polyphenism. *Adv. Insect Physiol.* **2017**, *53*, 167–285.
6. Song, H.; Foquet, B.; Mariño-Pérez, R.; Woller, D.A. Phylogeny of locusts and grasshoppers reveals complex evolution of den-sity-dependent phenotypic plasticity. *Sci. Rep.* **2017**, *7*, 6606. [CrossRef] [PubMed]
7. Harvey, A.W. A reclassification of the *Schistocerca americana* complex (Orthoptera: Acrididae). *Acrida* **1981**, *10*, 61–77.
8. Song, H.; Cigliano, M.M.; Lange, C.E. South American Locust. *Schistocerca cancellata* (Serville, 1838) (Acrididae). In *Encyclopedia of Pest Orthoptera of the World*; Lecoq, M., Zhang, L., Eds.; China Agricultural University Press: Beijing, China, 2019; pp. 198–203.
9. Pocco, M.E.; Cigliano, M.M.; Foquet, B.; Lange, C.E.; Nieves, E.L.; Song, H. Density-Dependent Phenotypic Plasticity in the South American Locust, *Schistocerca cancellata* (Orthoptera: Acrididae). *Ann. Èntomol. Soc. Am.* **2019**, *112*, 458–472. [CrossRef]
10. Simpson, S.J.; Despland, E.; Hägele, B.F.; Dodgson, T. Gregarious behavior in desert locusts is evoked by touching their back legs. *Proc. Natl. Acad. Sci. USA* **2001**, *98*, 3895–3897. [CrossRef]
11. Köhler, P. Relación entre biosfera y ecosistemas de la langosta migratoria. *Fund. Inst. Miguel Lillo Miscelánea* **1979**, *66*, 37.
12. Piou, C.; Zagaglia, G.; Medina, H.E.; Trumper, E.; Brizuela, X.R.; Maeno, K.O. Band movement and thermoregulation in *Schistocerca cancellata*. *J. Insect Physiol.* **2021**, 104328. [CrossRef] [PubMed]
13. Cigliano, M.M.; Pocco, M.; Lange, C.E. Acridoideos (Orthoptera) de Importancia Agroeconómica en la República Argentina. In *Biodiversidad de Artrópodos Argentinos*; Juñent, R., Ed.; Editorial INSUE—UNT: San Miguel de Tucumán, Argentina, 2014; Volume 3, pp. 11–36.
14. Dirsh, V.M. Morphometrical Studies on Phases of the Desert Locust (*Schistocerca gregaria*, Forskal). *Anti-Locust Bull.* **1953**, *16*, 1–34.
15. Uvarov, B.P. Grasshoppers and locusts: A handbook of general acridology. In *Anatomy, Psysiology, Development Phase Polymorphism Introduction to Taxonomy*; Cambridge University Press: Cambridge, UK, 1966; Volume 1.
16. Le Gall, M.; Overson, R.; Cease, A. A Global Review on Locusts (Orthoptera: Acrididae) and Their Interactions with Livestock Grazing Practices. *Front. Ecol. Evol.* **2019**, *7*, 263. [CrossRef]
17. COPR (Centre for Overseas Pest Research). *The Locust and Grasshopper Agricultural Manual*; COPR: London, UK, 1982.
18. Köhler, P. Ecología de la zona central y gregarización de la langosta en la República Argentina. *IDIA Supl.* **1962**, *7*, 7–108.

19. Waloff, Z.; Pedgley, D.E. Comparative biogeography and biology of the South American locust, *Schistocerca cancellata* (Serville), and the South African desert locust, *S. gregaria flaviventris* (Burmeister) (Orthoptera: Acrididae): A review. *Bull. Entomol. Res.* **1986**, *76*, 1–20.
20. López, R.M.; Copa Bazán, A.F. Prospección y Ciclo Biológico de la Langosta Voladora Schistocerca cancellata (Orthoptera: Acridi-dae) en el Departamento de Santa Cruz, Bolivia, 2017. Tesis de Grado, UAGRM, Presente en la Biblioteca Rafael Peña de la Facultad de Ciencias Agrícolas con N° de registro T-2563, 2018; p. 73.
21. Unterladstaetter, R. *Malezas del Oriente Boliviano;* CIAT: Santa Cruz de la Sierra, Bolivia, 2008; p. 542.
22. Talal, S.; Cease, A.; Farington, R.; Medina, H.E.; Rojas, J.; Harrison, J. High carbohydrate diet ingestion increases post-meal lipid synthesis and drives respiratory exchange ratios above 1. *J. Exp. Biol.* **2021**, *224*, jeb240010. [CrossRef]
23. White, T.C.R. *The Inadequate Environment: Nitrogen and the Abundance of Animals;* Springer: Berlin/Heidelberg, Germany, 1993.
24. Cease, A.J.; Elser, J.J.; Ford, C.F.; Hao, S.; Kang, L.; Harrison, J.F. Heavy Livestock Grazing Promotes Locust Outbreaks by Lowering Plant Nitrogen Content. *Science* **2012**, *335*, 467–469. [CrossRef]
25. Word, M.L.; Hall, S.; Robinson, B.E.; Manneh, B.; Beye, A.; Cease, A.J. Soil-targeted interventions could alleviate locust and grasshopper pest pressure in West Africa. *Sci. Total. Environ.* **2019**, *663*, 632–643. [CrossRef]
26. Le Gall, M.; Word, M.L.; Thompson, N.; Manneh, B.; Beye, A.; Cease, A.J. Linking land use and the nutritional ecology of herbivores: A case study with the Senegalese locust. *Funct. Ecol.* **2019**, *34*, 167–181. [CrossRef]
27. Hunter, D.M.; McCulloch, L.; Wright, D.E. Lipid accumulation and migratory flight in the Australian plague locust, *Chortoicetes terminifera* (Walker) (Orthoptera: Acrididae). *Bull. Èntomol. Res.* **1981**, *71*, 543–546. [CrossRef]
28. Herreid, C.F.; Full, R.J. Cockroaches on a treadmill: Aerobic running. *J. Insect Physiol.* **1984**, *30*, 395–403. [CrossRef]
29. Rogowitz, G.; Chappell, M. Energy metabolism of eucalyptus-boring beetles at rest and during locomotion: Gender makes a difference. *J. Exp. Biol.* **2000**, *203*, 1131–1139. [CrossRef] [PubMed]
30. Simpson, S.J.; Raubenheimer, D. The Geometric Analysis of Nutrient-Allelochemical Interactions: A Case Study Using Locusts. *Ecology* **2001**, *82*, 422. [CrossRef]
31. Zanotto, F.P.; Simpson, S.J.; Raubenheimer, D. The regulation of growth by locusts through post-ingestive compensation for variation in the levels of dietary protein and carbohydrate. *Physiol. Èntomol.* **1993**, *18*, 425–434. [CrossRef]
32. Cease, A.J.; Harrison, J.F.; Hao, S.; Niren, D.C.; Zhang, G.; Kang, L.; Elser, J.J. Nutritional imbalance suppresses migratory phenotypes of the Mongolian locust (*Oedaleus asiaticus*). *R. Soc. Open Sci.* **2017**, *4*, 161039. [CrossRef] [PubMed]
33. Le Gall, M.; Word, M.L.; Thompson, N.; Beye, A.; Cease, A.J. Nitrogen Fertilizer Decreases Survival and Reproduction of Fe-male Locusts by Increasing Plant Protein to Carbohydrate Ratio. *J. Animal Ecol.* **2020**, *89*, 2214–2221. [CrossRef] [PubMed]
34. Sánchez, N.E.; Wittenstein, E.; De Wysiecki, M.L.; Lange, C.E. Life History Parameters of the Gregarious Phase of the South American locust, *Schistocerca cancellata* (Serville) (Orthoptera: Acrididae), under Laboratory Conditions. *J. Orthoptera Res.* **1997**, 121. [CrossRef]
35. Uvarov, B.P. Biological and ecological basis of locust phases and their practical application. In Proceedings of the Fourth Inter-national Locust Conference, Cairo, Egypt, 22 April 1936; Government Press: Cairo, Egypt, 1936; Volume 7, p. 16.
36. Berryman, A.A. The theory and classification of outbreaks. In *Insect Outbreaks;* Barbosa, P., Schultz, J.C., Eds.; Academic Press: New York, NY, USA, 1987; pp. 1–30.
37. Akimenko, V.V.; Piou, C. Two-compartment age-structured model of solitarious and gregarious locust population dynamics. *Math. Methods Appl. Sci.* **2018**, *41*, 8636–8672. [CrossRef]
38. Magor, J.; Lecoq, M.; Hunter, D. Preventive control and Desert Locust plagues. *Crop. Prot.* **2008**, *27*, 1527–1533. [CrossRef]
39. Behmer, S.T.; Joern, A. Insect Herbivore Outbreaks Viewed through a Physiological Framework: Insights from Orthoptera. In *Insect Outbreaks Reviited;* Barbosa, F., Letourneau, D., Agrawaal, A., Eds.; Academic Press: Cambridge, MA, USA, 2012; pp. 3–29. [CrossRef]
40. Dwyer, G.; Dushoff, J.; Yee, S.H. The combined effects of pathogens and predators on insect outbreaks. *Nature* **2004**, *430*, 341–345. [CrossRef]
41. Hunter, D.M.; Elder, R.J. Rainfall sequences leading to population increases of *Austracris guttulosa* (Walker) (Orthoptera: Acrididae) in arid north-eastern Australia. *Aust. J. Èntomol.* **1999**, *38*, 204–218. [CrossRef]
42. Barrera, M.; Turk, S. Estado actual de la langosta *Schistocera cancellata paranensis* (Burm.) en la Republica Argentina: Nuevos aportes a su bioecología. *Acta Zool. Lilloana* **1983**, *27*, 15–29.
43. De Wysiecki, M.L.; Lange, C.E. La langosta *Schistocerca cancellata* Serville (Orthoptera: Acrididae: Cyrtacanthacridinae) en Argentina: Biologia, ecologia, historia y control. En: Barrientos Lozano, L. y P. In *Manejo Integrado de la Langosta Centroamericana (Schistocerca piceifrons piceifrons, Walker) y Acridoideos Plaga en América Latina;* Sierra, A., Ed.; Instituto Tecnológico de Ciudad Victoria: Tamaulipas, México, 2005; pp. 151–156. ISBN 970-18-7628-8.
44. Daguerre, J.B. Nuestros actuales conocimientos sobre la langosta. III. Equilibrio biológico. *Rev. Soc. Entomol. Argent.* **1938**, *10*, 65–69.
45. Daguerre, J.B. Informe de la Comisión Investigadora del Este. In *Memoria de la Comisión Central de Investigaciones sobre la Langosta correspondiente al año;* Ministerio de Agricultura de la Nación: Buenos Aires, Argentina, 1939; pp. 107–142.
46. Fernández, M.L.; Copa Bazán, A.F. Evaluación de Hongos Entomopatógenos Nativos Para el Control Biológico de la Langosta Voladora *Schistocerca cancellata* (Orthoptera: Acrididae) en el I.I.A. "El Vallecito", Santa Cruz, Bolivia, 2017. Tesis de Grado, UAGRM, Presente en la Biblioteca Rafael Peña de la Facultad de Ciencias Agrícolas Con N° de Registro T-2570, 2018; p. 63.

47. Greathead, D.J. Natural enemies of tropical locusts and grasshoppers: Their impact and potential as biological control agents. In *Biological Control of Locusts and Grasshoppers*; Lomer, C.J., Prior, C., Eds.; CABI: Wallingford, UK, 1992; pp. 105–121.
48. Mullié, W.C.; Keith, J.O. The Effects of Aerially Applied Fenitrothion and Chlorpyrifos on Birds in the Savannah of Northern Senegal. *J. Appl. Ecol.* **1993**, *30*, 536. [CrossRef]
49. D'Hérelle, F. Sur la propagation, dans la République Argentine, de l´épizootie des sauterelles du Mexique. *C. R. Acad. Sci. Paris Ser. D* **1912**, *154*, 623–625.
50. Deveson, E.; Martinez, A. Locusts in Southern Settler Societies: Argentine and Australian Experience and Responses, 1880–1940. In *Environmental History in the Making: Volume I: Explaining*; Vaz, E., Joanaz de Melo, C., Costa Pinto, L.M., Eds.; Environmental History; Springer International Publishing: Cham, Switzerland, 2017; pp. 259–286. ISBN 978-3-319-41085-2.
51. Lange, C.E.; Wittenstein, E. Susceptibilidad de la langosta *Schistocerca cancellata* (Orthoptera: Acrididae) a diferentes entomopatógenos. *Rev. Soc. Entomol. Argent.* **1998**, *57*, 19–22.
52. Plischuk, S.; Pocco, M.E.; Quintana, S.; de Wysiecki, M.L.; Lange, C.E. A new symbiont associated with the South American locust *Schistocerca cancellata*. In Proceedings of the 13th International Congress of Orthopterology, Agadir, Morocco, 24–28 March 2019.
53. Pelizza, S.A.; Medina, H.E.; Elíades, L.A.; Pocco, M.E.; Stenglein, S.A.; Lange, C.E. Virulence and enzymatic activity of three new isolates of *Beauveria bassiana* (Ascomycota: Hypocreales) from the South American locust *Schistocerca cancellata* (Orthop-tera: Acrididae). *J. King Saud Univ.-Sci.* **2020**, *32*, 44–47. [CrossRef]
54. Pelizza, S.A.; Ferreri, N.A.; Elíades, L.A.; Galarza, B.; Cabello, M.N.; Russo, M.L.; Vianna, F.; Scorsetti, A.C.; Lange, C.E. Enzymatic activity and virulence of *Cordyceps locustiphila* (Hypocreales: Cordycipitaceae) on the South American locust *Schistocerca cancellata* (Orthoptera: Acrididae). *J. King Saud Univ.-Sci.* **2021**, *33*, 101411. [CrossRef]
55. Zhang, L.; Hunter, D. Management of locusts and grasshoppers in China. *J. Orthoptera Res.* **2017**, *26*, 155–159. [CrossRef]
56. Lange, C.E.; Mariottini, Y.; Plischuk, S.; Cigliano, M.M. Naturalized, newly-associated microsporidium continues causing epizootics and expanding its host range. *Protistology* **2020**, *14*, 32–37. [CrossRef]
57. Pocco, M.E.; De Wysiecki, M.L.; Lange, C.E. Infectivity of *Paranosema locustae* (Microsporidia) against gregarious-phase South American locust (Orthoptera) when treated en masse. *J. Invertebr. Pathol.* **2020**, *177*, 107504. [CrossRef] [PubMed]
58. Fu, X.J.; Hunter, D.M.; Shi, Y.P. Effect of Paranosema (Nosema) locustae (Microsporidia) on morphological phase transfor-mation of *Locusta migratoria manilensis* (Orthoptera: Acrididae). *Biocontrol Sci. Technol.* **2010**, *20*, 683–693. [CrossRef]
59. Li, A.-M.; Yue, Y.; Zhang, Y.-X.; Zhang, L.; Zhang, K.Q.; Shen, J.; Shen, S.Q.; Shi, Y.P. Effects of *Paranosema locustae* (Micro-sporidia) on the development and morphological phase transformation of *Locusta migratoria* (Orthoptera: Acrididae) through modulation of the neurotransmitter taurine. *J. Integr. Agric.* **2020**, *19*, 204–210. [CrossRef]
60. Tranchini, E.M. *Politicas Agrarias y Comportamientos Sociales: El caso de la Plaga de Langosta en la Región Pampeana*; Facultad de Humanidades y Ciencias de la Educación, Uni-versidad Nacional de La Plata: La Plata, Argentina, 1995; p. 97.
61. Libonati, V.J. La Langosta: Su Historia en la Argentina (Síntesis) Buenos Aires: Casartelli y Fiol. 1928; p. 159.
62. Lizer y Trelles, C.C. *La Lucha Moderna Contra la Langosta en el País*; Academia Nacional de Agronomía y Veterinaria: Buenos Aires, Argentina; Universidad Nacional de La Plata: La Plata, Argentina, 1940; p. 33.
63. Uvarov, B.P. A Revision of the Genus *Locusta*, L. (= *Pachytylus*, Fieb.), with a New Theory as to the Periodicity and Migrations of Locusts. *Bull. Entomol. Res.* **1921**, *12*, 135–163. [CrossRef]
64. Predtechensky, S.A. The annual cycle of the Desert locust (*Schistocerca gregaria* Forsk.), its migrations and periodicity in Persia and adjacent countries of tropical and subtropical Asia. *Bull. Plant Prot. Entomol.* **1935**, *12*, 5–135.
65. Waloff, Z. Seasonal breeding and migration of the Desert locust (*Schistocerca gregaria* Forskål) in Eastern Africa. *Anti-Locust Mem.* **1946**, *1*, 1–74.
66. Kennedy, J.S. The migration of the Desert locust (*Schistocerca gregaria* Forsk.). I. The behaviour of swarms. II. A theory of long-range migrations. *Biol. Sciences Rev.* **1951**, *235*, 163–290.
67. Piou, C.; Bacar, M.E.H.J.; Ebbe, M.A.O.B.; Chihrane, J.; Ghaout, S.; Cisse, S.; Lecoq, M.; Halima, T.B. Mapping the spatiotemporal distributions of the Desert Locust in Mauritania and Morocco to improve preventive management. *Basic Appl. Ecol.* **2017**, *25*, 37–47. [CrossRef]
68. Deveson, E.D.; Drake, V.A.; Hunter, D.M.; Walker, P.W.; Wang, H.K. Evidence from traditional and new technologies for northward migrations of Australian plague locusts (*Chortoicetes terminifera*) (Walker) (Orthoptera: Acrididae) to western Queensland. *Austral Ecol.* **2005**, *30*, 920–935. [CrossRef]
69. Köhler, P. Informe de la Comisión Investigadora del Oeste. *Memoria. Comn. Cent. Invest. Langosta Corresp. Al Año* **1936**, *1936*, 17–106.
70. Archibald, E.D. Locusts and Sun-Spots. *Nature* **1878**, *19*, 145–146. [CrossRef]
71. Swinton, A.H. Locust and sunspots. *Science* **1881**, *2*, 255. [CrossRef]
72. Uichanco, L.B. Secular Trends of Locust Outbreaks in the Philippines and their apparent Relation with Sunspot Cycles. *Philip-Pine Agric.* **1936**, *25*, 321–354.
73. Cheke, R.A.; Young, S.; Wang, X.; Tratalos, J.A.; Tang, S.; Cressman, K. Evidence for a Causal Relationship between the Solar Cycle and Locust Abundance. *Agronomy* **2020**, *11*, 69. [CrossRef]
74. Maldonado-Bruzzone, R. Informe de la primera comisión exploradora. In *Memoria de la Comisión Central de Investigaciones Sobre la Langosta*; Ministerio de Agricultura de la Nación: Buenos Aires, Argentina, 1936; pp. 13–68.

75. Hunter, D.M. Locusts in the World. In *Olfactory Concepts of Insect Control-Alternative to Insecticides*; Picimbon, J.F., Ed.; Springer: Cham, Switzerland, 2019; Volume 1, pp. 30–48.
76. Bruch, C. Investigaciones sobre la langosta, experimentos en cautividad. *Mems. Comn. Cent. Invest. Langosta* **1936**, *1936*, 143–190.
77. CIPA, Comité Interamericano Permanente Antiacridiano, Memoria de la Conferencia Internacional de Lucha Antiacridiana realizada en Santa Cruz de la Sierra del 30 de enero al 1o de febrero, Ministerio de Agricultura, Ganadería y Colonización, Di-rección General de Agricultura Departamento de Sanidad Vegetal "Entomología". *La Paz Bolivia.* **1949**, 38.
78. Maldonado-Bruzzone, R. La formación de mangas de *Schistocerca cancellata* (Serville), Informe sobre una gira al oriente boli-viano, Año IV, Serie A, N° 35, Ministerio de Agricultura de la Nación, Buenos Aires. *Argentina* **1948**, 22.
79. Decreto Supremo No. 3081, 2017, Bolivia. Available online: https://www.lexivox.org/norms/BO-DS-N3081.html (accessed on 2 August 2021).
80. GICSV. Informe Regional Langosta Sudamericana Octubre. 2020. Available online: http://apps.iica.int/GICSV/programas/SanidadVegetal/archivos/Langosta/_Informe%20langostas%20GICSV%2010_2020.pdf (accessed on 2 August 2021).
81. Metsul "Guerra aos Gafanhotos" Que "Vai Ser Dura a Parada". 2020. Available online: https://metsul.com/guerra-aos-gafanhotos-que-a-parada-vai-ser-dura/ (accessed on 25 April 2021).
82. Lorier, E.; Zorbino, M.S. La langosta voladora *Schistocerca cancellata* (SERVILLE, 1838) (ORTHOPTERA, ACRIDIDAE, CYRTACANTHACRIDINAE) en Uruguay. *Bol. Soc. Zool. Urug.* **2020**, *29*, 52–65.
83. Gunn, D.L. Nomad encompassed. The development of preventive control of the red locust (*Nomadacris sepermfaciata* (Serville) by International Red Locust Control Service. *J. Ent. Soc. South Afr.* **1960**, *23*, 65–125.
84. Therville, C.; Anderies, J.; Lecoq, M.; Cease, A. Locusts and People: Integrating the Social Sciences in Sustainable Locust Management. *Agronomy* **2021**, *11*, 951. [CrossRef]
85. Daguerre, J.B. Estado actual de la langosta voladora. *Revta Soc. Ent. Argent.* **1970**, *32*, 115–116.
86. Therville, C.; Anderies, J.M.; Medina, H.E.; Overson, R.; Trumper, E.V.; Cease, A.J. *Synthesis of the Governance Workshop on the South American Locust*; Technical Report; Arizona State University: Tempe, AZ, USA, 2020; pp. 1–18.
87. Lecoq, M. Le Criquet pèlerin. Enseignements de la dernière invasion et perspectives offertes par la biomodélisation. In *La Lutte Anti-Acridienne*; Essaid, A., Ed.; AUPELF-UREF: Montreal, QC, Canada; John Libbey Eurotext: Paris, France, 1991; pp. 71–98.
88. Gay, P.-E.; Lecoq, M.; Piou, C. Improving preventive locust management: Insights from a multi-agent model. *Pest Manag. Sci.* **2017**, *74*, 46–58. [CrossRef] [PubMed]
89. Gay, P.; Lecoq, M.; Piou, C. The limitations of locust preventive management faced with spatial uncertainty: Exploration with a multi-agent model. *Pest Manag. Sci.* **2019**, *76*, 1094–1102. [CrossRef] [PubMed]
90. Medina, H.E. Langosta. Explosión Demográfica de la Plaga Que Nunca se fue. XII Encuentro de Monitoreo y Manejo de Plagas, Enfermedades y Malezas, Ciudad de Córdoba, 2016. Expuesto en Forma Oral, Publicado en Resumen de la Jornada. Available online: https://www.researchgate.net/publication/308892241_Langosta_Explosion_demografica_de_la_plaga_que_nunca_se_fue (accessed on 17 December 2021).
91. Medina, H.E. Emergencia Langostas 2020-2021-SENASA Argentina. 2020. Available online: https://geonode.senasa.gob.ar/maps/1806 (accessed on 10 August 2020).
92. Medina, H.E.; Massola, M. Informe de Gestión. Programa Nacional de Langostas y Tucuras. 2020. Available online: https://www.argentina.gob.ar/sites/default/files/anuario_2020_programa_nacional_de_langostas_y_tucuras_v_1.2.pdf (accessed on 10 August 2020).
93. Morello, J.; Pengue, W.; Rodríguez, A. Un Siglo de Cambios de Diseño del Paisaje: El Chaco Argentino. Primeras Jornadas Argentinas de Ecología del Paisaje. 2005, pp. 1–31. Available online: https://repositorio.cepal.org/bitstream/handle/11362/22011/S81102169_es.pdf?sequence=1 (accessed on 17 December 2021).
94. Volante, J.; Alcaraz-Segura, D.; Mosciaro, M.; Viglizzo, E.; Paruelo, J.M. Ecosystem functional changes associated with land clearing in NW Argentina. *Agric. Ecosyst. Environ.* **2012**, *154*, 12–22. [CrossRef]
95. Vallejos, M.; Volante, J.N.; Mosciaro, M.J.; Vale, L.M.; Bustamante, M.L.; Paruelo, J.M. Transformation dynamics of the natural cover in the Dry Chaco ecoregion: A plot level geo-database from 1976 to 2012. *J. Arid. Environ.* **2015**, *123*, 3–11. [CrossRef]
96. Piou, C.; Gay, P.-E.; Benahi, A.S.; Ebbe, M.A.O.B.; Chihrane, J.; Ghaout, S.; Cisse, S.; Diakite, F.; Lazar, M.; Cressman, K.; et al. Soil moisture from remote sensing to forecast desert locust presence. *J. Appl. Ecol.* **2018**, *56*, 966–975. [CrossRef]
97. Cressman, K. Climate Change and Locusts in the WANA Region. In *Climate Change and Food Security in West Africa and North Africa*; Sivakumar, M.V.K., Selvaraju, R.L.R., Hamdan, I., Eds.; Springer: Cham, Switzerland, 2013; Chapter 7; pp. 131–143.
98. Lecoq, M. Desert locust management: From ecology to anthropology. *J. Orthoptera Res.* **2005**, *14*, 179–186. [CrossRef]
99. Cisse, S.; Ghaout, S.; Mazih, A.; Babah Ebbe, M.A.O.; Benahi, A.S.; Piou, C. Effect of vegetation on density thresholds of adult desert locust gregarization from survey data in Mauritania. *Èntomol. Exp. Appl.* **2013**, *149*, 159–165. [CrossRef]
100. Cisse, S.; Ghaout, S.; Mazih, A.; Ebbe, M.A.O.B.; Piou, C. Estimation of density threshold of gregarization of desert locust hoppers from field sampling in Mauritania. *Èntomol. Exp. Appl.* **2015**, *156*, 136–148. [CrossRef]
101. Lazar, M.; Piou, C.; Doumandji-Mitiche, B.; Lecoq, M. Importance of solitarious desert locust population dynamics: Lessons from historical survey data in Algeria. *Èntomol. Exp. Appl.* **2016**, *161*, 168–180. [CrossRef]
102. Kayalto, M.; Hassani, M.I.; Lecoq, M.; Gay, P.-E.; Piou, C. Cartographie des zones de reproduction et de grégarisation du criquet pèlerin au Tchad. *Cah. Agric.* **2020**, *29*, 14. [CrossRef]

103. Hunter, D.M. Advances in the control of locusts (Orthoptera: Acrididae) in eastern Australia: From crop protection to preventive control. *Aust. J. Entomol.* **2004**, *43*, 293–303. [CrossRef]
104. Williams, T.; Arredondo-Bernal, H.C.; Rodríguez-Del-Bosque, L.A. Biological Pest Control in Mexico. *Annu. Rev. Èntomol.* **2013**, *58*, 119–140. [CrossRef] [PubMed]
105. Food and Agriculture Organization of the United Nations. Red Locust disaster in Eastern Africa prevented. In *Biopesticides being Used on a Large Scale*; FAO: Rome, Italy, 2009; Available online: http://www.fao.org/news/story/en/item/21084/icode/ (accessed on 14 August 2021).
106. Stokstad, E. In Somalia, an unprecedented effort to kill massive locust swarms with biocontrol. *Science* **2020**. [CrossRef]

agronomy

MDPI

Article

New Technology for Desert Locust Control

Graham A. Matthews

Faculty of Life Sciences, Imperial College, Ascot SL5 7PY, UK; g.matthews@imperial.ac.uk

Abstract: Locust outbreaks usually begin in remote unpopulated areas following higher than average rainfall. The need to survey such areas has suggested that unmanned aerial vehicles (UAVs), often referred to as drones, might be a suitable means of surveying areas with suitable detection devices to survey areas and detect important locust concentrations. This would facilitate determining where sprays need to be applied at this early stage and would minimise the risk of swarms developing and migrating to feed on large areas of crops. Ideally, a drone could also spray groups of hoppers and adults at this stage. To date, tests have shown limitations in their use to apply sprays, although it has been suggested that using a fleet of drones might be possible. The use of biopesticide in these areas has the advantage of being more environmentally acceptable as the spray has no adverse impact on birds.

Keywords: locust; drone; unmanned aerial vehicle; early warning; preventive control; biopesticide

Citation: Matthews, G.A. New Technology for Desert Locust Control. *Agronomy* **2021**, *11*, 1052. https://doi.org/10.3390/agronomy11061052

Academic Editors: Michel Lecoq, Arianne Cease and Beatriz Gámiz

Received: 20 April 2021
Accepted: 21 May 2021
Published: 24 May 2021

1. Introduction

Since biblical times, vast numbers of desert locusts (*Schistocerca gregaria* Forskål, 1775) have periodically increased to such an extent that the plagues cause extensive damage to major crops. Locust upsurges occur infrequently and during a recession period, international organisations have not prioritized research, nor have governments in countries subject to locust plagues maintained in-country research, due to years of under-funding, so they are not prepared for locust surveillance and the control of locust swarms when they do occur.

Over the last 60 years, the onset of a plague has been detected at an early stage, but in 2019, the early onset of locusts was following cyclones in 2018 which resulted in heavy rain in the inhospitable deserts of Arabia. This allowed locusts to breed unseen in the wet sands. Strong winds in 2019 blew the growing swarms into the Yemen where they were undetected as the country was beset with a war.

Initially, the solitary insects typically occur at low densities across a recession area, but following a period of good rainfall, the locusts thrived and soon aggregated and formed swarms, which spread both eastwards to Iran, India and Pakistan, and westwards to East Africa, with Kenya experiencing its worst outbreak in 70 years. Back in the 1950s, despite vast areas being invaded by swarms, it was possible to use aircraft to spray insecticides, including dieldrin to protect crops and eventually reduce the plague [1] (Figure 1). Dieldrin was used to kill hopper bands that crossed barrier strips and as a very low dose in very fine sprays to kill swarms of desert locust. However, due to increased concern over persistent insecticides affecting non-target species, the use of dieldrin was banned at meeting organised by FAO in 1988 [2]. Following this meeting, the FAO Pesticide Referee Group was established to advise which insecticides could be used. At the same time, CABI organised the international LUBILOSA Programme project, which led to the production of a biopesticide, based on *Metarhizium acridum,* to control locusts. This major development provides control of locusts that does not harm people or the environment [3,4]. Today, we need to bring together new technology to improve the detection of locusts in remote areas and use biopesticide to control locusts before populations build up and swarms invade extensive farming areas. Once swarms arrive in new areas, the control of adult locusts is rapidly needed before crop damage occurs. Biopesticide acts slowly so insecticides that can achieve high mortality within 24 h are needed. Nevertheless, the least hazardous insecticides should be selected in terms of mammalian toxicity and environmental protection.

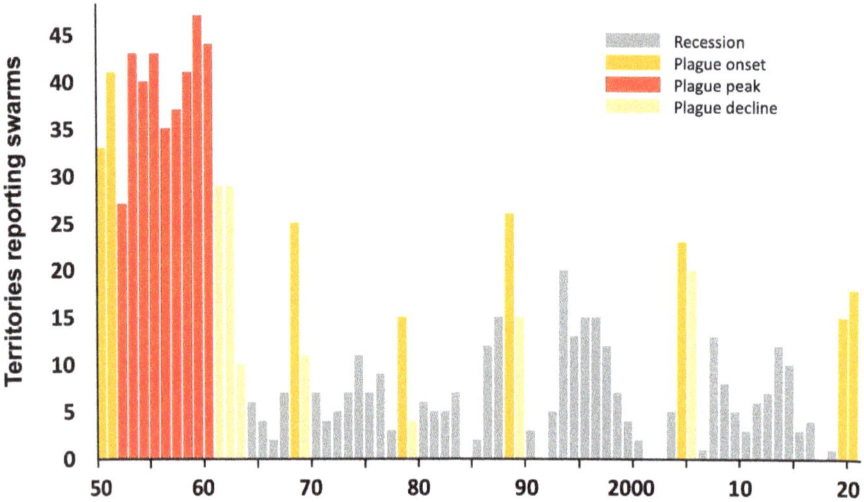

Figure 1. The number of countries in which desert locusts were detected since 1950, with fewer countries having an onset of major desert locust activity (i.e., major upsurges and plagues) since 1965. Control has been achieved sufficiently early to avoid a major plague, but in 2019/20 the locusts spread without early detection (updated from [5,6]).

2. The Initial Signs of a New Upsurge

During a recession, locusts survive in small numbers as solitary insects in arid areas in sparsely populated areas from the Atlantic Ocean to Northwest India, but particularly in the Middle East. After a period of above average rainfall that provides the vegetation, the locust population in these areas can increase rapidly, which translates into spectacular hopper band movements and swarms which migrate. Using phenomenological models using gridded monthly data, Tratalos et al. [7] suggested that desert locust dynamics are influenced by endogenous factors and rainfall, and that broad patterns of locust upsurges and declines can be forecast with some degree of success using data on only these factors.

Locust monitoring has relied on ground-based surveys, which have required individual searches in areas with a history of locust activity, albeit at long and irregular intervals to notify the national locust control units and share the information with neighbouring countries and international agencies [8]. As the population increases, and desert locusts often aggregate to lay eggs, and as they are extremely sensitive to density changes, this rapidly triggers their phase transformation, resulting in their movement in swarms to new areas. The phase transition is a continuous, cumulative, and easily reversible process that can take place within a short period (from 4 h to 32 h) in the desert locust [9,10]. Studies in China have indicated the presence of an aggregation pheromone in the migratory locust, *Locusta migratoria* (Linnaeus, 1758) [11], but as Vosshall [12] discusses, it is not clear whether pheromone traps could assist in detecting the presence of locusts, or whether a chemical could be found to block the receptor to prevent aggregation. The latter may not be desirable as it would not reduce locust populations but rather disperse large numbers of locusts over even larger areas, making the control of locust concentrations nearly impossible.

3. Detecting Where the Locusts Can Be Found after a Recession Period

Satellite imagery has been used for finding and mapping emerging vegetation in the desert. Data collected by the Moderate Resolution Imaging Spectroradiometer (MODIS) on NASA's Terra Satellite and the Normalized Difference Vegetation Index (NDVI) are measures of the health and greenness of vegetation based on how much red and near-infrared light is reflected. According to Tratalos and Cheke [13], there is evidence of a positive relationship between NDVI and the presence of grasshopper populations, probably because grasshoppers were found in areas with higher rainfall. This helps to monitor and forecast the desert locust [14] by assessing whether the ecological conditions are favourable to locust survival, breeding, and gregarization, but needs to be accompanied by thorough observations on the ground for making decisions regarding control interventions against the initial locust congregations. This significantly reduces the costs and contributes towards changing the paradigm of locust control from curative to preventive [15]. According to FAO, current satellites can provide continuous estimates of rain-producing clouds and ecological conditions, such as vegetation development and soil moisture, which are important factors for monitoring desert locust habitats and forecasting locust development. Related to soil moisture, observed differences in grass abundance and size contribute to locust gregarization [16]. The temporal, spectral and spatial characteristics of the sensor instruments onboard these earth observation satellites provide a wide range of sensing capabilities [17].

In 2018, the FAO initiated trials with a long-range fixed-wing drone to examine areas where locusts could be present by mapping areas of green annual vegetation. Ideally, the drone needs a range of at least 100 km or more and should be solar powered. Searching vegetation is possible with lightweight multispectral sensors with near-infrared (NIR) and/or red-edge bands, as well as hyperspectral sensors that provide spectral separability [18]. For locust work, the drone initially needs to detect green vegetation by flying at optimal heights to obtain imagery that can be processed easily and rapidly in the field, so the team can decide where to intensify survey efforts, again with the drone, to detect locust concentrations using optical images that could warrant control [19,20]. The drones need to be easy to operate both manually and automatically, as well as be robust, affordable and simple to maintain locally in locust-affected countries. As a result of three years of trials and refinement, the dLocust drone developed by HEMAV (Figure 2) became available for use in locust-affected countries in 2020 [21]. Imagery is processed in-flight so that when dLocust completes its long-distance survey, the results are immediately available and handed over to the eLocust3 tablet used by field teams for recording and transmitting survey and control data in real time [22,23]. In this way, the team does not need to carry an additional computer to the field. The eLocust3 tablet was also used to plan the dLocust flight and operate the drone. Using drones for surveillance with appropriate detection equipment will enable larger areas to be surveyed in contrast to using ground teams which may only be able to survey limited accessible areas.

Small multi-rotor drones with propellers positioned parallel to the ground may provide a more stable flight and vertical take-off ability, but these are liable to damage from flying locusts. At present, small drones are restricted to carry a limited quantity of spray (only 10 kg) and have a limited endurance of around 10–15 min productivity for locust control, due their small size and limited battery life. The drones need to be easy to control both manually and automatically, as well as be robust, affordable and easy to maintain locally in locust-affected countries.

Figure 2. Fixed wing drone for desert locust monitoring (source: FAO [24]).

4. Applying the Biopesticide in Remote Areas at the Initial Stage of Hopper Development

Studies have been carried out in West Africa to establish the importance of using the biopesticide known as "Green Muscle®", and based on the entomopathogenic fungus *Metarhizium acridum* [25,26]. The storage methods for fungi would be as dry conidia, perhaps with clay diluents, or in oils [27]. The LUBILOSA project focused on developing an oil formulation [28] suitable for ultra-low volume (ULV) application, a technique already well established for locust control. Early studies in South Africa in trials against Vth instar Brown locusts (*Locustana pardalina* Walker, 1870) used a large micro-light aircraft fitted with Micronair AU7000 atomisers to apply *Metarhizium flavoviride* isolate IMI 330189. Dry conidial powder was formulated in a paraffinic oil to apply approximately 2.0×10^{12} conidia per hectare with volume application rates of 1.0 and 2.5 L/ha with the atomiser blades set at three angles (25, 35 and 45°). Up to 98% mortality was obtained where locusts were in open top field enclosures, with the 2.5 L/ha apparently obtaining more consistent results [29]. As pointed out by Bateman [30], the key problems for further research and development would be the logistics and supply of consistently reliable formulations for application at a large scale, and the determination of mechanisms for effective dose transfer in the field. Since the development of the mycopesticide known as "Green Guard®", more than 100,000 ha have been treated with the FI-985 isolate of the fungus *Metarhizium anisoplae* var. *acridum*, since operational use began in Australia in 2000 [31].

At present, attempts to control locusts have been limited to using ground equipment with hand-carried spinning disc sprayers or using truck mounted equipment. Developments with ground equipment now include a ground positioning system (GPS) on truck-mounted sprayers to provide precision while applying ULV insecticide sprays along parallel tracks whilst also recording the position of the sprayer so that a record of the treatment is obtained and analysed.

The biopesticide, now marketed as Novacrid® from Elephant Vert (isolate EVCH077 of *Metarhizium acridum*), is a dry powder supplied in sachets, which is mixed with an appropriate oil before application at ULV rates of 1 L/ha and 0.5 L/ha. Oil formulations were tested in a range of commonly used spinning-disk sprayers, including the hand-held Micron Ulva-Plus and vehicle-mounted Ulva-Mast [3].

The enormous advantage of using the biopesticide for locust control is that it lacks adverse side-effects on biodiversity. In a study in Niger, Green Muscle® was sprayed operationally, using 107 g viable conidia per hectare, where the population of adult locusts, birds and vegetation greenness were previously simultaneously assessed along two transects from 12 days until 23 days after treatment [32]. Locusts started dying five days post-spray and the biopesticide reached its maximum effect one–two weeks after the spray, with 80% efficacy at day 21. After spraying, kestrels took significantly more of the larger female (75–80%) than smaller male (20–25%) locusts. This indicated that avian predation increased the impact of the biopesticide by removing more of the adult female locusts. No direct or indirect adverse side-effects were observed on non-target organisms including locust predators, such as ants and birds.

Where areas are accessible, ground equipment can spray up to 300 ha a day at a fraction of the cost and logistical effort. Similarly, aircraft can spray large areas at relatively low cost, but substantial logistics are required to support aerial operations.

Using a drone at present requires a trained operator, transportation for the drone with the spray formulation, an engine-powered generator and fuel for recharging the batteries. Drone battery technologies are expensive, and the number of charges is limited with existing technologies to less than an estimated 500 charges or only 75 h actual spraying before new batteries are required. Clearly, more research and development are needed to determine the most effective design, and operating procedures are needed to facilitate the control of hoppers prior to the expansion of the locust reaching plague status, with appropriate guidance and standard operating procedures (SOPs) for training to enable drones to be used effectively in certain situations to minimise the formation of swarms.

At present, there is little information available on how to treat desert locusts with drones in a safe and effective way. Solid scientific-based field testing is required. From there, guidelines and standard operating procedures (SOPs) can be developed, and training provided before using drones. The FAO is clear that these are very important steps that cannot be omitted.

5. Current Trials

5.1. Kenya

Drones were tentatively used in 2020 to determine whether their use would be effective. In Kenya, the main focus was to target hoppers and roosting locusts with better efficiency to develop SOPs for the optimal use of the technology. A DJI Agras T16 drone, fitted with a 16 L tank, six rotors, and programmed for operation in Auto and manual modes, was used to treat a 6.5 m swath applying 4.8 L/minute and treat 10 hectares. According to the manufacturer, the drone was fitted with either XR11001VS or XR110015VS fan nozzles which can apply 3.6 to 4.8 L/min, respectively.

The drone (Figure 3) can be operated at up to 3 km at a flight altitude of 2.5 m from the operator and spray at a maximum operational flight speed of 7 m/s. There were 16 nozzles operating from the height of 2.5 m above the canopy, so the drones automatically adjusted to maintain this height. As per the manual, the pilot controls the flight and the volume sprayed. The intention was to spray as per the recommendation on the pesticide label, so various flight parameters to determine which one results closest to the recommended dosage were tried with different heights and flying speeds.

Figure 3. Drone spraying locusts in trees in Kenya (source: CABI [33]).

5.2. India

The operation of drones was examined by deploying them initially for spot application, including high trees, dense plantation and inaccessible areas in association with ground control teams for effective control operation. Fifteen drones were each equipped with a 10 L tank and supplied with eight batteries so that 1 hectare could be treated in 15 min, and a drone could cover 12–15 hectares during a day. Flat fan nozzles were used for spraying a mixture of two insecticides, supplied as per EC formulations. They contained 5% lambdacyhalothrin and 2.8% deltamethrin. Insecticide in the range of 100–120 mL was mixed with 10 L of water. A total of 1753 L of the mixture was applied using 15 drones over 60 days, covering 7017 hectares in a total of 2007 h.

The flight height was 12.2–13.7 m. Flying speed depended on the whether the target was in a tree or in fields and ranged from 10 to 20 km/h. Swath width depended on wind speed and ranged from 1 to 5 m. The volume of spray applied was 10 L per ha. The mortality of locusts was between 50 and 90% at various locust stages.

No ULV formulations were sprayed and no ULV nozzles were used in drone spraying, but if used, the area coverage would increase 10 times.

The overall indications to date are that:

(a) The small multi-copter drones do not have an adequate lifting and endurance capacity to treat more than a small area;

(b) Most have hydraulic pressure nozzles that spray too high a volume so the area that can be treated is very limited. The power requirement to pump the volume of water would be better deployed by using ULV sprays applied with a rotary atomiser;

(c) Even when applying a ULV spray with rotary atomisers, the small payload (10–15 kg) carried by drones limits the area that can be treated with significantly higher operating costs than existing ground sprayers. Spraying small areas of trees on which locusts can be resting is one example as the target is not easily accessible with ground equipment;

(d) Most commercial drones are not designed for ULV products, as pumps, hoses and plastic materials are not compatible with the formulations being used;
(e) There is a question over durability in desert environments with sand and dust drawn across exposed electric motors from multi-rotor types;
(f) Battery cost and operating costs are high with limited duty cycle—400 × before they need to be changed;
(g) A larger payload drone (100 kg+) may offer better potential to fill the gap between using ground equipment and manned aircraft.

Ideally, locusts are controlled by the aerial application of ultra-low volume (ULV) sprays from *c.* 10 m height applied at one litre per hectare, using a spray with droplets of 120 μm volume median diameter (VMD) [34], while smaller droplets using 70–100 μm VMD sprays are applied with ground equipment. A narrower and more effective droplet spectrum is provided by rotary atomisation.

6. Conclusions

At present, the main use of drones will be to improve the surveillance of remote areas where locust populations can increase following more intense rainfall, especially when improvements in the effectiveness of deep learning and computer vision algorithms facilitate the efficiency of spotting the build-up of locusts to form swarms so that appropriate measures can be taken sooner [35]. Further research is needed to examine the effectiveness of drones which carry an increased payload of spray and are equipped with rotary nozzles to apply ultra-low volume sprays in remote areas and extend surveillance over larger areas, but also target sprays at hoppers to reduce the development of swarms. Where swarms have invaded new territories, rapid action insecticides applied as ULV sprays will continue to be needed. With concerns over climate change, surveillance and the ability to minimise the formation of swarms will be more important. In addition, in war or insecurity zones, any control of locusts will remain difficult and realistic mitigation measures will need to be carefully designed and implemented [36].

Funding: This research received no external funding.

Acknowledgments: The author gratefully acknowledges Keith Cressman (FAO, Desert Locust Information Service, Rome) for his guidance regarding recent work by FAO, as well as Ivan Rwomushana (CABI, Nairobi, Kenya) and K.L. Gurjar (Locust Unit, Faridabad, Haryana, India) for information about locust control and drone use in their respective countries.

Conflicts of Interest: The author declares no conflict of interest.

References

1. Bennett, L.V.; Symmons, P.M. *A Review of Estimates of the Effectiveness of Certain Control Techniques and Insecticides against the Desert Locust*; Anti-locust bulletin 50; Centre for Overseas Pest Research: London, UK, 1972.
2. FAO. *Report of the Meeting on the Use and Hazards of Dieldrin in Desert Locust Control, 19–20 October 1988*; FAO: Rome, Italy, 1988.
3. Lomer, C.J.; Bateman, R.P.; Johnson, D.L.; Langewald, J.; Thomas, M. Biological Control of Locusts and Grasshoppers. *Annu. Rev. Entomol.* **2001**, *46*, 667–702. [CrossRef]
4. FAO. Evaluation of Field Trial Data on the Efficacy and Selectivity of Insecticides on Locusts and Grasshoppers. In Proceedings of the Report to FAO by the Pesticide Referee Group, Seventh Meeting, Rome, Italy, 2–6 March 1998; FAO: Rome, Italy, 1998. Available online: http://www.fao.org/ag/locusts/common/ecg/1154/en/PRG7e.pdf (accessed on 27 March 2021).
5. Waloff, Z. *Some Temporal Characteristics of Desert Locust Plagues;* Anti-Locust Memoir No.1; Anti-Locust Research Centre: London, UK, 1976.
6. Cressman, K.; Stefanski, R. *Weather and Desert Locusts;* Technical Report WMO-No. 1175; World Meteorological Organization and Food and Agriculture Organization of the United Nations: Geneva, Switzerland, 2016.
7. Tratalos, J.A.; Cheke, R.A.; Healey, R.G.; Stenseth, N.C. Desert locust populations, rainfall and climate change: Insights from phenomenological models using gridded monthly data. *Clim. Res.* **2010**, *43*, 229–239. [CrossRef]
8. Latchininsky, A.V. Locusts and remote sensing: A review. *J. Appl. Remote Sens.* **2013**, *7*, 075099. [CrossRef]
9. Cullen, D.A.; Cease, A.J.; Latchininsky, A.V.; Ayali, A.; Berry, K.; Buhl, J.; De Keyser, R.; Foquet, B.; Hadrich, J.C.; Matheson, T.; et al. From molecules to management: Mechanisms and consequences of locust phase polyphenism. *Adv. Insect Phys.* **2017**, *53*, 167–285.

10. Ayali, A. The puzzle of locust density-dependent phase polyphenism. *Curr. Opin. Insect Sci.* **2019**, *35*, 41–47. [CrossRef]
11. Guo, X.; Yu, Q.; Chen, D.; Wei, J.; Yang, P.; Yu, J.; Wang, X.; Kang, L. 4-Vinylanisole is an aggregation pheromone in locusts. *Nature* **2020**, *584*, 584–588. [CrossRef]
12. Vosshall, L.B. Catching plague locusts with their own scent. *Nature* **2020**, *584*, 528–530. [CrossRef]
13. Tratalos, J.A.; Cheke, R.A. Can NDVI GAC imagery be used to monitor desert locust breeding areas? *J. Arid Environ.* **2006**, *64*, 342–356. [CrossRef]
14. Cressman, K. Satellites and GIS in desert locust monitoring worldwide: Lessons learned. In Proceedings of the 7th International IPM Symposium, Memphis, TN, USA, 27–29 March 2012.
15. Magor, J.I.; Lecoq, M.; Hunter, D.M. Preventive control and Desert Locust plagues. *Crop Prot.* **2008**, *27*, 1527–1533. [CrossRef]
16. Babah, M.A.O.; Sword, G.A. Linking locust gregarization to local resource distribution patterns across a large spatial scale. *Environ. Entomol.* **2004**, *33*, 1577–1583. [CrossRef]
17. Klein, I.; Oppelt, N.; Kuenzer, C. Application of Remote Sensing Data for Locust Research and Management—A Review. *Insects* **2021**, *12*, 233. [CrossRef]
18. Panday, U.S.; Pratihast, A.K.; Aryal, J.; Kayastha, R.B. A Review on Drone-Based Data Solutions for Cereal Crops. *Drones* **2020**, *4*, 41. [CrossRef]
19. Piou, C.; Jaavar Bacar, M.E.H.; Ould Babah Ebbe, M.A.; Chihrane, J.; Ghaout, S.; Cisse, S.; Lecoq, M.; Ben Halima, T. Mapping the spatiotemporal distributions of the Desert Locust in Mauritania and Morocco to improve preventive management. *Basic Appl. Ecol.* **2017**, *25*, 37–47. [CrossRef]
20. Piou, C.; Gay, P.E.; Benahi, A.S.; Babah Ebbe, M.A.O.; Chihrane, J.; Ghaout, S.; Cisse, S.; Diakite, F.; Lazar, M.; Cressman, K.; et al. Soil moisture from remote sensing to forecast desert locust presence. *J. Appl. Ecol.* **2019**, *56*, 966–975. [CrossRef]
21. Drones for Desert Locust Project—Hemav. Available online: https://www.youtube.com/watch?v=QmCznC6CbbE (accessed on 27 March 2021).
22. Cressman, K. The Use of New Technologies in Desert Locust Early Warning. *Outlooks Pest Manag.* **2008**, *19*, 55–59. [CrossRef]
23. FAO. eLocust3: Updated Field Data Entry & Transmission. Available online: http://www.fao.org/ag/locusts/en/activ/DLIS/eL3/index.html (accessed on 30 March 2021).
24. FAO. Drones for Locust Early Warning & Preventive Control. Available online: http://www.fao.org/ag/locusts/en/activ/DLIS/drones/index.html (accessed on 27 March 2021).
25. Langewald, J.; Kooyman, C.; Douro-Kpindou, O.-K.; Lomer, C.J.; Dahmoud, A.O.; Mohammed, H.O. Field treatment of Desert Locust (*Schistocerca gregaria* Forskål) hoppers in the field in Mauritania with an oil formulation of the entomopathogenic fungus *Metarhizium flavoviride*. *Biocontrol Sci. Technol.* **1997**, *7*, 603–611. [CrossRef]
26. Lomer, C.J.; Prior, C.; Kooyman, C. Development of *Metarhizium* spp. for the control of grasshoppers and locusts. *Mem. Entomol. Soc. Can.* **1997**, *171*, 265–286. [CrossRef]
27. Moore, D.; Caudwell, R.W. Formulation of entomopathogens for the control of grasshoppers and locusts. *Mem. Entomol. Soc. Can.* **1997**, *171*, 49–67. [CrossRef]
28. Bateman, R.P.; Carey, M.; Moore, D.; Prior, C. The enhanced infectivity of *Metarhizium flavoviride* in oil formulations to desert locusts at low humidities. *Ann. Appl. Biol.* **1993**, *122*, 145–152. [CrossRef]
29. Price, R.E.; Bateman, R.P.; Brown, H.D.; Butler, E.T.; Müller, E.J. Aerial spray trials against brown locust (*Locustana pardalina*, Walker) nymphs in South Africa using oil-based formulations of *Metarhizium flavoviride*. *Crop Prot.* **1997**, *16*, 345–351. [CrossRef]
30. Bateman, R.P. Methods of application of microbial pesticide formulations for the control of locusts and grasshoppers. *Mem. Entomol. Soc. Can.* **1997**, *171*, 69–81. [CrossRef]
31. Hunter, D.M. Credibility of an IPM Approach for Locust and Grasshopper Control: The Australian Example. *J. Orthoptera Res.* **2010**, *19*, 133–137. [CrossRef]
32. Mullié, W.C.; Cheke, R.A.; Young, S.; Ibrahim, A.B.; Murk, A.J. Increased and sex-selective avian predation of desert locusts *Schistocerca gregaria* treated with *Metarhizium acridum*. *PLoS ONE* **2021**, *16*, e0244733. [CrossRef] [PubMed]
33. CABI. Drones for Desert Locust Control in East Africa. Available online: https://www.cabi.org/projects/drones-for-desert-locust-control-in-east-africa/ (accessed on 27 March 2021).
34. Johnstone, D.R. Variations in insecticide dose received by settled locusts: A computer model for ultra-low volume Spraying. *Crop Prot.* **1991**, *11*, 183–194. [CrossRef]
35. Kumar, K.S.; Abdul Rahman, A. Early Detection of Locust Swarms Using Deep Learning. In *Advances in Machine Learning and Computational Intelligence*; Patnaik, S., Yang, X.S., Sethi, I., Eds.; Springer: Singapore, 2021; pp. 303–310. [CrossRef]
36. Showler, A.T.; Lecoq, M. Incidence and Ramifications of Armed Conflict in Countries with Major Desert Locust Breeding Areas. *Agronomy* **2021**, *11*, 114. [CrossRef]

Short Biography of Authors

Graham A. Matthews is a Professor Emeritus at Imperial College, Silwood Park (UK). His research within the International Pesticide Application Centre (IPARC) was on the application of pesticide technology, involving testing equipment suitable for vector control for the WHO and on applying pesticides for crop protection. Early in his career, he was involved in cotton pest research in Africa. Prof. G.A. Matthews was a member of the Pesticide Advisory Committee of the UK Department for Environment, Food and Rural Affairs from 2000 to 2004, and Chairman of the FAO Expert Panel on Locust Pesticides from 1989 to 2015. He is the author of numerous publications and books, including *Pesticide Application Methods* in 1979, now in its fourth edition, and most recently *Pesticides, Health, Safety and the Environment* (second edition—2016) and a *History of Pesticides* (2018).

agronomy

MDPI

Article

Regional Differences in Control Operations during the 2019–2021 Desert Locust Upsurge

Renata Retkute [1,*], Rebekah G. K. Hinton [1,2,*], Keith Cressman [3] and Christopher A. Gilligan [1]

1 Epidemiology and Modelling Group, Department of Plant Sciences, University of Cambridge, Downing Street, Cambridge CB2 3EA, UK; cag1@cam.ac.uk
2 Department of Civil and Environmental Engineering, University of Strathclyde, Glasgow G1 1XJ, UK
3 Food and Agriculture Organization of the United Nations, Viale delle Terme di Caracalla, 00153 Rome, Italy; Keith.Cressman@fao.org
* Correspondence: rr614@cam.ac.uk (R.R.); rgkh2@cam.ac.uk (R.G.K.H.)

Abstract: The desert locust remains a major threat to global food security. Control operations are a crucial tool to manage crisis; this research investigated the nature of control operations conducted between 2019–2021. Historical data on desert locust and control operations were obtained from the survey reports at the FAO Locust Hub and analysed with respect to survey reports, land cover types, cropland/rangeland extent and crop productivity data. We found that 16.1% of the grid cells with locust presence and 14.9% of the grid cells with control operations had a proportion of rangeland higher than 0.75; while 13.3% of the grid cells with locust presence and 13.2% of the grid cells with control operations had a proportion of croplands higher than 0.75, highlighting that locust presence and control operations were reported in both rangeland and cropland. Control operations continue to be used both to reduce overall locust numbers and to protect crops. Furthermore, through identifying which crops were most at risk, our analyses indicate that wheat production was under the highest strain during periods of increased locust infestations.

Keywords: desert locusts; control; crop loss; pastureland; land cover

Citation: Retkute, R.; Hinton, R.G.K.; Cressman, K.; Gilligan, C.A. Regional Differences in Control Operations during the 2019–2021 Desert Locust Upsurge. *Agronomy* **2021**, *11*, 2529. https://doi.org/10.3390/agronomy11122529

Academic Editors: Michel Lecoq and Arianne Cease

Received: 30 October 2021
Accepted: 7 December 2021
Published: 13 December 2021

1. Introduction

Desert locusts (*Schistocerca gregaria*) continue to pose one of the most severe threats to agricultural and rangeland crops, making their control a priority to food security in many regions [1]. Timely and precise early warning, and rapid response can be achieved through sustaining monitoring efforts [2], orienting surveys towards historical hotspots [3], increasing the areas accessed by field teams [4], integrating the roles of stakeholders [5], and by more efficient monitoring and control techniques [1]. Symmons [6] outlines three strategies in desert locust control: control during the upsurge for plague prevention, eliminating a plague through destroying all desert locusts, and allowing the plague to continue but protecting crops. The method of control applied and intensity of the control operation varies in accordance with the aim of the control operation and the stage in the desert locust outbreak in which it is applied. The United Nations Food and Agriculture Organisation (FAO) coordinates preventative and reactive control strategies through locust surveys and control operations, and providing early warning [7–12].

Prior to 1965, plagues of desert locust were common in East and West Africa, the Middle East and parts of Pakistan and India [13]. The availability of effective pesticides, together with an integrated programme for surveillance, monitoring and control, developed and advocated by FAO, has prevented local outbreaks and more widespread upsurges from developing into plagues in many regions since 1965 [14,15]. Pesticide application using ultra low volume (ULV) spraying formulations, originally developed in the 1950s and 1960s, remains one of the most extensive control methods [6,16]. Three types of sprayer platforms are utilised in a field, with each having specific control related characteristics:

portable ULV sprayers (slow work rate and up to 10 ha target size), vehicle-mounted sprayers (medium work rate and 1–100 ha target size), and aircraft-mounted sprayers (fast work rate and over 25 ha target size) [17]. The dispersal of the pesticide into small droplets for ULV application ensures efficient use of the pesticide and effective dispersal on the wind. Restrictions of resource and availability of aerial spraying mean that handheld and especially vehicle-mounted spraying to control desert locust remain common.

However, the control of desert locusts can have negative ecological consequences on non-target organisms [18–20]. Organophosphates and carbamates have historically been the most commonly used insecticides as they are fast acting and relatively non-persistent [16]. The non-persistence minimises their impact on non-target insects as they do not remain biologically active for long after application. Nevertheless, organophosphate and carbamate insecticides are broad spectrum compounds, exhibiting toxic effects on other arthropods that risk causing unintended ecological damage. Biopesticides represent an alternative method for the control of desert locusts, providing specific control [19]. For example, *Metarhizium anisopliae* var. acridum is a biopesticide specific to the family Acrididea, containing desert locusts [20]. However, despite the benefits in minimising the ecological consequences of desert locust control, *Metarhizium anisopliae* is often not used due to slow action, variable locust mortality and challenges in its production and storage [16]. Nevertheless, good results were achieved in Somalia where it was used exclusively during the 2019–2021 upsurge. Whilst pesticides with adverse ecological consequences remain in use, monitoring and managing the extent and location of control operations will be an important method in minimising negative ecological consequences for non-target insects.

In May and October 2018, the tropical cyclones Mekunu and Luban caused heavy rainfall and intense precipitation events in the Arabian peninsula, which provided suitable environmental conditions for breeding, multiplication and gregarization of the desert locust [21,22]. In January 2019, desert locust swarms spread to Yemen, Saudi Arabia and Iran, eventually crossing the Red Sea to East Africa [23]. From June 2018 to June 2020, desert locust swarms were observed in more than 22 countries [23].

The current research was designed to investigate the nature of control operations conducted during the 2019–2021 desert locust upsurge. The work is part of wider project, designed to evaluate the potential for epidemiological models, coupled with meteorological models for swarm dispersal, to optimise the deployment of control in order to minimise the impacts of desert locust on agricultural crops and rangeland. Here we describe an empirical analysis of control practices. Specifically, we investigate properties of control operations under the following aspects: survey reports, land cover types, cropland/rangeland presence and crop productivity. The primary objective is to assess the extent to which control occurs on crop and pastureland (i.e., control aimed directly at desert locusts to minimise economic loss) compared with bare ground and shrubs (i.e., control aimed directly at desert locusts and indirectly at preventing economic loss and hardship). Our analysis indicates that control operations in East Africa, the Middle East and an Eastern Region (comprising, Iran, Pakistan, India and Nepal) were qualitatively different in terms of the type of control used for locust management and how desert locust infestations affected the livelihoods of the communities in these regions.

2. Materials and Methods

2.1. Data

Historical data on desert locust presence and control operations were obtained from survey reports and archives at the FAO Locust Hub website [24]. We have extracted the records dated between 1 January 2019 and 1 June 2021.

We used land cover classification from Copernicus global map of land cover at 100 m resolution data (CLC100) [25,26]. Maps of an area fraction that represent the percentage of the pixel occupied by cropland and rangeland at 1 km resolution were downloaded from EU Science Hub [27]. The global crop production statistics data for 2010 [28] and data in

Sub-Saharan Africa for 2017 [29] (mapSPAM) were obtained from the International Food Policy Research Institute.

2.2. Analysis

We extracted Locust Hub reports for three regions: East Africa, the Middle East and the Eastern Region. East Africa region included Ethiopia (ET), Kenya (KE), Somalia (SO), Eritrea (ER), Djibouti (DJ), Uganda (UG), Tanzania (TZ), Sudan (SD&SU), South Sudan (SS), Egypt (EG) and D.R. Congo (CG). The country codes correspond to the "COUNTRYID" column in the Locust Hub database.

The Middle East region included Oman (OM), Saudi Arabia (SA), United Arab Emirates (AE&MU&TC), Yemen (YM&YE), Iraq (IQ), Jordan (JO), Bahrein (BH), Israel (IS), Kuwait) (KU&KW), Lebanon (LE) and Syria (SY).

The Eastern Region included India (IN), Pakistan (PK& PA), Iran (IR), and Nepal (NP). Our classification for the Eastern Region is based on observations that during recessions and even during upsurges, desert locust circulate between Iran, Pakistan and India. Only in a few situations do locusts invade the Eastern Region countries from outside, e.g., from Saudi Arabia to Iran, from Yemen/Somalia to India/Pakistan. This was also true during the current upsurge.

For each control report we extracted the country where it was applied and the type of control used. To align control data with 'Swarm' and 'Band' datasets from [24], we searched for reports that were reported within a 1 km radius and up to two days before the control operation. There remained a significant number of control reports within the criteria that did not correspond with the normal targets for control of 'Swarm' or 'Band' but with observations of the solitarious classifications, 'Adult' or 'Band'. It is likely that the observations of solitarious desert locusts may be inflated by misclassification of the remnants of former swarms and bands and by multiple reporting of the same insects, a common problem of crowd-sourced reporting of data. Accordingly control operations that were not matched to 'Swarms' and 'Bands' were assigned to a 'Not Assessed' ('NA') category.

Land cover type corresponding to the reported location of the control operation was extracted from CLC100 [25].

To align control reports with crop and pastureland maps, we calculated the number of control reports within each 1 km × 1 km grid cell for cropland/rangeland analysis and 10 km × 10 km grid cell for crop production analysis. All analysis and map preparations were performed using R Statistical Software (version 4.1.0; R Foundation for Statistical Computing, Vienna, Austria).

3. Results

3.1. Temporal Distribution of Locust Reports and Control Operations

The compiled dataset comprised more than 75,000 desert locust records with 42.5% reported in East Africa, 23.3% reported in the Middle East, and 34.2% reported in Eastern Region (Figure 1). Desert locusts can be either in solitarious (hoppers and adults) or gregarious (bands of hoppers and swarms) phases. In addition, remnant locusts from control operations against hopper bands and swarms were sometimes reported as gregarious hopper and adult individuals or groups. This is not unusual during upsurges. In terms of the types of desert locust classified within the reports, 19.7% were swarms, 14.3% were bands, 39.4% were solitarious adults and 26.6% were solitarious hoppers. There was a high degree of temporal heterogeneity in the numbers of reports. The highest weekly number of swarm reports (855) was reported in East Africa on the week commencing the 26 January 2021, the highest number of band reports (802) was also reported in East Africa on the week commencing the 9 June 2020. This coincides with the breeding periods.

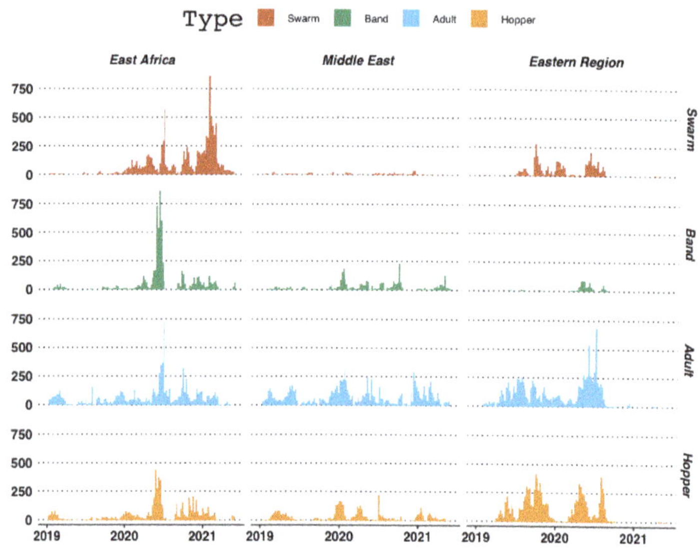

Figure 1. The weekly numbers of desert locust reports. Surveys are categorised as swarms, bands, adults or hoppers. Data from [24] (LocustHub).

The number of control operations performed in 2019–2021 reflects the enormous scale of the desert locust upsurge and deployment of resources to manage it (Figure 2). Overall, more than 39,000 control operations were conducted with 28.7% reported in East Africa, 23.8% reported in the Middle East and 47.5% reported in the Eastern Region. These consisted of 66.9% vehicle based control, 8.5% air based control and 4.8% handheld based control. About 19.8% of control reports were classified as "Unknown". This is likely to be control by teams using digital tools that report only basic information, e.g., control = Yes/No without indicating control method.

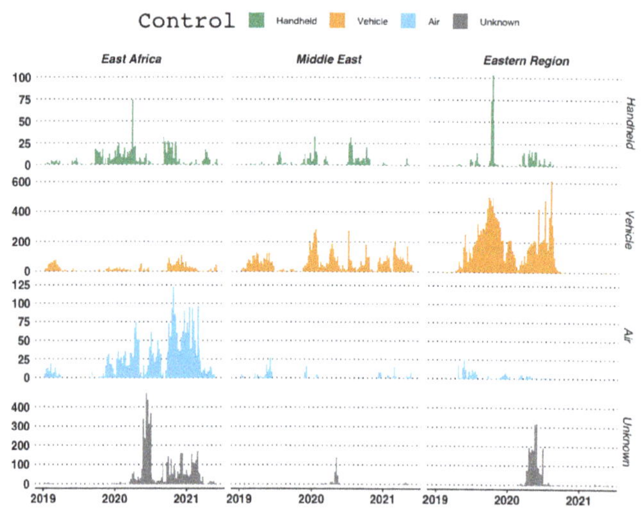

Figure 2. The weekly number of control operations. Control operations are categorised according to region: East Africa, the Middle East and the Eastern Region. Data from [24] (Locust Hub).

An increasing trend in the application of aerial control operations can be seen, with the peak reaching around 100 weekly aerial operations in November 2020 and February 2021. Most of the aerial control operations (86.2%) were conducted in East Africa. Further, within East Africa, 64.8% of aerial operations were conducted in Ethiopia, 4.7% in Kenya and 19.9% in Somalia. Overall, Kenya had the largest number of control reports (92.4%) classified as "Unknown".

The intensity of control operations differed amongst the regions (Figure 2). In all regions, the peaks correspond to breeding periods and are a result of an increase in locust numbers. In East Africa, the peak of control operations was in July 2020 (around 500 weekly control operations), followed by an approximately constant number of control operations (around 100 per week) from October 2020 till April 2021. In the Eastern Region, there were two peak periods reaching around 600 weekly operations in September–November 2019 and 700 weekly operations in May–August 2020. The desert locust situation was brought under control in the Eastern Region in Iran, Pakistan, and India in September 2020 [30]. In the Middle East, the intensity of control operations varied, a few periods (June 2019, January 2020, and May 2020) reached 300 operations.

In terms of pesticides applied during control operations, Malathion was frequently used in East Africa (18.5% of the reports) and the Eastern Region (65.8% of the reports). Malathion is an organophosphorus compound with a high kill rate for desert locust [31]. Malathion has low persistence in the environment as it breaks down quickly due to hydrolysis, photolysis and biodegradation [32]. The most frequently used pesticides in the Middle East were Deltamethrin (38.3% of the reports) and Fenitrothion (35.8% of the reports). Deltamethrin is a synthetic pyrethroid having low toxicity to birds and mammals [33]. Fenitrothion is an organophosphate, which has been shown to have a strong and long-lasting effect on non-target insect taxa [34]. Metarhizium, which is a biopesticide and can induce high mortality in *S. gregaria* [35], has been extensively applied in Somalia and used to treat at least 76.5% of all controlled area. Pesticide usage was ranked as follows according to area treated. In East Africa: Malathion (28.5%), Chlorpyrifos (17.5%), Metarhizium (6.0%), Fenitrothion (3.2%), Teflubenzuron (1.2%), Deltamethrin (0.3%), Triflumuron (<0.01%), Lambda-cyhalothrin (<0.01%). In the Eastern Region: Malathion (40.7%), Deltamethrin (38.8%), Lambda-cyhalothrin (11.3%), Fenitrothion (0.8%), Diflubenzuron (0.05%), Metarhizium (0.02%), Chlorpyrifos (<0.01%). In the Middle East: Fenitrothion (48.1%), Deltamethrin (33.8%), Malathion (10.4%), Chlorpyrifos (2.7%), Lambda-cyhalothrin (1.9%), Diflubenzuron (0.5%), Triflumuron (0.05%), Teflubenzuron (0.05%), Metarhizium (0.03%).

We found high correlations between the weekly number of desert locust reports and the weekly number of control operations: 0.94 in East Africa, 0.88 in the Middle East and 0.98 in the Eastern Region (Figure 3). However, there were differences between the three regions when we calculated the correlation coefficient between the number of swarm reports and control operations: 0.62 in East Africa, −0.02 in the Middle East and 0.74 in the Eastern Region. In East Africa, the highest correlation between the weekly number of different types of desert locust was between hopper reports and band reports (0.88). Low correlations between swarm reports and band reports were observed in all three regions: 0.01 in the Middle East region, 0.18 in East Africa and 0.27 in the Eastern Region.

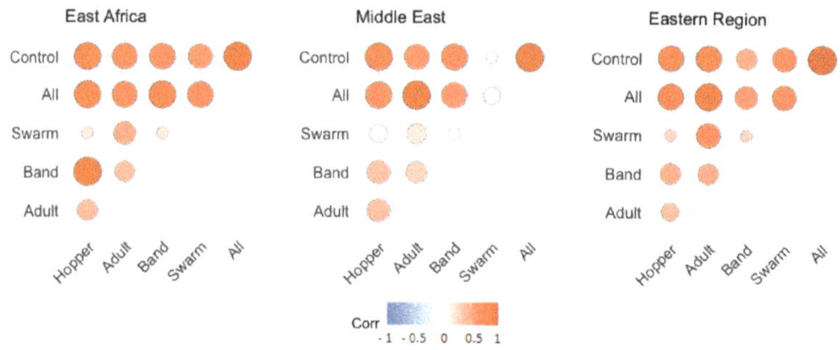

Figure 3. Correlation matrices between the weekly number of control operations and desert locust reports. Colour intensity and the size of the circle are proportional to the correlation coefficients. Data from [24] (Locust Hub).

3.2. Land Cover Characteristics

Figure 4 shows the frequency of land cover types for all regions and types of control. Bare parse vegetation, shrubs, cropland and herbaceous vegetation accounted for 96% of pixels with control operations. Bare/sparse vegetation is classified as lands with exposed soil, sand, or rocks and lands which never has more than 10% vegetated cover during any time of the year. Shrubs are classified as woody perennial plants with persistent and woody stems, or without any defined main stem, being less than 5 m tall [25].

The majority of vehicle based operations in the Middle East were reported over bare/sparse vegetation (83%), this is to be expected given that most of the region is desert. In the Eastern Region, a similar proportion of vehicle based control operations were reported over bare/sparse vegetation, shrubs and cultivated vegetation. In East Africa, 46% of aerial control were reported over land classified as shrubs, 20% over bare/sparse vegetation, and 14% over croplands. Overall, croplands (i.e., cultivated and managed vegetation/agriculture) accounted for 19% of the locations of control operations.

We could match 39.4% of control operations to swarm or band reports. The proportion of control operations that were not matched to swarms or bands (classified as 'NA' in Figure 4), could be explained by matching the control operations to solitarious locust reports (i.e., adult and hopper reports within a 1 km radius and up to two days before the control operation). This could be caused by erroneous classification of reports (for example, reporting remnant gregarious locusts from control operations as adults), as the standard operational procedure is that no sites with solitarious locust presence are subjected to control operations.

In East Africa, swarm reports preceded 35% of aerial operations conducted over bare/sparse vegetation, 65% of aerial operations conducted over shrubland, and 58% of aerial operations conducted over cropland.

3.3. Damage to Crop and Pasture Areas

We mapped the locations of control operations and desert locust reports against two of the factors determining livelihood: the proportion of grid cell area used as cropland and the proportion of grid cell area used as rangeland. We found that 13.3% of the grid cells with desert locust presence and 13.2% with control operations had a proportion of croplands higher than 0.75 (Figure 5). Similarly, 16.1% of the grid cells with desert locust presence and 14.9% with control operations had a proportion of rangeland higher than 0.75 (Figure 5).

To identify which crops are at the highest risk due to the desert locust upsurge, we used the data from the 2010 Spatial Production Allocation Model (mapSPAM) [28] to estimate crop production for the pixels with control operations. In East Africa, sugarcane (854 kt), other root crops (601 kt), maize (588 kt), sorghum (473 kt) and wheat (306 kt) were the five most vulnerable crops in terms of frequency of reported swarms and bands. In the Middle East, the top two crops potentially affected by desert locust were vegetables (1720 kt) and wheat (666 kt). In the Eastern Region, Wheat (4360 kt) and sugarcane (3923 kt) were the two most infested crops. Overall, wheat production would be under the highest strain (5151 kt) due to desert locust infestations.

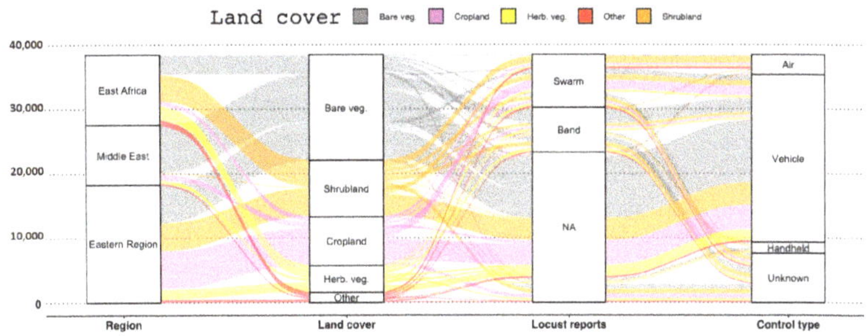

Figure 4. Relationship between control type, locust reports and land cover in the three regions. Y axis corresponds to the number of reported control operations. The most frequent land cover types are given by bare/sparse vegetation (Bare veg.), cropland, herbaceous vegetation (Herb. veg.) and shrubland. Locust reports refer to the gregarious categories 'Swarm' and 'Band' that were matched to reports of control (i.e., reports within a 1 km radius and up to two days before the control operation in the Locust Hub dataset [24]. The 'NA' ('Not Assessed') category refers to control operations that could not be matched to the gregarious classes.

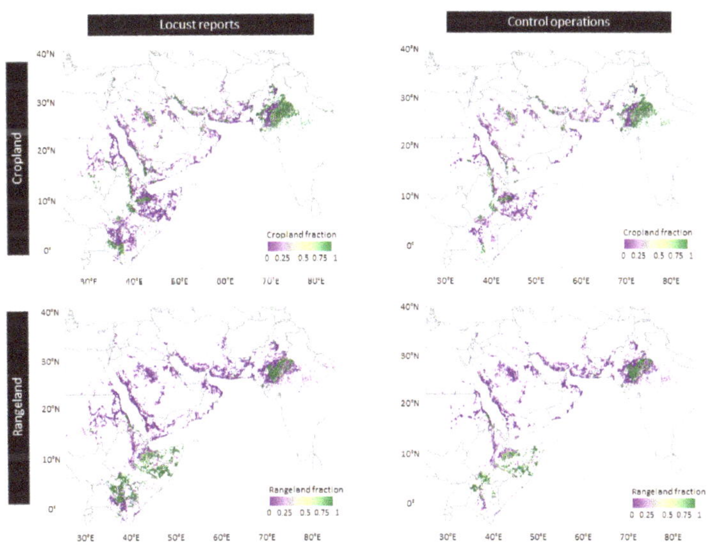

Figure 5. The fraction of land used as cropland and rangeland for pixels with locust reports and control operations. It should be noted that timing of locust infestations with crop production and pasture availability are not reflected in the above maps.

We used mapSPAM data for Sub-Saharan Africa (v2 2017) [29] to investigate crops that were at risk from desert locust presence in East Africa at a national level. In Ethiopia, the top five crops affected were other roots (451 kt), sorghum (437 kt), maize (418 kt), wheat (354 kt) and other cereals (339 kt). In Kenya, the largest effect was on maize (267 kt), vegetables (186 kt) and potato crops (143 kt). In Somalia, the most affected crops were sorghum (8 kt), maize (8 kt) and groundnut (5 kt). Overall, in the three countries, maize (694 kt), sorghum (454 kt) and other roots (453 kt) were the top three crops at the highest risk.

4. Discussion

Desert locust surveys and control operations are two management actions that have direct influence on desert locust dynamics [15]. More than 73,000 desert locust records and more than 38,000 control operations have been reported during the 2019–2021 desert locust upsurge. We found high correlation (0.92–0.98) between the weekly number of desert locust reports and the weekly number of control operations.

Control operations in the three regions were qualitatively different in terms of the type of control used for locust management. In East Africa 25.8% of control operations were aerial and 21.5% were vehicle based, while a variety of pesticides were used as control with the most frequent being Malathion (18.5%). In the Middle East 89.5% of control operations were executed by vehicle, and most frequently used pesticides were Deltamethrin (38.3% of the reports) and Fenitrothion (35.8% of the reports). In the Eastern Region 84.5% of control operations were vehicle based, and Malathion was most frequently used pesticide (65.8% of the reports).

Desert locust behaviour and dispersal patterns are complex processes that depend on environmental factors such as temperature [36], precipitation [22,37], the condition of vegetation [38–40], soil moisture [41,42], soil composition [41] and the wind [36]. These factors can create either stimulatory or inhibitory conditions for gregarisation, desert locust migration and breeding. For example, low-level wind-field convergence and concentration within the preferred habitat of desert locusts can affect the rate of gregarisation [43,44], whilst sandy soil is required for female desert locusts to breed [41]. Our analysis shows that the majority of control operations (79% in the Middle East, 28% in East Africa and 24% in the Eastern Region) were conducted over bare/sparse vegetation; this land type has high potential for the sufficient sand and clay content required for breeding. It is also the typical habitat of desert locust, which differentiates them from typical pests found in crops.

Despite improvements in both preventative control techniques and international collaboration in desert locust prevention, the desert locust remains a major threat to sustainable agriculture and rural livelihoods [7]. Desert locust infestations can aggravate food shortage situations [45]. We found that 13.3% of the grid cells with desert locust presence and 13.2% of the grid cells with control operations had a proportion of croplands higher than 0.75. We found that maize, sorghum and other roots were the top three crops at the highest risk in Ethiopia, Kenya and Somalia. Overall, our analysis of the Locust hub data indicate that wheat production is under the highest strain due to desert locust invasions in the 2019–2021 upsurge. This is in agreement with the estimate of a 12% decline in productivity of wheat due to the current desert locust upsurge in India [46].

Rangeland damage by desert locusts is also a threat to rural livelihoods. Rangelands provide local rural communities with natural resources for livestock production as grazing areas, as well as contributing to their livelihoods with fuel wood, poles, medicinal plants and wild foods. For example, for the most recent data available in 2019, Somalia was the largest producer of camels in East Africa, with the total number of camels recorded 7,243,792 [47], and a mean increase in camel population of 1.77%/year [48]. Rangelands also deliver important ecosystem services such as maintaining biodiversity and storing large amounts of carbon [49]. We found that 16.1% of the grid cells with desert locust presence and 14.9% of the grid cells with control operations had a proportion of rangeland higher than 0.75. Therefore, our analysis confirms that control operations continue to be used both in the reduction of desert locust populations and protection of crops to limit damage. Further work is required, however, to assess the cost effectiveness of surveillance and deployment of chemical control in reducing the increase and spread of locust populations during an upsurge, and also in managing agricultural and rangeland losses. It is possible that optimal deployment strategies of where to survey and where to apply chemical control may differ for minimising crop and other losses, compared with minimising insect dispersal. Moreover, the relative cost-effectiveness and availability of aerial, vehicle-based and hand-held control needs to be assessed as does the environmental and human health impacts of different pesticide options relative to the obvious needs for food security.

Pest outbreaks can strongly affect local economies with invasive insects costing an estimated minimum of US$70billion per year globally [50]. Epidemiological modelling has been regularly used as a tool to evaluate different strategies for crop protection from pests and diseases [51]. Typically, model development requires the partition of spatial data into layers such as 'host/habitat layer', 'environment layer', 'pest layer', and 'control layer' [52]. A spatially explicit model of pest entry and dispersal can be combined with these heterogeneous layers to identify the outcome of alternative control strategies. Furthermore, control strategies can incorporate time-varying economic constraints [53] or human and social dimensions [54]. We anticipate that the present study will help to develop a novel epidemiological modelling framework incorporating optimal control theory for deployment of scarce resources [55,56] to inform the surveillance and management of desert locust infestations.

Author Contributions: Conceptualization, R.R., R.G.K.H. and C.A.G.; methodology, R.R. and C.A.G.; software, R.R.; formal analysis, R.R.; investigation, R.R., R.G.K.H., K.C. and C.A.G.; resources, C.A.G.; data curation, K.C. and R.R.; writing—original draft preparation, R.R., R.G.K.H., K.C. and C.A.G.; writing—review and editing, R.R., R.G.K.H., K.C. and C.A.G.; visualization, R.R.; funding acquisition, C.A.G. All authors have read and agreed to the published version of the manuscript.

Funding: This research was funded by The UK Foreign, Commonwealth and Development Office. CAG also acknowledges support from the Bill and Melinda Gates Foundation.

Institutional Review Board Statement: Not applicable.

Informed Consent Statement: Not applicable.

Data Availability Statement: All data are publicly available from the sources listed in Materials and Methods.

Acknowledgments: The authors gratefully acknowledge Alexandre V. Latchininsky (FAO, Plant Production and Protection Division (NSP), Rome) for critical reading of the manuscript and his comments.

Conflicts of Interest: The authors declare no conflict of interest. The funders had no role in the design of the study; in the collection, analyses, or interpretation of data; in the writing of the manuscript, or in the decision to publish the results.

References

1. Zhang, L.; Lecoq, M.; Latchininsky, A.; Hunter, D. Locust and Grasshopper Management. *Annu. Rev. Entomol.* **2019**, *64*, 15–34. [CrossRef]
2. Meynard, C.N.; Lecoq, M.; Chapuis, M.P.; Piou, C. On the relative role of climate change and management in the current desert locust outbreak in East Africa. *Glob. Chang. Biol.* **2020**, *26*, 3753–3755. [CrossRef]
3. Gay, P.E.; Trumper, E.; Lecoq, M.; Piou, C. Importance of human capital, field knowledge and experience to improve pest locust management. *Pest Manag. Sci.* **2021**, *77*, 5463–5474. [CrossRef]
4. Gay, P.E.; Lecoq, M.; Piou, C. The limitations of locust preventive management faced with spatial uncertainty: Exploration with a multiagent model. *Pest Manag. Sci.* **2020**, *76*, 1094–1102. [CrossRef]
5. Gay, P.E.; Lecoq, M.; Piou, C. Improving preventive locust management: Insights from a multi-agent model. *Pest Manag. Sci.* **2018**, *74*, 46–58. [CrossRef]
6. Symmons, P. Strategies to combat the desert locust. *Crop Prot.* **1992**, *11*, 206–212. [CrossRef]
7. Lecoq, M. Desert locust threat to agricultural development and food security and FAO/International role in its control. *Arab J. Plant Prot.* **2003**, *21*, 188–193.
8. Lecoq, M. Desert locust management: From ecology to anthropology. *J. Orthoptera Res.* **2005**, *14*, 179–186. [CrossRef]
9. Cressman, K. Chapter 4.2—Desert Locust. In *Biological and Environmental Hazards, Risks, and Disasters*; Shroder, J.F., Sivanpillai, R., Eds.; Academic Press: Boston, MA, USA, 2016; pp. 87–105.
10. FAO. *Standard Operating Procedures (SOP) for Desert Locust Ground Survey*; FAO: Rome, Italy, 2021. Available online: http://www.fao.org/ag/locusts/common/ecg/359/en/SOPSurveyENv2021.pdf (accessed on 1 August 2021).
11. FAO. *Standard Operating Procedures (SOP) for Desert Locust Ground Control*; FAO: Rome, Italy, 2021. Available online: http://www.fao.org/ag/locusts/common/ecg/359/en/SOPControlE.pdf (accessed on 1 August 2021).

12. FAO. *Standard Operating Procedures (SOP) for Desert Locust Aerial Survey and Control*; FAO: Rome, Italy, 2021. Available online: http://www.fao.org/ag/locusts/common/ecg/359/en/SOPAerialE.pdf (accessed on 1 August 2021).
13. Pedgley, D. *Desert Locust Forecasting Manual (Volume 1 of 2)*; Centre for Overseas Pest Research: London, UK, 1981.
14. Lecoq, M. Recent progress in Desert and Migratory Locust management in Africa. Are preventative actions possible? *J. Orthoptera Res.* **2001**, *10*, 277–291. [CrossRef]
15. Magor, J.I.; Lecoq, M.; Hunter, D.M. Preventive control and Desert Locust plagues. *Crop Prot.* **2008**, *27*, 1527–1533. [CrossRef]
16. Dobson, H.M. *Desert Locust Guidelines Control*; Food and Agriculture Organization of the United Nations: Rome, Italy, 2001.
17. Dobson, H.M. *Desert Locust Guidelines 4. Control*; FAO: Rome, Italy, 2001. Available online: https://www.fao.org/ag/locusts/common/ecg/347_en_DLG4e.pdf (accessed on 1 August 2021).
18. Eriksson, H.; Wiktelius, S. Impact of chlorpyrifos used for desert locust control on non-target organisms in the vicinity of mangrove, an ecologically sensitive area. *Int. J. Pest Manag.* **2010**, *57*, 23–34. [CrossRef]
19. Matthews, G.A. New Technology for Desert Locust Control. *Agronomy* **2021**, *11*, 1052. [CrossRef]
20. Githae, E.W.; Kuria, E.K. Biological control of desert locust (Schistocerca gregaria Forskål). *CAB Rev. Perspect. Agric. Vet. Sci. Nutr. Nat. Resour.* **2021**, *16*, 13. [CrossRef]
21. Salih, A.A.M.; Baraibar, M.; Mwangi, K.K.M.; Artan, G. Climate change and locust outbreak in East Africa. *Nat. Clim. Chang.* **2020**, *10*, 584–585. [CrossRef]
22. Wang, L.; Zhuo, W.; Pei, Z.; Tong, X.; Han, W.; Fang, S. Using Long-Term Earth Observation Data to Reveal the Factors Contributing to the Early 2020 Desert Locust Upsurge and the Resulting Vegetation Loss. *Remote Sens.* **2021**, *13*, 680. [CrossRef]
23. Showler, A.; Ebbe, M.O.; Lecoq, M.; Maeno, K.O. Early intervention against desert locusts: Current proactive approach and the prospect of sustainable outbreak prevention. *Agron. J.* **2021**, *11*, 312. [CrossRef]
24. FAO Locust Hub. Available online: https://locust-hub-hqfao.hub.arcgis.com/ (accessed on 1 August 2021).
25. Buchhorn, M.; Lesiv, M.; Tsendbazar, N.E.; Herold, M.; Bertels, L.; Smets, B. Copernicus Global Land Cover Layers—Collection 2. *Remote Sens.* **2020**, *12*, 1044. [CrossRef]
26. Copernicus Global Land Cover Layers: CGLS-LC100 Collection 3. Available online: https://developers.google.com/earth-engine/datasets/catalog/COPERNICUS_Landcover_100m_Proba-V-C3_Global (accessed on 1 August 2021).
27. Perez-Hoyos, A.; Udias, F.; Rembold, F.; Kerdiles, H.; Gallego, J. Integrating multiple land cover maps through a multi-criteria analysis to improve agricultural monitoring in Africa. *Int. J. Appl. Earth Obs. Geoinf.* **2020**, *88*, 1020642. [CrossRef]
28. Yu, Q.; You, L.; Wood-Sichra, U.; Ru, Y.; Joglekar, A.K.B.; Fritz, S.; Xiong, W.; Lu, M.; Wu, W.; Yang, P. A cultivated planet in 2010: 2. the global gridded agricultural production maps. *Earth Syst. Sci. Data Discuss.* **2020**, *12*, 3545–3572. [CrossRef]
29. International Food Policy Research Institute. *Spatially-Disaggregated Crop Production Statistics Data in Africa South of the Sahara for 2017*; Harvard Dataverse, V2; International Food Policy Research Institute: Washington, DC, USA, 2020. [CrossRef]
30. Sultana, R.; Kumar, S.; Samejo, A.A.; Soomro, S.; Lecoq, M. The 2019–2020 upsurge of the desert locust and its impact in Pakistan. *J. Orthoptera Res.* **2021**, *30*, 145–154. [CrossRef]
31. Ahmed, H.; Gardiner, B.G. Activation of Malathion by the Locust Bodywall. *Nature* **1968**, *17*, 776–777. [CrossRef]
32. Tchounwou, P.B.; Patlolla, A.K.; Yedjou, C.G.; Moore, P.D. Environmental Exposure and Health Effects Associated with Malathion Toxicity. In *Toxicity and Hazard of Agrochemicals*; InTech: Vienna, Austria, 2015
33. Samways, M.J. Can Locust Control be Compatible with Conserving Biodiversity? In *Grasshoppers and Grassland Health*; NATO Science Series (Series 2. Environment Security); Lockwood, J.A., Latchininsky, A.V., Sergeev, M.G., Eds.; Springer: Dordrecht, The Netherlands, 2000; Volume 73.
34. Peveling, R.; Rafanomezantsoa, J.J.; Razafinirina, R.; Tovonkery, R.; Zafimaniry, G. Environmental impact of the locust control agents fenitrothion, fenitrothion esfenvalerate and triflumuron on terrestrial arthropods in Madagascar. *Crop Prot.* **1999**, *18*, 659–676. [CrossRef]
35. Blanford, S.; Thomas, M.B. Adult Survival, Maturation, and Reproduction of the Desert Locust Schistocerca gregaria Infected with the Fungus Metarhizium anisopliae var acridum. *J. Invertebr. Pathol.* **2001**, *78*, 1–8. [CrossRef] [PubMed]
36. FAO. *Standard Operating Procedures (SOP) Desert Locust Biology and Behaviour*; FAO: Rome, Italy, 2021. Available online: http://www.fao.org/ag/locusts/common/ecg/359/en/SOPBiologyENv2021.pdf (accessed on 1 August 2021).
37. Tratalos, J.; Cheke, R.; Healey, R.; Stenseth, N. Desert locust populations, rainfall and climate change: Insights from phenomenological models using gridded monthly data. *Clim. Res.* **2010**, *43*, 229–239. [CrossRef]
38. Pekel, J.F.; Ceccato, P.; Vancutsem, C.; Cressman, K.; Vanbogaert, E.; Defourny, P. Development and Application of Multi-Temporal Colorimetric Transformation to Monitor Vegetation in the Desert Locust Habitat. *IEEE J. Sel. Top. Appl. Earth Obs. Remote Sens.* **2011**, *4*, 318–326. [CrossRef]
39. Renier, C.; Waldner, F.; Jacques, D.M.; Ebbe, M.A.B.; Cressman, K.; Defourny, P. A Dynamic Vegetation Senescence Indicator for Near-Real-Time Desert Locust Habitat Monitoring with MODIS. *Remote Sens.* **2015**, *7*, 7545–7570. [CrossRef]
40. Maeno, K.O.; Ould Ely, S.; Ould Mohamed, S.; Jaavar, M.E.H.; Ould Babah Ebbe, M.A. Adult Desert Locust Swarms, Schistocerca gregaria, Preferentially Roost in the Tallest Plants at Any Given Site in the Sahara Desert. *Agronomy* **2020**, *10*, 1923. [CrossRef]
41. Ellenburg, W.E.; Mishra, V.; Roberts, J.B.; Limaye, A.S.; Case, J.L.; Blankenship, C.B.; Cressman, K. Detecting Desert Locust Breeding Grounds: A Satellite-Assisted Modeling Approach. *Remote Sens.* **2021**, *13*, 1276. [CrossRef]
42. Gomez, D.; Salvador, P.; Sanz, J.; Rodrigo, J.F.; Gill, J.; Casanova, J.S. Prediction of desert locust breeding areas using machine learning methods and SMOS (MIR SMNRT2) Near Real Time product. *J. Arid. Environ.* **2021**, *194*, 104599. [CrossRef]

43. WMO/FAO. *Weather and Desert Locusts, World Meteorological Organization and Food and Agriculture Organization of the United Nations*; World Meteorological Organization (WMO): Geneva, Switzerland, 2016.
44. Huis, A.V.; Cressman, K.; Magor, J.I. Preventing desert locust plagues: Optimizing management interventions. *Entomol. Exp. Appl.* **2007**, *122*, 191–214. [CrossRef]
45. Baro, M.; Deubel, T.F. Persistent hunger: Perspectives on vulnerability, famine, and food security in sub-Saharan Africa. *Annu. Rev. Anthropol.* **2006**, *35*, 521–538. [CrossRef]
46. Chatterjee, S. How Hard Did That Sting? Estimating the Economic Costs of Locust Attacks on Agricultural Production. *Appl. Econ. Perspect. Policy* **2020**, 1–26. [CrossRef]
47. FAOSTAT. 2021. Available online: http://www.fao.org/faostat/en/#data (accessed on 1 August 2021).
48. Faye, B. How many large camelids in the world? A synthetic analysis of the world camel demographic changes. *Pastoralism* **2020**, *10*, 25. [CrossRef]
49. Milton, S.; Barnard, P. Tenure, livelihoods and sustainable development- rangelands as systems for multiple land use and livelihood support. *Afr. J. Range Forage Sci.* **2003**, *20*, 202–209.
50. Bradshaw, C.; Leroy, B.; Bellard, C. Massive yet grossly underestimated global costs of invasive insects. *Nat. Commun.* **2016**, *7*, 12986. [CrossRef]
51. Cunniffe, N.J.; Gilligan C.A. Use of mathematical models to predict epidemics and to optimise disease detection and management. *Emerg. Plant Dis. Glob. Food Secur.* **2020**, *12*, 239–266.
52. Cunniffe, N.J.; Koskella, B.; Metcalf, C.J.; Parnell, S.; Gottwald, T.R.; Gilligan, C.A. Thirteen challenges in modelling plant diseases. Epidemics. *Epidemics* **2015**, *10*, 6–10. [CrossRef] [PubMed]
53. Forster, G.A.; Gilligan, C.A. Optimizing the control of disease infestations at the landscape scale. *Proc. Natl. Acad. Sci. USA* **2007**, *104*, 4984–4989. [CrossRef]
54. Therville, C.; Anderies, J.M.; Lecoq, M.; Cease, A. Locusts and People: Integrating the Social Sciences in Sustainable Locust Management. *Agronomy* **2021**, *11*, 951. [CrossRef]
55. Rowthorn, R.E.; Laxminarayan, R.; Gilligan, C.A. Optimal control of epidemics in metapopulations. *J. R. Soc. Interface* **2009**, *6*, 1135–1144. [CrossRef] [PubMed]
56. Klein, E.; Laxminarayan, R.; Smith, D.L.; Gilligan, C.A. Economic incentives and mathematical models of disease. *Environ. Dev. Econ.* **2007**, *12*, 707–732. [CrossRef]

agronomy

MDPI

Article

Long-Term Decline in Harvester Termites in Madagascar following Multiple Barrier Treatments with Fipronil against Migratory Locust

Ralf Peveling

Department of Environmental Sciences—Geoecology, University of Basel, 4056 Basel, Switzerland; ralf.peveling@unibas.ch

Abstract: Mounds of the harvester termite *Coarctotermes clepsydra* (Sjöstedt) (Isoptera: Termitidae, Nasutitermitinae) are a typical feature of savanna woodlands in Madagascar. With densities of up to 300 termitaria ha^{-1}, this species provides key ecosystem services and is an important food source for wildlife. Following large-scale aerial blanket and barrier treatments with the insecticide fipronil to control an outbreak of migratory locust in the late 1990s, evidence emerged that *C. clepsydra* and related food webs were adversely affected. However, neither the scale nor the duration of the effects were known. The present ex post study investigated the recovery of *C. clepsydra* populations subjected to multiple barrier treatments against hopper bands in 1998 and 1999 at estimated cumulative dose rates of 1.7–3.4 g fipronil ha^{-1}. At the time of the survey in 2007, both the density of occupied termitaria (30.2 versus 106.8 mounds ha^{-1}) and mound occupancy (24.3% versus 70.0%) were significantly lower in repeatedly sprayed so-called hotspots than in unsprayed areas. The overall adverse effect (mortality in sprayed areas corrected for mortality in unsprayed areas) was E = 64.4%. The main outcome of this study is a strikingly low resilience of *C. clepsydra* populations, which did not recover in hotspots within eight years, with likely repercussions on food webs. This study shows that the environmental benefits of barrier treatments are forfeited if the same areas are treated repeatedly during the same campaign. Recommendations are given for the mitigation of these risks.

Keywords: insecticide disturbance; barrier treatment; spray history; non-target effects; *Coarctotermes clepsydra*; recovery; resilience

Citation: Peveling, R. Long-Term Decline in Harvester Termites in Madagascar following Multiple Barrier Treatments with Fipronil against Migratory Locust. *Agronomy* **2022**, *12*, 310. https://doi.org/10.3390/agronomy12020310

Academic Editor: Marco Scortichini

Received: 10 October 2021
Accepted: 20 January 2022
Published: 25 January 2022

1. Introduction

Madagascar is a global biodiversity hotspot and has a high level of endemism in most taxonomic groups. Termite diversity, in contrast, is low [1], with only 56 species known in 2003, with the majority of them being wood feeders [2]. The grasslands and savanna woodlands in the semi-arid southwestern part of the island are home to just one mound-building species: the grass-feeding harvester termite *Coarctotermes clepsydra* (Sjöstedt) (Isoptera: Termitidae, Nasutitermitinae), whose conspicuous conical-shaped termitaria can reach densities of 100–300 mounds ha^{-1} [2,3]. *Coarctotermes* is one of three grass-feeding genera in the Nasutitermitinae and the only one in Madagascar [4]. Given their high biomass and exposed lifestyle during nocturnal grass-harvesting bouts, harvester termites are an important food source for both invertebrates [5] and vertebrates [6]. In Madagascar, they are also commonly collected for chicken feed [7,8].

Harvester termites share their habitat with the Malagasy migratory locust, *Locusta migratoria capito* Saussure, and the red locust, *Nomadacris septemfasicata* (Serville) (Orthoptera: Acrididae), and are therefore at risk of exposure to locust insecticides during control operations. During the last two major locust invasions, which lasted from 1997–2000 and 2013–2016, aerial control operations extended over 42,000 and 23,000 km^2, respectively [9,10], relying mainly on synthetic chemical insecticides and, during the 1997–2000 campaign, on barrier treatments with fipronil.

Barrier treatments target marching hopper bands, which cannot fly. The technique is based on barriers of sprayed vegetation lying hundreds of meters apart. It is considered environmentally benign because large tracts of land remain untreated [11,12]. While traversing the area, feeding hopper bands eventually pass through treated vegetation and die. During the 1997–2000 campaign, fipronil, a relatively persistent phenylpyrazole locust insecticide and also a potent termiticide [13,14], was the sole control agent for barrier treatments used at an operational scale [15]. Overall dose rates (i.e., rates applied over the total target area) of single barrier treatments ranged from 0.75–1.50 g active ingredient (a.i.) ha^{-1} at a 1000–500 m barrier spacing, respectively. In heavily locust-infested areas, treatments were applied repeatedly within the same season, which accordingly led to higher cumulative dose rates.

Earlier studies have shown detrimental effects on non-target fauna of both single barrier (1.0 g a.i. ha^{-1}) and blanket (3.2–4.0 g a.i. ha^{-1}) treatments with fipronil [16]. Harvester termites were found to be particularly vulnerable, with their decline leading to a decline in termite and/or ant-feeding reptiles and small mammals [3]. Field studies in Senegal and Australia corroborated the adverse effects of blanket treatments on termites over a range of dose rates (Senegal: 2.0–5.0 and 10.0–12.0 g a.i. ha^{-1}; Australia: 1.25 g a.i. ha^{-1}), though these studies did not investigate food chain effects [17–19]. In contrast, a subsequent study in Australia focusing on wood-eating termites, which owing to their more cryptic lifestyle are less exposed to insecticides, found no hazards associated with barrier treatments using fipronil at dosages per unit area ranging from 0.25–1.25 g a.i. ha^{-1} [20].

While the fipronil-induced decline in harvester termites in Madagascar was unequivocal, eventually leading to cancellation of the registration in 2006 (Arrêté n°4196/06, 23 Mars 2006), its duration and spatial scale were unknown. Rapid assessments in southwestern Madagascar in 2005 provided circumstantial evidence of low densities of live colonies of *C. clepsydra* in former barrier treatment areas while abandoned and to varying degrees degraded mounds were abundant (author's observation). The present ex post study followed from this evidence. It tested the hypothesis that densities and occupancies of *C. clepsydra* termitaria and morphological features differ among unsprayed (no barrier treatments recorded) and sprayed areas subjected to multiple ($\geq$2) barrier treatments from November 1998 to May 1999. The overall goal was to draw from the Madagascar experience and to inform risk assessments of fipronil and control agents with similar ecotoxicity worldwide.

2. Materials and Methods

This study was conducted in southwestern Madagascar. Based on spray protocols from aircraft equipped with differential global positioning systems, all barrier treatment areas (targets) were mapped on a monthly basis from November 1998 to May 1999 by the European Union (EU) Food Security Programme in Madagascar [15]. The main reason was to inform the campaign organization and to demonstrate progress in the control of the invasion. Original shape files of the Geographical Information System were lost after the campaign, but hard copies of the monthly spray maps and copies of the original spray protocols for 33 (out of more than 200) barrier treatments were retrieved by the author at the Delegation of the EU in Madagascar in 2005. The present study focused on an area covering 32,180 km^2 where intensive control operations had taken place (Figure 1). The area includes parts of the Sakaraha province (West) and the Haut Plateau de Horombé (East) and the national parks of Zombitse Vohibasia (West) and Isalo (East). Spray areas were, for the most part, adjacent to each other and sprayed only once, but a considerable proportion overlapped, resulting in multiple treatments ($\geq$2) within the same campaign. The present study focused on these hotspots. The underlying assumption was that if full recovery of harvester termites was achieved in hotspot areas, the same would hold for the entire barrier treatment area.

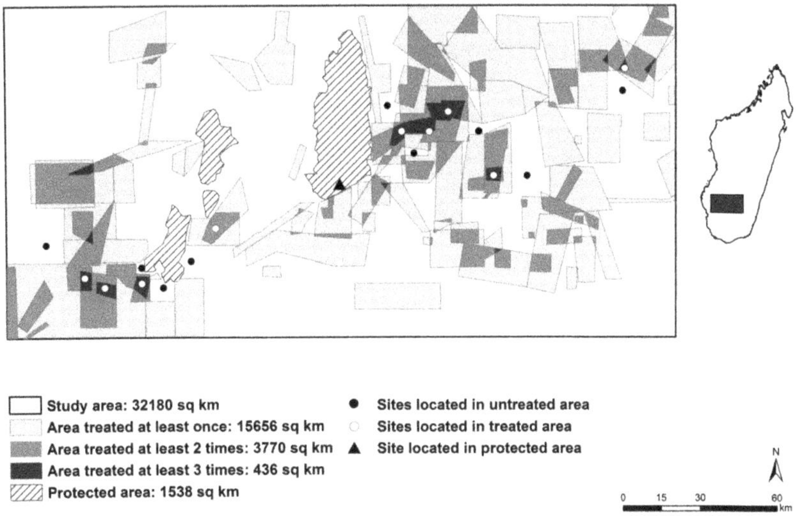

Figure 1. Study area and barrier treatment areas (targets) in southwestern Madagascar. Areas sprayed at least three times (dark grey) are included in areas sprayed at least twice (grey) and the latter are included in areas sprayed at least once (light grey). The design included 9 randomly selected pairs of sites—or blocks—located in untreated areas (black circles) and repeatedly treated hotspots (white circles), respectively. At each site, the densities, occupancies, and morphologies of termitaria were measured. At the single site located in Isalo national park (black triangle), only mound morphologies were measured.

According to the registered use of fipronil (Adonis® 7.5 UL) for barrier treatments in 1998, the maximum distance between barriers was 1000 m (Ministère d'Agriculture, Décision 166/98). No minimum distance was given at that time, but a decree issued by the Centre National Antiacridien in 2005 stipulated that the inter-barrier space should be 10 times as wide as the barrier (Note de service n°117-05/MAEP/CNA/Dir). According to available spray protocols (N = 33), the median barrier spacing was 875 m (range: 500–1000 m). At this distance, cumulative overall dose rates of fipronil correspond to ≈0.9, 1.7, 2.6, and 3.4 g a.i. ha^{-1}, respectively, at 1 single or 2, 3, and 4 overlapping barrier treatments. These values, however, are proxies because the actual combination of barrier spacing and the number of treatments was unknown for most targets.

A randomized paired-sample design was adopted, in which each site located in the center of a multiple barrier-treated hotspot was paired with a site located in an unsprayed area at a distance of 15–20 km (Figure 1). Sites were selected as follows: maps of barrier spray areas were scanned and imported into ArcGIS 9.4 (Esri Company, Redlands, CA, USA). They were georeferenced using the coordinates of targets for which spray protocols were available. Maps depicting monthly spraying operations were overlaid to identify overlapping areas of targets (Figure 1). Sampling sites were arbitrarily placed in the approximate center of hotspots. Paired unsprayed sites were arbitrarily placed in adjacent unsprayed areas at a minimum distance of 2–3 km from the boundary of the nearest sprayed area. For the purpose of the study presented in this paper, 9 pairs of sites—or blocks—were selected. The farthest distance between blocks was more than 200 km (Figure 1). Before starting the field work, the selected sites were verified to be located in typical locust and harvester termite savanna habitats by overlaying the barrier treatment map on the digital land use map of Madagascar (BD 500, Zone UTM 38 k, Institut Géographique et Hydrographique National de Madagascar, Antananarivo, Madagascar). The fire-resistant spear grass (*Heteropogon contortus*) is by far the most dominant grass species in these fire-prone savannas [21].

Each site was composed of 5 plots measuring 1000 m^2 each. The plots marked the center (original coordinates) and 4 corners of 1.4 × 1.4 km quadrats, with the corners located at a distance of 1 km at angles of 30°, 120°, 210°, and 300°, respectively, from the center point. Of the designated five plots per site, four were intended to be sampled depending on accessibility. Another site comprising six plots was investigated in Isalo national park. This was not planned before. Termitaria located there offered the opportunity to measure mound morphologies in presumably undisturbed conditions (no insecticide use, no harvesting of termites by humans).

The field work was conducted in March 2007. The designated plots were approached by vehicle as close as possible and further on foot, using the go-to function of the Global Positioning System. The median deviation from target waypoints was 7 m. A 17.84 m rope with wooden pegs fixed to both ends was used to delineate plots. One peg was driven into the ground to mark the center. The rope was pulled to its full length and the opposite peg circulated to outline the perimeter of a circle measuring 35.68 m in diameter and 1000 m^2 in surface area. All termitaria or remnants thereof lying within the circle were investigated. Even fully degraded or overgrown termitaria could be identified because grasses grew taller on the nutrient-enriched soil and subterranean mound chambers were still detectable. Mounds on the boundary were included if the center lay within the circle. The following parameters were recorded: minimum and maximum basal diameter, maximum height and number of domes, mound occupancy, and traces of human use such as regularly shaped marks from spades used to dig up termites. Neighboring termitaria with basal diameters >1 m apart were treated as different colonies. To verify occupancy, mounds were probed using picks. The search was extended to subterranean parts if no specimens emerged. Specimens of workers and soldiers were stored in 80% alcohol.

Data from the four plots per site (only three plots in one site) were pooled and extrapolated to one hectare. Mound density data were log-transformed and percent occupancy data (equivalent to survival rates) arcsine-transformed prior to two-factor analysis of variance (ANOVA) without replication, with treatment as the fixed and block as the random factor. This analysis is similar to a paired sample t-test.

The adverse effect E (mortality in repeatedly sprayed areas corrected for mortality in unsprayed areas) was calculated as a descriptive statistic according to Schneider-Orelli [22], where the percentage mortality per site was calculated as 100% minus the survival rate in this site. Effect calculations normally require pre-spray data from both sprayed and unsprayed sites. These were obviously not available as this study was conducted at 1 point in time 8 years post-spray. However, because mound structures persist for many years after the death of a colony, pre-spray densities (i.e., the status ante) of harvester termite colonies could be approximated.

The mound characters of occupied termitaria were analyzed using one-way ANOVA. Basal diameter was calculated as the mean of the minimum and maximum values. Data were log+1-transformed to achieve homogeneity of variances as needed. Analyses included pooled data from repeatedly sprayed, unsprayed, and control (national park) areas. Differences among means were tested using the Student–Newman–Keuls (SNK) test. A two-way contingency table was used to examine the association between treatment (repeatedly sprayed versus unsprayed) and termite-collecting efforts (traces of human use versus no traces). All tests were conducted using PASW Statistics 18 (SPSS Inc. Hong Kong, China).

3. Results

3.1. Density and Occupancy of Termitaria

The mean density of combined occupied and abandoned termitaria in repeatedly sprayed areas (93.5 ± 17.1 mounds ha^{-1}, mean ± SE) was significantly less than that recorded in unsprayed areas (156.6 ± 26.4 mounds ha^{-1}; $F_{1,8}$ = 9.1, p = 0.017). Likewise, the mean density of occupied termitaria was significantly less than in unsprayed areas (30.2 ± 15.1 ha^{-1} versus 106.8 ± 18.6 mounds ha^{-1}; $F_{1,8}$ = 17.9, p = 0.003) (Figure 2a). With ≤3 inhabited termitaria ha^{-1}, 4 out of 9 barrier treatment sites were virtually void of live harvester termite colonies though remnants of termitaria were frequently found. Mound occupancy was 24.3 ± 8.8% (mean ± SE) in repeatedly sprayed areas and 70.0 ± 4.7% in unsprayed areas ($F_{1,8}$ = 15.2, p = 0.005) (Figure 2b). The resulting adverse effect was E = 64.4 ± 12.7% (mean ± SE), meaning that at nearly 8 years post-spray, populations were suppressed by this percentage.

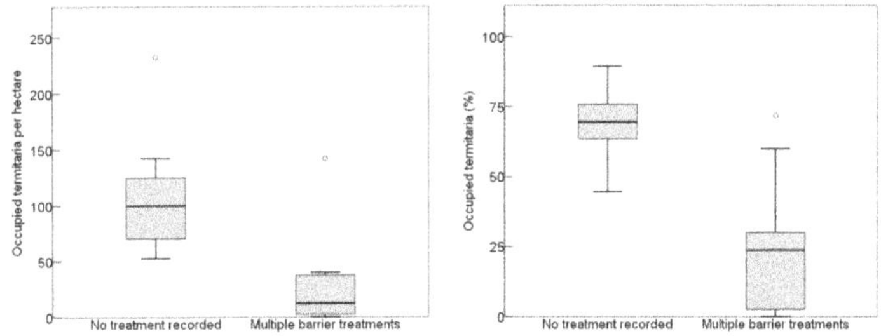

(a) (b)

Figure 2. Boxplots of the density (**a**) and percentage (**b**) of occupied termitaria in unsprayed (no treatment recorded) and repeatedly sprayed (multiple barrier treatments) areas (N = 9 in both treatments). Box plots show the median, lower/upper quartile, and smallest/largest observation and outliers (○). Differences among sprayed and unsprayed areas are statistically significant (see text).

3.2. *Mound Morphology*

The height of the occupied mounds was similar in sprayed (24.3 ± 1.8 cm; mean ± SE) and unsprayed (22.1 ± 0.9 cm) areas but differed significantly from the height recorded in undisturbed conditions in the Isalo national park (49.3 ± 4.5 cm; $F_{2,524}$ = 14.9, p < 0.001, SNK test significant at p < 0.05) (Figure 3a). Mound basal diameters were also similar among repeatedly sprayed (68.6 ± 3.9 cm) and unsprayed (70.7 ± 2.0 cm) areas but smaller than in undisturbed conditions (117.6 ± 8.1 cm; $F_{2,524}$ = 20.5, p < 0.001, SNK test significant at p < 0.05) (Figure 3b). The maximum number of domes per colony was 2. The mean number was 1.06 ± 0.04 in the national park and 1.07 in both unsprayed (SE = 0.01) and repeatedly sprayed (SE = 0.02) areas. These differences were not significant ($F_{2,455}$ = 0.02, p = 0.980).

3.3. *Human Use of Harvester Termites*

Of all the occupied mounds in unsprayed areas (N = 377), 33.4% showed traces of anthropogenic activity. In most cases, the apex (where termites concentrate in the morning) had been cut off, resulting in a flat- or bowl-shaped top after repair. In repeatedly sprayed areas (N = 119), the percentage was only 17.6%, which was significantly less than expected (29.6%; N = 496, $d.f.$ = 1, χ^2 = 10.8, p = 0.001).

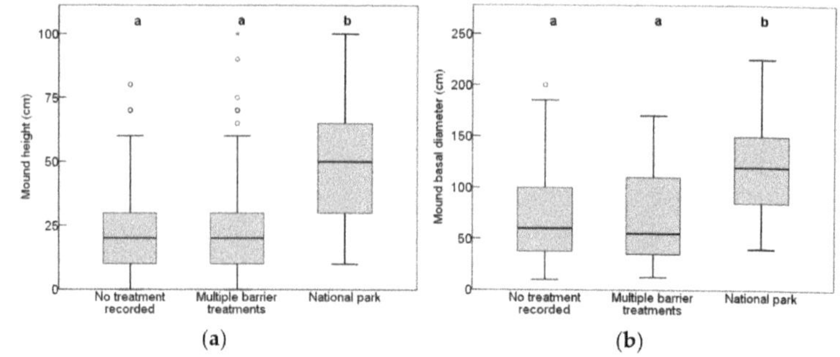

Figure 3. Boxplots of the height (**a**) and mean diameter (**b**) of occupied termitaria in unsprayed ($N = 377$; no treatment recorded) and repeatedly sprayed ($N = 119$; multiple barrier treatments) areas and in a national park ($N = 31$). Box plots show the median, lower/upper quartile, smallest/largest observation, outliers (○) and extreme cases (*). Means with different small letters are significantly different.

4. Discussion

The main outcome of the present research is a strikingly low resilience of *C. clepsydra* populations in hotspot areas. When exposed to 2–4 barrier treatments of fipronil within a single locust campaign, populations did not recover to pre-spray levels after 8 years. Large areas of intensive control were virtually cleared of *C. clepsydra*. This low resilience is likely due to the low dispersal capacity of termites. Dispersal distances generally do not exceed a few hundred meters [23] because alates are poor flyers [24]. Recolonizing vast deserted areas through repeated cycles of colony founding, growth, and dispersal would take a long time, depending on the density and spatial distribution of source colonies, and possibly also on the availability of abandoned but intact mounds for easier recolonization.

Another important outcome is that the environmental benefits of barrier over blanket treatments [11,12] are jeopardized if the same areas are sprayed repeatedly. The same applies if the barrier spacing is too narrow to provide truly unsprayed refugia from where recolonization can start. In an earlier study, even a 750 m barrier spacing resulted in a nearly 50% suppression in harvester termites after 2 years [3]. Considering that >30% of 33 barrier treatments for which spray protocols were available used a barrier spacing of only 500 m, adverse effects likely extended well beyond the 436 km^2 multiple barrier treatment hotspots studied in the present study. Extrapolating to the study area (Figure 1), 30% would correspond to about 3500 km^2. However, further research on *C. clepsydra* populations in single barrier treatment areas would have been required to test and quantify this assumption.

In any case, long-lasting reductions of harvester termites are expected to result in cascading effects across the trophic levels of affected ecosystems. This is because termites provide an array of ecological services, including decomposition of organic matter, nutrient cycling, enhancement of soil drainage, or bioturbation [25–28]. Moreover, termites influence the composition, structure, and spatial variation of vegetation [27,29,30] and the dynamics of ant communities [31], and mound-building termites play a keystone role by creating habitats for inquiline invertebrates [32]. The extent to which any of these processes was upset by the decline in *C. clepsydra* is not known. However, given its unique position as the only mound-building and grass-harvesting species of the savanna woodlands of Madagascar, perturbations were likely to be ecologically significant. At higher trophic levels, indirect effects of fipronil on wildlife are mediated mainly through loss in the quantity or quality of prey [33]. The above-mentioned study in Madagascar showed that the decline in harvester termites resulted in a reduced abundance of termite-feeding lizards and tenrecs [3]. A long-lasting decline in harvester termites over large areas would, therefore, affect the entire food web to which they belong.

In sub-Saharan Africa, termites and termite mounds are widely utilized by humans [8,34]. Examples include the consumption of protein-rich alates [35] or the use of mound material as fertilizer [36] or plaster for traditional houses [37]. In Madagascar and many other African countries, termites are frequently collected for chicken feed [7,8]. *C. clepsydra* colonies appear to recover well from recurrent damage to their mounds and loss in biomass, though this study showed that termitaria in both sprayed and unsprayed areas were smaller than in a protected area where the harvesting of termites is not allowed. An interesting finding was that inhabited termitaria in repeatedly sprayed areas were less frequently harvested than those in unsprayed areas. This suggests that it is less profitable to exploit a more widely dispersed resource because it requires more time and longer distances to find live colonies. It also shows that adverse effects on harvester termites directly translate into a negative impact on rural livelihoods.

An important methodological feature of this study was that inferences were made on pre-spray termitaria densities by combining occupied and abandoned termitaria. This was possible because remnants (e.g., subterranean mound chambers) or indications (e.g., circular grass patches different from surrounding grassland) of deserted mounds are detectable for many years. Moreover, the preponderance of deserted mounds in sprayed areas indicated that these had indeed been harvester termite habitats in the past.

However, the observed lower density of combined occupied and abandoned mound densities in repeatedly sprayed areas suggests that some fully weathered termitaria may have been undetected at the time of the field work. In this case, pre-spray densities and hence the adverse effect E would have been underestimated. There are only few studies on the persistence of abandoned mounds. A 10-year study on the growth of the Australian harvester termite *Drepanotermes perniger* (Froggatt) (Termitidae, Termitinae) showed that deserted mounds were not any more eroded than inhabited ones [38]. Lobry de Bruyn [39] hypothesized that mounds of *Drepanotermes tamminensis* (Hill) take at least 30 years to erode to ground level. However, the outer wall of *Drepanotermes* mounds is thicker and harder than that of *Coarctotermes* mounds (author's observation). Furthermore, mound erosion in southwestern Madagascar would be expected to be faster than in the Australian outback owing to much higher rainfall (≈750 mm).

Conversely, the absence of epigeous mounds was considered an indication of the absence of harvester termites. This was reasonable because successful recolonization would have resulted in the development of visible mounds. Colonies of the ecologically similar harvester termite *Trinervitermes trinervoides* (Sjöstedt) (Nasutitermitinae) in South Africa begin building mounds 2–3 years after colony founding [40]. In view of this, significant visible recolonization would have been expected in sprayed areas after eight years.

Another question is whether wildfires may have had an effect on the outcome of this study. It is estimated that 450,000 ha of savanna are burnt in Madagascar each year [41]. However, there is no evidence that wildfires destroy colonies of harvester termites. On the contrary, grasslands dominated by fire-resistant spear grass are typical habitats of harvester termites. This is in line with studies in South Africa [42], Brazil [43], and Australia [44]. If wildfires supposedly have no effect on harvester termites, the reverse is not the case. With the densities of colonies and concomitant grass removal rates reduced, a larger proportion of the standing crop would be devoured by wildfires instead of being decomposed and recycled by termites.

5. Conclusions

Systemic insecticides, such as fipronil, have been linked to a variety of direct and indirect effects on vertebrate and invertebrate non-target organisms and ecosystems [33,45]. Insecticides that have adverse effects on populations of non-target invertebrates over prolonged periods are classified as high-risk products [11,46]. The results of the present and previous studies [3,16–18] leave no doubt that fipronil belongs to this category and that even barrier treatments can have long-lasting adverse effects on termites and related food webs. Therefore, rigorous and effective risk mitigation measures must be applied.

Apart from using less environmentally harmful insecticides for barrier treatment in the first place (e.g., benzoylurea insect growth regulators), risk mitigation measures should at least include the following: (i) mapping of historical and current locations of locust populations as a basis for improved monitoring, control, and impact assessment; (ii) monitoring of the magnitude and duration of adverse effects; (iii) restricting the use of fipronil or insecticides with similar ecotoxicity until full recovery of the affected non-target fauna has been evidenced; (iv) ensuring that the barrier spacing is wide enough to provide truly unsprayed refugia; and (v) imperatively avoiding multiple treatments.

Funding: This research received no external funding.

Institutional Review Board Statement: Not applicable.

Informed Consent Statement: Not applicable.

Data Availability Statement: The data presented in this study are available on request from the corresponding author.

Acknowledgments: I am grateful to Christian Rejela and Christale Robelle Razafindrahova for supporting the field work and to Clarah Andriamalala and Palvanijaz Pirnijasov for conducting GIS analyses. The Centre National Antiacridien (CNA) of Madagascar is acknowledged for providing logistical support and the EU Food Security Programme together with the Projet de Lutte Antiacridienne dans l'Aire Grégarigène for mapping barrier treatments during the 1998/99 campaign. I am also grateful to Harold van der Valk for his encouragement to undertake this study against all odds and without external funding. The late Colin C.D. Tingle and his team members in Madagascar and the UK are acknowledged for spearheading research on the environmental impact of fipronil and other locust control agents in Madagascar.

Conflicts of Interest: The author declares no conflict of interest.

References

1. Paulian, R. The termites of Madagascar. In *Biology of Termites*; Krishna, K., Weesner, F.M., Eds.; Academic Press: New York, NY, USA, 1970; Volume 2, pp. 281–294.
2. Eggleton, P.; Davies, R. Isoptera, termites. In *The Natural History of Madagascar*; Goodman, S.M., Benstead, J.P., Eds.; The University of Chicago Press: Chicago, IL, USA, 2003; pp. 654–660.
3. Peveling, R.; McWilliam, A.N.; Nagel, P.; Rasolomanana, H.; Rakotomianina, L.; Ravoninjatovo, A.; Dewhurst, C.F.; Gibson, G.; Rafanomezana, S.; Tingle, C.C.D. Impact of locust control on harvester termites and endemic vertebrate predators in Madagascar. *J. Appl. Ecol.* **2003**, *40*, 729–741. [CrossRef]
4. Jones, D.T.; Eggleton, P. Global biogeography of termites: A compilation of sources. In *Biology of Termites: A Modern Synthesis*; Bignell, D.E., Roisin, Y., Lo, N., Eds.; Springer: Dordrecht, The Netherlands, 2011; pp. 477–498.
5. Traniello, J.F.A.; Leuthold, R.H. Behavior and ecology of foraging termites. In *Termites: Evolution, Sociality, Symbioses, Ecology*; Abe, T., Bignell, D.E., Higashi, M., Eds.; Kluwer: Dordrecht, The Netherlands, 2000; pp. 141–168.
6. Kok, O.B.; Hewitt, P.H. Bird and mammal predators of the harvester termite *Hodotermes mossambicus* (Hagen) in semi-arid regions of South Africa. *S. Afr. J. Sci.* **1990**, *86*, 34–37.
7. Zehrer, W. *Lutte Antiacridienne à Madagascar—Tome III—Ecotoxicologie*; DPV/GTZ, Ministère de l'Agriculture: Antananarivo, Madagascar, 2001.
8. Van Huis, A. Cultural significance of termites in sub-Saharan Africa. *J. Ethnobiol. Ethnomed.* **2017**, *13*, 8. [CrossRef] [PubMed]
9. Magor, J.I.; Lecoq, M.; Hunter, D.M. Preventive control and desert locust plagues. *Crop Prot.* **2008**, *27*, 1527–1533. [CrossRef]
10. Zhang, L.; Lecoq, M.; Latchininsky, A.; Hunter, D. Locust and grasshopper management. *Annu. Rev. Entomol.* **2018**, *64*, 15–34. [CrossRef]
11. Food and Agriculture Organization of the United Nations (FAO). *Evaluation of Field Trials Data on the Efficacy and Selectivity of Insecticides on Locusts and Grasshoppers*; Report to FAO by the Locust Pesticide Referee Group, 11th Meeting; FAO: Rome, Italy, 2021.
12. Van der Valk, H. *Environmental Impact of Barrier Treatments against Locusts—A Review of Field Studies*; Plant Production and Protection Division, Desert Locust Technical Series No. AGP/DL/TS/33; Food and Agriculture Organization of the United Nations: Rome, Italy, 2006.
13. Mao, L.; Henderson, G.; Scherer, C.W. Toxicity of seven termiticides on the Formosan and Eastern subterranean termites. *J. Econ. Entomol.* **2011**, *104*, 1002–1008. [CrossRef]
14. Rouland-Lefèvre, C. Termites as pests of agriculture. In *Biology of Termites: A Modern Synthesis*; Bignell, D.E., Roisin, Y., Lo, N., Eds.; Springer: Dordrecht, The Netherlands, 2011; pp. 499–517.

15. Lecoq, M. Recent progress in desert and migratory locust management in Africa. Are preventive actions possible? *J. Orthoptera Res.* **2001**, *10*, 277–291. [CrossRef]
16. Tingle, C.C.D.; Rother, J.A.; Dewhurst, C.F.; Lauer, S.; King, W.J. Fipronil: Environmental fate, ecotoxicology, and human health concerns. *Rev. Environ. Contam. Toxicol.* **2003**, *176*, 1–66. [CrossRef]
17. Danfa, A.; Ba, A.L.; Van der Valk, H.; Rouland-Lefèvre, C.; Mullié, W.C.; Everts, J.W. Long-term effects of chlorpyrifos and fipronil on epigeal beetles and soil arthropods in the semi-arid savanna of northern Senegal. In *Environmental Side-effects of Locust and Grasshopper Control*; Everts, J.W., Mbaye, D., Barry, O., Mullié, W., Eds.; Food and Agriculture Organization of the United Nations, CERES-Locustox Foundation, Ministry of Agriculture–Senegal: Dakar, Senegal, 2002; pp. 184–213.
18. Mullié, W.C.; Rouland-Lefèvre, C.; Sarr, M.; Danfa, A.; Ba, A.L.; Everts, J.W. *Impact of Adonis® 2 UL and 5 UL (fipronil) on Ant and Termite Communities in a Tropical Semi-Arid Savannah—A Two-Year Field Study in Northern SENEGAL*; Centre for Ecotoxicological Research in the Sahel, CERES-Locustox Foundation: Dakar, Senegal, 2003.
19. Steinbauer, M.J.; Peveling, R. The impact of the locust control insecticide fipronil on termites and ants in two contrasting habitats in northern Australia. *Crop Prot.* **2011**, *30*, 814–825. [CrossRef]
20. Maute, K.; French, K.; Story, P.; Bull, C.M.; Hose, G.C. Effects of two locust control methods on wood-eating termites in arid Australia. *J. Insect Conserv.* **2016**, *20*, 107–118. [CrossRef]
21. Rakotoarimanana, V.; Grouzis, M.; Le Floc'h, E. Influence du feu et du pâturage sur l'évolution de la phytomasse d'une savane à *Heteropogon contortus* de la région de Sakaraha (sud-ouest de Madagascar). *Tropicultura* **2008**, *26*, 56–60.
22. Schneider-Orelli, O. *Entomologisches Praktikum: Einführung in Die Land- und Forstwirtschaftliche Insektenkunde*; H.R. Sauerländer & Company: Aarau, Switzerland, 1947.
23. Bordereau, C.; Pasteels, J.M. Pheromones and chemical ecology of dispersal and foraging in termites. In *Biology of Termites: A Modern Synthesis*; Bignell, D.E., Roisin, Y., Lo, N., Eds.; Springer: Dordrecht, The Netherlands, 2011; pp. 279–320.
24. Eggleton, P. An introduction to termites: Biology, taxonomy and functional morphology. In *Biology of Termites: A Modern Synthesis*; Bignell, D.E., Roisin, Y., Lo, N., Eds.; Springer: Dordrecht, The Netherlands, 2011; pp. 1–26.
25. Lavelle, P.; Bignell, D.; Lepage, M.; Wolters, V.; Roger, P.A.; Ineson, P.; Heal, O.W.; Dhillion, S.P. Soil function in a changing world: The role of invertebrate ecosystem engineers. *Eur. J. Soil. Biol.* **1997**, *33*, 159–193.
26. Lobry de Bruyn, L.A.; Conacher, A.J. The role of termites and ants in soil modification: A Review. *Aust. J. Soil Res.* **1990**, *28*, 55–93.
27. Gabet, E.J.; Reichman, O.J.; Seabloom, E.W. The effects of bioturbation on soil processes and sediment transport. *Annu. Rev. Earth Planet. Sci.* **2003**, *31*, 249–273. [CrossRef]
28. Freymann, B.P.; De Visser, S.N.; Olff, H. Spatial and temporal hotspots of termite-driven decomposition in the Serengeti. *Ecography* **2010**, *33*, 443–450. [CrossRef]
29. Moe, S.R.; Mobæk, R.; Narmo, A.K. Mound building termites contribute to savanna vegetation heterogeneity. *Plant Ecol.* **2009**, *202*, 31–40. [CrossRef]
30. Okullo, P.; Moe, S.R. Termite activity, not grazing, is the main determinant of spatial variation in savanna herbaceous vegetation. *J. Ecol.* **2012**, *100*, 232–241. [CrossRef]
31. Palmer, T.M. Spatial habitat heterogeneity influences competition and coexistence in an African acacia ant guild. *Ecology* **2003**, *84*, 2843–2855. [CrossRef]
32. De Visser, S.N.; Freymann, B.P.; Schnyder, H. Trophic interactions among invertebrates in termitaria in the African savanna: A stable isotope approach. *Ecol. Entomol.* **2008**, *33*, 758–764. [CrossRef]
33. Gibbons, D.; Morrissey, C.; Mineau, P. A review of the direct and indirect effects of neonicotinoids and fipronil on vertebrate wildlife. *Environ. Sci. Pollut. Res.* **2015**, *22*, 103–118. [CrossRef] [PubMed]
34. Sileshi, G.W.; Nyeko, P.; Nkunika, P.O.Y.; Sekematte, B.M.; Akinnifesi, F.K.; Ajayi, O.C. Integrating ethno-ecological and scientific knowledge of termites for sustainable termite management and human welfare in Africa. *Ecol. Soc.* **2009**, *14*, 48. Available online: http://www.ecologyandsociety.org/vol14/iss1/art48 (accessed on 30 September 2021). [CrossRef]
35. Banjo, A.D.; Lawal, O.A.; Songonuga, E.A. The nutritional value of fourteen species of edible insects in southwestern Nigeria. *Afr. J. Biotechnol.* **2006**, *5*, 298–301.
36. Miyagawa, S.; Koyama, Y.; Kokubo, M.; Matsushita, Y.; Adachi, Y.; Sivilay, S.; Kawakubo, N.; Oba, S. Indigenous utilization of termite mounds and their sustainability in a rice growing village of the central plain of Laos. *J. Ethnobiol. Ethnomed.* **2011**, *7*, 24. [CrossRef] [PubMed]
37. Yamashina, C. Interactions between termite mounds, trees, and the Zemba people in the Mopane savanna in Northwestern Namibia. *Afr. Stud. Monogr.* **2010**, *40*, 115–128.
38. Watson, J.A.L.; Barret, R.A.; Green, J.P. Growth of mounds of Australian harvester termite, *Drepanotermes perniger* (Froggatt) (Termitinae). *Sociobiology* **1988**, *14*, 217–244.
39. Park, H.C.; Majer, J.D.; Hobbs, R.J. Contribution of the Western Australian wheatbelt termite, *Drepanotermes tamminensis* (Hill), to the soil nutrient budget. *Ecol. Res.* **1994**, *9*, 351–356. [CrossRef]
40. Adam, R.A.; Mitchell, J.D. Energetics and development of incipient colonies of the harvester termite, *Trinervitermes trinervoides* (Sjöstedt) (Termitidae, Nasutitermitinae). *Insectes Soc.* **2009**, *56*, 21–27. [CrossRef]
41. Rasambainarivo, J.H.; Ranaivoarivelo, N. *Madagascar Country Pasture/Forage Resource Profiles*; Food and Agricultural Organization of the United Nations: Rome, Italy, 2006.

42. Davies, A.B.; Eggleton, P.; van Rensburg, B.J.; Parr, C.L. The pyrodiversity–biodiversity hypothesis: A test with savanna termite assemblages. *J. Appl. Ecol.* **2012**, *49*, 422–430. [CrossRef]
43. DeSouza, O.; Albuquerque, L.B.; Tonello, V.M.; Pinto, L.P.; Reis Júnior, R. Effects of fire on termite generic richness in a savanna-like ecosystem ('Cerrado') of Central Brazil. *Sociobiology* **2003**, *42*, 639–649.
44. Abensperg-Traun, M.; Milewski, A.V. Abundance and diversity of termites (Isoptera) in unburnt versus burnt vegetation at the Barrens in Mediterranean Western Australia. *Austr. J. Ecol.* **1995**, *20*, 413–417. [CrossRef]
45. Pisa, L.; Goulson, D.; Yang, E.; Gibbons, D.; Sánchez-Bayo, F.; Mitchell, E.; Aebi, A.; van der Sluijs, J.; MacQuarrie, C.J.K.; Giorio, C.; et al. An update of the Worldwide Integrated Assessment (WIA) on systemic insecticides. Part 2: Impacts on organisms and ecosystems. *Environ. Sci. Pollut. Res.* **2021**, *28*, 11749–11797. [CrossRef] [PubMed]
46. European and Mediterranean Plant Protection Organization. Environmental risk assessment scheme for plant protection products. Chapter 9: Non-target terrestrial arthropods. *EPPO Bull.* **2003**, *33*, 131–139. [CrossRef]

agronomy MDPI

Article

Linseed Oil Affects Aggregation Behaviour in the Desert Locust *Schistocerca gregaria*—A Potential Swarm Disruptive Agent

Zainab Ali Saad Abdelatti [1,2] and Manfred Hartbauer [1,*]

1 Institute of Biology, University of Graz, Universitätsplatz 2, 8010 Graz, Austria; zienabentomologist@gmail.com

2 Department of Zoology, Faculty of Science, South Valley University, Qena 83523, Egypt

* Correspondence: manfred.hartbauer@uni-graz.at; Tel.: +43-316-380-5615

Received: 17 August 2020; Accepted: 21 September 2020; Published: 24 September 2020

Abstract: Gregarious desert locusts constitute very destructive agricultural pests. They aggregate and form collectively moving swarms that devastate vegetation and reduce crop production. To combat gregarious locusts, a bio-pesticide formulation that contains linseed oil as the main component was described recently. Since linseed oil is rich in fatty acids, some of which function as necromones that indicate injury or death in various insect species, we investigated the influence of linseed oil on the aggregation behaviour of sexually mature gregarious desert locusts. For this reason, we performed a series of aggregation experiments with six individuals of the same sex and brushed the wings of one individual (target individual) with linseed oil. The time the oil brushed target males spent close to any other individual was reduced in 76% of trials (average reduction of 18%), whereas the time target females spent in groups with members of the same sex did not alter. These results suggest that linseed oil may act as a bioactive agent that has the potential to disrupt swarm formation.

Keywords: locust swarm; bio-pesticide; gregarious locusts; aggregation behaviour; linseed oil; necromones

1. Introduction

Desert locusts (*Schistocerca gregaria* Forsskål, 1775) are considered to be among the most serious agricultural pests because of their polyphagous feeding behaviour, rapid reproduction rates and quick migration patterns [1,2]. Like other locust species, this species displays density-dependent phase polyphenism, which means that they can transform reversibly between two phases in response to population density: The solitary and gregarious phases. Individuals in either phase differ in terms of their morphology, physiology and behaviour [1,3]. Behavioural changes can occur quickly, appearing within just a few hours [4]. If the population density is low, locusts exist in a solitary phase and avoid each other, except when they are ready to mate. If the population density increases, even over a short time period, the behavioural transition to the gregarious phase is evoked [5,6]. This is the case, for example, after rainfalls that end long periods of drought, resulting in nymphs hatching from eggs laid in burrows in the ground [7]. As the population density increases, the locusts become social, aggregate and form marching bands on the ground (wingless hoppers); later, they form collectively moving swarms that can migrate over long distances [1]. Locust swarms can devastate entire fields and cause extensive crop damage over very short time periods. A small swarm of locusts contains thousands of individuals that spread out over several hundred square metres, but large swarms contain up to 80 million individuals per square kilometre. Since such swarms can cover a distance of 100 km per day [1], farmers regard gregarious locusts as one of the most destructive plagues on earth.

Locusts can sense swarm members using olfactory, visual and tactile cues [5,6,8]. Physical contact was found to be the most potent stimulus, causing solitary locusts to gregarise. Simpson et al. [8] discovered that touching the hind legs of others in a repetitive way induces phase transition. A patchy distribution of food plants also increases the probability of physical contact between locusts and boosts aggregation [9]. On the contrary, visual or olfactory stimulants have lesser or incompetent effects on phase transition in the desert locust [8,9], whereas the combination of them can lead to a behavioural gregarisation [10]. In addition, seeing other locusts for a longer period (24 h) can also mediate phase change, at least partially [10]. Agents that disrupt the formation of swarms have not yet been established on the market but may represent alternative measures that can be used against locust outbreaks.

The authors of this study recently developed a highly effective botanical pesticide formulation against two species of locusts that is mainly composed of linseed oil as a carrier oil and three essential oils [11]. A single spray treatment of locusts with this formulation killed all adults and nymphs of desert locusts (*Schistocerca gregaria*) and migratory locusts (*Locusta migratoria*) within 30 h [11,12]. In addition to this toxic effect, this botanical pesticide formulation may also exert a change in group formation behaviour after individuals have come in contact with linseed oil that contains 75–90% unsaturated fatty acids (50–55% linolenic acid, 15–20% oleic acid and 11–20% linoleic acid [13]). Yao et al. [14] described oleic and linoleic acid as necromones, substances that are associated with the injury and death of insects. These often evoke distinct behaviour patterns in other individuals of the same or different species. For example, eusocial and some semi-social species show necrophoric behaviour, including the removal of dead bodies from the nests (observed in bees, ants, spiders and aphids), burial (covering the dead with soil and/or other materials in ants and termites) and cannibalism (intraspecific necrophagy in ants and termites) [14–16]. In contrast, solitary and some sub-social species avoid dead or injured individuals (necrophobic behaviour) [14,16]. In this study, we performed aggregation experiments to test the following hypothesis: Linseed oil evokes a change in aggregation behaviour in gregarious desert locusts once one individual ("the target individual") has come into contact with this oil. To test this hypothesis, we evaluated the time the target individual spent in close proximity to others in a group of six individuals before and after brushing its wings with linseed oil. Moreover, we tested the behavioural responses of desert locusts towards a stationary linseed oil target, dead and crushed locust bodies.

2. Materials and Methods

2.1. Insect Species and Oil

Desert locusts (*Schistocerca gregaria*) were purchased in the gregarious phase from a breeding stock provided by the Buchner Company in Austria. Locusts were maintained in a crowded colony at the Institute of Zoology in Graz. About 100 locusts were kept in a glass terrarium with the dimensions of 60 × 30 × 30 cm. The light:dark cycle was 12:12 h, and the average temperature in the terrarium was 28 °C at night and 35 °C during the day. The relative humidity was 45–60%. The individuals used in the behavioural experiments were of the same age (mature: about four or five weeks after their last moult).

Organic linseed oil (*Linum usitatissimum*, Natur-Pur brand) was purchased from Spar Österreich. It was stored in a dark bottle at 4 °C during testing.

2.2. Insects Food

The locusts' diet consisted of organic wheat seedlings and organic wheat bran (DM-Drogeriemarkt, Karlsruhe, Germany). Pots of wheat seedlings (i.e., "cat grass") were purchased from Zoo Muser, a local pet shop. These pots of grass were watered daily and exposed to a light:dark schedule of 12:12 h. The wheat seedlings were replaced every two days in the terrarium.

2.3. Experimental Setup

Aggregation experiments were performed inside an anechoic chamber, using an arena with the dimensions of 85 × 65 × 45 cm. The floor of the arena was covered by a piece of paper that was replaced after each trial. We placed a heater (model: PF320LCD, ewt) to maintain a constant temperature of 30 °C inside the anechoic chamber. To illuminate the arena, we use two lamps, an 8 W inspection lamp (Electronic Montage Lamp, SLV Elektronik GmbH: Löhne, Germany) mounted on the top of the arena and a standard lamp (type: 160312, lamp/Bulb: 65× LED, 8 W, SLV Elektronik GmbH: Löhne, Germany). A top-view video camera (CB-38075, GKB: Taichung, Taiwan, China) was used to record the movement of locust individuals inside the arena.

2.4. Insect Isolation

Before performing aggregation experiments, we isolated gregarious desert locusts for three days to simulate swarm disruption that has to be expected after treatment of a swarm with the linseed oil based botanical pesticide. Therefore, we caged individuals in plastic boxes with dimensions of 9 × 9 × 6 cm and placed them at a distance of 2 cm from one another. Boxes were located near to a glass terrarium that contains crowded-reared locusts. All individuals used in experiments were taken from isolation boxes, behaved normally and were able to jump. The walls of insect boxes provided air exchange and olfactory communication among the individuals (ninety-eight pores with a diameter of 1 mm). These boxes were placed on sheets of soft kitchen paper towel to absorb the vibrations generated by the locusts. Additionally, white paper tape was applied to the walls of the plastic boxes to prevent visual contact. To study the aggregation ability of locusts three days after isolation, we evaluated the time individuals came close to any other individual in our aggregation experiments (see Table 1). During the isolation period, the same amount of grass was offered to all individuals. On the third day of isolation, a small piece of reflecting tape was mounted on the pronotum of one individual to mark the target locust. This reflecting tape was fixed in place using a small drop of super glue (Loctite, Henkel Central Eastern Europe GmbH, Vienna, Austria), which is harmless for insects such as bees and locusts.

2.5. Behavioural Experiments

Three days after their isolation, the locusts were transferred to the arena to study group formation during the light phase of their day cycle. In all trials, we studied the group formation of six individuals of the same sex to prevent premating and mating behaviour. All individuals were inspected for dual sexual characteristics (gynandromorph) to prevent an influence arising from mixing different sexes. The inter-individual distance between the target individual and any other individual was observed over a time period of 30 min before linseed oil application. Then, the wings of the target locust were brushed with a thin film of linseed oil (provided at room temperature), and the locusts were monitored for another 30 min. The movements of locusts were recorded under bright conditions using the top-view video camera. This camera was connected to a frame grabber (Pixelsmart Inc.: Lewiston, NY, USA) that captured images in intervals of 5 s, resulting in 12 images per minute. In total, 26 trials were performed with 13 female groups and 13 male groups. Since tactile stimulation may affect the target locust behaviour, we also performed sham operation experiments to exclude an influence of being handled and brushed. In sham operation trials, locusts were monitored over a time period of 30 min before treatment during the light phase of their day cycle. Then, the wings of the target locust were brushed with a clean brush, and the locust group was monitored for another 30 min after treatment. In total, 24 trials were performed with 6 female groups and 8 male groups. The ambient temperature and relative humidity were measured with a hand-held device during the trial. On average, the ambient temperature was 28.7 ± 0.9 °C (mean ± SD), and the mean relative humidity was 36.3 ± 2.9% (mean ± SD).

2.6. Behavioural Responses Towards Linseed Oil and Dead Insect Bodies

Experiments were performed on mature desert locusts of both sexes to study the attraction/avoidance behaviour after exposing locusts to linseed-oil-soaked paper and dead locust bodies. Five males or females were taken from the crowded-rearing arena and were placed in a test arena with a dimension of 48 × 24 × 28 cm. Two clean pieces of plastic foil (dimension of 8 × 3.5 cm) were placed on opposite corners of the arena. The movement of locust individuals was monitored for 30 min using a USB-camera that captured images in intervals of 5 s. Then, a filter paper saturated with 0.5 mL of linseed oil (dimension of 7 × 2.5 cm) was placed on a randomly selected plastic foil of the arena and locust individuals were monitored for another 30 min. Then, the filter paper was replaced by a clean plastic foil and the locusts were given 5 min pause before the next experiment was performed with dead insect bodies. For this purpose, two freshly killed locust individuals (a male and a female) were placed on the clean plastic foil opposite the linseed oil corner and the locust individuals were monitored for another 30 min. Locusts were killed by putting them into a freezer for one hour. In total, 11 trials were performed with 5 female groups and 6 male groups.

2.7. Behavioural Responses to Crushed Male Bodies

While intact dead bodies can be an indicator of contagion in wood lice, the crushed bodies of conspecifics can be a sign of an injury caused by a predator [14]. Therefore, we exposed male locusts to the crushed bodies of conspecific males in a similar setup. Two males were killed by freezing and then their bodies were crushed. A clean piece of plastic foil was placed in a randomly selected corner of the arena and the movements of locusts were monitored for 30 min by means of a top view USB-camera. Then, the foil was replaced by another foil to expose locusts to crushed conspecific males and to monitor locust movements for another 30 min. Ten trials were performed with groups of 5 males.

2.8. Data Evaluation and Statistical Analysis

The obtained video frames were imported into ImageJ (version 1.51j8, National institutes of Health: Bethesda, MD, USA) to measure the distances between the target locust and the next individual. Frame-by-frame distance measurements allow us to quantify the tendency to form groups. We regarded two or more individuals as a group when the inter-individual distance from the target individual was shorter than or equal to its body length (see red scale bar in Figure 1, on average: 5.8 cm for male groups and 6.2 cm for female groups). Such a small distance allowed individuals to see and smell each other. Gillet defined the grouping of locusts by the distance of two body-lengths to each other [17,18]. To analyse the group formation over a period of 30 min, 360 images were imported into ImageJ, and the time the target locust spent in groups (aggregation time) was evaluated by hand. The distance between the body of the target individual (see blue circle in Figure 1) and the closest individual was measured manually after spatially calibrating the image. The first five minutes at the beginning of frame grabbing (equals to 60 image) were excluded from the evaluation to remove possible behavioural influences related to insect handling. The time spent in groups was calculated using Equation (1):

$$Time\ spent\ in\ groups\ (minute) = \left(\frac{Sum\ of\ group\ formation \times 5}{60}\right) \quad (1)$$

The term "sum of group formation" refers to the number of frames in which the inter-individual distance of the target individual from any group member was shorter than or equal to its body length. To calculate the number of minutes individuals spent in a state of aggregation, we multiplied the "sum of group formation" by 5, which refers to the inter-frame interval of 5 s, and divided the result by 60. Percentages of aggregation relative to an observation period of 25 min were calculated using Equation (2):

$$Probability\ of\ aggregation\ (\%) = \left(\frac{Time\ spent\ in\ groups}{25} \times 100\right) \quad (2)$$

If the duration of group formation changed after the application of linseed oil or after sham operation by more than 10% (2.5 min), we considered this as a change in aggregation time. This threshold definition allowed us to discriminate between trials in which group formation either increased or decreased after linseed oil treatment.

To study the general tendency of the individuals to form groups after three days of isolation, the time any individual spent in groups of at least two individuals was quantified by measuring the distances between the closest parts of their bodies. We regarded two or more individuals as a group if their inter-individual distance was shorter than or equal to the average of the body lengths (on average: 5.6 cm for male groups and 6.2 cm for female groups) of all six locusts. The time any individual spent in groups was calculated according to Equation (1) (results are shown in Table 1).

To quantify the activity of the target individual before and after the application of linseed oil, the distance covered by the target individual was measured with the help of the MTrackJ plugin offered in ImageJ. The time spent in the target corner was measured by counting the frames in which any individual was within one body length to borders of the plastic foil. The time spent in the target corner was converted into minutes by using Equation (1).

The time spent by individuals in groups or in the target corner before and after the treatments was tested for statistically significant differences by performing a paired *t*-test. If data distribution deviated from a normal distribution, a Wilcoxon signed-rank test was performed. To test for statistically significant differences between the percentages of trials in which either an increased or decreased group formation time was observed after the application of linseed oil, we performed a *z*-test with Yates correction. All statistical tests were performed in Sigma Plot version 14 (Systat Software Inc.: Erkrath, Germany). The average amount of aggregation time and the proportion of aggregate formation relative to the observation period are given as the arithmetic mean ± standard deviation.

Figure 1. Snapshot of the top view camera. Blue circle: The target individual with the reflecting tape on the thorax. The red scale bar indicates the body length of the target individual, which was used as the minimum distance to define group formation.

3. Results

3.1. Group Formation after Isolation

To study the general tendency of individuals to form groups after three days spent in isolation, we quantified the amount of time (see Equation (1) during which **any individual** maintained a distance to other individuals that was equal to or smaller than the average body length of all locusts in a trial (Table 1). Neither males nor females changed the time spent close to any other individual before and after the wings of target individuals were brushed with linseed oil (females: 21.2 ± 2.2 min before vs. 22 ± 2.7 min after treatment, $p = 0.137$, paired t-test, $N = 13$; males: 23.0 ± 2.1 min before vs. 21.8 ± 1.9 min after treatment; $p = 0.112$, paired t-test, $N = 13$).

Table 1. Time any individual spent in groups of at least two individuals (including the target one) before and after the application of linseed oil. Males and females were tested in separate experiments. Total observation time = 25 min, Group size = 6 individuals, $N = 26$.

Experiments	Time of Forming Groups of at Least 2 Individuals (minute)	
	Before Treatment	After Treatment
Females	22.7	23.5
	23.2	24.4
	17.7	22.0
	21.3	21.2
	22.9	22.0
	22.5	23.8
	18.8	20.2
	25.0	25.0
	22.0	22.5
	20.5	22.8
	17.6	14.2
	21.4	21.8
	19.8	22.8
Average of females	**21.2 ± 2.2**	**22.0 ± 2.7**
Males	23.8	23.9
	24.7	22.8
	25.0	19.9
	20.2	20.9
	23.5	20.0
	20.9	19.1
	24.2	23.9
	23.8	21.8
	21.7	18.7
	24.5	22.1
	23.9	22.7
	24.9	23.4
	18.4	24.2
Average of males	**23.0 ± 2.1**	**21.8 ± 1.9**
Total average	**22.1 ± 2.3**	**21.9 ± 2.3**

3.2. *Linseed Oil Treatment of Male and Female Groups*

The average amount of time **target males** spent in groups with members of the same sex was significantly reduced by 4.6 ± 5.3 min (18.2 ± 21.2% of the observation time, Figure 2A, $p = 0.010$, paired t-test, $N = 13$) after brushing their wings with linseed oil. In 76.7% of trials, males reduced the time spent in aggregation by more than 10%. In these trials, the average aggregation time among males significantly decreased by 6.8 ± 3.2 min (27 ± 12.6% of observation time) after oil treatment (red horizontal stripes in Figure 2B, $p < 0.001$, paired t-test, $N = 10$). In only two trials did the aggregation time among males increase by 4.9 ± 1.9 min (20 ± 7.8% of observation time, blue vertical stripes in Figure 2B), and no change was observed in one trial (black area in Figure 2B). Furthermore, the proportion of trials exhibiting a decreased aggregation time after treatment was significantly higher compared to the proportion of trials showing an increase (Figure 2B, $p = 0.006$, z-test, $N = 13$). Before the application of the linseed oil, the target males sometimes mounted other males or were mounted by other males. Interestingly, male–male mounting involving the target individual was never observed after the target locust's wings were brushed with linseed oil. The average distance covered by the target males (activity) within 25 min did not change after linseed oil application (Table S1, $p = 0.818$, paired t-test, $N = 13$). Tracking brushed males revealed that, in 92.3% of trials, the target males were very active and moved towards other individuals in the arena, but other individuals were avoiding them.

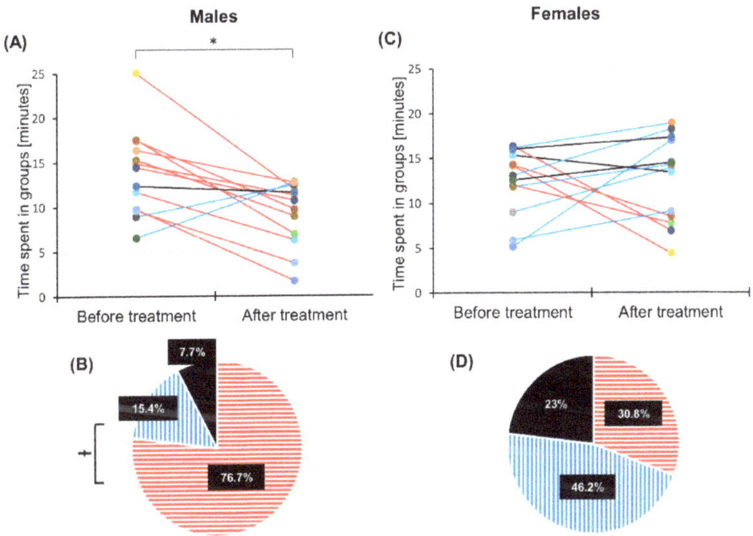

Figure 2. Sex differences in group formation after linseed oil treatment. Time that the target males (**A**) and target females (**C**) spent in groups. Red lines: Time spent in groups decreased. Blue lines: Time spent in group increased. Black lines: Time spent in groups did not change by more than 10%. Total observation time = 25 min, $N = 13$ experiments. (**B**,**D**) Percentage of experiments in which aggregation time decreased (red horizontal stripes), increased (blue vertical stripes) or was left unchanged (black area) after linseed oil treatment. * indicates a p-value < 0.05 (paired t-test, $N = 13$). ǂ indicates a significant difference between the percentage of experiments with a decreased and increased aggregation time related to the application of the linseed oil ($p < 0.05$, z-test, $N = 13$).

The average amount of time that the **target females** spent in groups with members of the same sex did not significantly differ before and after the application of the linseed oil (0.2 ± 6.2 min, Figure 2C, $p = 0.962$, paired t-test, $N = 13$). In six trials out of 13, the target females spent on average 5.0 ± 3.5 min longer in groups after oil treatment (blue vertical stripes in Figure 2D). However, the proportion of trials exhibiting an increased and decreased aggregation time did not differ significantly (Figure 2D, $p = 0.687$, z-test, $N = 13$). Egg pods laid by females were often found in the isolation boxes and one time inside the arena after the trial. The average distance covered by the target females did not change after linseed oil application (Table S1, $p = 0.814$, paired t-test, $N = 13$). Observations performed after linseed oil application revealed that, in 69.2% of trials performed with females, six females, including the target one, were active, behaved normally and moved towards each other. In the other 30.8%, the target females were less active and only walked along the borders of the arena.

3.3. Pooled Aggregation Time after Linseed Oil Treatment

In more than half of all trials performed with either males or females, the total amount of time that the target individual spent in groups of at least two individuals decreased by more than 10% after the application of linseed oil (Figure 3A,B). An increase in the amount of aggregation time, however, was found in only about one-third of the trials. Averaging over all trials and sexes, the time spent in an aggregation decreased by only 2.2 ± 6 min (8.7 ± 24.5% of the observation time), which indicates that the application of linseed oil did not significantly change the total aggregation time ($p = 0.077$, paired t-test, $N = 26$). However, in 53.8% of trials in which the target individuals spent less time in groups, the aggregation time was significantly reduced by 7 ± 2.9 min after the application of linseed oil (red striped segment in Figure 3B, $p < 0.001$, paired t-test, $N = 14$). In only 30.8% of trials, the target individuals significantly extended the average time spent in aggregation by 5 ± 3.1 min (blue-striped segment in Figure 3B, $p = 0.008$, Wilcoxon signed-rank test, $N = 8$) after treatment. In 15.4% of all trials, there was no change in the aggregation time of the target individuals of more than 10% (black segment in Figure 3B).

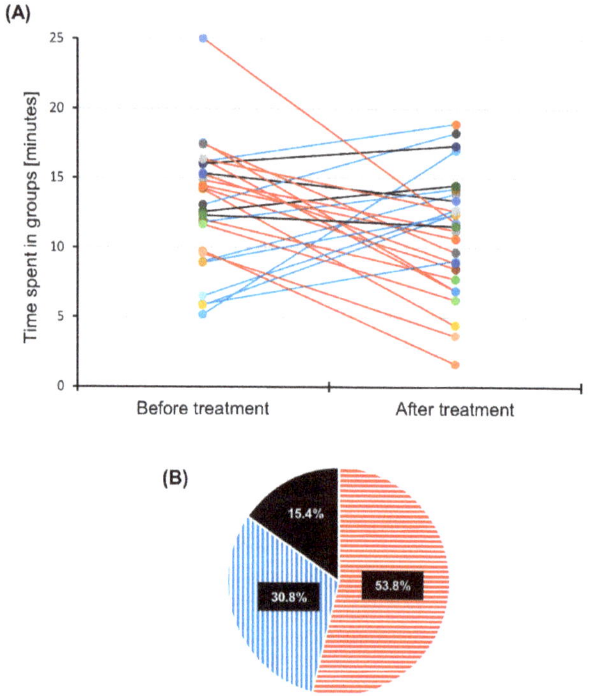

Figure 3. Time that the target individual spent in male and female groups. (**A**) Aggregation time of the target individual before and after treatment with linseed oil. Red lines: Time spent in groups decreased. Blue lines: Time spent in groups increased. Black lines: Time spent in groups did not change by more than 10%. (**B**) Percentage of experiments in which aggregation time decreased (red horizontal stripes), increased (blue vertical stripes) or was left unchanged (black area) after linseed oil treatment. Total observation time = 25 min, $N = 26$ experiments.

3.4. Sham Operation

To exclude an influence caused by the handling of the target individual, we quantified the total amount of time the target locust spent in groups of at least two individuals before and after the wings were brushed with a clean brush (results are shown in Table S2). Averaging over all trials and sexes, the time spent in an aggregation decreased by only 0.8 ± 6.2 min (2.7 ± 20.8% of the observation time), which shows that the clean brush did not change the total aggregation time significantly ($p = 0.538$, paired t-test, $N = 24$). The average amount of time target **males** spent in groups with members of the same sex was reduced by only 1.2 ± 6.2 min (3.9 ± 20.7% of the observation time) after sham operation. The average amount of time the target **females** spent in groups with members of the same sex in this experiment was reduced by only 0.4 ± 6.5 min (1.4 ± 21.7% of the observation time).

3.5. Group Exposure to Linseed Oil and Dead Bodies

The average amount of time males or females spent in the target corner did not change significantly after the addition of the filter paper containing linseed oil (Table 2, males: $p = 0.460$, paired t-test, $N = 6$; females: $p = 0.451$, paired t-test, $N = 5$).

Table 2. Time any individual spent in the target corner of the arena before and after the addition of the linseed oil. Males and females were tested in separate experiments. Total observation time = 25 min, Group size = 5 individuals, $N = 11$.

Experiments	Time Spent in the Target Corner of the Arena	
	Control	Linseed Oil
Females	20.4	20.4
	10.1	11.9
	15.8	18.9
	20.9	17.3
	16.8	21.4
Average of females	**16.8 ± 4.4**	**18.0 ± 3.7**
Males	14.6	14.6
	16.9	18.7
	17.9	15.1
	11.4	13.3
	16.0	15.3
	16.8	21.9
Average of males	**15.6 ± 2.3**	**16.5 ± 3.2**
Total average	**16.1 ± 3.3**	**17.2 ± 3.4**

The average amount of time males or females spent in the target corner did not change significantly after the addition of the dead bodies (Table 3, males: $p = 0.894$, paired t-test, $N = 6$; females: $p = 0.381$, paired t-test, $N = 5$).

Table 3. Time any individual spent in the target corner of the arena before and after the addition of the intact dead bodies. Males and females were tested in separate experiments. Total observation time = 25 min, Group size = 5 individuals, $N = 11$.

Experiments	Time Spent in the Target Corner of the Arena	
	Control	Dead Bodies
Females	7.5	23.8
	18.9	21.3
	13.8	11.8
	12.1	20.6
	14.8	8.9
Average of females	**13.4 ± 4.2**	**17.3 ± 6.5**

Table 3. *Cont.*

Experiments	Time Spent in the Target Corner of the Arena	
	Control	Dead Bodies
Males	11.2	6.7
	8.2	14.2
	14.8	14.4
	18.3	17.9
	19.0	20.8
	9.6	8.3
Average of males	**13.5 ± 4.6**	**13.7 ± 5.4**
Total average	**13.5 ± 4.2**	**15.3 ± 5.9**

3.6. Responses of Males Towards Crushed Bodies

In an observation period of 25 min, the average amount of time **males** spent in the target corner was significantly decreased by 6.4 ± 4.9 min (Table 4, $p = 0.003$, paired t-test, $N = 10$) after the addition of the crushed male bodies.

Table 4. Time any individual spent in the target corner of the arena before and after the addition of the crushed male bodies. Total observation time = 25 min, Group size = 5 individuals, $N = 10$.

Experiments	Time Spent in the Target Corner of the Arena	
	Control	Crushed Males
Males	11.6	9.3
	18.4	18.6
	15.7	2.1
	14.9	15.1
	20.9	12.9
	21.0	16.9
	21.2	12.8
	13.3	6.6
	16.3	8.1
	23.2	10.3
Average	**17.7 ± 3.9**	**11.3 ± 5.0**

4. Discussion

After isolating individuals for three days, we observed a general tendency for these individuals to form groups during the observation period (see Table 1), which is in contrast to solitary-reared locusts that are less active and freeze when they sense movement in the arena [5,10]. Furthermore, the amount of time any group member spent in groups was not influenced by brushing the wings of the target individuals with linseed oil (Table 1). Brushing the wings of the target locusts with linseed oil, however, affected the amount of time these individuals spent in proximity to other individuals in most trials (see Figure 3). In contrast, brushing the wings with a clean brush in the sham operation experiment did not change the average group formation time significantly (see Table S2). This suggests that the reduced group formation time observed in males is caused by linseed oil and less likely by physical stimulation. Brushing the wings of the target individuals with linseed oil did not change the average distance covered by the target locusts (see Table S1), which suggests that the reduced group formation time observed in males is not caused by an overall reduction in the activity of the target locust. Linseed oil is rich in unsaturated fatty acids, and some of these acids (especially oleic and linoleic acid) have been described as necromones that are released by injured and dead insects. These necromones are known to trigger distinct behaviour patterns in other individuals of the same or different species [14–16]. Therefore, we tested whether linseed oil could be used as an agent to control the aggregation behaviour of gregarious locusts by investigating the aggregation behaviour of members of both sexes in controlled experiments.

A reduced group formation time of males after the linseed oil application (see Figure 2A,B) may be the result of necrophobic avoidance behaviour, which may be absent in females most likely because of oviposition pheromone release. Males in this study showed a necrophobic avoidance and tended to move away from the target individual that released "the smell of death" in order to avoid possible infections that can be transmitted by swarm mates or indicates an injury from a predator attack [14]. This aversion to necromone fatty acids (especially oleic and linoleic acids) were reported in a large number of invertebrates [19]. However, locusts are also known to cannibalise other swarm mates in situations where food is scarce. We have not observed cannibalistic tendencies in our study because all individuals were fed with fresh grass during the isolation period. An interesting observation made in the course of this study was that the target individuals were no longer involved in mounting behaviours after being brushed with linseed oil. This male–male mounting behaviour is rather common in the absence of females, and our results indicate that the observed change in male behaviour was mediated by linseed oil. This result differs from that of Clancy et al. [20], who observed an increase in the frequency of male–male mounting (MMM) behaviour in desert locust males suffering from *Metarhizium acridum*, a fungal infection. This reduction in MMM is similar to the effect caused by "phenylacetonitrile" (PAN, also known as benzyl cyanide), which is a mature male volatile that prevents males from being mounted (homosexuality) by other males [5,21]. It is mainly released from the wings and the hind legs [22] and acts as a strong repellent for mature males to hide a female from other competing males (a courtship inhibition pheromone) [5,21]. Brushing the wings with linseed oil likely dilutes the concentration of PAN on the wings of the target male due to the high hydrophobicity of this pheromone. Therefore, we conclude that the repellent effect related to our treatment is mediated by fatty acids but less likely by PAN. However, Bashir et al. [23] proved that PAN has a solitarising effect (anti-gregarisation) on the hopper bands, and another recent study revealed that PAN serves as an antipredator defence in gregarious migratory locusts as it is converted into a hypertoxic cyanide (HCN) when they are under attack [24]. Since mature males exclusively produce aggregation pheromones that are attractive to members of both sexes [25–27], a reduction in male group formation caused by linseed oil may have important consequences for the formation and persistence of locust swarms.

We observed no significant difference in the average amount of time the target females spent in groups before and after the application of linseed oil (see Figure 2C,D). In almost half of all trials, there was a significant increase in the average amount of aggregation time by 5 ± 0.6 min. This result can be attributed to the attraction between female individuals mediated by the oviposition pheromone [25]. Since the target females were treated after the control recording ended, this might lead to an increase in the amount of oviposition pheromone released in the arena with time. This consequently increases the time the females spent in groups even in the presence of fatty acids necromones. This female–female attraction ensures the spatial aggregation of egg pods, which increases the survival rates and supports gregarious cohesion among members of the next generation. Furthermore, it has been also shown that ovipositing females of desert locusts aggregate responding to a pheromone produced by alive or dead individuals in all development stages [28]. In 30.8% of trials performed with females in this study, brushed target females stayed away, were less active and only walked on the borders of the arena. This may indicate that females carrying the smell of death seem to avoid being eaten by conspecifics because it is known that desert locusts show cannibalistic behaviour beginning in the fifth-instar stage, and especially adults may be regarded as an important source of proteins by conspecifics [29,30]. Bazazi et al. [31] studied collective motion and cannibalism in marching bands of desert locust nymphs and was able to show that individuals that were injured by conspecifics may suffer an increased risk of cannibalism. Since females have a higher demand for the proteins involved in egg development, they may respond to the linseed-oil-treated target individual in a different way than males.

To study possible attraction/avoidance behaviour of the desert locusts to necromones, we tested the time males and females spent in the presence of a stationary linseed oil target and dead bodies.

While we found no difference in the average amount of time males and females spent in the target corner after the addition of the linseed oil or dead bodies (see Tables 2 and 3), there was a significant reduction in this parameter in the experiment performed with crushed male bodies as males spent less time next to them (see Table 4). A response to freshly dead individuals is rather unlikely since the responses to fatty acid necromones increased over time [14,19,32]. According to our results, responses of some genera of isopods were very weak to intact dead bodies (disease) compared to crushed bodies or body extracts [14]. The latter indicates injury resulting from predation and swarm mates should avoid them.

Sex-specific responses to fatty acid necromones were also described in the cricket *Acheta domesticus*, where females respond to body extracts less than males, as females might be less risk averse because they seek out singing males and explore oviposition sites [19]. Another study investigated the repellent effect of various fatty acid necromones on cockroach males and females. In both sexes, the percentage of repellence was strictly dose dependent and the percentage of oleic acid repellence was more significant in cockroach males (70%) than females (43%) [33]. The results obtained from males in our study indicate that fatty acids from linseed oil may change the aggregation behaviour of desert locusts. A similar solitarising effect has been found in *S. gregaria* when nymphs were exposed to faeces of crowded locusts [34].

5. Conclusions

A novel botanical pesticide formulation uses linseed oil as the main component and was found to be highly effective against gregarious desert locusts as well as migratory locusts [11]. Since linseed oil contains unsaturated fatty acids that have been shown to act as necromone cues in different insect species, it was of interest to determine whether this oil affects group formation in gregarious desert locusts. Brushing the wings of single gregarious desert locust males with this oil significantly decreased the amount of aggregation time they spent in male groups in the majority of the trials (Figure 2). In contrast, most treated females either did not display alterations in aggregation time or the aggregation time increased (Figure 2). These sex differences in aggregation behaviour may be explained by the release of different pheromones produced by adult males and females in the gregarious phase. Since a reduction in the amount of aggregation time among males leads to a reduction in tactile stimuli that have been shown to be highly gregarizing [8,9], linseed oil seems to be a promising candidate agent for the control of aggregation behaviour in gregarious desert locusts, as it may even disrupt swarm formation once a certain percentage of individuals have come into contact with linseed oil in the course of botanical pesticide treatments. Therefore, fatty acids that act as necromones and are contained in linseed oil should be considered in pest management as previously suggested by Yao et al. [14]. Future studies need to reveal the origin of this behaviour by exposing males and females to different concentrations and types of necromones in controlled laboratory conditions.

Supplementary Materials: The following are available online at http://www.mdpi.com/2073-4395/10/10/1458/s1, Table S1: Distance covered by the target individual before and after the application of linseed oil. Males and females were tested in separate experiments. Total observation time = 25 min, Group size = 6 individuals, $N = 26$. Table S2: Time the target individual spent in groups before and after sham operation. Males and females were tested in separate experiments. Total observation time = 25 min, Group size = 6 individuals, $N = 14$.

Author Contributions: Z.A.S.A. and M.H. conceived and designed the research. Z.A.S.A. conducted the experiments. M.H. supervised the experiments. Z.A.S.A. analysed the data and drafted the manuscript. Both authors have read and agreed to the published version of the manuscript.

Funding: This research was funded by the Egyptian Ministry of Higher Education, Cultural Affairs and Mission Sector. Open Access Funding by the University of Graz.

Acknowledgments: Many thanks to Boris P. Chagnaud for his critical reading and valuable comments.

Conflicts of Interest: The authors declare no competing interests.

Data Availability: Data will be made accessible upon request.

Compliance with ethical standards: The experimental procedures were performed at the Institute of Biology (University of Graz). All experiments comply with the current Austrian and European Community laws for the ethical treatment of animals and are in line with the ASAB Guidelines for the Use of Animals in Research.

References

1. Steedman, A. *Locust Handbook. Overseas Development Natural Resources Institute*, 2nd ed.; Overseas Development Natural Resources Institute: London, UK, 1988.
2. Cressman, K. Desert Locust. In *Biological and Environmental Hazards, Risks, and Disasters.*; Shroder, J.F., Sivanpillai, R., Eds.; Elsevier: Amsterdam, The Netherland, 2016; pp. 87–105. [CrossRef]
3. Uvarov, B. *Grasshoppers and Locusts: A handbook of General Acridology*; Cambridge University Press: Cambridge, UK, 1966; Volume 1.
4. Gaten, E.; Huston, S.J.; Dowse, H.B.; Matheson, T. Solitary and gregarious locusts differ in circadian rhythmicity of a visual output neuron. *J. Biol. Rhythm.* **2012**, *27*, 196–205. [CrossRef] [PubMed]
5. Pener, M.P.; Simpson, S.J. Locust phase polyphenism: An update. *Adv. Insect Physiol.* **2009**, *36*, 1–286. [CrossRef]
6. Ma, Z.; Guo, X.; Lei, H.; Li, T.; Hao, S.; Kang, L. Octopamine and tyramine respectively regulate attractive and repulsive behavior in locust phase changes. *Sci. Rep.* **2014**, *5*. [CrossRef] [PubMed]
7. Latchininsky, A.V. Locusts and remote sensing: A review. *J. Appl. Remote Sens.* **2013**, *7*, 075099. [CrossRef]
8. Simpson, S.J.; Despland, E.; Hägele, B.F.; Dodgson, T. Gregarious behavior in desert locusts is evoked by touching their back legs. *Proc. Natl. Acad. Sci. USA* **2001**, *98*, 3895–3897. [CrossRef]
9. Roessingh, P.; Bouaichi, A.; Simpson, S.J. Gregarious behavior in desert locusts is evoked by touching their back legs., *Schistocerca gregaria*. *J. Insect Physiol.* **1998**, *44*, 883–893. [CrossRef]
10. Simpson, S.J.; McCaffery, A.R.; Hägele, B.F. A behavioural analysis of phase change in the desert locust. *Biol. Rev. Camb. Philos. Soc.* **1999**, *74*, 461–480. [CrossRef]
11. Abdelatti, Z.A.S.; Hartbauer, M. Plant oil mixtures as a novel botanical pesticide to control gregarious locusts. *J. Pest Sci.* **2020**, *93*, 341–353. [CrossRef]
12. Hartbauer, M.; Abdelatti, Z.A.S. Pesticidal compositions for pest control. P18417PCT/EP2019/056709 (WO 2019/179945 A1), 26 September 2019.
13. Bayrak, A.; Kiralan, M.; Ipek, A.; Arslan, N.; Cosge, B.; Khawar, K.M. Fatty acid compositions of linseed (*Linum usitatissimum* L.) genotypes of different origin cultivated in Turkey. *Biotechnol. Biotechnol. Equip.* **2010**, *24*, 1836–1842. [CrossRef]
14. Yao, M.; Rosenfeld, J.; Attridge, S.; Sidhu, S.; Aksenov, V.; Rollo, C.D. The ancient chemistry of avoiding risks of predation and disease. *Evol. Biol.* **2009**, *36*, 267–281. [CrossRef]
15. Wilson, E.O.; Durlach, N.I.; Roth, L.M. Chemical releaser of necrophoric behavior in ants. *Psyche* **1958**, *65*, 108–114. [CrossRef]
16. Sun, Q.; Zhou, X. Corpse management in social insects. *Int. J. Biol. Sci.* **2013**, *9*, 313–321. [CrossRef] [PubMed]
17. Gillett, S.D. Social determinants of aggregation behaviour in adults of the Desert locust. *Anim. Behav.* **1973**, *21*, 599–606. [CrossRef]
18. Gillett, S.D. Changes in the social behaviour of the desert locust, *Schistocerca gregaria*, in response to the gregarizing pheromone. *Anim. Behav.* **1975**, *23*, 494–503. [CrossRef]
19. Aksenov, V.; David Rollo, C. Necromone death cues and risk avoidance by the cricket *Acheta domesticus*: Effects of sex and duration of exposure. *J. Insect Behav.* **2017**, *30*, 259–272. [CrossRef]
20. Clancy, L.M.; Cooper, A.L.; Griffith, G.W.; Santer, R.D. Increased male-male mounting behaviour in desert locusts during infection with an entomopathogenic fungus. *Sci. Rep.* **2017**, *7*, 5659. [CrossRef] [PubMed]
21. Seidelmann, K.; Ferenz, H.J. Courtship inhibition pheromone in desert locusts, *Schistocerca gregaria*. *J. Insect Physiol.* **2002**, *48*, 991–996. [CrossRef]
22. Seidelmann, K.; Weinert, H.; Ferenz, H.-J. Wings and legs are production sites for the desert locust courtship-inhibition pheromone, phenylacetonitrile. *J. Insect Physiol.* **2003**, *49*, 1125–1133. [CrossRef]
23. Bashir, M.O.; Hassanali, A. Novel cross-stage solitarising effect of gregarious-phase adult desert locust (*Schistocerca gregaria* (Forskal)) pheromone on hoppers. *J. Insect Physiol.* **2010**, *56*, 640–645. [CrossRef]

24. Wei, J.; Shao, W.; Cao, M.; Ge, J.; Yang, P.; Chen, L.; Wang, X.; Kang, L. Phenylacetonitrile in locusts facilitates an antipredator defense by acting as an olfactory aposematic signal and cyanide precursor. *Sci. Adv.* **2019**, *5*, 1–13. [CrossRef] [PubMed]
25. Ferenz, H.J.; Seidelmann, K. Pheromones in relation to aggregation and reproduction in desert locusts. The Royal Entomological Society. *Physiol. Entomol.* **2003**, *28*, 11–18. [CrossRef]
26. Obeng-Ofori, D.; Njagi, P.G.N.; Torto, B.; Hassanali, A.; Amiani, H. Sex differentiation studies relating to releaser aggregation pheromones of the desert locust, *Schistocerca gregaria*. *Entomol. Exp. Appl.* **1994**, *73*, 85–91. [CrossRef]
27. Obeng-Ofori, D.; Njagi, P.G.N.; Hassanali, A. Aggregation pheromone complex of the desert locust, *Schistocerca gregaria* (Forskal) (Orthoptera: Acrididae): Current Status. *J. Ghana Sci. Ass.* **1998**, *1*, 69–83. [CrossRef]
28. Norris, M.J.; Richards, O.W. Aggregation response in ovipositing females of the desert locust, with special reference to the chemical factor. *J. Insect Physiol.* **1970**, *16*, 1493–1515. [CrossRef]
29. Whitman, D.W.; Blum, S.M.; Slansky, F., Jr. Carnivory in phytophagous insects. In *Functional Dynamics of Phytophagous Insects*; Ananthakrisnan, T.N., Ed.; Oxford & IBH: New Delhi, India, 1994; pp. 161–205.
30. Niassy, A. Interactions between *Schistocerca gregaria* (Forskal) and *Locusta migratoria* migratorioides (Reich & Farmaire) in Relation to Phase Polymorphism. Ph.D. Thesis, University of Ghana Legon, Accra, Ghana, 1996.
31. Bazazi, S.; Buhl, J.; Hale, J.J.; Anstey, M.L.; Sword, G.A.; Simpson, S.J.; Couzin, I.D. Collective Motion and Cannibalism in Locust Migratory Bands. *Curr. Biol.* **2008**, *18*, 735–739. [CrossRef]
32. Akino, T.; Yamaoka, R. Origin of oleic acid corpse recognition signal in the ant, *Formica japonica* Motschlsky (Hymenoptera: Formicidae). *Jpn. J. Appl. Entomol. Zool.* **1996**, *40*, 265–271. [CrossRef]
33. Rollo, C.D.; Czvzewska, E.; Borden, J.H. Fatty acid necromones for cockroaches. *Naturwissenschaften* **1994**, *81*, 409–410. [CrossRef]
34. Gillett, S.D.; Phillips, M. Faeces as a source of a locust gregarisation stimulus. Effects on social gregarisation and on cuticular colour of nymphs of the desert locust, *Schistocerca gregaria* (Forsk.). *Acrida* **1977**, *6*, 279–286.

agronomy

Article

Fungal-Based Biopesticide Formulations to Control Nymphs and Adults of the Desert Locust, *Schistocerca gregaria* Forskål (Orthoptera: Acrididae): A Laboratory and Field Cage Study

Waqas Wakil [1,2,*], Muhammad Usman Ghazanfar [3], Muhammad Usman [1], David Hunter [4] and Wangpeng Shi [5,*]

1 Department of Entomology, University of Agriculture, Faisalabad 38040, Pakistan; usmanbhattiuaf@gmail.com
2 Senckenberg German Entomological Institute, D-15374 Müncheberg, Germany
3 Department of Plant Pathology, College of Agriculture, Sargodha University, Sargodha 40100, Pakistan; usmanghazanfar1972@gmail.com
4 Locust and Grasshopper Control, McKellar, ACT 2617, Australia; davidmhunter100@gmail.com
5 Department of Entomology, College of Plant Protection, China Agricultural University, Beijing 100083, China
* Correspondence: waqaswakeel@hotmail.com (W.W.); wpshi@cau.edu.cn (W.S.)

Abstract: This is the first field study in which we have tested the efficacy of four different entomopathogenic fungal (EPF) formulations together in single study—i.e., Green Muscle, Green Guard, *Metarhizium anisopliae*, and an isolate of *Beauveria bassiana* (isolate WG-11)—against nymphs and adults of the desert locust, *Schistocerca gregaria* Forskål (Orthoptera: Acrididae). We conducted several different studies: (a) lethal bioassay against the 3rd, 5th, and adult stages under laboratory conditions; (b) sublethal effects on the reproduction, diet consumption, fecal production, and weight gain; (c) a greenhouse trial; and (d) a field cage trial. Under laboratory conditions, all EPF formulations caused significant mortality, and the highest efficacy was observed with Green Muscle, followed by Green Guard, *B. bassiana*, and *M. anisopliae*. Susceptibility was found to be greatest in 3rd-instar nymphs, followed by 5th instars, and then adults. Along with lethal effects, sublethal doses of EPF reduced the number of egg pods per female, total eggs per pod, and egg hatching, while extending nymphal developmental time and reducing adult longevity; again, Green Muscle performed better. Sublethal doses not only retarded reproduction, but also caused behavioral changes, including reductions in food consumption, fecal production, and weight gain. All EPF formulations not only produced significant mortality in laboratory conditions, but also performed very well under the greenhouse and field conditions. The maximum mortality against 3rd-instar (81.7% and 74.0%), 5th-instar (73.3% and 65.1%), and adult locusts (67.5% and 58.9%) was observed when using Green Muscle under greenhouse and field trials, respectively. The current study showed that all of the EPF formulations have the potential to reduce pest populations, and could be used in the integrated pest management program.

Keywords: desert locust; *Metarhizium acridum*; *Metarhizium anisopliae*; *Beauveria bassiana*; lethal effect; sublethal effect; greenhouse; field efficacy

Citation: Wakil, W.; Ghazanfar, M.U.; Usman, M.; Hunter, D.; Shi, W. Fungal-Based Biopesticide Formulations to Control Nymphs and Adults of the Desert Locust, *Schistocerca gregaria* Forskål (Orthoptera: Acrididae): A Laboratory and Field Cage Study. *Agronomy* **2022**, *12*, 1160. https://doi.org/10.3390/agronomy12051160

Academic Editor: Beatriz Gámiz

Received: 22 July 2021
Accepted: 11 August 2021
Published: 11 May 2022

1. Introduction

Locust plagues are considered a major constraint in sustainable agriculture production [1–3]. The desert locust, *Schistocerca gregaria* Forskål (Orthoptera: Acrididae), has produced extensive losses of different food crops in Africa and Asia [4–6], especially during locust upsurges. In 2019, *S. gregaria* entered the province of Baluchistan in Pakistan from Iran, and then spread to different parts of Sindh, and finally to all four provinces. It was estimated that Pakistan faced a loss of PKR 353 billion and 464 billion if 25% damage occurred to Rabi (sown in winter) and Kharif crops (sown in spring), respectively (https://www.dawn.com/news/1555487; accessed on 12 April 2021). There was apprehension that, if left uncontrolled, the number of locusts would increase 500-fold by June 2020 in East Africa [7]—potentially the worst infestation in past 100 years [8]. Climate change aggravates this situation due to high temperatures and increased precipitation, which accelerate locust growth and reproduction [9]. *Schistocerca gregaria* has three different developmental stages: egg, nymph, and adult. Adult females lay eggs in the form of egg pods in the soil at 5–10-cm depth [10]. Egg pods are 3–4 cm long, white in color, and shaped like bananas. Egg hatching occurs in 10–14 days when it is hot, or up to 65 days when it is cool, with each egg pod consisting of 80–160 eggs [11]. Hatched eggs produce nymphs known as hoppers that have 5–6 instars, and the developmental period from the first to the last instar is 28–48 days, followed by the adult stage, which can last for several months [12,13].

Management of locust plagues depends heavily on the use of chemical insecticides [14,15]. Prior to 1980, the organochlorine insecticide dieldrin was effectively used for many years to control this plague because of its persistence. However, due to their hazardous effects on animals and the environment [4], the use of organochlorines is now banned, and locust control mainly relies on organophosphate insecticides [16] that have low persistence, making them less effective than dieldrin [17]. Between 2003 and 2005, to combat the locust outbreak, over 13 million hectares were treated using 13 million liters of chemical insecticides [18]. During the 2019–20 locust outbreak in Pakistan, the Department of Plant Protection (DPP) sprayed 150,839 L of pesticides over an area of 300,595 hectares to suppress the locust population. The effects of non-judicious, intensive, and heavy use of chemical pesticides are probably underestimated, leading to environmental problems resulting from lethal effects on the whole ecosystem [19,20].

Knowing the hazardous effects of conventional insecticides, the development of safe alternatives is essential [21,22]. The utilization of natural enemies such as insect predators and parasitoids as control methods is not likely to be effective, due to the rapid growth and highly dispersed movement of locusts [4,15]. Green technology such as biological control uses microorganisms and birds to help to reduce the locust population [9], reducing the toxic impact of pesticides on the ecosystem. Microbial biopesticides could be an effective alternative to synthetic insecticides in locust management [6,23,24]. Microbial-based control methods, including the utilization of entomopathogenic fungi (EPF), have an extra benefit in that they invade the insect host via direct contact with the insect's integument instead of ingestion of the pathogen [23,25]. The Food and Agriculture Organization (FAO) has been working for the past 20 years in different countries on the development of biopesticides based on *Metarhizium acridum* (Driver and Milner) J.F. Bisch, Rehner and Humber (Hypocreales: Clavicipitaceae) to infect grasshoppers and locusts [18]. In field conditions, locusts can be exposed to *M. acridum* conidia in three ways: direct exposure to spray application, secondary pick-up of conidia from treated vegetation, or horizontal transmission from infected individuals during copulation and during aggregation phenomena [26]. While, there have been a number of studies on the efficacy of EPF such as *M. acridum* [15,26–30] and *Beauveria bassiana* (Bals.-Criv.) Vuill. (Hypocreales: Cordycipitaceae) [29,31] against *S. gregaria*, ours is the first to test four different formulations against *S. gregaria* in a single study.

In addition to lethal effects, EPF have sublethal effects that can disturb overall pest status [27,32,33]. Reduction in feeding was observed among acridid insect pests following exposure to sublethal doses of entomopathogens [32,34–38]. Reduced fecundity was observed in surviving adults after being exposed to sublethal doses [27,39,40]. Similar effects have also been observed in other insect pests, such as houseflies and mosquitoes [41–43]. This is the first study to test the effects of different EPF on the survival and longevity of different developmental stages of *S. gregaria*.

Many insect pests are able to thermoregulate their body temperature across different environmental conditions [44]. Acridid insects not only change their body temperature with respect to location, but also alter their body temperature (behavioral fevers) when infected with fungal conidia under field conditions. They assume basking postures or locations that elevate their body temperature above optimal temperatures for microbial growth [45,46]. Commonly, lower mortality was observed in the field conditions compared to the laboratory [47]. Consequently, greenhouse and field cage experiments were included in this study in order to determine whether locusts exposed to sunlight could thermoregulate and reduce EPF efficacy compared to the laboratory. The objective of this study was to evaluate the efficacy of two different commercial formulations of *M. acridum*—one of *M. anisopliae*, and one local strain of *B. bassiana* (isolate WG-11)—against 3rd- and 5th-instar nymphs and adults of *S. gregaria* in laboratory, greenhouse, and field trials.

2. Materials and Methods

2.1. Insect Collection and Rearing

Adults of *S. gregaria* were collected from a swarm in Bhakkar Punjab (Pakistan) and reared in the laboratory inside wooden cages (30 cm × 30 cm × 30 cm). The cages were made with mesh wires on three sides of cages and on the roof. The front side of each cage was fitted with a door that contained white cloth with sleeves for operations like insect introduction, diet supplementation, and cleaning. The bottom of the cage consisted of several small holes (10-cm diameter) for egg laying. Plastic cups filled with sand with appropriate moisture content were placed below the wooden cage for egg collection. The adults were provided with wheat seedlings and wheat bran as food. Three light bulbs (60 W) were placed on the top of the roof for heating, and were switched on for 6–8 h each day for thermoregulation, which enabled basking behavior, as would occur in the field. After egg laying, the small plastic cups (covered with mosquito wire mesh) were placed under the light bulb (60 W) for 6–8 h to accelerate embryonic development. Upon hatching, first-instar nymphs were released inside the cage for subsequent development [26].

2.2. Entomopathogenic Fungi and Their Culturing

The biopesticides used in this study included two commercial products based on *M. acridum*—Green Guard (BASF, CSIRO FI (=ARSEF 324)) and Green Muscle (Elephant Vert, IMI 330189 (=ARSEF7486))—one *Metarhizium anisopliae* (Metchnikoff) Sorokin (Hypocreales: Clavicipitaceae) (CQMCC No. 0877) from China, and a local strain of *B. bassiana* (isolate WG-11) originally isolated from soil samples collected from Lal Suhanra, Punjab, (Pakistan). Each fungal isolate was individually inoculated on a potato dextrose agar (PDA) Petri plates (100 × 20 mm), wrapped with Parafilm, and finally placed in an incubator at 25 °C with a 14:10 h (light:dark) photoperiod. After 7–10 days, fungi were harvested using a sterile scalpel, followed by the conidial suspension being placed inside a falcon tube (50 mL) containing 30 mL of 0.05% Silwet L-77 solution. Eight glass beads were added to the tube and the solution was vortexed for about 5 min for proper agitation, and the desired conidial doses were determined using hemocytometer. Conidial viability of each fungal isolate was determined by spreading 0.1 mL of solution at rate of 1×10^6 conidia/mL on Sabouraud dextrose agar with 1% yeast (SDAY) plates (two plates for each isolate) wrapped with Parafilm and placed in the incubator at 25 °C with a 14:10 h (light:dark) photoperiod for 16 h [48]. A coverslip was placed on each SDAY plate, and approximately 200 conidia were assessed for germination for one count, with a conidium considered to be germinated if the germ tube was longer than the conidia [48]. Two counts were taken from each plate, totaling four counts for each isolate [49,50].

2.3. Bioassay against 3rd- and 5th-Instar Nymphs and Adults of S. gregaria

The bioassays were conducted to assess the effect of different doses (1×10^5, 1×10^6, 1×10^7, and 1×10^8 conidia/mL) of entomopathogenic fungi against 3rd- and 5th-instar nymphs and adults (two weeks after last molt) of *S. gregaria* at the Department of Entomology, University of Agriculture Faisalabad, Punjab (Pakistan). Different doses of each fungal isolate were prepared in 0.05% Silwet L-77 solution. Both nymphal instars and adults were treated with a 1-mL suspension of entomopathogenic fungi using a 35-mL glass atomizer [51]. Control individuals were treated in the same fashion, but with 1 mL of 0.05% Silwet L-77. After application, individuals were transferred to (30 cm × 30 cm × 30 cm) cages and provided with wheat bran and wheat seedlings grown in pots [16]. Each dose of specific formulations/isolates represents a treatment, each treatment consisted of three replicates (total 45 individuals, 15/replication), and the experiment was conducted twice (using new individuals and new materials), giving a total of six replicates [16,28]. Mortality was determined at 24-h intervals, with the final count at 12 days post-application [52]. The environmental conditions were maintained at a 14:10 h (light:dark) photoperiod, 30 °C, and 60% relative humidity (RH) [51].

2.4. Effect of a Sublethal Dose on the Reproduction and Development of *S. gregaria*

The effect of a sublethal dose (1×10^4 conidia/mL) of each fungal isolate was examined against adults of *S. gregaria*. For this, a group of 30 adults (15 females and 15 males) was treated as described in the previous section. The cages were provided with 25-cm-diameter plastic cups filled with clean sand, and were provided with appropriate moisture to encourage egg laying. The total number of egg pods laid by each female, average number of eggs present per pod, and percentage of eggs hatching was determined. The egg pods were collected at the end of experiment from among the different treatments, and mean egg pods per female were determined using the number of females still alive on the day of first oviposition. The egg pods were dissected and the total number of eggs per pod assessed. Then, eggs were kept under optimal conditions of 27 °C, 60% RH, and total egg hatching was determined among the different treatments [53]. After egg hatching, a group of 30 first-instar nymphs was collected from each replicate and kept inside the cages (30 cm × 30 cm × 30 cm) and provided with fresh diet (wheat seedlings and wheat bran). The developmental duration of different stages (1st-, 2nd-, 3rd-, 4th-, and 5th-instar nymphs, and adults) was determined. Each treatment consisted of three replicates, and the whole experiment was repeated twice, with new individuals and materials for each repetition.

2.5. Effects of a Sublethal Dose on Diet Consumption, Weight Gain, and Frass Production of *S. gregaria*

The effects of a sublethal dosage (1×10^4 conidia/mL) on diet consumption, weight gain, and frass production of *S. gregaria* were determined. Fifth-instar nymphs were treated with a sublethal dose of EPF in a similar manner as in the above-mentioned bioassays. Individual 5th-instar nymphs were weighed and then transferred into a small plastic box and provided with a wheat seedling for feeding. Prior to feeding, diet weight was measured on weight balance and then transferred inside the small boxes for feeding. Daily unused diet was taken from the boxes and measured on the weight balance, and diet consumed was calculated by subtracting the remaining diet from the initial diet. During days 1–15 after inoculation, fecal pellets were collected daily from each cage (replicate). The fecal production per cage was oven-dried at 80 °C until constant weight (24 h) and then weighed. The fecal production per insect per day for each replicate was calculated by dividing the daily fecal production by the number of insects remaining alive on that date [34]. Insects were also weighed each day. Each treatment consisted of three replicates, with each replication containing 15 individuals (total 45 individuals per treatment), and the experiment was repeated twice, giving a total of six replicates.

2.6. Greenhouse Trial

The greenhouse experiment was conducted using nymphs (3rd- and 5th-instar) and adults of *S. gregaria*. The 3rd- and 5th-instar nymphs and adults were inoculated in a similar manner as in the laboratory bioassay, with 1×10^9 conidia/mL. After inoculation, each individual of each specific stage was released inside a plastic pot (12.5 cm H × 8 cm D) that contained wheat seedlings and covered with a cylindrical metal cage (30 cm H × 8 cm D). All pots with cages were kept inside trays provided with water and placed on the bench top at 1-m height. Each treatment consisted of three replicates, with 20 pots per replicate (for each stage), and each pot contained one individual of a specific stage. The experiment was repeated twice, giving a total of six replicates. Adult mortality was observed daily for 15 days [52]. In order to avoid scavenging by live locusts, their frass and dead were removed from the cages daily. Conditions were maintained at 30 °C, 16:8 h (light:dark) photoperiod, and 60% RH.

2.7. Field Trial

The effectiveness of different fungal isolates against 3rd- and 5th-instar nymphs and adults of *S. gregaria* was evaluated in the field cages (70 cm × 70 cm × 70 cm). These wooden cages with a door on the front were placed in the field, and 120 individuals of each stage were placed in each cage [51]. Each wooden cage served as replicate, and each treatment consisted of three replicates. As is common for field applications for Green Muscle [18,26], conidia were mixed with diesel and applied at a dose of 1×10^9 conidia/mL using a knapsack sprayer. The experiment consisted of five treatments, with three replicates (three cages) in each treatment. The whole experiment was repeated twice [26], with new individuals and materials each time. The data on the locust mortality were taken daily by removing the dead from each field cage. Untreated wheat seedlings were provided in the cages in the late afternoon so that the locusts had fresh food all night. Sometimes there was mortality in the first two days due to transport and handling; to compensate for this, any mortality that occurred in the first two days, in both the treated and untreated cages, was ignored. The experiment endpoint was day 15 after the introduction of insects and treatment application.

2.8. Statistical Analysis

In the laboratory trials, the mortality among the different developmental stages (3rd- and 5th-instar nymphs, and adults) was corrected with control mortality using Abbott's formula [54]. The values were subjected to two-way analysis of variance (ANOVA), and treatment means were separated by Tukey's honestly significant difference (HSD) test [55]. The mortality among different stages and intervals was found to be less than 5%. The effects on reproduction, diet consumption, frass production, and weight gain were subjected to analysis of variance (ANOVA). The mortality in greenhouse and field trials was subjected to analysis of variance (ANOVA), and treatment means were separated by Tukey's HSD test. All of the statistical analysis was conducted using Minitab 17 software [56].

3. Results

3.1. Bioassay against 3rd- and 5th-Instar Nymphs and Adults of S. gregaria

Against 3rd-instar nymphs, significant differences were observed between different EPF within each dose at 5 ($F_{3,23} \leq 6.05$; $p < 0.01$), 7 ($F_{3,23} \leq 7.92$; $p < 0.01$), and 12 ($F_{3,23} \leq 7.31$; $p < 0.01$) days post-treatment. Within each EPF, a significant dose response was observed ($F_{3,23} \leq 13.6$; $p < 0.01$ at 5 days; $F_{3,23} \leq 14.3$; $p < 0.01$ at 7 days; $F_{3,23} \leq 17.2$; $p < 0.01$ at 12 days) post-treatment. The mortality of the EPF at the 10^6 dose at 12 days ranged from 46.5 to 67.5%, while the 10^8 dose resulted in 65–95% mortality. The dose response was such that increasing the dose 100-fold (from 10^6 to 10^8) meant mortality was reached 5 days earlier. Green Muscle produced mortality of 68% at the 10^6 dose after 12 days, while similar mortality was observed at day 7 for the 10^8 dose (Figure 1a–c). For *M. anisopliae*, there was 47% mortality at 12 days at the 10^6 dose, and at 7 days at the 10^8 dose (Figure 1b,c). The time taken to reach > 50% mortality was 5 days after a 10^8 dose of Green Muscle, 7 days after a 10^7 dose of Green Guard, 7 days after a 10^8 dose of *B. bassiana*, and 12 days after a 10^7 dose of *M. anisopliae* was applied. By the end of the experiment on day 12, maximum mortality was observed with Green Muscle, followed Green Guard, *B. bassiana*, and *M. anisopliae*. Overall, mortality was greatest with Green Muscle and least with *M. anisopliae* at all time intervals and at all doses (Figure 1a–c).

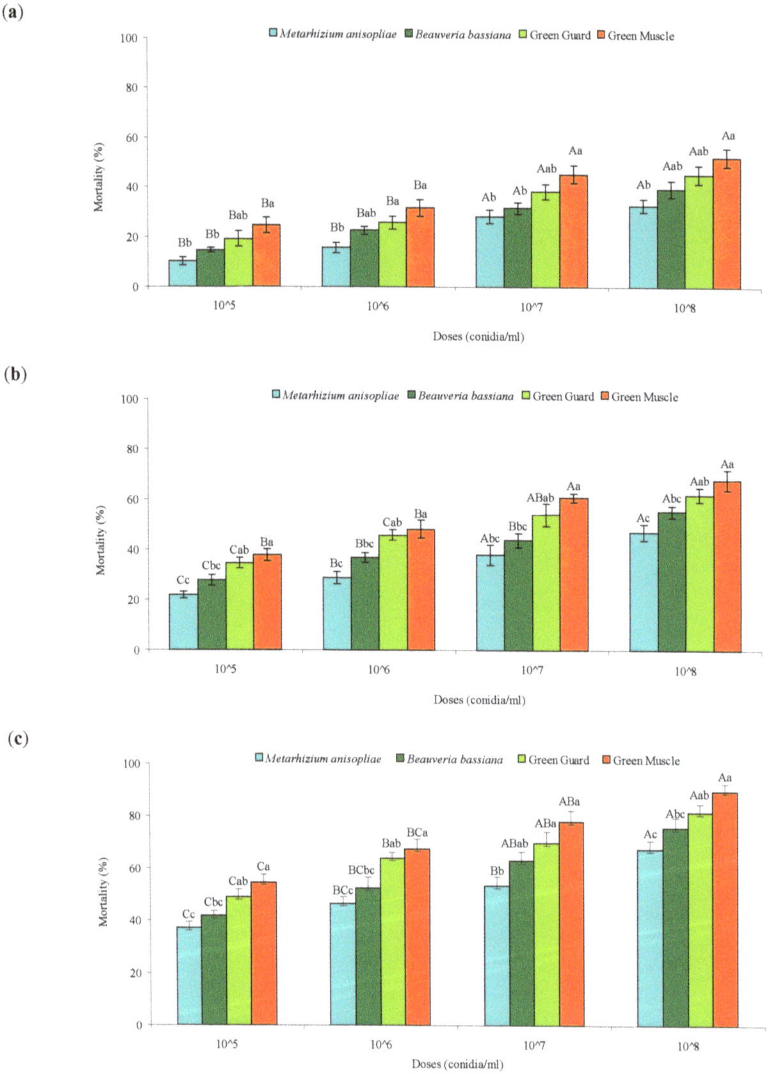

Figure 1. Mean mortality percentages (±SE) of 3rd-instar nymphs of *S. gregaria* after (**a**) 5 days, (**b**) 7 days, and (**c**) 12 days of treatment with various doses (10^5, 10^6, 10^7, and 10^8 conidia/mL) of Green Muscle, Green Guard, *B. bassiana*, or *M. anisopliae*. Mortality data were corrected for control mortality, which was <5%. For each subfigure: different lowercase letters indicate significant differences in mortality between each of the EPF; different uppercase letters indicate significant differences in mortality between doses (Tukey's HSD test at $p \leq 0.05$).

Clear treatment effects were observed against 5th-instar nymphs at different dose rates at 5 ($F_{3,23} \leq 15.1$; $p < 0.01$), 7 ($F_{3,23} \leq 22.4$; $p < 0.01$), and 12 ($F_{3,23} \leq 17.0$; $p < 0.01$) days post-treatment. All of the EPFs had a significant dose response ($F_{3,23} \leq 4.28$; $p < 0.01$ at 5 days; $F_{3,23} \leq 5.92$; $p < 0.01$ at 7 days; $F_{3,23} \leq 3.88$; $p < 0.01$ at 12 days). The dose response was such that increasing the dose 100-fold (from 10^6 to 10^8) meant mortality was reached just under 5 days earlier. The mortality due to the EPF at the 10^6 dose at 12 days ranged from 37.1 to 54.8%. For Green Muscle; there was 55% mortality at 12 days at a dose of 10^6, similar to the 61% mortality at 7 days at a dose of 10^8 (Figure 2b,c). Until day 5, no EPF was able to kill 50% of 5th-instar nymphs, while at 7 days of exposure only two EPF (Green Muscle and Green Guard) caused > 50% mortality at a level not statistically different from one another, with the greatest mortality from Green Muscle (Figure 2a–c).

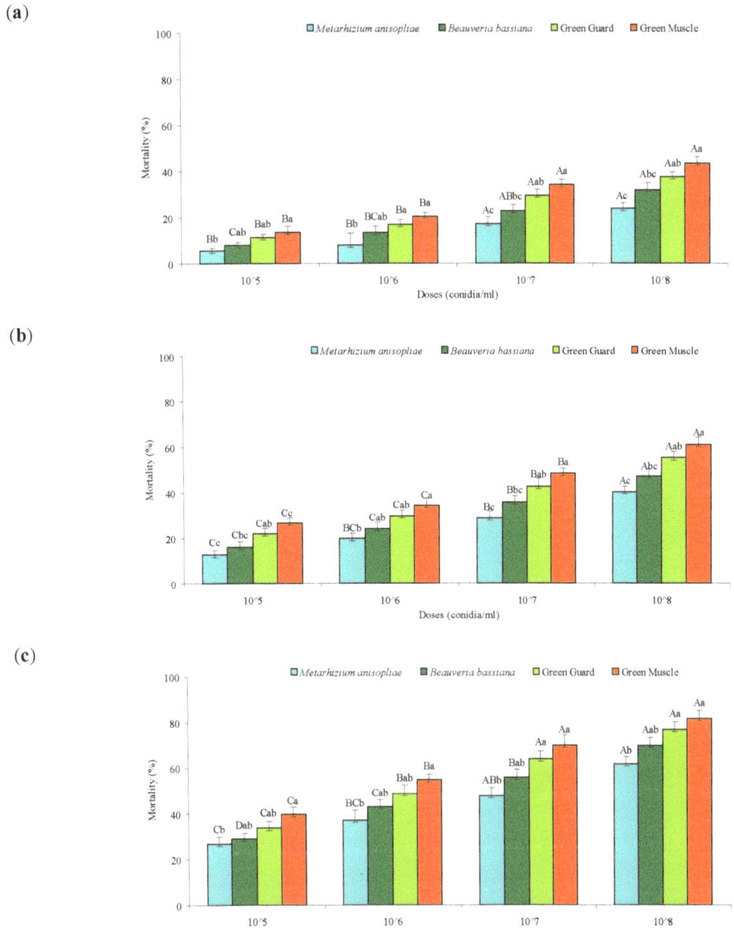

Figure 2. Mean mortality percentages (±SE) of 5th-instar nymphs of *S. gregaria* after (**a**) 5 days, (**b**) 7 days, and (**c**) 12 days of treatment with various doses (10^5, 10^6, 10^7, and 10^8 conidia/mL) of Green Muscle, Green Guard, *B. bassiana*, or *M. anisopliae*. Mortality data were corrected for control mortality, which was <5%. For each subfigure: different lowercase letters indicate significant differences in mortality between each of the EPF; different uppercase letters indicate significant differences in mortality between doses (Tukey's HSD test at $p \leq 0.05$).

Against adults, each dose rate of EPF produced significantly more mortality compared to the control treatment ($F_{3,23} \leq 7.23$; $p < 0.01$), ($F_{3,23} \leq 5.53$; $p < 0.01$), and ($F_{3,23} \leq 3.17$; $p < 0.01$) at 5, 7, and 12 days after treatment, respectively. A clear dose response in terms of mortality was observed in different EPF at ($F_{3,23} \leq 16.6$; $p < 0.01$) 5 days, ($F_{3,23} \leq 17.1$; $p < 0.01$) 7 days, and ($F_{3,23} \leq 20.6$; $p < 0.01$) 12 days post-application. Once again, the dose response was such that increasing the dose 100-fold (from 10^6 to 10^8) resulted in mortality being reached just under 5 days earlier. The mortality of the EPF at the 10^6 dose at 12 days ranged from 49% for Green Muscle to 34% for *M. anisopliae*—slightly less than the levels reached at 7 days with a 10^8 dose. No EPF caused 50% mortality at 5 days, and even at 7 days, only two EPF caused 50% mortality, and even then only at the highest conidial dose. After 12 days, all EPF caused > 50% mortality at the two highest conidial doses. As with treatments of 3rd- and 5th-instar nymphs, the greatest mortality was observed with Green Muscle, followed by Green Guard, *B. bassiana*, and *M. anisopliae* (Figure 3a–c).

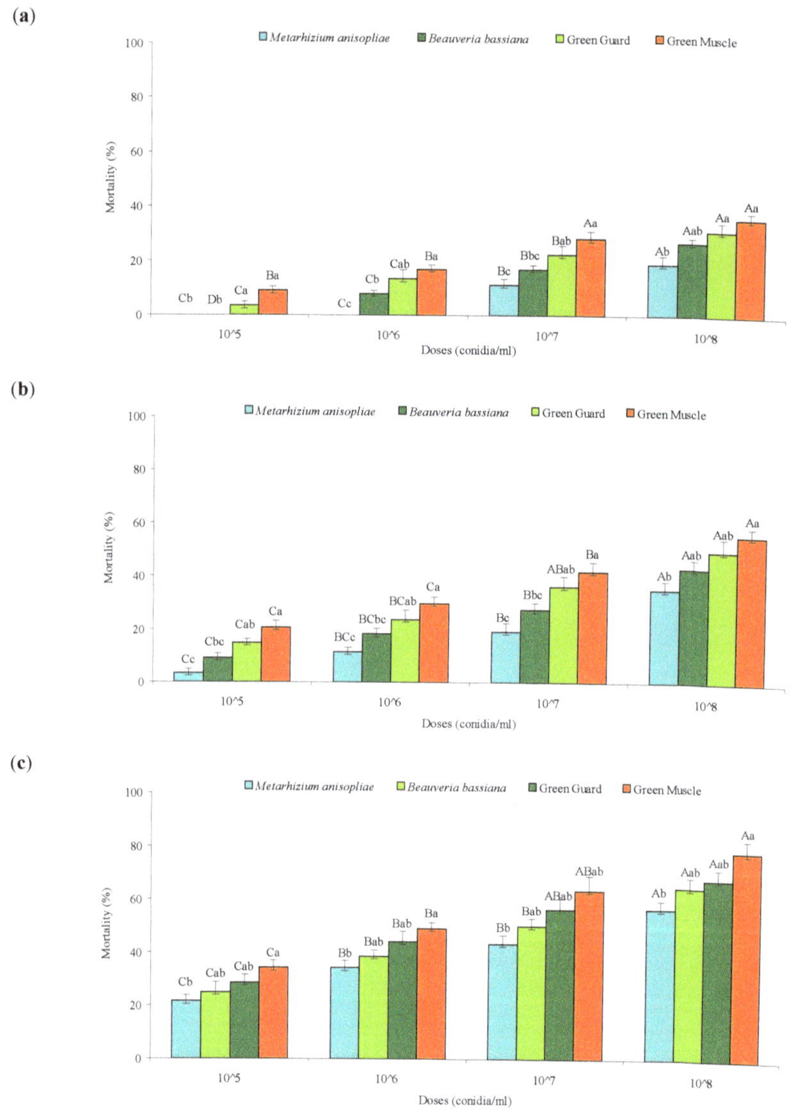

Figure 3. Mean mortality percentages (±SE) of *S. gregaria* adults after (**a**) 5 days, (**b**) 7 days, and (**c**) 12 days of treatment with various doses (10^5, 10^6, 10^7, and 10^8 conidia/mL) of Green Muscle, Green Guard, *B. bassiana*, or *M. anisopliae*. Mortality data were corrected for control mortality, which was <5%. For each subfigure: different lowercase letters indicate significant differences in mortality between each of the EPF; different uppercase letters indicate significant differences in mortality between doses (Tukey's HSD test at $p \leq 0.05$).

Overall, 3rd-instar nymphs were found to be more susceptible than 5th-instar nymphs or adults.

3.2. Effects of Sublethal Doses on the Reproduction and Development of *S. gregaria*

Sublethal doses of EPF seriously affected the reproduction capability of *S. gregaria*, and also retarded the growth of their offspring. The females treated with all EPF except *M. anisopliae* produced fewer egg pods compared to the untreated controls ($F_{4,29}$ = 28.1; p < 0.01), with no significant difference between the Green Muscle and Green Guard treatments (Table 1). In addition, there were fewer eggs per pod with EPF treatments ($F_{4,29}$ = 73.1; p < 0.01), with Green Muscle treatments leading to the lowest number of eggs per pod (Table 1). Of the eggs laid, hatching was lowest with the Green Muscle and Green Guard treatments, with both *B. bassiana* and *M. anisopliae* resulting in intermediate levels between the other EPF and the controls. Even though the offspring themselves were uninfected, the offspring of infected females had an increase in nymphal developmental time ($F_{4,29}$ = 7.20; p < 0.01) and a reduction in adult longevity compared to the untreated controls ($F_{4,29}$ = 41.2; p < 0.01).

Table 1. Effects of applying one sublethal dose (1 × 10^4 conidia/mL) of Green Muscle, Green Guard, *B. bassiana*, or *M. anisopliae* to *S. gregaria* adults on the number of egg pods/female, number of eggs/pod, egg hatching, development time for offspring of different instars, and survival of resulting adults. Data are means ± SE, and within each column and treatment, different letters indicate significant differences (Tukey's HSD test at $p \le 0.05$).

Treatment	No. of Egg Pods/Female	Eggs/Pod	Egg Hatching (%)	Development Time (days)					
				L1	L2	L3	L4	L5	Adult
Green Muscle	0.76 ± 0.11 c	23.13 ± 1.75 d	34.44 ± 4.99 c	8.63 ± 0.32 a	7.31 ± 0.17 a	7.51 ± 0.20 a	7.90 ± 0.26 a	10.21 ± 0.42 a	18.65 ± 1.17 d
Green Guard	1.06 ± 0.12 bc	34.54 ± 1.54 c	45.33 ± 4.89 bc	7.96 ± 0.23 ab	6.86 ± 0.20 ab	6.96 ± 0.21 ab	7.48 ± 0.21 ab	9.64 ± 0.32 ab	27.30 ± 2.16 cd
B. bassiana	1.41 ± 0.11 b	40.58 ± 1.34 bc	53.33 ± 3.56 b	7.22 ± 0.17 bc	6.20 ± 0.17 bc	6.71 ± 0.17 b	7.11 ± 0.20 abc	9.25 ± 0.27 abc	33.20 ± 1.81 bc
M. anisopliae	2.01 ± 0.16 a	47.78 ± 2.29 b	61.11 ± 3.18 b	6.46 ± 0.19 cd	5.60 ± 0.16 cd	5.85 ± 0.15 c	6.76 ± 0.28 bc	8.41 ± 0.28 bc	38.71 ± 2.59 b
Control	2.56 ± 0.15 a	67.36 ± 2.44 a	85.55 ± 4.00 a	6.04 ± 0.16 d	5.08 ± 0.11 d	5.01 ± 0.13 d	6.20 ± 0.21 c	8.06 ± 0.28 c	55.95 ± 2.75 a
F	28.3	73.1	21.6	22.1	28.0	30.7	7.20	7.30	41.2
p	<0.01	<0.01	<0.01	<0.01	<0.01	<0.01	<0.01	<0.01	<0.01

3.3. Effects of a Sublethal Dose on Diet Consumption, Weight Gain, and Frass Production of *S. gregaria*

Sublethal doses induced significant behavioral changes in fifth-instar *S. gregaria*, with lower diet consumption, frass production, and weight compared to controls; most effects were already seen at 3 days after infection, and were clearly evident by 6 days (Figure 4a–c). As the days progressed, diet consumption declined in the treated groups ($F_{4,29} \le 19.6$; p < 0.01), but increased in the untreated controls ($F_{4,29} \le 6.51$; p < 0.01). Except for day 12, no significant difference was observed in diet consumption between Green Muscle and Green Guard. A similar trend was observed in frass production, with significantly lower frass production in the treated groups compared to the controls ($F_{4,29} \le 104$; p < 0.01). Except for day 12, Green Muscle led to significantly lower frass production compared to the rest of the treatments. As days progressed, there was a decline in weight gain per day by treated locusts, while weight gain increased in the controls. Except at days 6 and 9, the effects of Green Muscle were not statistically significantly different from those of Green Guard (Figure 4a–c).

3.4. Greenhouse Trial

In the greenhouse trial, mortality in the controls was very low, and was <10% even at 15 days post-application, while mortality from all of the EPF was much higher (Figure 5). Green Muscle caused the highest mortality of third- and fifth-instar nymphs, but for adults, mortality was similar for Green Muscle and Green Guard. Mortality was less for the other two EPF, with mortality from *M. anisopliae* the lowest.

(a) (b) (c)

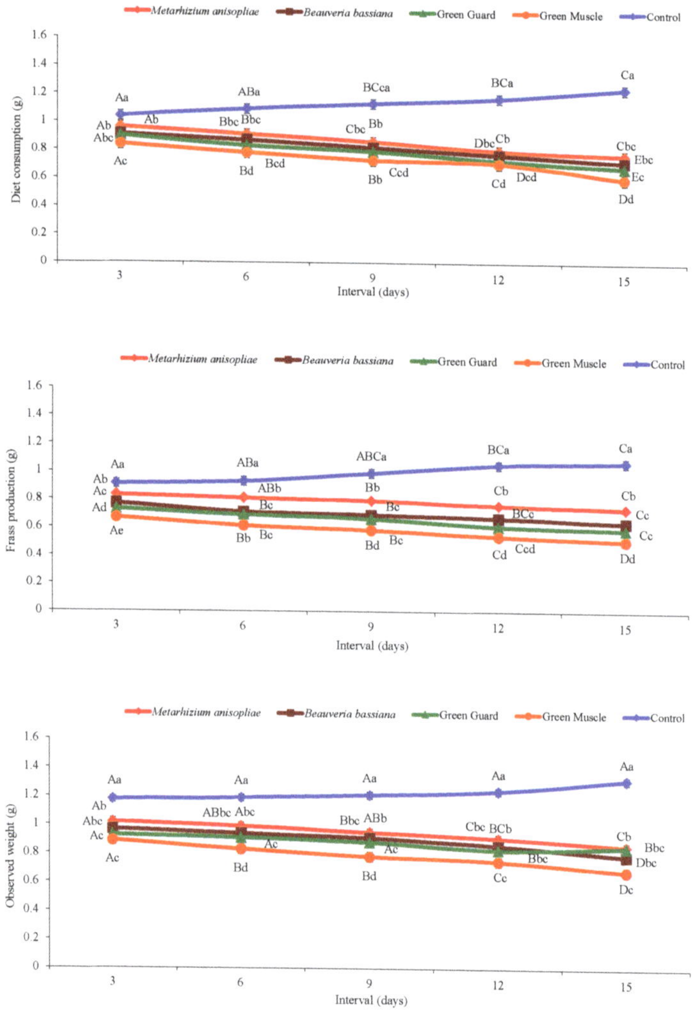

Figure 4. Means (±SE) for (**a**) diet consumption, (**b**) frass production, and (**c**) observed weight of fifth-instar *S. gregaria* treated with a sublethal dose (1×10^4 conidia/mL) of Green Muscle, Green Guard, *B. bassiana*, or *M. anisopliae*. For each subfigure: different lowercase letters indicate significant differences between different EPF; different uppercase letters indicate significant differences between different days (Tukey's HSD test at $p \leq 0.05$).

As with the laboratory assay, the greenhouse bioassay demonstrated that 3rd-instar nymphs were most susceptible, followed by 5th-instar nymphs, and then adults (Figure 5a–c).

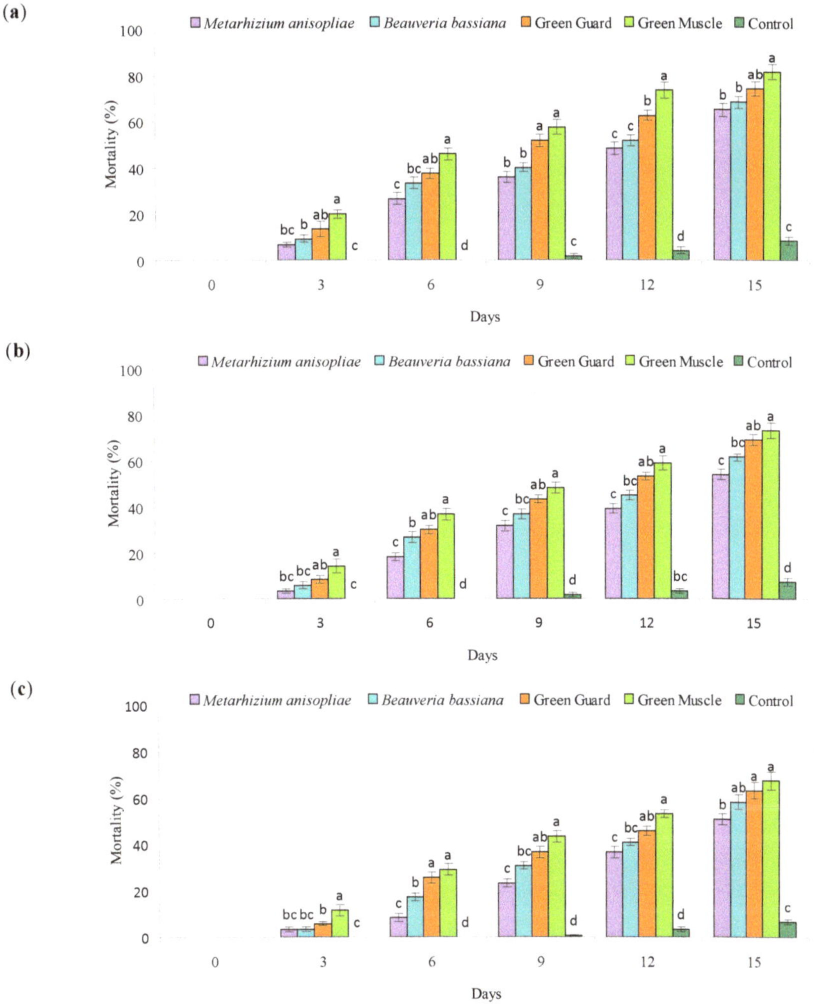

Figure 5. Mean mortality percentages (±SE) of (**a**) 3rd-instar, (**b**) 5th-instar, and (**c**) adult *S. gregaria* inoculated with 1×10^9 conidia/mL of Green Muscle, Green Guard, *B. bassiana*, or *M. anisopliae* in a greenhouse. Within each subfigure and between the treatments, different letters indicate significant differences (Tukey's HSD test at $p \leq 0.05$).

3.5. Field Trial

For EPF sprayed with a knapsack sprayer in field cages, all of the EPF produced significantly higher mortality ($F_{4,29} \leq 49.6$; $p < 0.01$ in 3rd-instar nymphs; $F_{4,29} \leq 69.9$; $p < 0.01$ in 5th-instar nymphs; and $F_{4,29} \leq 133.0$; $p < 0.01$ in adults) compared to untreated controls. Green Muscle caused the greatest mortality among all developmental stages, with mortality from Green Guard statistically slightly lower. As with the other assays, the field bioassay demonstrated that 3rd-instar nymphs were the most susceptible (Figure 6a–c).

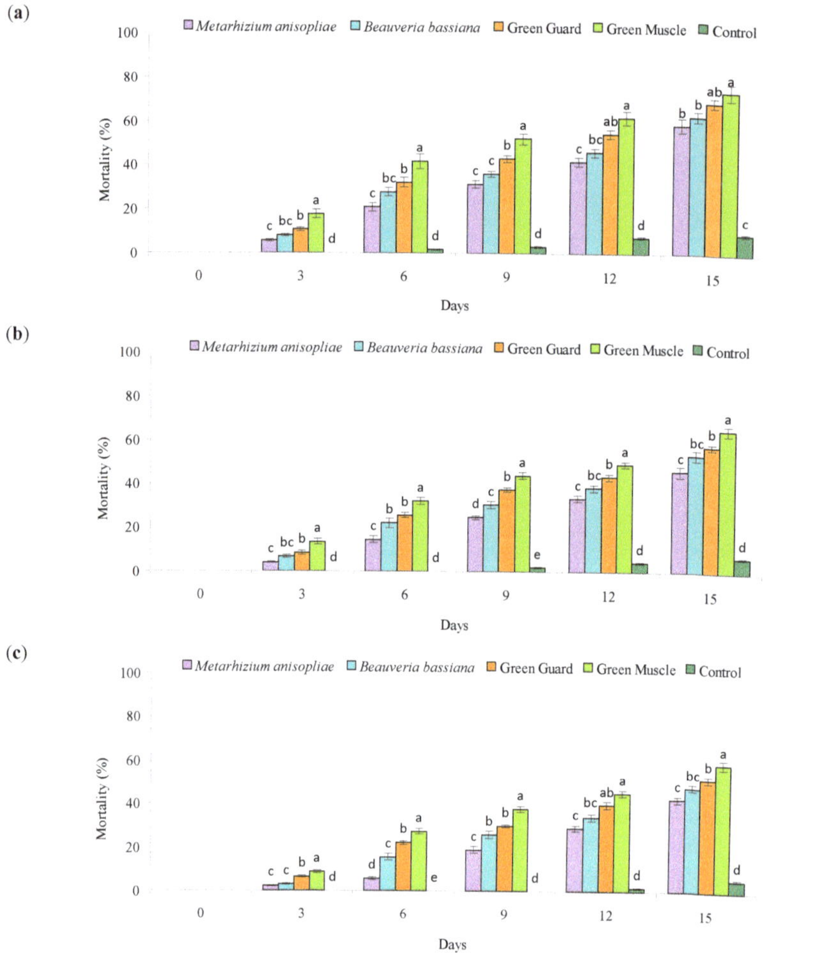

Figure 6. Mean mortality percentages (±SE) of (**a**) 3rd-instar, (**b**) 5th-instar and (**c**) adult *S. gregaria* sprayed with 1×10^9 conidia/mL of Green Muscle, Green Guard, *B. bassiana*, or *M. anisopliae* using a knapsack sprayer in field cages. Within each subfigure and between the treatments, different letters indicate significant differences (Tukey's HSD test at $p \leq 0.05$).

4. Discussion

In the present study, laboratory, greenhouse, and field trial results showed that all of the formulated products (Green Muscle, Green Guard, and *M. anisopliae* (CQMCC No. 0877)) and the local isolate of *B. bassiana* caused significant mortality of *S. gregaria*. Green Muscle performed best, with the greatest mortality, followed by Green Guard, *B. bassiana*, and *M. anisopliae*. Third-instar nymphs were more susceptible than fifth-instar nymphs or adults. Experiments in greenhouse and field conditions showed that all of the tested EPF have the potential to reduce pest populations under natural conditions. Previous studies showed significant efficacy of Green Muscle, Green Guard, and *B. bassiana* against different species of locust and grasshopper, and the present study showed that all of the EPF caused significant mortality of various stages of *S. gregaria*. As in our studies, 3rd-instar nymphs were found to be more susceptible to EPF by both Bashir and El Shafie [30], who treated desert locusts of various stages with *M. acridum*, and Youssef et al. [31], who treated desert locusts with *B. bassiana* and *Entomophthora* sp. Additive and synergistic interaction was observed when lower and higher doses of *M. acridum* and *Paranosema locustae* were tested against 5th-instar *S. gregaria* nymphs under laboratory conditions [28].

The current study is the first report on sublethal effects of EPF on the reproduction ability, fertility, and subsequent developmental stages of *S. gregaria*. Even though the applied dose was 1/10th of a dose causing 40% mortality, and 1/10,000th of the dose causing 80% mortality, infected females produced fewer egg pods and fewer eggs per pod, and fewer of the eggs that were produced actually hatched. Increased developmental duration was observed in offspring nymphs, while reduction in adult longevity was also detected. Effects on reproduction have been found in the past, but these were at a dose that was eventually lethal. Fewer egg pods per female and a lower total number of eggs per pod were observed in *S. gregaria* and in the lamenting grasshopper *Eyprepocnemis plorans* (Charpentier) when infected with *B. bassiana* and *M. acridum* [39], and in *Dichroplus maculipennis* (Blanchard) and *Ronderosia bergi* (Stål) infected with *B. bassiana* [40]. A 21–53% reduction in egg pods per female was observed for Moroccan locusts *Dociostaurus maroccanus* (Thunberg) (Orthoptera: Acrididae) infected with *M. acridum* (IMI 380189) and *B. bassiana* [53]. Fungal infection seems to utilize host resources to such an extent that there are less resources available for egg maturation [33,57]. Contrary to our findings was a report by Blanford and Thomas [27], who did not observe any sublethal effect of *M. acridum* (IMI 330189) on *S. gregaria* reproduction. Furthermore, we have for the first time detected the effect of EPF on the fertility or hatching of eggs of this species. Lower egg fertility was observed in *D. maroccanus* when infected with fungal isolates [53].

Sublethal EPF infections not only reduced reproductive capability, but also led to behavioral changes, including reduction in food consumption, reduced fecal production, and reduced weight. Upon infection with EPF, reduction in feeding was observed, similar to the observations of Moore et al. [35] of a reduction in diet consumption in both *S. gregaria* and the Central American locust *S. piceifrons* starting three days post-infection. Likewise, Mohammadbeigi and Port [34] observed a reduction in the diet consumption of *Uvarovistia zebra* (Uvarov) (Orthoptera: Tettigoniidae) infected with fungi. In the present study, infected individuals had lower fecal production compared to controls. A reduction in fecal production in *D. maroccanus* was observed from all of the fungal treatments except EABb 90/2-Dm [53]. Reduction in weight was observed among the different EPF compared to controls but, contrary to our findings, a previous study [30] reported no significant difference in the weight of controls and individuals treated with neem oil and *M. acridum*.

In this study, we observed significant efficacy of EPF not only in the laboratory, but also in the greenhouse and field experiments, despite several environmental factors that may restrict the efficacy of these pathogens under natural conditions [47]. A most important factor is ultraviolet (UV) radiation, which directly reduces pathogen survival, and also enables locusts to increase their body temperature and produce immunity against fungal infection [58]. The current study revealed that in spite of thermoregulation, all of the EPF have the ability to cause mortality, indicating their potential as biocontrol agents in the field. Our results are consistent with [15], where it was found that Green Muscle was effective in controlling locust adults, with some added advantages compared to chemicals. As with the greenhouse experiment, in field conditions these results show that EPF have great efficacy against *S. gregaria*. Future research will focus on investigating the compatibility of EPF with other control agents, including entomopathogenic nematodes, chemical insecticides, and botanical extracts, as these have been found to be effective against a variety of insect pests [30,49,50,59–61].

5. Conclusions

The present research shows that Green Muscle, Green Guard, *B. bassiana*, and *M. anisopliae* have the potential to control locusts under field conditions. In addition, there are sublethal effects on their reproductive ability and behavior, with reduction in diet consumption, frass production, and weight observed. Future research needs to explore the combination of EPF with other control agents—including chemical insecticides, microsporidia, and botanical extracts—as a means of integrated pest management of locusts in the field.

Author Contributions: Conceptualization, W.W.; research design, W.W.; methodology, W.W., M.U.G., D.H. and W.S.; data curation, W.W.; formal analysis, W.W. and M.U.; performed research, W.W. and M.U.G.; visualization, W.W.; writing—review and editing, W.W., M.U.G., D.H. and W.S.; supervision, W.W.; funding acquisition, W.W. All authors have read and agreed to the published version of the manuscript.

Funding: This study was supported by the National Disaster Management Authority (NDMA), Government of Pakistan, Islamabad, Pakistan.

Institutional Review Board Statement: Not applicable.

Informed Consent Statement: Not applicable.

Data Availability Statement: Data are contained within the article.

Acknowledgments: The authors are thankful to Atif Kamal and Murtaza Quddusi (BASF Pakistan Pvt. Ltd.) for providing Green Guard, Yann Miège (Elephant Vert, Morocco) for Green Muscle, and Zhongkang Wang (Chongqing Chongda Bio-Tech Development Ltd. Co., China) for *M. anisopliae* (CQMCC No. 0877). We are thankful to Tariq Butt (Swansea University, Swansea, UK) and Syed Farooq Shah (Razbio Ltd., UK) for their technical support in this study. This study was financially supported by the National Disaster Management Authority (NDMA), Government of Pakistan, Islamabad, Pakistan.

Conflicts of Interest: The authors declare no conflict of interest.

References

1. Millist, N.; Abdalla, A. Benefit-cost analysis of Australian plague locust control operations for 2010–11. In *ABARES Report Prepared for the Australian Plague Locust Commission*; Australian Bureau of Agricultural and Resource Economics and Sciences: Canberra, Australia, 2011; p. 22.
2. Latchininsky, A.V. Locusts and remote sensing: A review. *J. Appl. Rem. Sens.* **2013**, *7*, 075099. [CrossRef]
3. Zhang, L.; Hunter, D.M. Management of locusts and grasshoppers in China. *J. Orthop. Res.* **2017**, *26*, 155–159. [CrossRef]
4. Steedman, A. *The Locust Handbook*; Overseas Development National Resources Institute: London, UK, 1988.
5. Showler, A.T. Desert locust, *Schistocerca gregaria* (Forskål) (Orthoptera: Acrididae), campaign in Tunisia, 1988. *Agricul. Syst.* **1993**, *42*, 311–325. [CrossRef]
6. Zhang, L.; Lecoq, M.; Latchininsky, A.; Hunter, D. Locust and grasshopper management. *Ann. Rev. Entomol.* **2019**, *64*, 15–34. [CrossRef]
7. Anonymous. Locust alarm, scientific rigour and a debate over risky disease experiments. *Nature* **2020**, *577*, 602–603. [CrossRef]
8. Sharmila, D. Locust swarms in east Africa could be "a catastrophe". *Lancet* **2020**, *395*, 547.
9. Peng, W.; Ma, N.L.; Zhang, D.; Zhou, Q.; Shing, X.Y.; Khoo, C.; Yang, H.; Guan, R.; Chen, H.; Zhang, X.; et al. A review of historical and recent locust outbreaks: Links to global warming, food security and mitigation strategies. *Environ. Res.* **2020**, *191*, 110046. [CrossRef] [PubMed]
10. Le Gall, M.; Overson, R.; Cease, A. A global review on locusts (Orthoptera: Acrididae) and their interactions with livestock grazing practices. *Front. Ecol. Evo.* **2019**, *7*, 263. [CrossRef]
11. Shrestha, S.; Thakur, G.; Gautam, J.; Acharya, N.; Pandey, M.; Shrestha, J. Desert locust and its management in Nepal: A review. *J. Agricul. Nat. Res.* **2021**, *4*, 1–28.
12. Symmons, P.M.; Cressman, K. Desert Locust Guidelines: Biology and Behaviour. Food and Agriculture Organization of the United Nations: Rome, Italy, 2001; pp. 1–25. Available online: http://www.fao.org/ag/locusts/oldsite/PDFs/DLG1e.pdf (accessed on 16 August 2021).
13. FAO. *Weather and Desert Locusts WMO-No. 1175*; World Meteorological Organization and Food and Agriculture Organization of the United Nations: Rome, Italy, 2016; pp. 1–48. Available online: http://www.fao.org/3/i6152en/i6152en.pdf (accessed on 22 August 2021).
14. FAO. *Desert Locust Bulletin No. 484*; Food and Agriculture Organization of the United Nations: Rome, Italy, 2019. Available online: http://www.fao.org/ag/locusts/common/ecg/2462/en/DL484e.pdf (accessed on 20 August 2021).
15. Mullié, W.C.; Cheke, R.A.; Young, S.; Ibrahim, A.B.; Murk, A.J. Increased and sex-selective avian predation of desert locusts *Schistocerca gregaria* treated with *Metarhizium acridum*. *PLoS ONE* **2021**, *16*, e0244733. [CrossRef]
16. Abdelatti, Z.A.S.; Hartbauer, M. Plant oil mixtures as a novel botanical pesticide to control gregarious locusts. *J. Pest. Sci.* **2020**, *93*, 341–353. [CrossRef]
17. Arthurs, S. Grasshoppers and locusts as agricultural pests. In *Encyclopedia of Entomology*; Capinera, J.L., Ed.; Springer: Dordrecht, The Netherlands, 2008; pp. 1690–1694.

18. Lecoq, M. Integrated pest management for locusts and grasshoppers: Are alternatives to chemical pesticides credible? *J. Orthop. Res.* **2010**, *19*, 131–132. [CrossRef]
19. Alavanja, M.C.R.; Bonner, M.R. Occupational pesticide exposures and cancer risk: A review. *J. Toxicol. Environ. Health* **2012**, *15*, 238–263. [CrossRef]
20. Köhler, H.R.; Triebskorn, R. Wildlife ecotoxicology of pesticides: Can we track effects to the population level and beyond? *Science* **2013**, *341*, 759–765. [CrossRef] [PubMed]
21. Nicolopoulou-Stamati, P.; Maipas, S.; Kotampasi, C.; Stamatis, P.; Hens, L. Chemical pesticides and human health: The urgent need for a new concept in agriculture. *Front. Public Health* **2016**, *4*, 1–8. [CrossRef] [PubMed]
22. Roussi, A. The battle to contain gigantic locust swarms. *Nature* **2020**, *579*, 330. [CrossRef]
23. Lacey, L.A.; Shapiro-Ilan, D.I. Microbial control of insect pests in temperate orchard systems: Potential for incorporation into IPM. *Annu. Rev. Entomol.* **2008**, *53*, 121–144. [CrossRef]
24. Lacey, L.A.; Grzywacz, D.; Shapiro-Ilan, D.I.; Frutos, R.; Brownbridge, M.; Goettel, M.S. Insect pathogens as biological control agents: Back to the future. *J. Invertebra. Pathol.* **2015**, *132*, 1–41. [CrossRef] [PubMed]
25. Goettel, M.S.; Eilenberg, J.; Glare, T.R. Entomopathogenic fungi and their role in regulation of insect populations. In *Comprehensive Molecular Insect Science*; Gilbert, L.I., Iatrou, K., Gill, S., Eds.; Elsevier: Amsterdam, The Netherlands, 2005; Volume 6, pp. 361–406.
26. Atheimine, M.; Bashir, M.; Ely, S.; Kane, C.; Mohamed, S.; Babah, M.; Benchekroun, M. Efficacy and persistence of *Metarhizium acridum* (Hypocreales: Clavicipitaceae) used against desert locust larvae, *Schistocerca gregaria* (Orthoptera: Acrididae), under different vegetation cover types. *Inter. J. Trop. Insect Sci.* **2014**, *34*, 106–114. [CrossRef]
27. Blanford, S.; Thomas, M.B. Adult survival, maturation, and reproduction of the desert locust *Schistocerca gregaria* infected with the fungus *Metarhizium anisopliae* var acridum. *J. Invertebra. Pathol.* **2001**, *78*, 1–8. [CrossRef]
28. Tounou, A.-K.; Kooyman, C.; Douro-Kpindou, O.-K.; Poehling, H.-M. Interaction between *Paranosema locustae* and *Metarhizium anisopliae var. acridum*, two pathogens of the desert locust, *Schistocerca gregaria* under laboratory conditions. *J. Invertebra. Pathol.* **2008**, *97*, 203–210. [CrossRef]
29. Milat-Bissaad, F.Z.; Bounaceur, F.; Halouane, F.; Behidj, N.; Chebouti, N.; Doumandji-Mitiche, B. Effect of entomopathogenic fungi *Beauveria bassiana* and *Metarhizium anisopliae* var *acridum* on the haemolymph of the desert locust *Schistocerca gregaria*. *Tunis. J. Plant Protect.* **2011**, *6*, 127–132.
30. Bashir, E.M.; El Shafie, H.A. Laboratory evaluation of the effects of neem (*Azadirachta indica*) oil and *Metarhizium anisopliae* against some immature stages of the desert locust *Schistocerca gregaria* (Forskal) (Orthoptera: Acrididae). *J. Agricul. Vet. Sci.* **2014**, *18*, 116–126.
31. Youssef, N.A. Effect of certain entomopathogenic fungi and nematode on the desert locust *Schistocerca gregaria* (Forskal). *Ann. Agricul. Sci.* **2014**, *59*, 125–131. [CrossRef]
32. Arthurs, S.; Thomas, M.B. Effects of a mycoinsecticide on feeding and fecundity of the brown locust *Locustana pardalina*. *Biocontrol Sci. Technol.* **2000**, *10*, 321–329. [CrossRef]
33. Roy, H.E.; Steinkraus, D.C.; Eilenberg, J.; Hajek, A.E.; Pell, J.K. Bizarre interactions and endgames: Entomopathogenic fungi and their arthropod hosts. *Annu. Rev. Entomol.* **2006**, *51*, 331–357. [CrossRef]
34. Mohammadbeigi, A.; Port, G. Effect of infection by *Beauveria bassiana* and *Metarhizium anisopliae* on the feeding of *Uvarovistia zebra*. *J. Insect. Sci.* **2015**, *15*, 88. [CrossRef]
35. Moore, D.; Reed, M.; Le Patourel, G.; Abraham, Y.J.; Prior, C. Reduction of feeding by the desert locust, *Schistocerca gregaria*, after infection with *Metarhizium flavoviride*. *J. Invertebr. Pathol.* **1992**, *60*, 304–307. [CrossRef]
36. Schmidt, F.G.V.; Conceição, P.J.; Benito, N.P.; Lopes, R.B. Susceptibility of three orthopteran species to infection by *Metarhizium acridum* (Hypocreales: Clavicipitaceae). *Int. J. Trop. Insect Sci.* **2018**, *38*, 1–5. [CrossRef]
37. Seyoum, E.; Moore, D.; Charnley, A.K. Reduction in flight activity and food consumption by the desert locust, *Schistocerca gregaria* Forskål (Orth., Cyrtacanthacrinae), after infection with *Metarhizium flavoviride*. *J. Appl. Entomol.* **1994**, *118*, 310–315. [CrossRef]
38. Thomas, M.B.; Blanford, S.; Lomer, C.J. Reduction of feeding by the variegated grasshopper, *Zonocerus variegatus*, following infection by the fungal pathogen, *Metarhizium flavoviride*. *Biocontrol Sci. Technol.* **1997**, *7*, 327–334. [CrossRef]
39. El-Maghraby, M.M.A.; Gomaa, E.A.; Metaweh, H.H.; Abdelatef, G.M. Susceptibility of *Schistocerca gregaria* (Forskål) and *Euprepocnemis plorans* (Charpentier) to *Metarhizium anisopliae* var *acridum* (Metchnikoff) Soroken, *Beauveria bassiana* (Bals.) Vuill. and *Nosema locustae* Canning. *Egypt. J. Biol. Pest. Control.* **2009**, *19*, 55–61.
40. Pelizza, S.A.; Mariottini, Y.; Russo, M.L.; Cabello, M.N.; Lange, C.E. Survival and fecundity of *Dichroplus maculipennis* and *Ronderosia bergi* (Orthoptera: Acrididae: Melanoplinae) following infection by *Beauveria bassiana* (Ascomycota: Hypocreales) under laboratory conditions. *Biocontrol Sci. Technol.* **2013**, *23*, 701–710. [CrossRef]
41. Blanford, S.; Shi, W.; Christian, R.; Marden, J.H.; Koekemoer, L.L.; Brooke, B.D.; Coetzee, M.; Read, A.F.; Thomas, M.B. Lethal and pre-lethal effects of a fungal biopesticide contribute to substantial and rapid control of malaria vectors. *PLoS ONE* **2011**, *6*, e23591. [CrossRef]
42. Anderson, R.D.; Blanford, S.; Thomas, M.B. House flies delay fungal infection by fevering: At a cost. *Ecol. Entomol.* **2013**, *38*, 1–10. [CrossRef]
43. Acharya, N.; Rajotte, E.G.; Jenkins, N.E.; Thomas, M.B. Potential for biocontrol of house flies, *Musca domestica*, using fungal biopesticides. *Biocontrol Sci. Technol.* **2015**, *25*, 513–524. [CrossRef]

44. Roxburgh, L.; Pinshow, B.; Prange, H.D. Temperature regulation by evaporative cooling in a desert grasshopper, *Calliptamus barbarous*. *J. Therm. Biol.* **1996**, *21*, 331–337. [CrossRef]
45. Ouedraogo, R.M.; Cusson, M.; Goettel, M.S.; Brodeur, J. Inhibition of fungal growth in thermoregulating locusts, *Locusta migratoria*, infected by the fungus *Metarhizium anisopliae* var acridum. *J. Invertebra. Pathol.* **2003**, *82*, 103–109. [CrossRef]
46. Ouedraogo, R.M.; Goettel, M.S.; Brodeur, J. Behavioral thermoregulation in the migratory locust: A therapy to overcome fungal infection. *Oecologia* **2004**, *138*, 312–319. [CrossRef]
47. Foster, R.N.; Jaronski, S.; Reuter, K.C.; Black, L.R.; Schlothauer, R.; Harper, J.; Jech, L.E. Simulated aerial sprays for field cage evaluation of *Beauveria bassiana* and *Metarhizium brunneum* (Ascomycetes: Hypocreales) against *Anabrus simplex* (Orthoptera: Tettigoniidae) in Montana. *Biocontrol Sci. Technol.* **2011**, *21*, 1331–1350. [CrossRef]
48. Inglis, G.D.; Enkerli, J.; Goettel, M.S. Laboratory techniques used for entomopathogenic fungi: Hypocreales. In *Manual of Techniques in Invertebrate Pathology*; Lacey, L.A., Ed.; Academic Press: London, UK, 2012; pp. 189–253.
49. Usman, M.; Gulzar, S.; Wakil, W.; Wu, S.; Piñero, J.C.; Leskey, T.C.; Nixon, L.J.; Oliveira-Hofman, C.; Toews, M.D.; Shapiro-Ilan, D. Virulence of entomopathogenic fungi to *Rhagoletis pomonella* (Diptera: Tephritidae) and interactions with entomopathogenic nematodes. *J. Econ. Entomol.* **2020**, *113*, 2627–2633. [CrossRef]
50. Gulzar, S.; Wakil, W.; Shapiro-Ilan, D.I. Combined effect of entomopathegns against *Thrips tabaci* Lindeman (Thysanoptera: Thripidae): Laboratory, greenhouse and field trials. *Insects* **2021**, *12*, 456. [CrossRef]
51. Pelizza, S.A.; Medina, H.; Ferreri, N.A.; Elíades, L.A.; Pocco, M.A.; Stenglein, S.A.; Lange, C.E. Virulence and enzymatic activity of three new isolates of *Beauveria bassiana* (Ascomycota: Hypocreales) from the South American locust *Schistocerca cancellata* (Orthoptera: Acrididae). *J. King Saud. Univer.* **2020**, *32*, 44–47. [CrossRef]
52. Dakhel, W.H.; Latchininsky, A.V.; Jaronski, S.T. Efficacy of two entomopathogenic fungi, *Metarhizium brunneum*, strain F52 alone and combined with *Paranosema locustae* against the migratory grasshopper, *Melanoplus sanguinipes*, under laboratory and greenhouse conditions. *Insects* **2019**, *10*, 94. [CrossRef]
53. Valverde-Garcia, P.; Santiago-Alvarez, C.; Thomas, M.B.; Maranhão, E.A.A.; Garrido-Jurado, I.; Quesada-Moraga, E. Sublethal effects of mixed fungal infections on the Moroccan locust, *Dociostaurus maroccanus*. *J. Invertebra. Pathol.* **2019**, *161*, 61–69. [CrossRef]
54. Abbott, W.S. A method of computing the effectiveness of an insecticide. *J. Econ. Entomol.* **1925**, *18*, 265–267. [CrossRef]
55. Sokal, R.R.; Rohlf, F.J. *Biometry*; Freeman: New York, NY, USA, 1995.
56. Minitab. *Minitab 17 Statistical Computer Software*; Minitab Inc.: State College, PA, USA, 2010.
57. Kaur, S.; Harminder, P.K.; Kaur, K.; Kaur, A. Effect of different concentrations of *Beauveria bassiana* on development and reproductive potential of *Spodoptera litura* (Fabricius). *J. Biopest.* **2011**, *4*, 161–168.
58. Inglis, G.D.; Johnson, D.L.; Goettel, M.S. Effects of temperature and thermoregulation on mycosis by *Beauveria bassiana* in grasshoppers. *Biol. Control.* **1996**, *7*, 131–139. [CrossRef]
59. Wakil, W.; Ghazanfar, M.U.; Riasat, T.; Qayyum, M.A.; Ahmed, S.; Yasin, M. Effects of interactions among *Metarhizium anisopliae*, *Bacillus thuringiensis* and chlorantraniliprole on the mortality and pupation of six geographically distinct *Helicoverpa armigera* field populations. *Phytoparasitica* **2021**, *41*, 221–234. [CrossRef]
60. Wakil, W.; Tahir, M.; Al-Sadi, A.M.; Shapiro-Ilan, D. Interactions between two invertebrate pathogens: An endophytic fungi and externally applied bacterium. *Front. Microbiol.* **2020**, *11*, 522368. [CrossRef]
61. Wakil, W.; Yasin, M.; Shapiro-Ilan, D. Effect of single and combined applications of entomopathogenic fungi and nematodes against *Rhynchophorus ferrugineus* (Olivier). *Sci. Rep.* **2017**, *7*, 1–11. [CrossRef]

agronomy

Review

Nosema locustae (Protozoa, Microsporidia), a Biological Agent for Locust and Grasshopper Control

Long Zhang [1,2,*] and Michel Lecoq [3]

1 Shandong Academy of Agricultural Sciences, Jinan 250100, China
2 Department of Grassland Resources and Ecology, College of Grassland Science and Technology, China Agricultural University, Beijing 100193, China
3 CIRAD, UMR CBGP, F-34398 Montpellier, France; mlecoq34@gmail.com
* Correspondence: locust@cau.edu.cn

Abstract: Effective locust and grasshopper control is crucial as locust invasions have seriously threatened crops and food security since ancient times. However, the preponderance of chemical insecticides, effective and widely used today, is increasingly criticized as a result of their adverse effects on human health and the environment. Alternative biological control methods are being actively sought to replace chemical pesticides. *Nosema locustae* (Synonyms: *Paranosema locustae, Antonospora locustae*), a protozoan pathogen of locusts and grasshoppers, was developed as a biological control agent as early as the 1980s. Subsequently, numerous studies have focused on its pathogenicity, host spectrum, mass production, epizootiology, applications, genomics, and molecular biology. Aspects of recent advances in *N. locustae* show that this entomopathogen plays a special role in locust and grasshopper management because it is safer, has a broad host spectrum of 144 orthopteran species, vertical transmission to offspring through eggs, long persistence in locust and grasshopper populations for more than 10 years, and is well adapted to various types of ecosystems in tropical and temperate regions. However, some limitations still need to be overcome for more efficient locust and grasshopper management in the future.

Keywords: locust; grasshopper; biological control; *Nosema locustae*; application; epizootics

Citation: Zhang, L.; Lecoq, M. *Nosema locustae* (Protozoa, Microsporidia), a Biological Agent for Locust and Grasshopper Control. *Agronomy* **2021**, *11*, 711. https://doi.org/10.3390/agronomy11040711

Academic Editor: Andrea Sciarretta

Received: 23 February 2021
Accepted: 2 April 2021
Published: 8 April 2021

1. Introduction

Locust and grasshopper (L&G) outbreaks, often resulting in huge plagues, have been a serious threat to global food security since ancient times [1]. Traditional control of L&G consists mainly of the application of chemical pesticides, which often results in many side effects. These include toxic chemical residues on food, adverse health effects on humans and nontarget animals, and environmental pollution. One of the most promising alternatives to chemical pesticides is biological control. Although there are many natural enemies of L&G [2–4], only a few have been developed as biological control agents or potential agents, including the microsporidian *Nosema locustae* (synonym: *Antonospora locustae, Paranosema locustae*) and the fungus *Metarhizium acridum*, both of which have been quite widely used in the control of L&G.

N. locustae, a unicellular eukaryote (Figure 1) with an obligate intracellular lifestyle [5–7], was the first to be commercially developed for L&G control since its discovery and characterization of its potential as a biological control agent against these species [8–10]. Due to its slow action and various other constraints, *N. locustae* was considered to be of limited application [11,12]. However, as the only microsporidian agent for L&G control, *N. locustae* has been studied extensively [1,13–17]. In recent years, there has been a renewed interest in *N. locustae*, mainly due to work in China, where it is produced in large quantities and used extensively, and Argentina, where its long-term persistence appears to reduce the frequency and intensity of grasshopper outbreaks [18]. Recent advances in areas such as mass production, formulation, application, and epizootiology have further promoted the application of this pathogen. Numerous studies on its host spectra, molecular biology, and evolution have provided a solid fundamental basis for the use of *N. locustae* for L&G management.

Figure 1. Spores of *N. locustae* under scanning electron microscopy (photo by Long Zhang).

2. Host Spectrum

N. locustae was first identified in a laboratory population from the migratory locust *Locusta migratoria* (Linnaeus, 1758) by Elizabeth Canning in 1953 [19]. This pathogen occurs under natural conditions and has been found in various areas of the United States (Montana, Northern Dakota, Minnesota, Oregon, Wyoming, Colorado, Arizona, and Idaho), Canada (Saskatchewan, Ontario) and South America (western Pampas and northwestern Patagonia, Argentina), as well as Asia (Rajasthan and Vidarbha, India; Inner Mongolia, Hainan and Qinghai, China) and Africa (Karoo, South Africa) [20]. The highest level of infection was reported of about 5.5% in *Melanoplus sanguinipes* (Fabricius, 1798) in Idaho between 1963 and 1967 [21].

N. locustae has a wide host range restricted to orthopteran insects. Henry [22] in 1969 provided an initial list of 55 North American species that he knew to be susceptible. Brooks [15] published in 1988 a new list of worldwide susceptible orthopterans, expanding the host range to 95 species. The last review was done in 2005 by Lange [20], who arrived at a total of 121 species. Since then, various authors have been added to this list (Table 1), and we now reach a total of 144 susceptible orthopterans.

Table 1. List of orthopteran species infected by *N. locustae* but not included in Lange [20].

Species	Inoculation Infection (Caging)	Field Trial Infection	References
	Asia		
Bryodemella holdereri (Krauss, 1901)	X		[23]
Calliptamus italicus (Linnaeus, 1758)	X		[24]
C. abbreviatus Ikonnikov, 1913	X		[23]
Ceracris kiangsu Tsai, 1929	X	X	[25]
Chondracris rosea (De Geer, 1773)	X	X	[26]
Chorthippus dubius (Zubovski, 1898)		X	[27]
C. brunneus (Thunberg, 1815)	X		
Damalacantha vacca (Fischer von Waldheim, 1846)	X		[23]
Deracanthella aranea (Fischer von Waldheim, 1833)	X		
Dociostaurus kraussi (Ingenitskii, 1897)	X		[24]
Fruhstorferiola tonkinensis (Willemse, 1921)		X	[28]
Gampsocleis sedakovii (Fischer von Waldheim, 1846)	X		[23]
Haplotropis brunneriana Saussure, 1888	X		
Oedaleus decorus (Germar, 1825)	X		[24]
Arcyptera meridionalis Ikonnikov, 1911	X	X	
Sphingonotus mongolicus Saussure, 1888	X		[23]
	Africa		
Acrotylus blondeli Saussure, 1884		X	
Acrotylus patruelis (Herrich-Schäffer, 1838)		X	[29]
Aiolopus thalassinus (Fabricius, 1781)		X	
Anacridium melanorhodon (Walker, 1870)		X	
	America		
Amblytropidia australis Bruner, 1904		X	[30]
Dichroplus vittigerus (Blanchard, 1851)		X	

Although *N. locustae* has a broad host spectrum in Orthoptera, it is incapable of affecting non-orthopteran insects. It has been shown that the honey bee (*Apis mellifera*) and the lepidoterans *Heliothis zea* and *Agrotis ipsilon* were not susceptible [15]. Menapace et al. [31] reported that honey bees were not infected, even when fed high doses of spores. The American cockroach *Periplaneta americana* and the spider *Butalus occidentalis* were also found to be non-susceptible [32]. *N. locustae* has been demonstrated to be safe for vertebrates. Brooks [15] summarized the assessment tests on primary skin irritation, acute dermal toxicity, acute inhalation toxicity and pathogenicity, subacute oral toxicity, acute oral toxicity, acute pathogenicity, and possible hazards of *N. locustae* to vertebrates, including rabbits, guinea pigs, rainbow trout, giant toads, mallards, ring-necked pheasant, mice, and rats. No significant effects were observed. The toxicity of *N. locustae* at 20 million spores/mL to certain non-target organisms, such as *Coturnix japonica*, *Apis mellifera*, *Bombyx mori*, *Daphnia magna*, and *Brachydanio rerio*, was examined in China. The results showed that there was no risk to honey bees per os at the maximum exposure dose tested, no deaths in the five animal species tested in the contact toxicity experiments, and *N. locustae* was relatively safe for nontarget beneficial organisms in the environment [33].

3. Pathogenicity

N. locustae, as an obligate parasite, reproduces in host target cells. Its infection involves the polar tube of the spore to inject its plasma into the target cells [5]. It causes high mortality in L&G. The locust's main target organ is the host's adipose tissue (fat body) [9]. *N. locustae* penetrates fat body cells and produces meronts, sporonts, sporoblasts, and spores. In the migratory locust, *Locusta migratoria migratorioides* (Reiche & Fairmaire, 1849), the younger the nymphs, the more susceptible they are, but even newly emerged adults are still susceptible [9]. Studies by Tounou et al. [34] on the effects of *N. locustae* on the desert locust, *Schistocerca gregaria* (Forskål, 1775) and the Senegalese grasshopper, *Oedaleus senegalensis* (Krauss, 1877), showed that *N. locustae* has high pathogenicity on the young nymphal instars of these two species. The median survival time for first, second, third, fourth, and fifth nymphal instars was 6, 9, 10, 14, and 15 days, respectively, when locusts were inoculated with 1 × 10^7 spores on 10 g wheat bran in groups. Similar results were obtained with the Senegalese grasshopper using the same treatment method, with median survival times for instars 1, 3, and 5 being 5, 9, and 15 days, respectively. The median survival time therefore increased with the age of the locust and with decreasing inoculation doses. For instance, third-instar desert locust nymphs were inoculated in groups with 5.62 × 10^6 or 3.16 × 10^4 spores with a median survival time of 14 and 16 days, respectively. However, cumulative mortality increases with the increase of spore concentration and decreases with the increasing age of the nymphs. *N. locustae* caused high mortality in young desert locust nymphs. Mortality was 100% in first and second nymphal instars inoculated in groups with 1 × 10^7 spores of *N. locustae*, as well as in first nymphal instars with 1 × 10^6 spores. With third and fourth nymphal instars inoculated with 1 × 10^7 spores, cumulative mortality was above 90%. Similar results were obtained with the Senegalese grasshopper, which was inoculated with 1 × 10^7 spores: 100% mortality was observed in the first instar but 88.5% mortality in the third instar. In the fifth instar, mortality was only 66.3% in desert locust and 70% in the Senegalese grasshopper. These results suggest that the appropriate period for the application of *N. locustae* is mainly between the first and fourth instars, i.e., a possible application period of about 20–28 days, taking into account an average duration of 5–7 days per nymphal instar.

Inoculation of the 1st–5th nymphal instars of *Chondracris rosea* (De Geer, 1773) with *N. locustae* resulted in 78.2% to 100% mortality in field caging experiments by Liu and Chen [26]. The LC_{50} was approximately 3.88 × 10^5 spores/mL for the 1st–3rd instars and 3.98 × 10^6 spores/mL for the 4th–5th instars. In a caging experiment on a large forest site, the same authors observed that *C. rosea* mortality was greater than 91.1% when they sprayed *N. locustae* at a rate of 5 × 10^7 spores/mL. At 1 × 10^8 spores/mL, mortality reached 100% 25 days after treatment [26]. Chen et al. [25] conducted a laboratory experiment with five concentrations of *N. locustae* spores (1 × 10^4, 1 × 10^5, 1 × 10^6, 1 × 10^7, and 1 × 10^8 spores/mL) to treat early stages (1st–2nd instar) of the yellow-spined bamboo locust (*Ceracris kiangsu*). Mortalities were 16.6, 32.9, 29.2, 34.0, and 83.7%, respectively, increasing with spore concentrations. They also found that yellow-spined bamboo locust mortality was 85.1% after the application of a suspension of *N. locustae* spores at a concentration of 5 × 10^7 spores/mL in field trials in the forest. Zang et al. [35] reported that mortality of *Oxya chinensis* (Thunberg, 1815) was about 65.4–68.1% 30 days after the field treatment of third-instar nymphs sprayed with 1.5 × 10^{10}, 2.25 × 10^{10} spores/ha, respectively. The infection rate in survivors was approximately 40%.

Concurrent use of *N. locustae* and *Metarhizium* spp. has shown additive effects on locusts and grasshoppers. When fifth-instar nymphs of *S. gregaria* were inoculated first with *N. locustae* at doses between 1 × 10^4 and 1 × 10^6 spores on wheat bran in groups, and then 10 days later with *M. anisopliae* at doses between 1 × 10^2 and 1 × 10^4 spores/nymph, the median survival times ranged from 3 to 9 days, and the shortest duration was only 3 days at doses of 1 × 10^6 spores of *Nosema* and 1 × 10^4 spores of *Metarhizium* [29]. When the locusts were inoculated in groups with *N. locustae* at doses of 1 × 10^5 and 1 × 10^6 spores, and *Metarhizium* at 1 × 10^3 and 1 × 10^4/nymph, the mortality was at 90% and 97% over 3 days, and both reached 100% 10 days after inoculation. Mortality was only 12.2% in the control, showing a synergistic effect between these two agents. Similar results were obtained with *N. locustae* directly mixed with *Metarhizium* spp., with the oriental migratory locust (*L. migratoria manilensis* Meyen, 1835). Mortality was higher with the mixture compared to treatments with each agent applied separately. The effects of a mixture of *M. acridum* and *N. locustae* on third nymphal instars under laboratory conditions showed that at a ratio of 1:1 (*M. acridum* at 3.90 × 10^6 spores/g locust body weight and *N. locustae* at 3.99 × 10^6 spores/g locust body weight), locust mortality was about 96.7% 24 days after inoculation, showing an additive effect of these two agents [36]. When 2nd–3rd-instar nymphs of *L. migratoria* were inoculated with 1 × 10^6 spores/nymph of *N. locustae*, and after 3, 6 and 9 days with *Metarhizium* at 1 × 10^7 conidia/mL, an additive effect was only observed in nymphs inoculated with *Nosema* and then 9 days later with *Metarhizium* [37]. After examining the stage of *Nosema* development on Days 3, 6, and 9 after inoculation, they found that it was not until the *Nosema* spore maturation stage that locusts were most susceptible to fungal infection.

Lv et al. [38] identified 4 defensins from migratory locust palp transcriptomes, named LmigDEF1 (78 amino acids), LmigDEF3 (78 amino acids), LmigDEF4 (69 amino acids), and LmigDEF5 (67 amino acids) with theoretical isoelectric points (pI)/molecular weights (Mw, kDa) of 6.48/8.29, 6.56/8.37, 8.23/7.19, and 8.27/6.93, respectively. The expression patterns of LmigDEF1, LmigDEF3, and LmigDEF5 in the fat body and salivary glands were examined by qRT-PCR after the locusts were inoculated with *N. locustae*. Results indicated that all three defensins varied over time in the fat body and salivary glands after *Nosema* infection, the transcript level of the LmDEFs being at their lowest in the fat body over the 10 days. This indicates that *Nosema* infection reduced the locust's immune response via the defensins and may explain why the coordinated use of *Nosema* and *Metarhizium* results in higher locust mortality [29,36]. In particular, for about four days when *Nosema* were sporulating after inoculation, the locusts were more easily infected by the fungus due to the lower level of the three types of defensins in the fat body [37]. In addition, Chen et al. [39] demonstrated the key role of *N. locustae* sporulation in locust mortality. They identified a spore wall protein, AlocSWP2 from *N. locustae*, containing four cysteines. AlocSWP2 has been detected in the wall of mature spores, sporoblasts, and sporonts during sporulation in the host body by immunocytochemistry localization experiments. AlocSWP2 was detected in the fat body of infected locust only on Day 9 after inoculation using RT-PCR. The survival percentage of infected locusts that received a dsRNA injection of AlocSWP2 on Days 15, 16, and 17 after inoculation of *Nosema* spores was significantly higher than that of infected locusts without dsRNA treatment. Similarly, the number of spores in locusts infected with *Nosema* and treated with RNAi of AlocSWP2 was significantly lower than that in infected locusts without RNAi of this gene. This indicates that this *N. locustae* spore wall protein is involved in sporulation, contributing to host mortality.

L. migratoria migratorioides infected with *N. locustae* demonstrated reduced sustainable flight capacity [9]. Zhang et al. [40] confirmed this with the oriental migratory locust (*L. migratoria manilensis*). Flight capacity of infected and healthy adult locusts, 5 to 15 days after emergence, was determined using a flying mill for 18 h. On average, the flight distance of healthy versus infected locusts was 14,279 vs. 864 m; flight speed, 1.23 vs. 0.51 m/s, flight time, 3 vs. 0.33 h; maximum flight distance, 72,538 vs. 1544 m; and maximum sustained flight time, 6.7 h vs. 0.1 h. The decrease in flight capacity may be due to the fact that *N. locustae* destroyed the fat bodies reducing the supplement of glyceride and fat as energy resources. In *L. migratoria manilensis* infested with *N. locustae*, the glyceride content decreased rapidly, while lipase activity increased in both hemolymph and total fat [41].

Locusts and grasshoppers infected with *N. locustae* have reduced fertility [13,42]. Reduced vitellogenin was observed in fourth-instar nymphs of *L. migratoria manilensis* inoculated with *N. locustae*: vitellogenin levels in the fat body, hemolymph, and ovaries were very low compared to the control [43]. The maximum vitellogenin level in infected vs. healthy locusts, respectively, were 4.663 vs. 18.655 mg/mL in the fat body, 2.627 vs. 7.603 mg/mL in the hemolymph, and 4.927 vs. 73.367 mg/mL in the ovaries. This explains why the reproductive capacity of infected locusts is low.

The disease caused by *N. locustae* is transmitted vertically in eggs and egg pods [5]. Infected females have been reported to lay eggs containing spores of *N. locustae* [42]. Raina et al. [44] reported vegetative stages of *N. locustae* in the yolk of the oocyte and spores in the eggs of *L. migratoria*. When fourth-instar nymphs were inoculated with a dose of 1.5×10^6 spores, the infected parents laid eggs and the prevalence of infection was 100% in the next generation. The disease was vertically transmitted up to 14 generations, and mortality due to vertical transmission at times reached more than 90%. Parents of *S. gregaria* and *O. senegalensis* infected with *N. locustae* produced progeny with a 50% infection rate, indicating high vertical transmission in these species [34].

4. Genomics and Molecular Biology

Understanding the molecular biology of *N. locustae* is fundamental in determining its biology and molecular interactions with the host and improving applications in its use for pest control. A complete genomic sequence of *N. locustae* has recently been revealed by Chen et al. [45], which is a major achievement. Sequencing of its genome yielded 3,170,203 nucleotides, encoding for 1857 predicted genes, of which 1755 are single-exon genes, and 102 are multiple exon genes. A total of 17 scaffolds, ranging from 88.763 to 388.82 kb, were identified and assigned to 17 chromosomes. Genomic and protein sequences from *N. locustae* and several other single-celled organisms were used to study evolutionary relationships in genetic synteny and collinearity using the MCScanX method. Results showed that within the microsporidia, most genes exhibit good collinearity [45].

In addition to genomic sequencing, Chen et al. [45] performed an analysis of locust midgut transcripts versus locust fat body transcripts from *N. locustae* infected locusts. They found that the abundant expression of locust antimicrobial peptides and other defense genes, such as peroxiredoxin and amine oxidase in the midgut, may explain the lower number of microsporidian spores in the midgut of the host. In the fat body, however, the large number of *N. locustae* spores present may be related to the fact that several locust phenol oxidases and peroxisome proliferator-activated receptors have been inhibited to allow it to escape the host immune response. This revealed the interactions of *N. locustae* and the host at the transcript molecular level, in particular why *N. locustae* can reproduce massively in the locust fat body and not in the gut.

Identification of the genes and proteins of the polar tube is certainly useful in understanding the interactions between *N. locustae* and its hosts at the molecular level. Two polar tube protein genes, PTP1 and PTP2, have been identified, and it has been suggested that PTP1 plays a key role in interactions with the host cell surface [46]. Two other orthologous polar tube proteins, named AlPTP2b and AlPTP2c, exhibit elastomeric characteristics. The AlPTP2b and AlPTP2c genes encode for proteins of 568 and 599 amino acids with deduced molecular weights of 55,399 and 56,664 Da, respectively. These proteins are highly conserved (84.2% identity), larger than the previously reported AlPTP2 (287 amino acids) [47].

Fast et al. [48] identified a gene for the TATA box binding protein (TBP) of *N. locustae*. The predicted amino acid sequence of the TBP gene consists of 259 amino acids. In the phylogenetic analysis of TBP, the authors emphasized that TBP from *N. locustae* is close to that of fungi and supported previous studies on the evolution of microsporidia with tubulins, HSP70, and the greater subunit of RNA polymerase II (RPB1) proteins [49–53]. However, in a study on the large subunit (LSU) rRNA, Peyretaillade et al. [54] proposed that the origin of microsporidia was not specifically linked to a particular group of eukaryotes. In general, there are only a few studies on proteins and genes of *N. locustae* [55], and even rarer studies on their functions, especially pathogen–host interactions at the molecular level.

5. Mass Production and Products

The production of *N. locustae* spores is done in vivo. The rearing of infected hosts is the main process for mass production (Figure 2). Grasshoppers *Melanoplus bivittatus* (Say, 1825) have been used as hosts in the United States. Henry et al. [10] pointed out that several factors influence production yield. Cage size and the number of individuals in each cage, light in the cage, and time of harvest, as well as factors such as grasshopper species and sex, are important. The general process is as follows: grasshopper nymphs are reared to the 4th–5th instars, inoculated with *N. locustae* spore suspension, and then reared until they die. The cadavers are collected and crushed, screened, and centrifuged to obtain a high concentration of spores, which are stored between −10 and −20 °C. In the United States, after several improvements in mass production techniques, a yield of 1×10^9 to 3×10^9 spores per grasshopper has been achieved. However, the use of *Melanoplus differentialis* (Thomas, 1865) as a host resulted in up to 7.1×10^9 spores per individual [56]. The time of harvest is an important factor in influencing the spore yield; the further away from the inoculation date, the more spores are obtained [17].

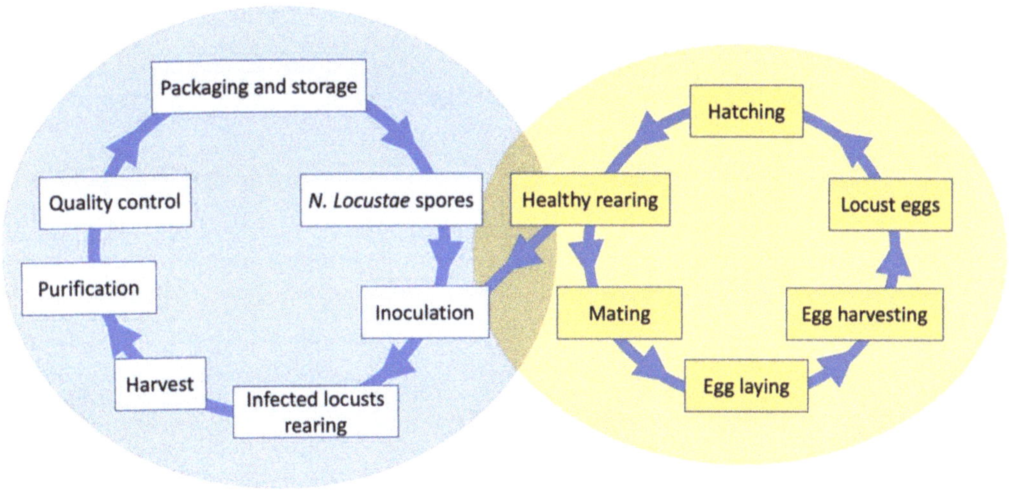

Figure 2. Mass producing process of *N. locustae* spores.

In China, *L. migratoria manilensis* has been selected as a host to produce spores of *N. locustae*. In this species, there is no diapause, and individuals can be used year-round. Several factors influence the efficiency of mass production, including the inoculation concentration, the stage of development for inoculation, and the harvesting time [57]. A higher concentration of spores in the inoculation suspension resulted in rapid locust death, and a lower concentration increased the time to harvest and reduced the number of spores. The best inoculation concentration was found to be 1×10^6 spores/mL, the stage for inoculation the fourth instar, and the harvest time about 30–40 days after inoculation. The average spores/individual yield was approximately 6×10^9, the highest being about 9.9×10^9. In China, several insectariums have been established for the mass production of *N. locustae* since the 1990s. Currently, the average spore yield per individual can reach 22×10^9 to 34×10^9. There are about 5–6 harvests per year, and the total yield can reach 2×10^{15} spores for each insectarium [17].

The products are usually formulated either as bran bait, aqueous suspension, or water-based suspension. They are prepared from high concentrations of spores stored at low temperature, diluted directly with water or mixed with wheat bran in an appropriate ratio. In China, a new water-based suspension has been developed. The spores are mixed with xanthan gum, sorbic acid, and other environmentally friendly additives. The suspension is more stable and homogeneous and can be stored for about one year at room temperature [58]. It is better adapted to the needs of users and should be widely applied. To date, several commercial products have been registered such as Nolo Bait™, Semaspore™, and Grasshopper Attack™ in the United States since the 1980s [12]. In China, *Nosema locustae* products for locust control have been developed since the 1990s and are currently distributed by Beijing JiaJing Biotechnology Ltd. (Beijing, China). They are based on a highly pathogenic strain (AL2008L-04) that significantly improves product efficacy, a high-yield production technology, and an aqueous suspension formulation that can be stored at room temperature [59].

6. Application and Epizootics

The three main components whose interactions can lead to a reduction in locust and grasshopper populations and epizootics are *N. locustae,* its hosts, and its environments. *N. locustae* has been used to control *L. migratoria manilensis* since the 1990s. A field trial was conducted on Hainan Island, China, in areas close to rice fields [60]. The locust population consisted of first to third instars with pretreatment densities of 0.9, 1.5, and 2.2 individuals/m^2 in the three treatment plots and 0.6 in the control plot. The plots were treated at 50×10^9, 99×10^9, and 150×10^9 spores/ha. At 25 days after treatment, the locust population density was reduced by 70.2%, 74.0%, and 78.9%, respectively, in the treated plots but increased by 58.1% in the control plot. The prevalence of *N. locustae* infection in survivors was 29.4%, 31.1%, and 31.8% at 30 days and 40.6%, 40%, and 35.7% at 40 days after treatment in the treated plots. The prevalence of infection persisted the following summer (one year) in the three treated plots, respectively, at 23.5%, 13%, and 9.1% levels.

In the United States, in Montana, studies indicated that the prevalence of *N. locustae* during the treatment season was variable, but mainly around 30%, with the highest being 50% when the spore application rate was 1.4×10^{10} in 1.12 kg of wheat bran/ha after three applications [10,61]. In Canada, in a grasshopper population consisting mainly of third instars of *Melanoplus sanguinipes*, *M. packardii* (Scudder, 1878), and *Camnula pellucida* (Scudder, 1862) treated with 2.5×10^9 or 5.0×10^9 spores in 1.68 kg wheat bran per ha, an infection rate of approximately 50% was observed 4–5 weeks after treatment [62]. Toward the end of the season, about 12 weeks after treatment, this rate was about 95–100%. Zhang et al. [63] reported the prevalence of *N. locustae* in the years following the application of spore-based baits to mixed grasshopper populations—consisting of *Oedaleus asiaticus* Bey-Bienko, 1941, *Myrmeleotettix palpalis* (Zubovski, 1900), *Angaracris rhodopa* Fischer von Waldheim, 1836, *A. barabensis* (Pallas, 1773) and *Dasyhippus barbipes* (Fischer von Waldheim, 1846)—in Inner Mongolia rangelands, a region where temperatures are below −10 °C in winter and 20 to 30°C in summer and rainfall is about 200 mm per year. In two areas treated in 1991 and 1992 with 7.5×10^9 spores/1.5 kg wheat bran bait/ha, locust infection rates of *N. locustae* were 14.5% and 19.8% in the second year after treatment. In a third area treated in 1988 with *N. locustae* baits, the observed infection rates were 6.8% and 23% two and six years after treatment. Mortality rates of 60–80% in mixed grasshopper populations were achieved following applications of *N. locustae,* in Inner Mongolia rangeland, at a rate of 3×10^{10} spores/ha for two consecutive years [64]. *N. locustae* was even used in the highlands (3270–3350 m ASL) in Qinghai province of China, and the disease could still be transmitted and persisted for a long time among grasshoppers [65], up to 10 years with an infection rate exceeding 50% in some years [27]. In three locations in Canada in two consecutive years (1988–1989), Johnson and Dolinski [66] found that *N. locustae* persisted in grasshopper populations. They suggested that the activity of *N. locustae* was not inhibited despite severe weather conditions, hot in summer and very cold in winter, with air temperature down to −39 °C.

N. locustae was reportedly applied in a forest to control the yellow-spined bamboo locust (*Ceracris* kiangsu) [25]. In a 1.5 ha area of bamboo sprayed with 50 kg of *N. locustae* spore suspension at a concentration of 5×10^7 spores/mL, mortality of *C. kiangsu* was 85% 15 days after treatment. *N. locustae* was also used in 2018 to control the yellow-spined bamboo locust in bamboo forest ecosystems in Phongsaly province, Lao [67]. Nymphs were in the third and fourth instars at a density of over 100 individuals/m^2. Treatment was carried out on May 13 at an application rate of 2×10^7 spores/mL. After 17 days, many dead locusts were observed in the treated area, where the average density was less than 10 individuals/m^2, a reduction of more than 90% compared to the untreated control area. *N. locustae* spores were observed in all dead locusts that were collected from the treated plots and examined individually under a microscope in the laboratory. The average grade of infection was three out of five, corresponding to severe disease. In contrast, the infection rate of locusts collected from the control area was zero. In addition, 46% of surviving locusts in treated plots were infected with *N. locustae*, and two and a half months after treatment, the density was very low, about 0–2 individuals/m^2. It was concluded that *N. locustae* could be an effective agent to control high densities of yellow-spined bamboo locusts in bamboo forests.

In Africa, Tounou et al. [68] conducted field trials to assess the effects of *N. locustae* and *Metarhizium anisopliae* against various grasshopper species: *Pyrgomorpha cognata* Krauss, 1877, *Acrotylus blondeli* Saussure, 1884 (both predominant species), and *Oedaleus senegalensis*. Both agents were mixed with wheat bran as bait with *M. anisopliae* alone, *N. locustae* + *M. anisopliae*, *N. locustae* spores alone, or *N. locustae* + sugar. The same treatments were carried out with *O. senegalensis* in the early stages of development and resulted in mortalities of 64–85%. The population density during the three weeks of monitoring decreased by 44.7 ± 6.9% in the *N. locustae* plot, 52.8 ± 8.4% in the *N. locustae* + sugar plot, 73.7 ± 5.5% in the *M. anisopliae* plot, and 89.1 ± 1.8% in the *N. locustae* + *M. anisopliae* plot. The prevalence of *N. locustae* in adult grasshoppers surviving at 28 days after application was 48.1 ± 2.3, 28.9 ± 4.8, and 27.4 ± 3.7% in the three treatments with *N. locustae*. The results suggest that these two biological control agents have the potential to control the early stages of grasshoppers in Africa. These conclusions differ from those of Lima et al. [69], who found no significant difference between treated and untreated plots and concluded that *N. locustae* could not be used to control *O. senegalensis* in the Cape Verde archipelago. However, they did not specify the developmental stage of the treated grasshoppers, and it is possible that efficacy was less on the fifth nymphal instar or on adults.

Long-term epizootics of *N. locustae* have been investigated in Argentina by Lange et al., who have shown that this pathogen can persist for many years after its introduction [30,70,71]. These authors, in their field survey, found that the prevalence of *N. locustae* ranged from 1.8 to 41% in 9 of 13 sites between 1995 and 2003. In 1990, in another field survey in an area treated 11 years earlier, Lange and Azzaro [71] observed infection rates of 2.9%, 3.5%, and 3.6% by *N. locustae* spores in three grasshopper species, *Dichroplus elongatus* Giglio-Tos, 1894, *D. maculipennis* (Blanchard, 1851), and *Scotussa lemniscata* (Stål, 1861). Lange et al. [30] examined grasshoppers collected from sites treated with *N. locustae* in Argentina in the late 1970s, early 1980s, and mid-1990s and found that the highest prevalence of *N. locustae* in grasshopper populations was about 50%.

7. Conclusions and Prospects

N. locustae is becoming increasingly biologically understood, as well as for its potential in L&G control. Recent studies have shown that it can be applied in different ways to achieve control objectives and plays a more important role than ever in locust and grasshopper management programs. It has a fairly broad host spectrum, and at least 144 orthopteran species are susceptible. However, it is very safe for non-orthopteran insects and other non-target animals [17,33]. It is transmitted orally and is therefore probably less sensitive to environmental factors than fungal-type agents. This is certainly the reason why good results have been recorded at a 3000–4000 m altitude on the Tibetan plateau in China and in high-temperature regions such as Hainan province in China or Lao and Vietnam. *N. locustae* has been demonstrated to be efficient in the control of L&G in rangelands, crop fields, and forest ecosystems. Its disease can persist for many years after application in L&G populations and can be part of their long-term management.

When L&G are at low density, a low dose of *N. locustae* will slowly kill them as a long-term control method. However, recent field trials have indicated that *N. locustae* can also be used at high locust densities, as has been demonstrated to control second and third instars of the yellow-spined bamboo locust. *N. locustae* also exhibits a synergistic effect when used in admixture with the fungus *Metarhizium* spp. and can potentially weaken the host immune response by reducing its defensins. The simultaneous use of *N. locustae* and *Metarhizium* spp. could be encouraged to control high-density locust outbreaks.

L&G mortality caused by *N. locustae* is dose-dependent; the higher the dose used, the higher the mortality. However, *N. locustae* spores are produced restrictively in vivo. The mass rearing of locusts to produce spores is expensive; therefore, increasing the spore yield per locust or grasshopper is a bottleneck for mass production and large-scale application. Understanding the molecular mechanisms of *N. locustae* and its interactions with its host may be valuable to explore ways to mass-produce *N. locustae* spores in order to improve the traditional in vivo mass production method or to find an alternative, as well as to increase the effects on L&G. Genome sequencing and transcriptomic analysis of *N. locustae* genes and proteins have provided a valuable basis for future studies to improve the genetics of *N. locustae* and screen for high-virulence strains [72]. The development of new formulations and other technologies to maintain high spore vitality at room temperature will also be useful in promoting the wider application of *N. locustae*.

Author Contributions: L.Z. conceived the article and drafted the manuscript with substantial input from M.L. All authors have read and agreed to the published version of the manuscript.

Funding: This work is supported by a grant from Shandong Academy of Agricultural Sciences, and a grant from the Ministry of Agriculture of China (15206030).

Acknowledgments: Authors would like to thank anonymous reviewers' comments and suggestions.

Conflicts of Interest: The authors declare no conflict of interest.

Disclaimer: Mention of trade names or commercial products in this publication is solely for the purpose of providing specific information and does not imply recommendation or endorsement by the authors.

References

1. Zhang, L.; Lecoq, M.; Latchininsky, A.; Hunter, D. Locust and Grasshopper Management. *Annu. Rev. Entomol.* **2019**, *64*, 15–34. [CrossRef] [PubMed]
2. Greathead, D.J. Natural Enemies of Tropical Locusts and Grasshoppers: Their Impact and Potential as Biological Control Agents. In *Biological Control of Locusts and Grasshoppers*; Lomer, C.J., Prior, C., Eds.; CAB International: Wallingford, UK, 1992; pp. 105–122.
3. Lomer, C.J.; Prior, C. *Biological Control of Locusts and Grasshoppers*; CAB International: Wallingford, UK, 1992.
4. Goettel, M.S.; Johnson, D.L. (Eds.) Microbial Control of Grasshoppers and Locusts. *Mem. Entomol. Soc. Can.* **1997**, *129*, 400. [CrossRef]
5. Solter, L.F.; Becnel, J.J.; Oi, D.H. Microsporidian Entomopathogens. In *Insect Pathology*, 2nd ed.; Vega, F.E., Vega, H.K., Eds.; Academic Press: Cambridge, MA, USA, 2012; pp. 221–263. [CrossRef]
6. Keeling, P.J.; Fast, N.M. Microsporidia: Biology and Evolution of Highly Reduced Intracellular Parasites. *Annu. Rev. Microbiol.* **2002**, *56*, 93–116. [CrossRef] [PubMed]
7. Keeling, P.J.; McFadden, G.I. Origins of Microsporidia. *Trends Microbiol.* **1998**, *6*, 19–23. [CrossRef]
8. Canning, E.U. The Life Cycle of *Nosema locustae* Canning in *Locusta migratoria migratorioides* (R. & F.) and its Infectivity to other Hosts. *J. Inver. Pathol.* **1962**, *4*, 237–247.
9. Canning, E.U. The Pathologenicity of *Nosema locustae* Canning. *J. Insect Pathol.* **1962**, *4*, 248–256.
10. Henry, J.; Tiahrt, K.; Oma, E. Importance of Timing, Spore Concentrations, and Levels of Spore Carrier in Applications of *Nosema locustae* (Microsporida: Nosematidae) for Control of Grasshoppers. *J. Invertebr. Pathol.* **1973**, *21*, 263–272. [CrossRef]
11. Lockwood, J.A.; Ewen, A.B. Biological Control of Rangeland Grasshoppers and Locusts. In *The Bionomics of Grasshoppers, Katydids and their Kin*; Gangwere, S.K., Muralirangan, M.C., Muralirangan, M., Eds.; CAB International: Wallingford, UK, 1997; pp. 421–442.
12. Lockwood, J.A.; Bomar, C.R.; Ewen, A.B. The History of Biological Control with *Nosema locustae*: Lessons for Locust Management. *Int. J. Trop. Insect Sci.* **1999**, *19*, 333–350. [CrossRef]
13. Henry, J.E.; Oma, E.A. Pest Control by *Nosema locustae*, a Pathogen of Grasshoppers and Crickets. In *Microbial Control of Pests and Plant Diseases (1970–1980)*; Burges, H.D., Ed.; Academic Press: New York, NY, USA, 1981; pp. 573–586.
14. Henry, J.E. Natural and Applied Control of Insects by Protozoa. *Annu. Rev. Entomol.* **1981**, *26*, 49–73. [CrossRef]
15. Brooks, W.M. Entomogenous Protozoa. In *CRC Handbook of Natural Pesticides*; Apple Academic Press: Palm Bay, FL, USA, 2019; Volume 5, pp. 1–149.
16. Lomer, C.J.; Bateman, R.P.; Johnson, D.L.; Langewald, J.; Thomas, M.B. Biological control of Locusts and Grasshoppers. *Annu. Rev. Entomol.* **2001**, *46*, 667–702. [CrossRef]
17. Johnson, D.L. Nosematidae and other Protozoa as Agents for Control of Grasshoppers and Locusts: Current Status and Prospects. *Memoirs Entomol. Soc. Can.* **1997**, *129*, 375–389. [CrossRef]
18. Lange, C.E.; Sokolova, Y.Y. The Development of the Microsporidium *Paranosema* (*Nosema*) *locustae* for Grasshopper Control: John Henry's Innovation with Worldwide Lasting Impacts. *Protistology* **2017**, *11*, 170–174. [CrossRef]

19. Canning, E.U. A New Microsporidian, *Nosema locustae* n.sp., from the Fat Body of the African Migratory Locust, *Locusta migratoria migratorioides* R. & F. *Parasitol.* **1953**, *43*, 287–290. [CrossRef]
20. Lange, C.E. The Host and Geographical Range of the Grasshopper Pathogen *Paranosema* (*Nosema*) *locustae* Revisited. *J. Orthoptera Res.* **2005**, *14*, 137–141. [CrossRef]
21. Henry, J.E. Epizootiology of Infections by *Nosema locustae* Canning (Microsporidia: Nosematidae) in Grasshoppers. *Acrida* **1972**, *1*, 111.
22. Henry, J.E. Extension of the Host Range of *Nosema locustae* in Orthoptera. *Ann. Entomol. Soc. Am.* **1969**, *62*, 452–453. [CrossRef]
23. Dong, Y. Preliminary Experiments on Application of *Nosema locustae* to Control Grasshoppers in Inner Mogolia Rangland. Master's Thesis, Agricultural University, Beijing, China, 1989.
24. Wang, L.; Yan, Y.; Dong, Y.; Lin, Z. Observations of a Microsporidian Parasite *Nosema locustae* Canning in the Oriental Migratory Locust (*Locusta migratoria manilensis* (Meyen)) and other Hosts from Inner Mogolia & Xinjiang. *Acta Agric. Univ. Pekin.* **1987**, *13*, 459–462.
25. Chen, R.; Liu, Q.; Huang, H. Application of Insect Pathogenic Microbials to Control *Ceracris kiangsu* Tsai. *Nat. Enem. Insects* **2002**, *24*, 123–127.
26. Liu, Q.; Chen, R. A Preliminary Study on *Chondracris rosea* Control by Inoculating *Nosema locustae*. *Guangdong For. Sci. Technol.* **1996**, *12*, 30–36.
27. Ma, D.; Bai, S.; Li, Z. Studies on Sustainable Controlling Rangeland Grasshopper with *Nosema locustae* in Caidamu of Qinghai Province. *J. Shandong Agric. Univ. Nat. Sci.* **2005**, *36*, 199–205.
28. Chen, Y.; Chen, Z.; Luo, R.; Chen, W.; Lu, S.; Li, Y. A Study on Application of *Nosema locustae* to Control *Fruhstorferiola tonkinesis*. In Proceedings of the 8th Conference of Chinese Entomology Society, Hebi, China, 11–13 October 2007.
29. Tounou, A.-K.; Kooyman, C.; Douro-Kpindou, O.-K.; Poehling, H.-M. Interaction between *Paranosema locustae* and *Metarhizium anisopliae* var. acridum, two Pathogens of the Desert Locust, Schistocerca gregaria under Laboratory Conditions. *J. Invertebr. Pathol.* **2008**, *97*, 203–210. [CrossRef] [PubMed]
30. Lange, C.E.; Mariottini, Y.; Plischuk, S.; Cigliano, M.M. Naturalized, Newly-Associated Microsporidium Continues Causing Epizootics and Expanding its Host Range. *Protistology* **2020**, *14*, 32–37. [CrossRef]
31. Menapace, D.M.; Sackett, R.; Wilson, W.T. Adult Honey Bees are not Susceptible to Infection by *Nosema locustae*. *J. Econ. Entomol.* **1978**, *71*, 304–306. [CrossRef]
32. Krall, S.; Knausenberger, W. *Efficiency and Environmental Impact for Biological Control of Nosema locustae on Grasshoppers in Cape Verde, a Synthesis Report*; Deutsche Gesellschaft fur Technische Zusammenarbeit (GTZ): Eschborn, Germany, 1992; p. 16.
33. Yuan, S.; Jiang, J.; Zhang, L.; Wu, S.; Liu, X.; An, X.; Li, G. Toxic Effects of *Nosema locustae* on Environmental Non-Target Bene-ficial Organisms. *Mod. Agrochem.* **2020**, *19*, 39–43.
34. Tounou, A.; Kooyman, C.; Douro-Kpindou, O.; Gumedzoe, Y.; Poehlingn, H. Laboratory Assessment of the Potential of *Paranosema locustae* to Control Immature Stages of *Schistocerca gregaria* and *Oedaleus senegalensis* and Vertical Transmission of the Pathogen in Host Populations. *Biocontrol Sci. Technol.* **2011**, *21*, 605–617. [CrossRef]
35. Zang, J.; Cao, J.; Li, G.; Li, C.; Yan, Y. Biological Control of *Oxya chinensis* by using *Nosema locustae*. *Chin. J. Boil. Control* **2001**, *17*, 126–128. Available online: http://www.zgswfz.com.cn/EN/Y2001/V17/I3/126 (accessed on 3 March 2021).
36. Ding, X.; Zhang, L. Virulence of *Metarhizium anisopliae* and *Nosema locustae* Against the Nymphs of Locusta migratoria manilensis. *J. Beijing Univ. Agric.* **2009**, *24*, 9–14.
37. Tokarev, Y.S.; Levchenko, M.V.; Naumov, A.M.; Senderskiy, I.V.; Lednev, G.R. Interactions of two Insect Pathogens, *Paranosema locustae* (Protista: Microsporidia) and *Metarhizium acridum* (Fungi: Hypocreales), during a Mixed Infection of *Locusta migratoria* (Insecta: Orthoptera) Nymphs. *J. Invertebr. Pathol.* **2011**, *106*, 336–338. [CrossRef]
38. Lv, M.; Mohamed, A.A.; Zhang, L.; Zhang, P.; Zhang, L. A Family of CSαβ Defensins and Defensin-Like Peptides from the Migratory Locust, *Locusta migratoria*, and Their Expression Dynamics during Mycosis and Nosemosis. *PLoS ONE* **2016**, *11*, e0161585. [CrossRef]
39. Chen, L.; Li, R.; You, Y.; Zhang, K.; Zhang, L. A Novel Spore Wall Protein from *Antonospora locustae* (Microsporidia: Nosematidae) Contributes to Sporulation. *J. Eukaryot. Microbiol.* **2017**, *64*, 779–791. [CrossRef]
40. Zhang, L.; Yan, Y.H.; Li, G.B.; Cao, Y.Z. The Effect of *Nosema locustae* Infection on Flying Ability of Locust (*Locusta migratoria manilensis*). *Acta Agrestia Sin.* **1995**, *3*, 324–327.
41. Chen, J.; Shen, J.; Song, D.; Zhang, L.; Yan, Y. Effect of *Nosema locustae* on the Content of Fat in *Locusta migratoria manilensis*. *Acta Entomol. Sin.* **2000**, *43*, 109–113.
42. Zhang, L. Epizootic and Transmission of *Nosema locustae* in the Populations of Grasshoppers and Locusts. Ph.D. Thesis, Beijing Agricultural University, Beijing, China, 1994.
43. Chen, J.; Shen, J.; Song, D.; Zhang, L.; Yan, Y. Effect of *Nosema locustae* on the Content of Vitellogenin of *Locusta migratoria manilensis*. *Acta Entomol. Sin.* **2002**, *45*, 170–174.
44. Raina, S.K.; Das, S.; Rai, M.M.; Khurad, A.M. Transovarial Transmission of *Nosema locustae* (Microsporida: Nosematidae) in the Migratory Locust *Locusta migratoria migratorioides*. *Parasitol. Res.* **1995**, *81*, 38–44. [CrossRef] [PubMed]
45. Chen, L.; Gao, X.; Li, R.; Zhang, L.; Huang, R.; Wang, L.; Song, Y.; Xing, Z.; Liu, T.; Nie, X.; et al. Complete Genome of a Unicellular Parasite (*Antonospora locustae*) and Transcriptional Interactions with its Host Locust. *Microb. Genom.* **2020**, *6*, e000421. [CrossRef]

46. Polonais, V.; Prensier, G.; Metenier, G.; Vivarès, C.P.; Delbac, F. Microsporidian Polar Tube Proteins: Highly Divergent but Closely Linked Genes Encode PTP1 and PTP2 in Members of the Evolutionarily Distant Antonospora and Encephalitozoon Groups. *Fungal Genet. Biol.* **2005**, *42*, 791–803. [CrossRef]
47. Polonais, V.; Belkorchia, A.; Roussel, M.; Peyretaillade, E.; Peyret, P.; Diogon, M.; Delbac, F. Identification of Two new Polar Tube Proteins Related to Polar Tube Protein 2 in the Microsporidian *Antonospora locustae*. *FEMS Microbiol. Lett.* **2013**, *346*, 36–44. [CrossRef]
48. Fast, N.M.; Logsdon, J.M.; Doolittle, W.F. Phylogenetic Analysis of the TATA Box Binding Protein (TBP) Gene from *Nosema locustae*: Evidence for a Microsporidia-Fungi Relationship and Spliceosomal Intron Loss. *Mol. Biol. Evol.* **1999**, *16*, 1415–1419. [CrossRef]
49. Edlund, T.D.; Li, J.; Visvesvara, G.S.; Vodkin, M.H.; McLaughlin, G.L.; Katiyar, S.K. Phylogenetic Analysis of β-Tubulin Sequences from Amitochondrial Protozoa. *Mol. Phylogenetics Evol.* **1996**, *5*, 359–367. [CrossRef]
50. Keeling, P.J.; Doolittle, W.F. Alpha-Tubulin from Early-Diverging Eukaryotic Lineages and the Evolution of the Tubulin Family. *Mol. Biol. Evol.* **1996**, *13*, 1297–1305. [CrossRef]
51. Germot, A.; Philippe, H.; Le Guyader, H. Evidence for Loss of Mitochondria in Microsporidia from a Mitochondrial-Type HSP70 in *Nosema locustae*. *Mol. Biochem. Parasitol.* **1997**, *87*, 159–168. [CrossRef]
52. Hirt, R.P.; Healy, B.; Vossbrinck, C.R.; Canning, E.U.; Embley, T. A Mitochondrial Hsp70 Orthologue in *Vairimorpha necatrix*: Molecular Evidence that Microsporidia Once Contained Mitochondria. *Curr. Biol.* **1997**, *7*, 995–998. [CrossRef]
53. Hirt, R.P.; Logsdon, J.M.; Healy, B.; Dorey, M.W.; Doolittle, W.F.; Embley, T.M. Microsporidia are Related to Fungi: Evidence from the Largest Subunit of RNA Polymerase II and other Proteins. *Proc. Natl. Acad. Sci. USA* **1999**, *96*, 580–585. [CrossRef]
54. Peyretaillade, E. Microsporidian *Encephalitozoon cuniculi*, a Unicellular Eukaryote with an Unusual Chromosomal Dispersion of Ribosomal Genes and a LSU rRNA Reduced to the Universal Core. *Nucleic Acids Res.* **1998**, *26*, 3513–3520. [CrossRef]
55. Burri, L.; Williams, B.A.P.; Bursac, D.; Lithgow, T.; Keeling, P.J. Microsporidian Mitosomes Retain Elements of the General Mitochondrial Targeting System. *Proc. Natl. Acad. Sci. USA* **2006**, *103*, 15916–15920. [CrossRef]
56. Henry, J.E. Effect of Grasshoppers Species, Cage Density, Light Intensity, and Method of Inoculation on Mass Production of *Nosema locustae* (Microsporida: Nosematiodae). *J. Econ. Entomol.* **1985**, *78*, 1245–1250. [CrossRef]
57. Wang, L.; Yu, X. Mass Production and Application of *Nosema locustae* Against Grasshoppers. *Acta Agrestia Sin.* **1994**, *2*, 49–54.
58. Zhang, J. Research and Application of New Green Control Technology Extension Mode of Locusts in China. China agricultural university. Ph.D. Thesis, China Agricultural University, Beijing, China, 2015.
59. Beijing JiaJing Biotecnology Ltd. Available online: http://www.jiajingbiotech.com (accessed on 26 March 2021).
60. Zhang, L.; Shi, W.; Yan, Y.; Lin, B.; Wu, M.; Ye, S.; Li, D. Relationships between Application Rate of *Nosema locustae* and Infection of *Locusta migratoria manilensis* in Hainan Province. *J. China Agric. Univ.* **2001**, *6*, 90–95.
61. Henry, J. Experimental Application of *Nosema locustae* for Control of Grasshoppers. *J. Invertebr. Pathol.* **1971**, *18*, 389–394. [CrossRef]
62. Ewen, A.B.; Mukerji, M. Evaluation of *Nosema locustae* (Microsporida) as a Control Agent of Grasshopper Populations in Saskatchewan. *J. Invertebr. Pathol.* **1980**, *35*, 295–303. [CrossRef]
63. Zhang, L.; Yan, Y.; Wang, G.; Zhang, Z.; Pan, J.; Yang, Z. A Preliminary Survey on Epizootics of Infection of *Nosema locustae* Among Grasshoppers in Rangeland. *Acta Agrestica Sin.* **1995**, *3*, 223–229.
64. Zhang, F.; Bai, Y.; Wang, L.; Wang, X.; Zhang, L. A Preliminary Study on Strategies of Management of Grasshoppers in crop-Grassland Ecosystem in Inner Mongolia. *Entomol. J. East China* **2007**, *16*, 113–118.
65. Gong, A.; Liu, X.; Jiang, X.; Zhang, L. Transmission of *Nosema locustae* Disease in Grasshopper Populations in Qinghai Grassland. *Chin. J. Biol. Control* **2003**, *19*, 118–121.
66. Johnson, D.L.; Dolinski, M. Attempts to Increase the Prevalence and Severity of Infection of Grasshoppers with the Entomopathogen *Nosema locustae* Canning (Microsporida: Nosematidae) by Repeated Field Application. *Mem. Entomol. Soc. Can.* **1997**, *129*, 391–400. [CrossRef]
67. Phithalsoun, S.; Zhang, L. An Effective Biological Control Method of Yellow-Spined Bamboo Locust (*Ceracris kiangsu*) has been de-Veloped in Lao PDR and Vietnam. *Metaleptea* **2018**, *38*, 15–16.
68. Tounou, A.K.; Kooyman, C.; Douro-Kpindou, O.K.; Poehling, H.M. Combined Field Efficacy of *Paranosema locustae* and *Metarhizium anisopliae* var. acridum for the Control of Sahelian Grasshoppers. *BioControl* **2007**, *53*, 813–828. [CrossRef]
69. Lima, M.L.L.; Brito, J.M.; Henry, J.E. Biological Control of Grasshoppers in the Cape Verde Islands. In *Biological Control of Locusts and Grasshoppers*; Lomer, C.J., Prior, C., Eds.; CAB International: Wallingford, UK, 1992; pp. 287–295.
70. Lange, C. Long-term Patterns of Occurrence of *Nosema locustae* and *Perezia dichroplusae* (Microsporidia) in Grasshoppers (Orthoptera: Acrididae) of the Pampas, Argentina. *Acta Protozool.* **2003**, *42*, 309–315.
71. Lange, C.E.; Azzaro, F.G. New Case of Long-Term Persistence of *Paranosema locustae* (Microsporidia) in Melanopline Grasshoppers (Orthoptera: Acrididae: Melanoplinae) of Argentina. *J. Invertebr. Pathol.* **2008**, *99*, 357–359. [CrossRef]
72. Zhou, X.; Zhang, L. Selection of *Antonospora locustae* (Protozoa: Microsporidae) with Higher Virulence Against *Locusta migratoria manilensis* (Orthoptera:Acrididae). *Biocontrol Sci. Technol.* **2009**, *19*, 421–427. [CrossRef]

agronomy

Article

Locusts and People: Integrating the Social Sciences in Sustainable Locust Management

Clara Therville [1,*], John M. Anderies [1,2], Michel Lecoq [3] and Arianne Cease [1,4]

1 School of Sustainability, Arizona State University, Tempe, AZ 85287, USA; m.anderies@asu.edu (J.M.A.); acease@asu.edu (A.C.)
2 School of Human Evolution and Social Change, Arizona State University, Tempe, AZ 85287, USA
3 CIRAD, UMR CBGP, F-34398 Montpellier, France; mlecoq34@gmail.com
4 School of Life Sciences, Arizona State University, Tempe, AZ 85287, USA
* Correspondence: clara.therville@gmail.com; Tel.: +33-6-7959-9705

Abstract: Locust outbreaks have impacted agricultural societies for millennia, they persist today, and humans aim to manage them using preventative strategies. While locusts have been a focus for natural sciences for more than a century, social sciences remain largely underrepresented. Yet, organizational, economic, and cultural variables substantially impact these management strategies. The social sciences are one important means through which researchers and practitioners can better understand these issues. This paper examines the scope and purpose of different subfields of social science and explores how they can be applied to different issues faced by entomologists and practitioners to implement sustainable locust research and management. In particular, we discuss how environmental governance studies resonate with two major challenges faced by locust managers: implementing a preventative strategy over a large spatial scale and managing an intermittent outbreak dynamic characterized by periods of recession and absence of the threat. We contend that the social sciences can help facilitate locust management policies, actions and outcomes that are more legitimate, salient, robust, and effective.

Keywords: environmental governance; social variables; locusts; social sciences

Citation: Therville, C.; Anderies, J.M.; Lecoq, M.; Cease, A. Locusts and People: Integrating the Social Sciences in Sustainable Locust Management. *Agronomy* **2021**, *11*, 951. https://doi.org/10.3390/agronomy11050951

Academic Editor: Nikolaos Papadopoulos

Received: 18 April 2021
Accepted: 7 May 2021
Published: 12 May 2021

Publisher's Note: MDPI stays neutral with regard to jurisdictional claims in published maps and institutional affiliations.

1. Locust Science Needs Social Sciences

It is evident to all researchers and practitioners concerned by locusts that these insects have always been of great concern to people socially, culturally, economically, and politically. Locusts are currently considered by the FAO as "one of the most destructive migratory pests in the world"(http://www.fao.org/locusts/en/, accessed on 10 May 2021): there are at least 20 different agriculturally important locust species affecting the economies of most continents, a large range of more or less developed countries, and the livelihoods of millions during outbreaks [1,2]. In addition to impacting vulnerable agricultural communities in the short term, a locust plague can have long-term consequences such as spurring migration to urban areas, decreased access to education for children, exacerbating tensions between pastoralists and farmers, and overexploitation of natural resources [3,4]. Control campaigns can cost many millions of dollars [5], and the chemical insecticides used can have serious side effects on human health and the environment [6,7]. Locusts are also an object of political power. For example, Peloquin (2013) demonstrated how the French colonial empire designed the management of the desert locust crisis in 1943 to be perceived as a transnational and constructive federation against a common enemy (locusts), but in ways that strengthened its authority during a geopolitical legitimacy crisis [8].

Throughout history, humans have predominantly considered locusts as a threat and have tried to protect themselves from them. Until about the last one hundred years, locusts were described as literally falling from the sky and anticipating invasions was impossible [9]. The urgency to find solutions to this agricultural challenge led governments to support scientific research, locust ecology in particular, to locate main breeding areas (outbreak areas) and develop more efficient survey and control methods [10]. It was the discovery by Uvarov of locust phase polyphenism [11,12] that directed field research towards locating outbreak areas from the 1930s, and thus provided a rationale for preventive strategies around the world [10]. This discovery induced a paradigm shift from old reactive approaches focusing on crop protection in a context of complete uncertainty to anticipative preventive strategies and risk management systems [13–15]. Many entomologists around the world thus devoted their careers to exploring the best potential biological and ecological solutions to deal with the locust problem [10]. Since then, tremendous scientific work, especially in the natural sciences, as well as technological advances for example in communication, remote sensing imagery and climatic and weather data collection were developed to manage locusts [13,14]. The current preventive strategy relies on the surveillance of areas prone to producing gregarious populations, which ultimately help people make decisions for control according to locust densities and location of infestations, using either chemical, biological or ecological control [13,14]. However, this strategy remains questioned, as illustrated by the debate that followed the 2003–2005 desert locust invasion [16,17]. Locust outbreaks continue to arise and some authors wonder if sustainable locust management is possible [18].

This observation has led some scientists and practitioners to insist on the social and organizational variables that may impact locust management [19]. Research has described this impact on our capacity to understand and manage locusts since the end of the 19th century for different locust species in different part of the world (see, e.g., [10,20]). These studies point out the social constraints to maintaining an efficient preventive strategy through time and through large spatial scales. Recent reviews highlight a common and significant constraint for sustainable locust management: the reduction of resources during recession times [14]. As one of the most well-known species, the case of the desert locust provides old testimonies of this challenge. Roy, an historical expert in the fight against locust invasions in West Africa, looked back at the conditions that made the great invasion of 1987–1989 possible through the degradation of an existing preventive control strategy [21]. He highlights the impact of the loss of knowledge during recession times, induced by the dismissal or retirement of field staff and researchers. Past and current outbreaks of the desert locust have also been explained by armed conflicts and insecurity that have led some countries to lack capacity to sustain expertise, monitoring, and control [22,23]. In 2005, Lecoq insisted on the need for stronger risk-management systems, and put an emphasis on governance, funding, flexibility and a better understanding of the role of locust-control stakeholders [19]. Fifteen years later, the identified highest priorities to face desert locust outbreaks sounds familiar [22]: "(a) to ensure that the political and socio-economic conditions are in place so that vulnerable human populations can adapt to new large-scale threats and (b) to maintain a culture of long-term risk assessment with constant necessary means".

The social sciences are one means through which researchers and practitioners could come to understand the social variables that impact locust management. However, paradoxically, while locust management is seriously constrained by organizational, economic, and coordination issues [14], social sciences are underused in locust research. There are numerous recent studies on locust biology and on the use of remote sensing technologies to improve locust management systems (reviewed in Cullen et al. 2017) [24]. We do not question the fact that further research is needed on genetics, ecology, population dynamics, or remote sensing to improve our knowledge of locusts and management strategies, especially under a changing climate. However, more than fifteen years after the call to switch locust management "from ecology to anthropology" [19], social sciences are still overlooked in locust studies. That being said, counter-examples can be found. For instance, economics has already provided useful arguments to defend preventive strategies for locusts. In the case of the Australian plague locust [25], the benefit–cost ratio for all the control campaigns of the 2010–2011 plague is estimated at 19:1, indicating that every dollar invested in locust control generates $19 in avoided losses.

This call towards social sciences also meets other calls towards the reframing of risk management, crop pest management, insect science, and locust management in particular, using social-ecological, sustainability, and transdisciplinary approaches [26–28]. A social-ecological approach is a way to emphasize the interlinked social and ecological dynamics and the cross-scale and cross-level social-ecological complexity involved in managing the environment [29,30]. For example, Cease et al. [31] described the locust-grassland-human system as a coupled human and natural system, implying complex feedback that connect geographically distinct people and places across time (ecological connections with locust migration or socio-economic connections through markets). Considering locusts as a "wicked" problem [32] also induces new ways of doing science, capable of dealing with problems characterized by high stakes, uncertainties, values in dispute, and urgent decisions [33]. In such a renewed frame, research not only has to be scientifically robust, but should also be more involved in the resolution of societal problems and promote the participation of civil society. This requirement justifies an extension of the peer community involved in the resolution of environmental problems, including social scientists, and the mobilization of both interdisciplinarity (within academia but across disciplinary boundaries) and transdisciplinarity—also referred to as cross-sectoral (across the professional boundaries between researchers and practitioners, including managers and agriculturists in this case) [34].

In response to calls to make social sciences mainstream in locust research and management, this paper provides a succinct guide and overview of the social sciences for present and future locust managers, for natural scientists leading or participating in multi-stakeholder processes and interested in strengthening the role of social sciences, and for social scientists interested in locust research. We consider that the failure of social science to be mainstreamed stems in part from a lack of clearly articulated objectives and values associated with the social sciences. This article firstly identifies the distinct contributions that different fields of the social sciences can make to understanding and improving locust management. Then, we detail the main social challenges faced in locust management and how different areas of social science can shed light on them. In particular, we mention other examples of risk studies (e.g., shocks such as earthquakes or floods), institutional approaches used in these cases, and advancements surrounding the resilience of complex social-ecological systems. We conclude with a discussion of several key considerations for better engaging with the social sciences to improve locust management.

2. What Are the Social Sciences?

The social sciences are a set of tools and ideas that focus on how humans make decisions, create and maintain social relations, and how these individual processes lead to macro-level patterns of organization, such as interest groups, cultures, or nation states. Locust management involves decisions of individual farmers, managers, and policy actors, as well as the relationships among these actors (e.g., information exchange, coordination of activities). These decisions and relationships lead to macrolevel patterns (e.g., organizations to manage locust outbreaks across the spatio-temporal expanse of locust outbreaks). Thus, social science is a useful tool to understand the human and social dimensions of environmental issues in general [35], and of locust management in particular. Yet their absence is commonly reported in environmental studies, either dealing with global environmental change research [36], biodiversity conservation [35], or pest management [26,37]. This can be observed in the locust community through the limited number of social scientists participating in locust symposia and in locust review papers [24]. The barriers impacting the integration of social sciences and natural sciences has been particularly explored in the field of conservation sciences [38,39]. These papers highlight how, among many natural scientists and practitioners, there remains a lack of awareness about the social sciences. This knowledge void interferes with the ability to engage with the social sciences in a constructive manner that can inform management practices [35,40]. The multiple challenges that are known to interfere with the integration can be conceptual (e.g., confusion over the role of social sciences, different expectations, and/or disagreements on frameworks, definitions or concepts) or structural (e.g., lack of social scientists, compartmentalized organization, and/or lack of funding or opportunities for interdisciplinary collaborations) [38,41]. Another common challenge is the instrumentalization of social scientists to better package or market ecological knowledge [38], that devalues the knowledge they produce. Conversely, the critical approach developed by some social scientists on the "politics of nature", highlighting the role of ecology in power relations and injustices [39], has led to misunderstandings and even to a mistrust between social and natural scientists. Resolving these challenges requires a mutual respect and a basic understanding of respective approaches: both social and natural scientists should be able to identify the different disciplines embedded in the other realm, and be aware of the philosophical principles and theoretical assumptions of each of these disciplines [42]. On the other hand, each scientist should be able to make their discipline more digestible for scientists originating from other disciplines, through a more straightforward use of language, direct communication and accessibility of research findings [39].

Without going into detail—see Moon and Blackman [42] or Bennett et al. [35] for a more complete understanding of the social sciences—we provide here some elements of social science ideas and theory relevant to locust management. Social sciences are used to study five broad categories of the social realm: (i) social phenomena (e.g., markets, governance, politics, culture, demographics), (ii) social practices and processes (e.g., social organization, decision-making, knowledge exchange, collaboration), (iii) social attributes (e.g., social capital, trust, memory), (iv) actors, roles, and positions (e.g., bridging organizations, leadership), or (v) individual attributes (e.g., values, beliefs, knowledge, motivations, preferences, behaviors) [35,43]. They include disciplines such as sociology, anthropology, political science, geography, economics, history, psychology, and other more applied social sciences such as education, development, or communication science. Each discipline has specific topical strengths (e.g., governance for political science, social interactions and social facts for sociology, communication science to inform communication strategies during a crisis) and relies on established and specific bodies of social theories. Specific disciplines of the social sciences have included environmental concerns, such as environmental psychology or environmental economics. To tackle complex social-ecological problems, multiple interdisciplinary fields have emerged as well such as the resilience of social-ecological systems [29], environmental social science [44], political ecology [45] or environmental humanities [46]. All of them highlight the interconnectedness between the biophysical and the social and are supposed to facilitate the engagement between natural and social scientists.

Social sciences can be conducted on issues at different scales, from individual to local to global. The individual scale might be adopted to study perceptions, attitudes, or behaviors. For example, what are the preferences of farmers regarding the use of insecticides [47]? At the global scale, scientists might study governance models and the impact of decentralization in locust control [48]. The applications of social sciences vary from predicting, understanding, emancipating, or deconstructing the issue of interest [42]. They can focus on understanding social phenomena, testing or developing theories, and/or giving critical insights on a situation to overcome injustices. More applied social science fields such as communication science can help to think about the role played by information and communication technologies to promote coordination and appropriate responses in the face of extreme events [49], e.g., locust plagues. Methodology can be qualitative (e.g., interviews, focus groups, discourse analysis), quantitative (e.g., closed-ended questionnaires, cost-benefit analysis), participatory (e.g., participatory action research, photovoice), and/or forward-thinking (e.g., scenario planning, economic modeling) [35]. Alongside this "social science" panel, the arts and humanities, which designate knowledge fields related to culture in general (including history, philosophy, cultural and science studies or literature) should also be mentioned as an important complement to both social sciences and natural sciences. They question our values, our representations, our responsibilities [46], and provide sensitive means to address these fundamental issues.

3. An Overview of Existing Social Science Contributions to Locust Management

Although poorly represented, social sciences are not completely absent from the locust research and management world and some contributions can be found. They can be differentiated depending on their disciplinary approach and on their applications, from more analytical to more applied (Figure 1). Figure 1 is an attempt to summarize these distinctions and highlight some of the major contributions of the social sciences applied to locust management. The proposed categories are not necessarily exclusive from one another, nor as clear as they appear in Figure 1 since this should rather be seen as a continuum.

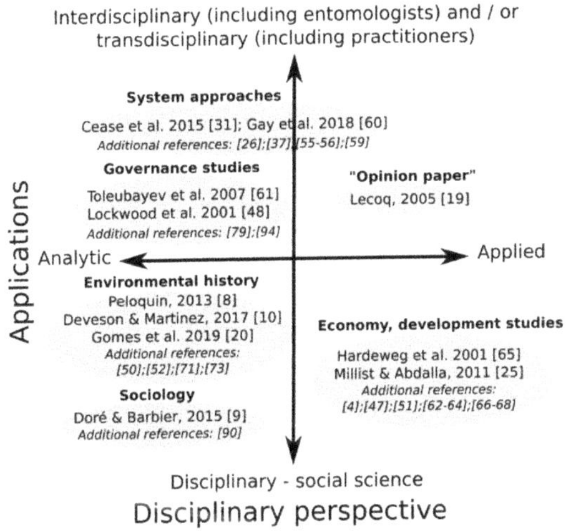

Figure 1. A framework to summarize some of the main types of studies in locust science that include social sciences, depending on their disciplinary approach and on their applications, from more analytical to more applied. In additional references, some papers may not focus on locusts but are representative of a given approach, e.g., [50].

Diverse disciplinary approaches can be found in the social science literature on locusts: contributions can be embedded in one subdiscipline of the social sciences, such as sociology [9] or economics [25]; they can emerge from social science fields that are specifically interested in environmental issues, such as environmental history [10] or political ecology [51]; finally, they can come from interdisciplinary and transdisciplinary groups of authors who are willing to address the social variables impacting locust management [31,48]. Such papers may involve both entomologists/natural scientists alongside social scientists (interdisciplinary approach) and practitioners (transdisciplinary approach) who sometimes have a strong interest in social sciences and can even be double-hatted at the same time, being a practitioner or entomologist and trained as a social scientist. These contributions also reflect diverse possible applications: while some of them aim to describe and understand social phenomenon from an analytical point of view [9], others are clearly oriented towards action: they urge some changes in the ways we are thinking, managing and governing locusts outbreaks [19,48], are assessing the economic feasibility of current management strategies [25] or are exploring the possibility to use renewed instruments in locust management, such as payments for ecosystem services [52] or compensation funds [47].

Along this framework (Figure 1), four main categories of studies can be identified. First, studies embedded in an analytical approach and rooted in classical fields of the social sciences such as sociology or history can be found in the lower left corner. Interestingly, history is a particularly well represented discipline with contributions focusing on the desert locust [8], the South American locust [10], the Australian plague locust [53] and the

Moroccan locust [20]. These studies can focus on a specific event in history, such as the 1943 conference convened to discuss ongoing locust plagues in Northwestern Africa [8], or retrace a longer history of structuration of locust management of over 50 years [10,20]. Lessons learned include perspectives on historical distributions of locust invasions, explanations regarding the ineffectiveness of the measures taken, including a lack of transboundary coordination [20], or the strong interactions between political power and scientific research, leading entomologists to be recognized as public scientists [10]. By learning lessons from the past, environmental history offers enormous potential to inform contemporary locust management. According to Gomes et al. [20], histories can help to change the world and *"Neglecting the history of locusts [...] is a step towards forgetting the presence of the species..."*, which is a key concern in locust management, as we will discuss later.

Second, studies located in the upper left corner of the figure were written by interdisciplinary and transdisciplinary groups of authors, even if they were still embedded in a particular analytical perspective. This includes what we have called social-ecological systems or "SES"approaches, which draw on theory from ecology, economics, and political science to focus on understanding the complex dynamics generated by multi-level interactions (interactions on multiple temporal and spatial scales linked across different levels of social organization) between inter-related social and ecological systems. These studies have been strongly influenced by systems theory and resilience studies and used concepts such as coupled human and natural systems, telecoupling, scales and scale mismatch, feedback loops or complex adaptive systems [30,54,55] to discuss the complex, and sometimes long-distance social-ecological interactions within the locust system. Ecologists can be pushed towards such an approach when studying the interactions between locusts and their environment, which thus includes the impact of human practices (crop farming, grazing) on the surrounding environment. For example, Cease et al. [31], Le Gall et al. [56], Word et al. [57] and others before them [58,59], demonstrated there can be a substantial impact of overgrazing or the type of crops and soil management regime on locust dynamics. This opens a space for discussion to include land management practices as a way to manage locusts, and thus to include farmers and their social environment (e.g., markets, policies) as key players. Within this family of systems studies, Gay et al. [60,61] focus on the ability of the current preventive system to manage desert locust plagues. Using multi-agent systems that include both human (e.g., field teams, national control units, funding institutions) and non-human agents (e.g., locusts), they explore the role of funding institutions' awareness, budget cyclicity, and the lack of access to some areas due to insecurity. This last topic has also been recently explored by Showler and Lecoq [23] who demonstrated the various effects of armed conflicts in the case of the desert locust over the last 35 years. Lastly, some studies develop a particular interest for collective action and governance in general. For example, Toleubayev et al. [62] focuses on the locust management system in Kazakhstan since the formation of the Soviet State and its disintegration after the collapse of the Soviet Union. Similarly, Lockwood et al. [48] analyzed the relative importance of various stakeholders and interests in three case studies (Wyoming in the United States, Eritrea, and Irkutsk in Russia). They observed the changing role of central governments with the deployment of bottom-up approaches and decentralization, and the consequences of locust plagues for farmers in different contexts. From these observations, the authors questioned the historical dependence on distant governments. In the same vein, in the upper right corner of Figure 1, Lecoq [19] proposed what we identify as an opinion paper, which suggested that *"a new approach to locust issues (...) using techniques derived from sociological and anthropological sciences"* is needed. After retracing the history of recent desert locust plagues, Lecoq acknowledged the role played by organizational problems and by the lack of involvement in prevention systems of various stakeholders such as farmers' organizations: *"the locust is no longer the real problem, humans are the real problem. Every time there has been an outbreak over the last 50 years, the main problem has been human organization..."*. He thus called towards the development of a risk management system, a new governance strategy involving a larger panel of stakeholders, and a better understanding of the role of each of them.

Studies in the lower right corner of Figure 1 used a disciplinary approach, mostly rooted in economic or in socio-economic and development studies with the practical objective of evaluating current management practices and proposing credible alternatives where appropriate. In particular, desert locust campaigns have been submitted to socio-economic evaluations under the impetus of the FAO in the 1990s [63]. These studies aimed to identify who is affected by the desert locust and to what extent, if the current management strategy and control campaigns are appropriate, and what are the costs and the other possible policy responses. Such studies are recognized to be difficult to carry out because of the enormous area that must be studied, the heterogeneity of the potential damages of the pest, the multiple variables that can impact crop production, and the irregular occurrence of outbreaks [64]. Some initial studies criticized the effectiveness of control strategies which are presented as not very cost effective economically, socially, or environmentally [64,65]. Complementary studies were developed later to better integrate the stakes and perceptions of diverse groups at the farm and households levels, and proposed alternatives such as compensation, insurance policies, farmers' adaptation strategies or biological control [47,66–68]. Lastly, some recent studies demonstrated the long term impact of locust plagues, in particular on health and education, and the lasting damages the plagues caused during early childhood into adulthood [4,69].

To conclude, social science studies interested in locusts are as diverse as natural science studies [24]. As illustrated in Figure 2, they are included in a diversity of approaches (e.g., environmental history, systems approach, sociology, economics, and development studies). They mobilize a diversity of methodologies (e.g., document analysis, quantitative data analysis, economic evaluation, ethnographic approach, reliance on expertise knowledge). They can pursue different objectives, from policy function and structure or efficiency analysis to the inclusion of people's perceptions and social-ecological analysis. They can focus on different scales, from individuals to nations and beyond, they can support existing management practices or adopt a critical approach and try to push towards alternative solutions. While some disciplines are particularly well represented, such as environmental history or economics, others are less so, such as communication, philosophy, sociology, anthropology, and psychology. Their inputs could be valuable to locust science in questioning ethics and epistemologies (see, e.g., the recent work of Lockwood and Sardo [70]), or in observing the issues in the day-to-day management of locusts, during crisis and recession times. For example, communication science could contribute to the development of communication plans to ensure trust over the long run and support coordination during outbreaks.

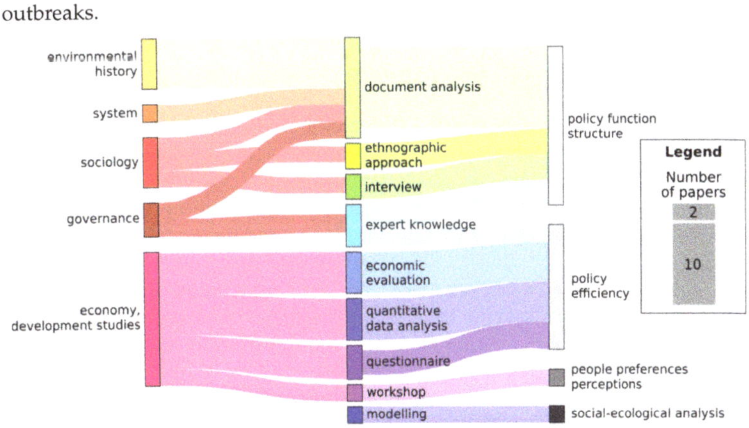

Figure 2. Main approaches, methodologies and objectives represented in the studies included in Figure 1 and specifically focusing on locusts (n = 25). Only the boxes and flows represented in two or more papers are represented on this figure. The flows on the left represent which methods are used

by each approach. The flows on the right represent which methods are used to pursue the different objectives. The size of the flows is proportional to the number of papers using this represented flow. For more detail, see Table S1.

4. Social Sciences and Tackling Locust Management Challenges

In this section, we highlight how social sciences can help to tackle two of the main challenges of locust management: (1) ensuring coordination and a multi-level governance structure from a local to a very large scale, and (2) maintaining the management system during recession times.

4.1. Ensuring Multi-Level and Large-Scale Coordination

Most countries recognize that effective locust management requires some form of large scale coordination because control programs are beyond the capabilities of individual landholders alone [14]. At a landscape scale, locusts readily move from one farm to another, so their treatment is often considered a community responsibility. Cooperation at a local scale can be implemented more or less informally or formally, for example informally at a village/town scale or formally through cooperative programs at the county or state level, as in Wyoming, USA [37]. Combating locust outbreaks has also led governments to take responsibility since the latter part of the 1800s, as is the case in Australia or in Argentina [10], or the US in response to the Rocky Mountain locust outbreaks [71] that led to the emergence of some of the first national plant health policies [72].

For species such as the desert locust, which is associated with one of the most ancient organized preventive systems, the great mobility of the swarms imposes the necessity of bilateral, regional and international cooperation [73]. As illustrated by historical studies, some evidence from large-scale collective mobilization against locusts can already be found in Antiquity, with the existence of local locust control specialists and the use of armies to destroy the insects or help with the harvest [74]. The internationalization of the locust problem started as early as the beginning of the 20th century when a large plague in Algeria contributed to the awareness of the international community of the fight against the locust plague [50]. However, the first attempt to find a transboundary solution at a truly international scale only emerged in the 1920s with the first international conference on locust control held in Rome under the auspices of the International Institute for Agriculture and which resulted in an "International Convention for the control of locusts and grasshoppers" [75]. Since that time, it was considered as hopeless to study the locust problem on a narrow territorial basis and cooperation has not ceased to develop to the present day. Many efforts have been pursued to ensure better transboundary coordination. In the 1930s the process of formation of a locust scientific community and the institutionalization of the international scientific fight against the locust took place by means of five international locust conferences, held in Rome (1931), Paris (1932), London (1934), Cairo (1936) and Brussels (1938), resulting in a first conception of a preventive control program for a few locust species by regular monitoring of the outbreak areas [76]. Cooperation became a true reality in the 1950s, with the creation of the desert locust control committee (DLCC) under the aegis of the FAO in 1954, mandated by its Member States to ensure the coordination of monitoring and control activities of the desert locust [73]. Since the creation of the DLCC, FAO has continued to play an active role to promote international locust control cooperation, on the desert locust as well as for many other locust species (i.e., migratory, Moroccan, South American, and red locusts). In present days, the DLCC includes representatives from all the countries affected by the desert locust as well as those which take part in the locust control campaigns, such as various donor countries helping to finance control campaigns. The DLCC is linked in the field by three regional Commissions (created under the auspices of the FAO and at the request of the affected countries): the commission for the Western Area, CLCPRO (the "*Commission de lutte contre le Criquet Pèlerin dans la Région Occidentale*" created in 2002 to coordinate and finance a regional solidarity on both sides of the Sahel [51,77,78]), the Central Region (CRC), and the eastern region in Southwest Asia (SWAC), as well as by an inter-state organization: the Desert Locust Control Organization for Eastern Africa (DLCO-EA). These Commissions are mandated to promote all activities, research and training necessary to ensure effective preventive control and capacity to deal with any invasion of the desert locust [73]. However, coordination still presents some weaknesses, even if solidarity and mutual support are well established. In their evaluation of the 2003–2005 desert locust campaign for instance, Brader et al. note that "*A weakness is the lack of clarity concerning the different roles of different actors, and how these relate to each other*" [3].

Given this context, a challenge that the social sciences can help to address is to understand if, when, and how collaboration is effective [79] from local to the international levels. Because of their mobility, locusts represent a collective action problem: they reveal a situation of mutual interdependence between the actors where outcomes depend on the actions of all of these actors. Invasive species, crop diseases, locusts, and other migratory pests have been considered by some authors as a "public bad" [62], a "collective bad" [80], or as a "public good" for their controlling system [81]. These terms all refer to collective action theory [82]. Many authors have noted that desert locust control should be considered as a regional or even international "public good" [83,84], meaning that it is provided as a service accessible to all, including vulnerable people, and supported by those who can afford it and are willing to pay for it, e.g., the general public through governments, donor countries or otherwise. As acknowledged by Toleubayev regarding this "common enemy" [62]: "*agricultural producers are not able to control locusts outside their private plots. This is why many countries treat the control of migratory and highly destructive pests as a public service, comparable with emergency services such as the fire brigade and the police*". Compared to other archetypal examples of commons such as small-scale forests or irrigation systems, locusts are special since they pose a problem that evades and exceeds the spatial reach and extent of control of the nation state. They thus raise transboundary governance issues as do bluefin tuna fisheries, another archetypal example used to study collective action in large-scale social-ecological systems [85,86]. These transboundary cases have a high number of stakeholders, the temporal and spatial extent of biophysical processes go beyond and are permeable to geopolitical boundaries, and there is often a limited understanding of the complexities of ecological and societal dynamics. Therefore, collective action and collaborative environmental governance are essential [79], which underscores the importance of stronger inclusion of collective action theories [79,82] in the design of governance regimes. In particular, the notion of mismatch can be useful regarding locust governance. According to several authors, many of the problems encountered by societies in managing natural resources arise because of a mismatch between the scale of management and the scale of the ecological processes being managed [30,55,87]. This has been referred to as "scale mismatches", the "scale challenge", institutional "fit" or "interplay", a challenge which is particularly relevant for transboundary issues such as locusts.

Raising coordination issues and the scale challenge, including at a transnational level, is not new in studying locusts. For example, through an historical approach [20], Gomes et al. demonstrated that transnational collaboration was considered fundamental to fight against locusts since the beginning of the 20th century. Through the example of the Iberic peninsula, the authors highlighted the attempts and difficulties in implementing such a coordinationeffort. Conflicting interests and lack of means led the Spanish actors to blame each other for losses, and the Portuguese actors to blame the inaction of their Spanish neighbors as they observed locust swarms frequently travelling across the border. While a widespread opinion throughout the twentieth century was that Iberians should fight the locusts together, and some evidence support this idea of cooperation, Gomes et al. concluded that "*cooperation and coordination were rare or non-existent*". In a second example, Lockwood et al. [48], recognized that scale and scale mismatch are fundamental problems in assessing the sustainability of controlling locusts. By focusing on what they called the "geopolitical scale", the authors insisted on the problem of the historical dependence of local agents on distant governments that often have the resources, knowledge, and expertise. In parallel, the authors observed a paradoxical trend towards decentralization in locust control characterized by a decline in external support with the aim of increasing responsibilities at the local level. With the notions of "social scale" and "interest scale", they highlighted the diversity of motivations, gains, and losses in different contexts and at several governance levels, for example between farmers, the general public, agrochemical industries, and governments. In a final example, Peloquin [8,51] provided an example of a critical study regarding the concept of scale mismatch. Similar to other political ecologists, rather than taking the mismatch between institutions and ecosystems as given, he critically investigated why and how given socio-ecological dynamics become adopted as management mandates by agencies, and what particular technological or institutional arrangement is favored by these configurations. Taking the example of locusts, he illustrated how the French colonial Empire prioritized organizing against the desert locust during the Second World War because the spatial extent of the insect provided an ideal context in which to reinvent the spatiality and legitimacy of the French Empire as a transnational and constructive federation during a geopolitical crisis [8]. Similarly, Roy describes how the existence of two colonial empires on the African continent—French in the West, English in the East—explained for decades the lack of cooperation at the scale of the continent regarding the management of the desert locust [21].

4.2. Maintaining Vigilance during Recession Times

The second main challenge is to maintain the infrastructures needed for sustainable locust management during recession times. Locusts are temporal, alternating absence with peaks of high activity that must be rapidly managed. Major outbreaks can be separated by several decades, depending on the species and on the maintenance and efficiency through time of the management system. Such long-term environmental problems raise the issue of funding, support provision, and knowledge maintenance over substantial time frames [79]. For locust managers, an acknowledged challenge is the maintenance of an operational control system and a set of elements: (i) available and functional material for survey and control (e.g.,vehicles, planes, insecticides) and (ii) available and trained staff during recession times. Funding is of course essential to ensure the maintenance of these two crucial elements.

Some studies on locusts have tackled this issue. As illustrated in Figure 3, Lecoq was one of the first to explicitly refer to an institutional cyclicity in the case of locusts [88]. Making an analogy with locust phase polyphenism, he characterized an alternation between two institutional phases: a motivation phase associated with important funding in the face of an emergency situation, and an oblivion phase during recession times when "*the memory of the invasion fades, funding is reduced below an operational level, specialists are dispersed, and motivation wanes*". This "vicious" cycle [60] thus refer to a situation in which one is locked in a complex chains of events that reinforce themselves through a feedback loop and that has detrimental results. It is typified by the desert locust but can also be observed in Madagascar for the migratory locust [14] and in the South American locust [89]. When looking back at the desert locust invasion of 1987–1989 [21], Roy notes that when the severity of the threat diminished, vigilance is decreased. Under other constraints and faced with other emergencies, government authorities gradually considered locust control action a secondary interest. Funding from government partners was deferred, then reduced or even suspended, leading to a decrease in operational potential, staff and non-renewed equipment that degrades. Through an insidious process, managers worried about the possible loss of their positions and tended to hide the seriousness of the decline in field staff and resources from the government on which they depended. Roy described the loss of knowledge induced by the dismissal or retirement of field staff and researchers as "*one of the most devastating effects (...) as equipment can be renewed and insecticide stocks replenished in a matter of months, but the loss of these experienced and dedicated field staff is an irretrievable loss—it will take many years to train their successors, but the trainers themselves have become scarce*". Roy and Lecoq thus asked this question which remains topical [21,88]: how to get out of this vicious cycle? How can an operational structure be maintained against an intermittent scourge? We add this complementary question: how to identify emerging institutional weaknesses before they reveal themselves mostly during a time of crisis?

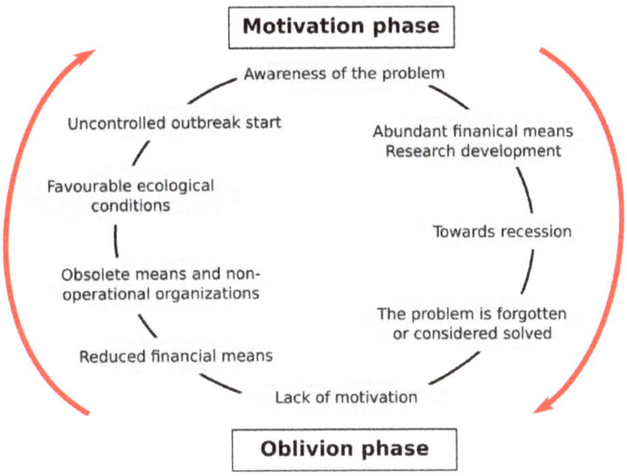

Figure 3. The phase theory applied to institutions: a representation of the vicious cycle alternating a motivation phase and an oblivion phase in locust management. Adapted from Lecoq, 1991 [88].

The French sociologist Doré and colleagues studied the maintenance of vigilance during recession times [9,90]. In a time when the efficiency of the preventive control system was questioned, their work was an attempt to understand the failures of the preventive system that led to the 2003–2005 desert locust invasion. More than a lack of trained human resources, Doré et al. highlighted the shortcomings of international involvement which, by seeking to rationalize expenditures, created a slow and under-mobilization of logistical and human resources. Bilateral and regional assistance mechanisms can mitigate these shortcomings to some extent. For example, the authors discussed the necessity to integrate the complexity of organizational and governance issues in a multi-level actor system, and to no longer think of international action only as the game of war against locusts, but also as a permanent biovigilance, which equitably distributes risks among the players [90]. Stopping the erosion of vigilance thus relies on the perpetuation of a difficult coordination among heterogeneous actors: solitarious desert locusts, the information network that connects the managers, and research projects play a central role in this maintenance over time. These multiple components allow the managers to have more continuity and free themselves from having to act in a crisis mode [9]. While Gay et al. also illustrated the role of funding institutions' awareness as a facilitating factor for cyclic locust plagues, they argued that loss of skilled field agents was of primary concern and their recommendations thus included plague simulation exercises [61]. Most locust response systems are subject to this cyclic loss, but awareness has been building for several years and locust managers are now trying to take better account of these issues. For instance, in the case of the desert locust, CLCPRO has implemented risk management plans including simulation exercises between outbreaks, a monitoring and evaluation plan to enable actors in the system to know in real time and at the country level, the state of equipment, staff or pesticides, and a communication plan (http://www.fao.org/clcpro/fr/, accessed on 10 May 2021). More recently, a financing system designed to address the various levels of desert locust infestations began to be developed [91]. There are also successful examples such as the Australian Plague Locust Commission (APLC) in Australia. The APLC has been funded continuously by four Australian member states from 1974. This funding supports the maintenance of their structures and keeps them prepared to immediately react in the event of an alert [92]. We can speculate that this stability is explained both by the fact that the relatively high frequency of Australian plague locust outbreaks helps to maintain the motivation of the actors, and that it concerns a single country that is politically stable and has a high GDP.

These spatio-temporal challenges of governing transient socio-environmental problems are not specific to locusts and can be analyzed using the concept of "temporal misfit" or "temporal-scale mismatch" between biophysical systems and institutions [30,93]. A temporal misfit can be defined as a case in which an institution is formed too early or too late to cause desired ecosystem effects, or if conflicting time frames exist between policy-makers and those of the environment. In the literature, such temporal governance challenges can be found for a diversity of risks and disasters, ranging from epidemics [94] and wildfires [95] to nuclear risk [96]. As illustrated for governing epidemics [94], the dominance of an outbreak narrative can have drawbacks such as marginalizing narratives that emphasize long-term structural, land use, and environmental changes, and local knowledge. The risk of loss of knowledge and expertise has also been observed in the case of nuclear risk [96]. Many nuclear employees and experts who designed, constructed, and currently operate existing nuclear power plants are now approaching retirement. A consequence is the potential loss of a substantial amount of critical nuclear energy knowledge. In response, nuclear organizations conduct knowledge loss risk assessments, evaluate the consequences of the loss of critical knowledge and skills, develop action plans to retain this knowledge, and use this knowledge to improve the skills and competencies of new and existing workers. Knowledge loss may be related to cyclicity, as described for example in bank lending behavior [97]. In this case, the institutional memory hypothesis speculates that cyclicity may be driven by a deterioration in the ability to recognize problems: institutions may tend to forget the lessons they learned as time passes since their last learning experience with the problem. Studies on risk perceptions have shown that personal experience of a natural hazard and trust in authorities has a substantial impact on risk perception. However, perceiving a risk is not sufficient to guarantee that an actor will take action; action depends on motivation, responsibility, and ability to respond [98].

In the case of locusts, these elements question the dominance of an outbreak narrative, the frequency of learning experiences, and the fact that they may be growing increasingly apart in time due to the implementation of effective preventive strategies. A common trap with preventive strategies is that when you are successful in prevention, there seems to be no need for prevention: the perception of the risk disappears, and with it the feeling that investment in prevention systems must be maintained. Bödin argues that the management of such transient environmental problems requires rapid responses, and thus the mobilization of relevant actors organized in ad hoc collaborative networks [79]. In such conditions, studies on collaborative governance have demonstrated that more-centralized networks with specific actors acting as leaders by distributing and coordinating tasks were more favorable [99], if actors already agree on what needs to be done, share inter-personal and inter-organizational trust, and are willing to comply with the identified responses. This of course requires underlying, dense and longer lasting collaborations, even during the absence of locust plagues.

5. Conclusions: Engaging Social Sciences to Improve Sustainable Locust Management

Social sciences are a vital component, along with the natural sciences, for effective and sustainable locust management, including decision-making, planning, implementation, and management both during crisis and recession times. In this paper, our objective has been to clarify the role played by social sciences and inspire uptake of the social sciences in locust research and management. We have highlighted the diversity of possible applications and contributions, as well as some of the existing contributions to respond to the main challenges faced by locust managers around the world. We have also pointed out some shortcomings, such as the limited questioning of locust management narratives and underlying epistemologies. Examples of studies on other large-scale and erratic socio-environmental problems such as epidemics, hurricanes, and wildfires could shed new light and perspectives on the present and future challenges faced by locust managers. In particular, some long-term environmental and social changes may challenge the way locusts were previously managed. These changes include not only land-use and climate change and rising environmental concerns, but also renewed environmental governance models that include more bottom-up and participative approaches, and increasing calls for social justice [70]. Such changes call for new ways of interacting and living with nature and with locusts. They question the dominant narratives focusing predominantly on techno-scientific control, mobilizing war rhetoric or continuing to present locusts as a biblical plague. The social sciences can help take a step towards making salient what we often take for granted such as taking into account only techno-scientific parameters, highlight the impact of variable social aspirations and political divisiveness, and invent the future of sustainable locust management.

Supplementary Materials: The following are available online at https://www.mdpi.com/article/10.3390/agronomy11050951/s1, Table S1: Detailed categorization of the papers referenced in Figure 1 and graphically illustrated in Figure 2 (n = 25).

Author Contributions: Conceptualization, C.T., J.M.A. and A.C.; validation, J.M.A. and A.C.; writing—original draft preparation, C.T.; writing—review and editing, C.T., J.M.A., M.L. and A.C.; visualization, C.T.; supervision, J.M.A. and A.C.; project administration, A.C.; funding acquisition, A.C. All authors have read and agreed to the published version of the manuscript.

Funding: This research was funded by the Foundation for Food and Agricultural Research, grants no. 593561, by the Global Locust Initiative and by the Swette Center for Sustainable Food Systems.

Acknowledgments: Research reported in this publication is supported by the Foundation for Food &Agriculture Research, Grant ID: 593561, the Global Locust Initiative and the Swette Center for Sustainable Food Systems at Arizona State University. The content of this publication is solely the responsibility of the authors and does not necessarily represent the official views of the Foundation for Food &Agriculture Research.

Conflicts of Interest: The authors declare no conflict of interest. Given their role as Guest Editor, Michel Lecoq & Arianne Cease had no involvement in the peer-review of this article and had no access to information regarding its peer-review.

References

1. Symmons, P.; Cressman, K. *Desert Locust Guidelines 5. Campaign Organization and Execution;* Food and Agriculture Organization of the United Nations: Rome, Italy, 2001.
2. Pener, M.P.; Simpson, S.J. Locust Phase Polyphenism: An Update. In *Advances in Insect Physiology;* Simpson, S.J., Pener, M.P., Eds.; Academic Press: Cambridge, MA, USA, 2009; Volume 36, pp. 1–272.
3. Brader, L.; Djibo, H.; Faye, F.; Ghaout, S.; Lazar, M.; Luzietoso, P.; Babah, M.O. *Towards a More Effective Response to Desert Locusts and Their Impacts on Food Security, Livelihoods and Poverty;* FAO: Rome, Italia, 2006.
4. de Vreyer, P.; Guilbert, N.; Mesplé-Somps, S. *The 1987–89 Locust Plague in Mali: Evidences of the Heterogeneous Impact of Income Shocks on Education Outcomes;* No. DT/2012/06, WorkingPapers; DIAL (Développement, Institutions et Mondialisation): Paris, France, 2012; 48p.
5. Belayneh, Y.T. Acridid pest management in the developing world: A challenge to the rural population, a dilemma to the international community. *J. Orthopt. Res.* **2005**, *14*, 187–195. [CrossRef]
6. Peveling, R. Environmental conservation and locust control—Possible conflicts and solutions. *J. Orthopt. Res.* **2001**, *10*, 171–187. [CrossRef]
7. Everts, J.W.; Ba, L.; Krall, S.; Peveling, R.; Diallo, D.B. Environmental effects of locust control: State of the art and perspectives. In *New Strategies in Locust Control;* Springer Science and Business Media LLC: Berlin/Heidelberg, Germany, 1997; pp. 331–336.
8. Péloquin, C. Locust swarms and the spatial techno-politics of the French Resistance in World War II. *Geoforum* **2013**, *49*, 103–113. [CrossRef]
9. Doré, A.; Barbier, M. Maintenir la vigilance. *Rev. Anthropol. Connaiss.* **2015**, *9*, 189. [CrossRef]
10. Deveson, E.; Martinez, A. Locusts in Southern Settler Societies: Argentine and Australian Experience and Responses, 1880–1940. In *Environmental History in the Making;* Vaz, E., Joanaz de Melo, C., Costa Pinto, L.M., Eds.; Springer International Publishing: Cham, Switzerland, 2017; Volume 1, pp. 259–286.
11. Uvarov, B.P. A Revision of the Genus Locusta, L. (= Pachytylus, Fieb.), with a New Theory as to the Periodicity and Migrations of Locusts. *Bull. Èntomol. Res.* **1921**, *12*, 135–163. [CrossRef]

12. Uvarov, B. Biological and Ecological Basis of Locust Phases and Their Practical Application. In Proceedings of the 4th International Locust Conference, Cairo, Egypt, 23 April 1936.
13. Showler, A.; Ebbe, M.O.B.; Lecoq, M.; Maeno, K. Early Intervention against Desert Locusts: Current Proactive Approach and the Prospect of Sustainable Outbreak Prevention. *Agronomy* **2021**, *11*, 312. [CrossRef]
14. Zhang, L.; Lecoq, M.; Latchininsky, A.; Hunter, D. Locust and Grasshopper Management. *Annu. Rev. Èntomol.* **2019**, *64*, 15–34. [CrossRef] [PubMed]
15. Pantenius, C. *Towards Better Desert Locust Risk Prevention Systems in the Central Region*; FAO: Rome, Italy, 2015; 24p.
16. Magor, J.; Lecoq, M.; Hunter, D. Preventive control and Desert Locust plagues. *Crop. Prot.* **2008**, *27*, 1527–1533. [CrossRef]
17. Symmons, P. A Critique of "Preventive Control and Desert Locust Plagues". *Crop Prot.* **2009**, *28*, 905–907. [CrossRef]
18. Enserink, M. Can the War on Locusts Be Won? *Science* **2004**, *306*, 1880. [CrossRef] [PubMed]
19. Lecoq, M. Desert locust management: From ecology to anthropology. *J. Orthopt. Res.* **2005**, *14*, 179–186. [CrossRef]
20. Gomes, I.; Queiroz, A.I.; Alves, D. Iberians against locusts: Fighting cross-border bio-invaders (1898–1947). *Hist. Agrar. Rev. Agric. Hist. Rural.* **2019**, *78*, 127–159. [CrossRef]
21. Roy, J. *Histoire d'un Siècle de Lutte Anti-Acridienne en Afrique. Contributions de la France*; L'Harmattan: Paris, France, 2001.
22. Meynard, C.N.; Lecoq, M.; Chapuis, M.; Piou, C. On the relative role of climate change and management in the current desert locust outbreak in East Africa. *Glob. Chang. Biol.* **2020**, *26*, 3753–3755. [CrossRef]
23. Showler, A.T.; Lecoq, M. Incidence and Ramifications of Armed Conflict in Countries with Major Desert Locust Breeding Areas. *Agronomy* **2021**, *11*, 114. [CrossRef]
24. Cullen, D.A.; Cease, A.J.; Latchininsky, A.V.; Ayali, A.; Berry, K.; Buhl, J.; De Keyser, R.; Foquet, B.; Hadrich, J.C.; Matheson, T.; et al. Chapter Seven—From Molecules to Management: Mechanisms and Consequences of Locust Phase Polyphenism. In *Advances in Insect Physiology*; Verlinden, H., Ed.; Academic Press: Cambridge, MA, USA, 2017; Volume 53, pp. 167–285.
25. Millist, N.; Abdalla, A. *Benefit–Cost Analysis of Australian Plague Locust Control Operations for 2010–11*; Australian Bureau of Agricultural and Resource Economics and Sciences: Canberra, Australia, 2011; 22p.
26. Magarey, R.D.; Chappell, T.M.; Trexler, C.M.; Pallipparambil, G.R.; Hain, E.F. Social Ecological System Tools for Improving Crop Pest Management. *J. Integr. Pest Manag.* **2019**, *10*, 10. [CrossRef]
27. De Marchi, B.; Ravetz, J.R. Risk Management and Governance: A Post-Normal Science Approach. *Futures* **1999**, *31*, 743–757. [CrossRef]
28. Dangles, O.; Pérez, V.C. Editorial overview: Devastating locust swarms and pandemics: The same pressing need for sustainability science. *Curr. Opin. Insect Sci.* **2020**, *40*. [CrossRef] [PubMed]
29. Berkes, F.; Folke, C. *Linking Social and Ecological Systems: Management Practices and Social Mechanisms for Building Resilience*; Cambridge University Press: Cambridge, UK, 1998.
30. Cash, D.W.; Adger, W.N.; Berkes, F.; Garden, P.; Lebel, L.; Olsson, P.; Pritchard, L.; Young, O. Scale and Cross-Scale Dynamics: Governance and Information in a Multilevel World. *Ecol. Soc.* **2006**, *11*, 8. [CrossRef]
31. Cease, A.J.; Elser, J.J.; Fenichel, E.P.; Hadrich, J.C.; Harrison, J.F.; Robinson, B.E. Living with Locusts: Connecting Soil Nitrogen, Locust Outbreaks, Livelihoods, and Livestock Markets. *Bioscience* **2015**, *65*, 551–558. [CrossRef]
32. Balint, P.J.; Stewart, R.E.; Desai, A.; Walters, L.C. *Wicked Environmental Problems: Managing Uncertainty and Conflict*; Island Press: Washington, DC, USA, 2011.
33. Funtowicz, S.O.; Ravetz, J.R. The Emergence of Post-Normal Science. In *Science, Politics and Morality*; Springer Science and Business Media LLC: Berlin/Heidelberg, Germany, 1993; pp. 85–123.
34. Lang, D.J.; Wiek, A.; Bergmann, M.; Stauffacher, M.; Martens, P.; Moll, P.; Swilling, M.; Thomas, C.J. Transdisciplinary research in sustainability science: Practice, principles, and challenges. *Sustain. Sci.* **2012**, *7*, 25–43. [CrossRef]
35. Palsson, G.; Szerszynski, B.; Sörlin, S.; Marks, J.; Avril, B.; Crumley, C.; Hackmann, H.; Holm, P.; Ingram, J.; Kirman, A.; et al. Reconceptualizing the 'Anthropos' in the Anthropocene: Integrating the social sciences and humanities in global environmental change research. *Environ. Sci. Policy* **2013**, *28*, 3–13. [CrossRef]
36. Bennett, N.J.; Roth, R.; Klain, S.C.; Chan, K.; Christie, P.; Clark, D.A.; Cullman, G.; Curran, D.; Durbin, T.J.; Epstein, G.; et al. Conservation social science: Understanding and integrating human dimensions to improve conservation. *Biol. Conserv.* **2017**, *205*, 93–108. [CrossRef]
37. Lewis, W.J.; Van Lenteren, J.C.; Phatak, S.C.; Tumlinson, J.H. A total system approach to sustainable pest management. *Proc. Natl. Acad. Sci. USA* **1997**, *94*, 12243–12248. [CrossRef]
38. Fox, H.E.; Christian, C.E.; Nordby, J.C.; Pergams, O.R.W.; Peterson, G.D.; Pyke, C.R. Perceived Barriers to Integrating Social Science and Conservation. *Conserv. Biol.* **2006**, *20*, 1817–1820. [CrossRef]
39. Büscher, B.; Wolmer, W. Introduction: The Politics of Engagement between Biodiversity Conservation and the Social Sciences. *Conserv. Soc.* **2007**, *5*, 1–21.
40. Brosius, J.P. Common ground between anthropology and conservation biology. *Conserv. Biol.* **2006**, *20*, 683–685. [CrossRef] [PubMed]
41. Sievanen, L.; Campbell, L.M.; Leslie, H.M. Challenges to Interdisciplinary Research in Ecosystem-Based Management. *Conserv. Biol.* **2012**, *26*, 315–323. [CrossRef] [PubMed]
42. Moon, K.; Blackman, D. A Guide to Understanding Social Science Research for Natural Scientists. *Conserv. Biol.* **2014**, *28*, 1167–1177. [CrossRef]

43. Alexander, S.M.; Armitage, D. A Social Relational Network Perspective for MPA Science. *Conserv. Lett.* **2015**, *8*, 1–13. [CrossRef]
44. Cox, M. A basic guide for empirical environmental social science. *Ecol. Soc.* **2015**, *20*, 20. [CrossRef]
45. Robbins, P. Critical Introductions to Geography. In *Political Ecology: A Critical Introduction*; Blackwell Publishing: Oxford, UK, 2004.
46. Rose, D.B.; Van Dooren, T.; Chrulew, M.; Cooke, S.; Kearnes, M.; O'Gorman, E. Thinking Through the Environment, Unsettling the Humanities. *Environ. Humanit.* **2012**, *1*, 1–5. [CrossRef]
47. Abou-Ali, H.; Belhaj, M. *Cost Benefit Analysis of Desert Locusts' Control: A Multicountry Perspective*; Working paper 0801; Economic Research Forum: Giza, Egypt, 2008.
48. Lockwood, J.A.; Showler, A.; Latchininsky, A.V. Can we make locust and grasshopper management sustainable? *J. Orthopt. Res.* **2001**, *10*, 315–329. [CrossRef]
49. Gaspar, R.; Yan, Z.; Domingos, S. Extreme natural and man-made events and human adaptive responses mediated by information and communication technologies' use: A systematic literature review. *Technol. Forecast. Soc. Chang.* **2019**, *145*, 125–135. [CrossRef]
50. Künckel D'Herculais, J. *Invasion des Acridiens, Vulgo Sauterelles, en Algérie (1893–1905)*; Imprimerie Administrative et Commerciale Giralt, Gouvernement Général de l'Algérie: Alger, Algeria, 1905.
51. Peloquin, C. Unruly Nature and Technological Authority: Governing Locust Swarms in the Sahel. Ph.D. Thesis, University of Arizona, Tuscon, AZ, USA, 2014.
52. Byrne, A.T.; Hadrich, J.C.; Robinson, B.E.; Han, G. A factor-income approach to estimating grassland protection subsidy payments to livestock herders in Inner Mongolia, China. *Land Use Policy* **2020**, *91*, 104352. [CrossRef]
53. Deveson, E.D. The Search for a Solution to Australian Locust Outbreaks: How Developments in Ecology and Government Responses Influenced Scientific Research. *Hist. Rec. Aust. Sci.* **2011**, *22*, 1–31. [CrossRef]
54. Liu, J.; Hull, V.; Batistella, M.; DeFries, R.; Dietz, T.; Fu, F.; Hertel, T.W.; Izaurralde, R.C.; Lambin, E.F.; Li, S.; et al. Framing Sustainability in a Telecoupled World. *Ecol. Soc.* **2013**, *18*, 18. [CrossRef]
55. Cumming, G.S.; Cumming, D.H.M.; Redman, C.L. Scale Mismatches in Social-Ecological Systems: Causes, Consequences, and Solutions. *Ecol. Soc.* **2006**, *11*, 11. [CrossRef]
56. Le Gall, M.; Overson, R.; Cease, A. A Global Review on Locusts (Orthoptera: Acrididae) and Their Interactions with Livestock Grazing Practices. *Front. Ecol. Evol.* **2019**, *7*, 7. [CrossRef]
57. Word, M.L.; Hall, S.J.; Robinson, B.E.; Manneh, B.; Beye, A.; Cease, A.J. Soil-targeted interventions could alleviate locust and grasshopper pest pressure in West Africa. *Sci. Total. Environ.* **2019**, *663*, 632–643. [CrossRef] [PubMed]
58. Skaf, R. Le Criquet Marocain au Proche-Orient et sa grégarisation sous l'influence de l'homme. *Bull. Soc. Ecol.* **1972**, *3*, 247–325.
59. Benfekih, L.; Chara, B.; Doumandji-Mitiche, B. Influence of Anthropogenic Impact on the Habitats and Swarming Risks of Dociostaurus maroccanus and Locusta migratoria (Orthoptera, Acrididae) in the Algerian Sahara and the Semiarid Zone. *J. Orthopt. Res.* **2002**, *11*, 243–250. [CrossRef]
60. Gay, P.; Lecoq, M.; Piou, C. The limitations of locust preventive management faced with spatial uncertainty: Exploration with a multi-agent model. *Pest Manag. Sci.* **2020**, *76*, 1094–1102. [CrossRef]
61. Gay, P.-E.; Lecoq, M.; Piou, C. Improving preventive locust management: Insights from a multi-agent model. *Pest Manag. Sci.* **2018**, *74*, 46–58. [CrossRef]
62. Toleubayev, K.; Jansen, K.; Van Huis, A. Locust Control in Transition: The Loss and Reinvention of Collective Action in Post-Soviet Kazakhstan. *Ecol. Soc.* **2007**, *12*. Available online: http://www.ecologyandsociety.org/vol12/iss2/art38/ (accessed on 10 May 2021).
63. Joffe, S. *Economic and Policy Issues in Desert Locust Management: A Preliminary Analysis*; FAO: Rome, Italy, 1998.
64. Krall, S. Desert Locusts in Africa: A Disaster? *Disasters* **1995**, *19*, 1–7. [CrossRef] [PubMed]
65. Joffe, S.R. *Desert Locust Management*; The World Bank: Washington, DC, USA, 1995.
66. Hardeweg, B. *A Conceptual Framework for Economic Evaluation of Desert Locust Management Interventions*; Institute of Horticultural Economics: Hannover, Germany, 2001.
67. Emana, B. *Socio-Economics of Desert Locust Control in Sudan. A Micro Level Case Study. Final Draft Report*; Food and Agriculture Organization of the United Nations: Rome, Italy, 2002.
68. De Groote, H.; Douro-Kpindou, O.-K.; Ouambama, Z.; Gbongboui, C.; Müller, D.; Attignon, S.; Lomer, C. Assessing the Feasibility of Biological Control of Locusts and Grasshoppers in West Africa: Incorporating the Farmers' Perspective. *Agric. Hum. Values* **2001**, *18*, 413–428. [CrossRef]
69. Manthey, S. *Curse of the Good Rains the Long-Run Impacts of Locust Infestations. Evidence from the 1986–89 Locust Plague in the Sahel*; Lund University: Lund, Sweden, 2020; 77p.
70. Lockwood, J.; Sardo, M. A Swarm of Injustice: A Sociopolitical Framework for Global Justice in the Management of the Desert Locust. *Agronomy* **2021**, *11*, 386. [CrossRef]
71. Lockwood, J.A. An Excerpt of Locust: The Devastating Rise and Mysterious Disappearance of the Insect that Shaped the American Frontier. *Am. Entomol.* **2004**, *50*, 222–227. [CrossRef]
72. Catenaccio, C. (Senasa, Buenos Aires, Argentina) South American Locusts: A Crisis That Creates Opportunities. Personal communication, 2018.
73. Lecoq, M. Desert Locust Threat to Agricultural Development and Food Security and FAO/International Role in Its Control. *Arab. J. Plant Prot.* **2003**, *21*, 188–193.

74. Desanges, J. Témoignages antiques sur le fléau acridien. *Pub. Acad. Inscript. Belles Lett.* **2006**, *17*, 221–235.
75. Buj, A. Nature et environnement. International experimentation and control of the locust plague: Africa in the first half of the 20th century. In *Les Sciences Hors d'Occident au 20ème Siècle*; Waast, R., Chatelin, Y., Bonneuil, C., Eds.; Orstom: Paris, France, 1995; pp. 93–105.
76. Uvarov, B.P. *Locust as a World Problem*; Commission Française d'Études des Calamités with the Support of Union Internationale de Secours: Paris, France, 1937; pp. 376–382.
77. Lecoq, M. *Projet de Restructuration des Organismes Chargés de la Surveillance et de la Lutte Contre le Criquet Pèlerin en Région Occidentale. Justifications et Propositions*; FAO/AGP-P; Food and Agriculture Organization of the United Nations: Rome, Italy, 1999.
78. FAO. *Accord Portant Création d'Une Commission de Lutte Contre le Criquet Pèlerin dans la Région Occidentale, en Vertu de l'Article XIV de l'Acte Constitutif de la FAO*; Food and Agriculture Organization of the United Nations: Rome, Italy, 2000.
79. Bodin, Ö. Collaborative environmental governance: Achieving collective action in social-ecological systems. *Science* **2017**, *357*, eaan1114. [CrossRef] [PubMed]
80. Damtew, E.; Van Mierlo, B.; Lie, R.; Struik, P.; Leeuwis, C.; Lemaga, B.; Smart, C. Governing a Collective Bad: Social Learning in the Management of Crop Diseases. *Syst. Pract. Action Res.* **2020**, *33*, 111–134. [CrossRef]
81. Graham, S.; Metcalf, A.L.; Gill, N.; Niemiec, R.; Moreno, C.; Bach, T.; Ikutegbe, V.; Hallstrom, L.; Ma, Z.; Lubeck, A. Opportunities for better use of collective action theory in research and governance for invasive species management. *Conserv. Biol.* **2018**, *33*, 275–287. [CrossRef]
82. Ostrom, E. *Governing the Commons*; Cambridge University Press: Cambridge, UK, 1990.
83. Thomson, A.; Miers, H. *Assessment of the Socio-Economic Impact of Desert Locust and Their Control*; UK Department for International Development: London, UK, 2002; p. 37.
84. Food and agriculture organization of the United Nations. *Programme Evaluation of The Emergency Prevention System for Transboundary Animal and Plant Pests and Diseases (EMPRES): Desert Locust*; Food and Agriculture Organization of the United Nations: Rome, Italy, 2002.
85. Berkes, F. From Community-Based Resource Management to Complex Systems: The Scale Issue and Marine Commons. *Ecol. Soc.* **2006**, *11*, 11. [CrossRef]
86. Fleischman, F.D.; Ban, N.C.; Evans, L.S.; Epstein, G.; Garcia-Lopez, G.; Villamayor-Tomas, S. Governing Large-Scale Social-Ecological Systems: Lessons from Five Cases. *Int. J. Commons* **2014**, *8*, 428–456. [CrossRef]
87. Young, O.R. *The Institutional Dimensions of Environmental Change: Fit, Interplay, and Scale*; MIT Press: London, UK, 2002.
88. Lecoq, M. Lecoq, M. Le Criquet pèlerin. Enseignements de la dernière invasion et perspectives offertes par la biomodélisation. In *La Lutte Anti-Acridienne*; AUPELF-UREF, John Libbey Eurotext: Paris, France, 1991; pp. 71–98.
89. Medina, H.; Cease, A.; Trumper, E. The Resurgence of the South American Locust (*Schistocerca cancellata*). *Metaleptea* **2017**, *37*, 17–21.
90. Doré, A.; Barbier, M.; Lecoq, M.; Babah, M.A.O. Prévention des invasions de criquets pèlerins: Analyse socio-technique d'un dispositif de gestion du risque. *Cah. Agric.* **2008**, *17*, 457–464.
91. Deshormes, A. *Institutional Study to Enhance the Roles and Responsibilities of the Desert Locust Control Commissions Established under Article XIV. Financial Governance. Final Report*; Food and Agriculture Organization of the United Nations: Rome, Italy, 2011.
92. Adriaansen, C.; Woodman, J.D.; Deveson, E.D.; Drake, V.A. The Australian Plague Locust—Risk and Response. In *Biological and Environmental Hazards, Risks, and Disasters*; Elsevier: Amsterdam, The Netherlands, 2016; pp. 67–86.
93. Rosenschöld, J.M.A.; Honkela, N.; Hukkinen, J.I. Addressing the temporal fit of institutions: The regulation of endocrine-disrupting chemicals in Europe. *Ecol. Soc.* **2014**, *19*, 19. [CrossRef]
94. Leach, M.; Scoones, I.; Stirling, A. Governing epidemics in an age of complexity: Narratives, politics and pathways to sustainability. *Glob. Environ. Chang.* **2010**, *20*, 369–377. [CrossRef]
95. Boin, A.; 'T Hart, P. Organising for Effective Emergency Management: Lessons from Research 1. *Aust. J. Public Adm.* **2010**, *69*, 357–371. [CrossRef]
96. Boyles, J.E.; Kirschnick, F.; Kosilov, A.; Yanev, Y.; Mazour, T. Risk Management of Knowledge Loss in Nuclear Industry Organisations. *Int. J. Nucl. Knowl. Manag.* **2009**, *3*, 125–136. [CrossRef]
97. Berger, A.N.; Udell, G.F. The Institutional Memory Hypothesis and the Procyclicality of Bank Lending Behavior. *J. Financ. Intermed.* **2004**, *13*, 458–495. [CrossRef]
98. Wachinger, G.; Renn, O.; Begg, C.; Kuhlicke, C. The Risk Perception Paradox-Implications for Governance and Communication of Natural Hazards. *Risk Anal.* **2012**, *33*, 1049–1065. [CrossRef] [PubMed]
99. Lubell, M.; Jasny, L.; Hastings, A. Network Governance for Invasive Species Management. *Conserv. Lett.* **2016**, *10*, 699–707. [CrossRef]

agronomy

MDPI

Article

A Swarm of Injustice: A Sociopolitical Framework for Global Justice in the Management of the Desert Locust

Jeffrey A. Lockwood [1,*] and Michael Christopher Sardo [2]

[1] Department of Philosophy and Religious Studies, University of Wyoming, 1000 E. University, Laramie, WY 82071, USA

[2] Department of Politics, Occidental College, 1600 Campus Road, Los Angeles, CA 90041, USA; msardo@oxy.edu

* Correspondence: lockwood@uwyo.edu

Abstract: In recent years, scientists and managers have advocated for the integration of the social sciences (particularly political science and economics) and the humanities (particularly moral philosophy) with the natural sciences (particularly entomology and ecology) in developing a full understanding of locust-management programs. In this paper, we pursue such a synthesis by using the desert locust (*Schistocerca gregaria*) as an exemplar case. After an overview of this insect's biology, ecology, and management, we provide a brief summary of the standard, moral theories (utilitarianism, deontology, and virtue ethics) and consider their shortcomings with regard to developing a framework for understanding the socioeconomic complexity of locust management. Next, we address some of the models of global justice and focus on two fundamental questions: Who is a moral agent with regard to desert locust management, and how should we justly distribute the responsibilities among agents during preventive and reactive modes? After identifying the agents, we use a fourfold set of principles to construct a framework for locust management consistent with global justice and apply this conceptual system to two hypothetical scenarios. We conclude with some observations from political philosophy that offer progress toward a comprehensive and applicable theory for locust management in the context of global justice.

Keywords: locust plagues; preventative and reactive programs; moral agents; capacity-and-capability model; social connections model; responsibility; global justice

Citation: Lockwood, J.A.; Sardo, M.C. A Swarm of Injustice: A Sociopolitical Framework for Global Justice in the Management of the Desert Locust. *Agronomy* **2021**, *11*, 386. https://doi.org/10.3390/agronomy11020386

Academic Editors: Michel Lecoq and Arianne Cease

Received: 22 January 2021
Accepted: 19 February 2021
Published: 21 February 2021

Publisher's Note: MDPI stays neutral with regard to jurisdictional claims in published maps and institutional affiliations.

1. Introduction

Locusts (Orthoptera: Acrididae) are among the most devastating pests in human agriculture. This common name for these insects is derived from the Latin *locus ustus*, meaning "burnt place," to describe the condition of the land after a swarm has passed. These insects have caused serious damage to crops and forage across the globe, and their depredations have become the basis for legends, myths, and—in recent times—staggeringly complex, multinational control programs. Aside from pandemics, no biological phenomenon afflicts a larger expanse of the Earth than locust plagues. For example, the desert locust, *Schistocerca gregaria*, spans 60 countries, generating cooperation (and conflict) across three continents—and arguably on all six inhabited continents if we include the role of international donors during insect-driven, humanitarian crises.

Human efforts to manage locust outbreaks entail the integration of virtually all fields of academic inquiry. Understanding the insects themselves requires the study of biology (evolution, genetics, biochemistry, physiology, and anatomy), while their ecology demands an understanding of atmospheric science, chemistry, soil science, hydrology, and botany, among other fields. The movement of locusts across borders necessitates the inclusion of political science and economics, along with cultural and anthropological research. Grasping the complex international aspects of locust management also requires the study of history, including the legacy of colonialism. And the conflicts and harms arising from locust infestations raise important questions with regard to environmental ethics and global justice.

Such complexity calls for an overarching framework for understanding the locust–human nexus, which integrates interdisciplinary scholarly research and real-world practices. While universities, government agencies, and international organizations "talk the talk" of interdisciplinary collaboration, locusts provide an urgent test of our willingness and ability to "walk the walk". In particular, we argue that the scale and magnitude of locust swarms raise crucial normative questions of global justice that cannot be answered by natural scientific inquiry alone. Such questions include the scope of moral and political obligations, whether transnational inequalities constitute a question of justice, and the proper agents responsible for implementing the standards of global justice. However, while the global justice literature often engages environmental questions, such as natural resource inequalities and climate change, locust plagues have garnered little attention in that field. In this paper, we hope to make a modest contribution—along with those who have already taken the first, small steps—toward a desperately needed collaboration of the natural sciences, social sciences, and humanities to address an ongoing and worsening cause of human suffering.

In this paper, we review the ecological and economic relationships between humans and locusts, with particular attention to why the challenges with respect to pest management cannot be mitigated solely in terms of the natural sciences, and how these shortcomings are a matter of justice. We briefly review the triad of standard ethical theories (utilitarianism, deontology, and virtues) and make the case that features of locust population dynamics require us to apply theories of global justice. From this argument, we then address the two fundamental questions of who an agent of justice is and how responsibility should be fairly distributed among these agents. We conclude with the application of the most compelling theoretical frameworks to simplified but realistic case studies to illustrate how the principles of global justice could inform our policies and actions.

2. Background: Entomology, Society, and Politics

2.1. Locusts and Humans

2.1.1. Locust Biology

Locusts are those species of grasshoppers that exhibit behavioral, anatomical, and physiological "phase changes" associated with aggregation and migration under crowded conditions. At high population densities, these insects undergo a transformation from the solitarious to the gregarious phase, during which they shift from avoiding one another to seeking close proximity with conspecifics. In the gregarious condition, the immature hoppers (nymphs) aggregate in dense bands that march across the habitat, and the adults form swarms that fly long distances [1–3]. Tactile and olfactory cues are used by the insects to assess population densities and thereby initiate the reversible phase transformations [4].

While some locust species are chronic hazards to agriculture and require treatment in most years, many of the most serious pests exhibit erratic population dynamics with periods of low densities punctuated by irruptive population growth requiring emergency interventions to avoid disastrous crop losses [5–7]. In this paper, we will focus attention on the prototypical, widespread, and damaging species, the desert locust, which extends across 29 million km^2 from India to Mauritania [7].

Like many other locust species, *S. gregaria*'s population dynamics are driven by weather. In particular, precipitation is primarily responsible for indirectly triggering the gregarious phase by generating abundant vegetation in the desert habitats where the solitarious locusts chronically persist. In conventional terminology, a period of low population density in nonagricultural habitats is called a recession. An outbreak is characterized by localized, small swarms with most populations remaining solitarious. During an upsurge, continuing regional rains allow locusts to increase in density across multiple generations with more frequent swarming. Perhaps 20% of upsurges develop into plagues with swarms infesting an entire region causing extreme damage to crops and pastures [8].

2.1.2. Locust Management

The management of the desert locust involves two, distinct strategies: the chronic, localized management of incipient outbreaks to prevent large-scale population increases; and the acute, reactive, large-scale treatment of hopper bands and adult swarms when prevention fails. The challenge in shifting from reactive to preventive modes strategy is the so-called "vicious cycle" [9–11] in which funding is abundant during plagues due to an urgent need to mitigate suffering, while between these humanitarian crises, there is diminished support for surveillance and tactical treatments.

During plagues of the desert locust, millions of people relying on subsistence farming may require food aid [7]. The socioeconomic damage is further manifest as pastoralists sell their livestock at deflated prices to meet acute needs. As such, the harm to agriculturalists can extend for years as they incur debt or lose their land entirely. In addition, there is a negative impact to child education and health that persists well beyond the locust plague [12,13]. Reactive-control programs rely heavily on broad-spectrum chemical insecticides applied by aircraft and ground sprayers. Positioning equipment and supplies in pursuit of fast-moving swarms is a significant logistical challenge. Biological insecticides have made some inroads, but because these pathogenic products are largely restricted to locusts, the market is erratic and requires dramatic, rapid increases in production during an upsurge or plague. Moreover, once a plague has subsided, these materials cannot be used for other pests, and storage stability is limited. While some broad-spectrum chemical insecticides can find agricultural uses after a reactive program, large stocks of obsolete, environmentally undesirable or banned insecticides have accumulated in locust-afflicted countries [14].

A strategy of preventing locust plagues through locating and treating hopper bands and adult swarms during the outbreak phase [15,16] is considered to be the most economically efficient use of resources [17–19] (for a dissenting view, see [5]). Satellite imagery combined with geographic information systems is used to find green vegetation that signals the possibility of breeding conditions in desert areas, as locusts encounter abundant food and moist soil for their eggs [20]. But ultimately, scouts must go into the field to assess infestations and apply localized treatments—this a weak link in that the locusts often arise in places that are extremely remote and locations with military hostilities or minefields from previous conflicts [14]. The actual efficacy of preventive programs in most countries within outbreak areas can be called into question by consideration of the frequent upsurges and plagues that have developed in the last 35 years.

2.2. The Desert Locust in a Sociopolitical Context

2.2.1. Need for New Approaches

The ecological and social challenges arising from preventive and reactive locust-management programs are too interdependent to be addressed through the natural sciences alone [21]. Although biologists have pursued studies of transboundary coordination, risk management, stakeholder strategies, and economic cost-benefit analyses, social scientists have rarely been involved, with a few laudable exceptions (e.g., [10,22–25]). As argued by Zhang et al., "The sustainability of locust and grasshopper control must be better assured by considering not only scientific and technical aspects, but also socioeconomic mechanisms involved in the management of these pests" [7] p. 27. However, there is a further consideration. A sound system of locust management must be not only efficacious but moral, as unjust demands on countries and regions are sociopolitically unsustainable.

The Food and Agriculture Organization (FAO)'s proposal for an improved financing system illustrates the problem of analyzing desert locust management without regard to ethical considerations [26]. The prospectus includes a plan for funding during various periods of locust population dynamics, but there is no consideration of who ought to be responsible for providing support or how this responsibility should be fairly allocated. In this example, as in many others, there is a valuable, descriptive critique (i.e., what is being done) but little or no normative analysis (i.e., what should be done).

In a similar sense, intense conflicts over the goals of desert locust management are exemplified by the 2020 exchange between Tom, US Ambassador and Permanent Representative, United Nations Rome Based Agencies, and Metlerkamp, research fellow at the Environmental Learning and Research Center Rhodes University. Although competing socioeconomic ideologies were presented, neither side in this debate went beyond simple, normative assertions (i.e., that industrial or agroecological values were to be preferred) [27,28]. As such, the competing and compelling ethical foundations were largely obscured by political rhetoric.

The current locust plague reflects ecological and political conditions [29], including armed conflicts among weak states in the Middle East and eastern Africa, which precluded effective preventive programs following heavy rains [30]. The early outbreaks went undetected, allowing swarms to develop [31]. Using a spatially explicit multiagent model, Gay et al. [11] found that a preventive program can be nullified with just 5% of the territory having limited access. The study concluded that plagues are most likely to originate in weak states with larger areas inaccessible to pest managers. Add to this the reluctance of donor nations to provide funds to countries that engaged in terrorism and human-rights abuses (e.g., Sudan), and the potential for locust outbreaks to emerge undetected and unimpeded is significant [14]. And this is not a new problem, as the 1986–1989 plague was substantially attributable to a lack of early intervention due to armed conflict in the primary locust breeding areas of Eritrea and Sudan [14].

Natural disasters have been analyzed using various risk governance, operational, and funding models (e.g., [32–34]), but such theories rarely consider locust plagues. Although locust control is increasingly viewed as being the management of a natural hazard [7], the desert locust presents several unique features that must be considered in developing any viable, socioecological analysis. These qualities include ecological dynamics, geographic scale, geopolitical complexity, and socioeconomic factors.

2.2.2. Distinctive Features of Locust Plagues

With respect to ecological dynamics, since 1860 there have been 10 upsurges lasting 1–4 years (mean = 2 years) and nine plagues lasting for 1–22 years (mean = 10 years) [8]. This means that recession periods are the norm and can last more than a decade. During recessions, international funds, along with regional, national, and local resources are shifted to urgent problems rather than outbreak-prevention programs for locusts. However, it is important to recognize that past ecological patterns may not be reliable indicators of future dynamics, as climate change is likely to exacerbate the unpredictability and intensity of plagues [35,36]. With rising temperatures, insect development is accelerated (locusts typically take about a month from hatching to fledging) [4] and flight performance (speed and distance) may be enhanced [8]. Extreme weather events make forecasting locust dynamics increasingly difficult. For example, despite a general trend of increased droughts in Africa [37], the current plague developed during unusually heavy rains from two tropical cyclones in the Arabian Peninsula in just 5 months [29,38]—phenomena linked to the rapid warming of the Indian Ocean [36].

In terms of geographic scale, the desert locust afflicts 10% of the human population [39]; only pandemics have a wider scope. The current plague involves nine countries (Ethiopia, India, Iran, Kenya, Pakistan, Somalia, South Sudan, Uganda, and Yemen) with swarms moving into parts of six other nations [40]. Food shortages may soon resemble those in 2003–2005, when crop loss exceeded 80% in Burkina Faso, Mali, and Mauritania despite the treatment of 13 million ha at a cost of USD 500 million [41,42]. Upsurges and plagues produce mismatched scales in which the potential for harm is widespread and control operations must be dispersed across enormous areas to track swarms moving up to 150 km/day, but damage is highly concentrated wherever locusts descend to feed. With both reactive and preventive management, the costs accrued in one place yield benefits elsewhere. Individualizing costs and collectivizing benefits becomes a serious challenge in terms of funding and cooperation.

The spatial scope of the desert locust leads to geopolitical complexity. A brief overview provides a sense of the international efforts [26]. The Food and Agriculture Organization (FAO) of the United Nations has established three Regional Commissions for Locust Control (western, central, and eastern, totaling 56 countries) that obligate member states to develop monitoring and control programs through national centers or ministries of agriculture. The FAO also hosts: (1) the Desert Locust Control Committee, which is composed of national representatives and coordinates monitoring and control plans; (2) the Desert Locust Information Service, which provides data, analyses, and forecasts; and (3) the Emergency Centre for Transboundary Plant Pests, which mobilizes emergency aid provided by donors such as USAID (USA), DGIS (the Netherlands), CIRAD (France), CIDA (Canada), SIDA (Sweden), ODA (UK), and GTZ (Germany) [14]. Different countries and cultures give rise to various challenges, such as the implementation of control programs during Ramadan, the acceptability of locusts as human food, the reluctance to accept foreign aid, and the movement of swarms across borders, which erodes fragile relations between hostile countries [43].

Socioeconomic factors exacerbate the spatiotemporal challenges. Standard economic analyses of desert locust-management programs have been criticized for omitting the perspectives of nomads, the value of food security, the costs of environmental damage, the sociological constraints in different countries, and the humanitarian benefits of pest management [14]. In large part, these insects are found in impoverished nations where colonial legacies have created weak national governments with continuing dependence on donors for the costs of preventing or controlling locust plagues [43]. In turn, perverse economic incentives may delay a poor nation's response to outbreaks in order to generate more foreign aid, given the opportunities for lucrative contract services during a plague. If the afflicted people are subsistence agriculturalists, then other nations, and even the urban centers within an infested country, are not directly harmed and assistance becomes less urgent, particularly if locust control competes with more dire threats (e.g., the current plague is unfolding at the same time as the COVID-19 pandemic).

In sum, it is evident that the ecological, geographical, geopolitical, and socioeconomic factors that shape the course of desert locust population dynamics and human responses, contribute in clear but complex ways to the unjust distribution of both the costs of preventive and reactive programs and the harms resulting from the failure of these interventions.

3. Locust Swarms as a Question of Global Justice

3.1. Locusts and Justice

3.1.1. The Limits of Individualistic Ethical Theories

Traditional theories of ethical and moral obligation identify principles that determine the proper moral action for individuals to take. Utilitarian principles emphasize the consequences of action, arguing that the moral action is that which produces the greatest good for the greatest number [44], while deontological principles defend absolute duties that must be followed regardless of their consequences [45]. The third traditional school of thought, virtue ethics, takes the position that we should cultivate character traits through practice so that individuals will develop their fullest, human potential [46].

These ethical positions can fruitfully expand the scope of one's moral obligations. For example, Singer has argued, because pleasure and pain are universal, the obligation to reduce suffering transcends national boundaries [47]. There is no difference, in his framework, between letting a child drown in front of you for fear of damaging your expensive clothing and purchasing expensive clothing in the first place, when that money could be used to alleviate starvation and disease in other countries. Therefore, he concludes, "If it is in our power to prevent something bad from happening, without thereby sacrificing anything of comparable moral importance, we ought, morally, to do it" [47] p. 231. A similar argument could be reached through deontological reasoning, contending that we have an absolute duty to preserve human dignity regardless of consequences. However, as

O'Neill (1993) [48] contends, such duties are "imperfect," meaning that while we have an absolute duty not to directly impair the dignity of another person ourselves, promoting the dignity of distant others would be beneficent, but not morally obligatory. While virtue ethics presents a more difficult case, as such theories emphasize the particularity of ethical communities and traditions [49], Van Hooft argues that cosmopolitanism can be considered a virtue in a globalized world, in which individuals view themselves as citizens of the world and see their moral obligations extending globally [50]. Therefore, whether taken as a moral obligation to prevent suffering, an imperfect duty to promote human dignity, or as an fulfilling the virtue of cosmopolitanism, the traditional ethical "triad" can provide good moral reasons why individuals should seek to provide aid—whether preventative or reactive—to those suffering from locust swarms.

While a detailed analysis of ongoing debates between these rival theories is well beyond the scope of this article, we note that there are significant challenges in applying these principles to desert locust management. For example, utilitarian calculations could in fact argue that locust plagues (we will use this term henceforth to include upsurges) are not worthy of moral consideration because resources could be more fruitfully put toward assisting refugees, battling climate change, or providing vaccinations against infectious disease. A similar line of argument was implicit in Tom's impassioned argument for using chemical insecticides to suppress the current desert locust plague, along with shifting global agriculture to an industrial model so as to produce abundant food at a low cost to feed as many people as possible [27]. Similar challenges face deontological approaches, since humanitarian assistance would qualify as an imperfect duty: we ought to practice charity but we do not have a duty to donate to every person or nation requesting our assistance. Thus, for a locust-control program, deontology offers no way of determining how the general duty to assist others in need should be instantiated, nor is it the case that anyone is unethical for not providing aid to locust afflicted people if one discharges one's charitable duty in other ways. The result is an impotent demand that "somebody should do something," which is a common response during humanitarian crises. Virtue theorists fare little better, given widely divergent conceptions of the meaning of the virtues across different cultures. What constitutes the "good life"—or the full realization of human potential—for one culture or community may not for another, raising concerns of paternalism or cultural imperialism.

Additionally, moving from individual moral obligation to questions of collective action and responsibility creates further challenges. We cannot simply scale up these systems to make sense of the ethical obligations of collectives (groups play a vital role concerning the responsible and effective management of the desert locust) for two important reasons. First, there are collective duties or responsibilities that a society is obligated to meet, but that no individual is blameworthy for failing to fulfill (e.g., a society ought to provide education and healthcare to people, but no particular person is ethically required to become a teacher or doctor). Second, because collectives are internally diverse with respect to power, influence, privilege, and causal relation to any particular harm, the question of allocating responsibility becomes vital. To take Singer's example of a child drowning in a river: If a dozen people see the child struggling, the problem of who should act gives rise to questions such as, "Who is trained in lifesaving?" "Does anyone have a special relationship with the child?" "Did someone cause the dire situation?" and "Can we form a human chain to reach the child and thereby reduce individual risk?" Given the cost of acting (jumping into the river to save a drowning child or donating funds to save hungry farmers), there is an incentive to wait for others who are more willing and able to intervene [47].

As such, cases of shared responsibility require a framework for justice. This is not to say that justice is devoid of utilitarian or deontological elements, but that the standard theories are not the principled foundations for collective, moral, and political action. Our more fundamental concern, however, is that relying upon these moral theories in isolation overlooks the nature of locust plagues as an injustice. Given the complex cultural dynamics within which the prevention, management, and harms of locust plagues are embedded, individuals trying to fulfill their moral obligations—whether to alleviate suffering, promote human dignity, or practice the virtues—find that their actions are mediated by social, political, and economic forces beyond their control. This can be a disempowering realization, leading otherwise well-intentioned individuals to lament that there is nothing they can do to solve the problem. More fundamentally, these ethical approaches tend to treat human suffering in a vacuum, as an exogenous fact or tragedy that must be morally remedied like a tornado devastating a farm community. Our contention, however, is that locust plagues are better understood as injustice arising through the interaction of natural and human processes.

3.1.2. Injustice: Beyond Bad Luck

Not every inequality connotes an injustice. Some people are taller or younger than others and some countries have better soil or more minerals than others, and these unequal qualities are not unfair. Likewise, in the context of natural phenomena, inequalities may arise because someone is struck by lightning or a region is more prone to earthquakes. Given the sentiment expressed by Hayek that justice "ought to be confined to the deliberate treatment of men by other men" [51] p. 163, it is tempting to treat locust plagues, and other natural disasters, as misfortunes, tragedies, or bad luck, rather than injustices. Such thinking misunderstands the nature of justice and the particular injustice of natural disasters in fundamental ways.

Justice is often understood in distributive terms. Such a conception has a long pedigree, dating to Aristotle's argument that justice entails treating equals equally and unequals unequally, when such inequalities are morally relevant (e.g., sentience or nonsentience rather than male or female). Sandel helpfully summarizes this conception as how a society "distributes the things we prize—income and wealth, duties and rights, powers and opportunities, offices and honors," or ensuring that society "gives each person his or her due" [52] p. 19. To understand locust plagues as injustices, under this framework, is to highlight the inequitable distribution of harm they generate.

Rawls' conception of "justice as fairness" is perhaps the most well-known and influential account [53]. This theory begins with a thought experiment: Imagine that you are behind a "veil of ignorance" such that you do not know your particular characteristics (nationality, ethnicity, wealth, gender, etc.) and you must develop the principles that will apply when you reenter the world. Rawls maintained that in this "original position," rational individuals will select three impartial rules of governance: the Liberty Principle granting all individuals a right to basic freedoms, the Equality Principle granting each person a right to the same opportunity of acquiring resources as others with the same natural abilities, and the Difference Principle allowing inequalities as long as they benefit the worst-off individuals. Rawls' goal is to nullify "the accidents of natural endowment and the contingencies of social circumstances" that are "arbitrary from a moral point of view" [53] p. 15. Such thinking can be applied to the question of locust plagues, by considering the contingencies of birth (e.g., geography) that make one more susceptible to suffering harm. Resources should be distributed to benefit those worst off in society, to nullify the negative effects of these contingencies.

Justice should not only be understood in terms of material distributions, however. Following Young (1990 [54] p. 37), injustices are more fundamentally understood as social and institutional conditions that create "oppression, the institutional constraint on self-development, and domination, the institutional constraint on self-determination" [54] p. 37. Goodhart similarly describes injustices as "deformit[ies] in social relations [that] create, sustain, or contribute to subordination" [55] p. 143. In this conception, the true injustice of locust plagues lie in the social, political, and economic institutions and structures that create unequal patterns of vulnerability, adaptive capacity, and suffering in the first place. In either Rawls' or Young's conception, locust plagues should not be written off as mere tragedies of nature to be remedied, but as injustices generating moral and political responsibilities for two crucial reasons.

First, human influences may make natural events more frequent, severe, or prolonged. Hurricanes, floods, droughts, and wildfires occur without human intervention, but anthropogenic climate change is worsening these disasters. In terms of acridid population dynamics, rangeland grasshopper outbreaks are natural phenomena in the western US, but the use of broad-spectrum insecticides that eliminate parasites and predators increases the duration and intensity of these infestations [56]. With regard to desert locusts, upsurges have been attributed to both the failure of countries to implement preventive management methods [14,29] and the triggering of population increases as a consequence of climate change [35,36].

Second, we must consider the allocation of the economic and social costs of natural disasters, which may be unjustly distributed even with random events. For example, the damage of an earthquake or hurricane may be far greater for people lacking the resources to access healthcare or rebuild houses. However, even if humans are partially or wholly responsible for a natural disaster, there is no injustice if those who allowed the problem to develop (e.g., by neglecting to take steps to prevent harm) or benefited from fostering the adverse environmental conditions (e.g., by accumulating comforts and conveniences that emit greenhouse gases), accrue the costs. However, with phenomena such as climate change and locust plagues, the costs are primarily borne by those lacking wealth and power—impoverished countries, former colonies, unstable governments, and subsistence agriculturalists who did not neglect to act or benefit from inaction [7,14,30,43].

Therefore, locust plagues should be understood as what Shue calls a compound injustice, which occurs "when an initial injustice paves the way for a second, as when colonial exploitation weakens the colonized nation to such an extent that the colonizer can impose unequal treaties upon it even after it gains independence" [57] p. 4. In the case of locust plagues, existing inequalities in the global political-economic order—themselves legacies of colonialism [58]—leave the already-vulnerable both more susceptible and less equipped to effectively respond, manage, and adapt to locust plagues.

To treat locust plagues as mere bad luck, would, following Shklar, impose a "sense of tragic inevitability upon events that are in fact entirely (or at least substantially) amenable to purposive human alteration" [59] p. 70. This fatalism then obscures questions of responsibility and ultimately reifies existing inequalities as natural. Similarly, Erskine contends that retrospective ascriptions of blame after international, natural disasters can distract from addressing how responsibilities should be distributed in the future [60]. Who has a moral duty and what constitutes a just allocation of such responsibilities are as vital as questions about logistics, financing, and politics. Dealing with prospective responsibilities to avert future crises is more important than investing energy into assigning retrospective culpability when nothing (or too little) was done. Vague references to the failure of the "international community" are singularly unhelpful. It is the purpose of this paper to open a conversation as to how we might prospectively identify the moral agents and just distributions of responsibilities in the context of a sound sociopolitical and ecological understanding of desert locust management.

3.1.3. The Challenge of Global Justice

The normative concerns with desert locust management become even more challenging when placed in their global context. As Zhang et al. argue, "Robust mechanisms need to be put in place to ensure continuous financial support at national and international levels so that treatment programs can be put in place in a timely manner as part of successful strategies of sustainable preventive management" [7] p. 26. If one's concern is justice, desert locust management inevitably raises questions of how to give "each person his or her due." Who should provide this financial support: domestic governments, foreign governments, international organizations, or nongovernmental organizations? To whom should such financial support flow? How should different projects be prioritized? What do citizens of one nation-state owe to citizens of distant ones? If justice concerns not only distributive questions, but also, as Young and Goodhart contend, questions of social structure, institutional domination, and oppression, these questions are even more vexing [54,55]. We must ask what obligations do nation-states (or their citizens) have to reform unjust institutions in other nation-states, or would such actions be themselves unjust? If the institutional structure in question is the global political–economic order, including the structure of global trade, international patent law, and humanitarian assistance, what obligations can individuals realistically and meaningfully have with respect to such a structure? The debate over such questions in the literature on global justice is voluminous, and our treatment will necessarily be schematic. However, our contention is both that this body of literature can be fruitfully used to study the ethical challenges of desert locust management, and that the question of desert locust management can add further richness to the relatively abstract discussion of global justice.

Much of the debate concerns whether justice is even a meaningful concept at the global level, or if the scope and boundaries of justice should be restricted to more formal political communities. We begin with a consideration of cosmopolitan views and then address statist views.

Cosmopolitans, as we have earlier noted, argue that both membership in a political community and physical proximity are both morally arbitrary when evaluating one's moral obligations. Emblematic of this line of argument is Singer's insistence that there is no morally relevant difference between letting a child drown within eyesight and letting a child starve to death on the other side of the world [47]. One has a moral obligation to alleviate suffering, wherever it occurs.

In an increasingly globalized and interconnected world, it becomes difficult to sustain the idea that moral obligations stop at national boundaries, even if one rejects Singer's utilitarian reasoning. While Rawls believed that his principles of justice could not be applied beyond the boundaries of autonomous nation-states [53,61], Beitz argues that this formulation fails to recognize that the interdependency of modern societies undermines national autonomy and creates a new basis for justice [62]. Because a state's endowment of natural resources are contingent—and thus morally arbitrary in a Rawlsian sense—Beitz contends that "principles of distributive justice must apply in the first instance to the world as a whole, then derivatively to nation-states" [62] p. 383.

Other approaches focus less on Rawlsian contractualism, but on the conditions for living a dignified life. Shue argues that all human beings are entitled to the protection of certain basic rights, which constitute "everyone's minimum reasonable demands upon the rest of humanity" [63] p. 19, while Nussbaum instead focuses on human capabilities that are necessary for flourishing (e.g., bodily health and integrity, self-determination, and free expression), to which all human beings are entitled regardless of geographic location [64]. Insofar as locust plagues cause human suffering, deprive individuals their basic rights of subsistence and security, or undermine their ability to flourish, they generate moral obligations to provide aid regardless of political boundaries.

Statists object to such arguments despite their appeal to moral intuitions about universality and equality. Nagel argues that justice refers specifically to duties shared by co-citizens, by virtue of living under a common sovereign [65]. His argument is not merely pragmatic—that without a world government is difficult if not impossible to enforce the demands of cosmopolitan justice—but that living under a common sovereign who wields violence in one's own name creates special ethical bonds among compatriots that give rise to the demands of justice. While we have basic humanitarian obligations to all people to alleviate extreme distress, the demands of justice, which concern questions of equality and political relationships, are properly bounded by political membership. Miller critiques cosmopolitan theories of justice for imposing uniform sets of rights and duties over and against the diversity of communities around the world [66]. Cosmopolitanism excludes "the possibility that there exist deep cultural differences between and among societies which the members of each find valuable, and which they want to see protected by political means" [66] p. 84. While these statists may accept that providing humanitarian assistance in response to locust plagues may be morally good, they would reject that the demands of justice generate equal responsibilities to victims of distant locust plagues as they do to those suffering from hunger or poverty within their own political communities.

Eschewing the binary choice between cosmopolitans and statists, we follow theoretical projects that navigate between the extremes of the overwhelming burden of cosmopolitan justice on one hand and the moral isolationism of statist conceptions of justice on the other. Rejecting the logic that the demands that justice must either be restricted to one's own political community or have universal reach, Forst argues that many questions of transnational justice involve multiple overlapping relationships of power, domination, and inequality at local, national, and transnational scales [67]. Shifting perspective from abstract ideals of justice towards actual victims of injustice, "... reveals that theirs is a situation of multiple domination: most often they are dominated by their own (hardly legitimate) governments, elites, or warlords, which in turn are both working together and are (at least partly) dominated by global actors... The various contexts of justice—local, national, international, and global—are connected through the kind of injustice they produce, and a theory of justice must not remain blind to this interconnectedness" [67] (pp. 166–167).

Locust plagues can best be understood as involving multiple contexts of justice—from domestic concerns of failed preventative management and socioeconomic inequalities that leave the vulnerable at greatest risk, to transnational economic and political forces that weaken state capacity to address these compound challenges, to the broader contexts of global climate change and the lasting legacies of colonialism on economic development. Given this complexity, the proper question of justice is not, "Do I have any moral responsibility to assist those suffering from locust plagues?" Instead, in an attempt to lay the foundation for what we hope will be a continuing discourse on moral agency and the just allocation of responsibilities pertaining to locusts, we consider two essential questions: Who are the moral agents that can be held accountable for achieving justice, and how should the responsibilities for reactive and preventive programs be fairly distributed among these agents?

4. Moral Agents in Locust Management

4.1. Who Is Responsible?

Individuals are paradigmatic moral agents, having the capacity to understand and respond to ethical obligations. The ability to be morally responsible is often taken as a hallmark of moral personhood, distinguishing human persons from other animals. As Kant writes, "A person is a subject whose action can be imputed to him" [68] p. 16. However, other types of agents are also treated as capable of bearing moral responsibility, including collectives [69], states [70], and international institutions [60]. As O'Neill argues, theories of justice that do not identify which agents are responsible for realizing them are necessarily incomplete and generate considerable theoretical and practical challenges [70]. For example, the Universal Declaration of Humans Rights adopted by the United Nations in 1948 asserts that individuals have certain rights, but fails to specify who has the corresponding duties. The Declaration contends that every individual and "organ of society" should strive to advance human rights, but tacitly identifies states as the primary agents without any explicit rationale.

When considering who should be responsible for reactive and preventive programs for the desert locust, are the proper subjects of moral accountability subsistence farmers, global citizens, rural communities, scientific organizations, local governments, regional governments, ministries of agriculture, afflicted nations, foreign governments, relief organizations, corporations, and international bodies [7,14,30,43]? The result is a cosmopolitan conception of individualized rights enmeshed within a vaguely statist view of moral duties. While an Ethiopian farmer has a right to avoid starvation due to desert locust damage, who is morally obligated to treat the devastating plague is not evident.

One might intuitively contend that whoever is causally responsible for a harm is a moral agent. Indeed, responsibility is often, as Young summarizes, assigned "to a particular agent (or agents) whose actions can be shown to be causally connected to the circumstances for which responsibility is sought" [71] p. 116. However, causal responsibility is neither necessary nor sufficient for moral responsibility. One can cause harm to another person purely by accident and so accrue no moral responsibility, and one can be obligated to provide aid to a suffering individual without having caused the harm. The desert locust case demonstrates this point: No single agent is uniquely causally responsible for a locust plague, but that does not eliminate all responsibility to provide preventative and reactive aid. We contend that causal contribution is best understood as a factor in allocating responsibility rather than assigning agency. Therefore, we consider two broad approaches to identifying the moral agents responsible for global justice, with particular attention to the management of the desert locust: the capacity-and-capability model and the social connectionist model. Before addressing these, however, a crucial distinction must be made.

Not every assemblage of moral individuals constitutes a new moral agent [60,72]. A mere grouping of people (or collection of social units, such as nations) can be termed "aggregate collectivity," insofar as the group has no different or greater obligations than the sum of its parts (e.g., the spectators at a sporting event or the countries comprising Asia). This contrasts with "conglomerate collectivity," in which the assemblage has an identity and continuity over time, organizational and decision-making procedures, and an ability to take purposive action through deliberative goals. Such institutional agents can possess novel or emergent moral responsibilities beyond those of their members, and might be exemplified by the United Nations or the Global Locust Initiative (2020) [73]. It is sometimes useful to also distinguish primary from secondary agents of justice, the former exerting control over the latter (e.g., the UN versus the UNFAO, or the US Congress versus the US Agency for International Development) [70].

4.2. *Contemporary Models*

4.2.1. Capacity-and-Capability Model

O'Neil contends that institutional or collective agents of justice must possess two features: the capacity to make moral decisions and the capability to affect these decisions [70]. Along these lines, Erskine contends that institutional agents have capacity insofar as they can engage in moral deliberation and thereby take on collective responsibility [60]. For example, the FAO's Desert Locust Control Committee, the government of Kenya, and Dow AgroSciences all have the potential to assess their organization's collective duty to respond to a locust plague. Each of these groups possesses the temporal continuity, analytical processes, and goal-setting mechanisms for conglomerate collectivity.

Capacity is a necessary but insufficient condition of being an agent of justice, as good intentions without commensurate actions might satisfy a deontological standard, but most views of justice require a consequentialist element as well. As such, agents must also possess the capability to actualize the outcome of their moral reasoning—to deploy resources in actual circumstances [74]. In essence, this requirement reflects the ethical principle that "ought implies can"—that to say one should act in some manner entails that the agent has the ability to do so. However, as Caney argues in the context of climate change, there are considerable limitations to ascribing responsibility on the basis of capacity alone, as it ignores how injustices themselves produce inequalities in capacity and can absolve those who have contributed to an injustice if they now lack effective capacity [75].

Additionally, capability is a continuous, rather than discrete, feature of an institutional agent. As Sardo argues, using discrete qualities of agents as criteria for responsibility overlooks the social, economic, and political structures within which agents are embedded and that generate differences in capacity [76]. For example, an agent that has fostered dependency and thereby diminished the capability of another agent may thereby become responsible for either providing resources or developing autonomy and capability in the reliant agent (e.g., a European nation in relation to its former African colony). Therefore, Sardo argues that in the case of complex and compound structural injustices, such as climate change and, as we contend, locust plagues, responsibility should be ascribed on the basis of social-relational criteria, rather than individualistic ones [76]. One such theory of relational responsibility is Young's "social connections model" [71,77].

4.2.2. Social Connection Model

Although cosmopolitan principles of justice would apply to all moral agents, O'Neil raises the concern that ceding the obligation to act on behalf of afflicted people to global institutions could degenerate into a kind of centralized tyranny [70]. One approach to avoiding such a situation would be to disperse the responsibility among a plurality of agents. Such a strategy can be found in Young's critique and reconceptualization of Rawls's understanding of the basic structure of society, in which she makes the case that modern life entails vital, mutual interests extending far beyond the borders of one's nation into a system of international connectivity [71,77].

The social connections model is based on the recognition that people participate in political institutions as a response to individuals being socioeconomically interdependent, and that these connections often arise without regard to political boundaries. According to this view, truly global moral obligations, such as human rights, are not the proper realm of justice, as universals do not arise from, nor are they contingent upon, social relationships. However, all agents who contribute by their actions to economic and political systems that yield unjust distributions of resources among affected countries have a responsibility to remedy these moral failings. This responsibility is different than the standard notion of liability insofar as many current injustices are the product of sociohistorical events giving rise to so-called structural injustices that are immoral in a way that is distinct from the wrongful action of either individuals (who may participate in and benefit from the structures while not having created them or intended harm) or states (which may not willfully diminish the well-being of the repressed people). As such, it is not possible to trace specific actions by a US citizen or the French government, for example, to the suffering of a chronically impoverished Somali farmer during a locust infestation.

The social connections model, therefore, can fruitfully negotiate between the cosmopolitan and statist conceptions of justice and responsibility. It recognizes the realities of economic interdependence without holding individuals responsible for rectifying every injustice everywhere on Earth. In the case of desert locust plagues, it does not blame individuals for the suffering caused by these disasters. However, individuals can be held responsible in a forward-looking sense to act not merely to provide humanitarian aid, but to work to reform the national and transnational institutions that unequally distribute the harm of natural disasters such as locust plagues. This responsibility is not universal, but adheres to those who participate in and benefit from the same structures and institutions that lead to the compound injustice of locust plagues.

4.3. Inventory of Moral Agents in Desert Locust Management

Our goal is not to settle theoretical debates between different theories of responsibility, but to draw on both the capacity and capability (C&C) model and the social connections (SC) model to identify responsible agents in the context of pest management. This yields a realistically complex suite of parties given the nature and scope of desert locust infestations.

4.3.1. Multinational Agents

The unspecified "global community" is an aggregate collective lacking the necessary qualities for moral responsibility. In terms of the desert locust, the dominant, conglomerate collective is the UN and its secondary agent, the FAO. This collective, along with its downstream organizations (Regional Commissions, Desert Locust Control Committee, and Emergency Centre for Transboundary Plant Pests) all have substantial capacities and capabilities, as well as elements of social connectivity.

4.3.2. National Agents

At the national level, it is important to consider both strong and weak states. The former unambiguously meet the criteria of the C&C model [70]. The executive and legislative branches of these governments constitute the primary agents, with various development agencies being secondary agents (e.g., the French Agricultural Research Centre for International Development).

The agency of weak states is a matter of degree according to the C&C model. So-called quasi-states, exemplified by several countries in the Middle East and sub-Saharan Africa, are exceptionally dependent on foreign creditors and lack institutions to constrain or outlast the individuals holding power [60]. As such, they exhibit aggregate, more than conglomerate, collectivity. In comparison, weak states have greater autonomy, although still depending to a substantial degree on international aid, which can impede their sovereignty [78]. Miller contends that the wellbeing of weak states is largely a function of external, economic, and historic conditions that would entail the obligation of strong states via the SC model [79]. A country having habitats that give rise to desert locust outbreaks is an ecological accident, which might be analogized to a person having a genetic handicap that entails dependency and hence generates a moral duty for others to provide accommodations.

4.3.3. Nonstate Agents

There are at least three nonstate agents of justice. First, NGOs of various types have capacity but often limited capabilities, although they actively cultivate social connectivity. Examples of these agents include international aid organizations (e.g., CARE and Mercy Corps) and university-based organizations (e.g., the Global Locust Initiative and Association for Applied Acridology International). Next, corporations meet the criteria of both the C&C and SC models. Although desert locust irruptions are too erratic for the agrichemical industry to bother shaping locust-management policies and markets (producers of locust-specific biological insecticides have a greater interest), their products are important to preventive and reactive programs. Just as it is sociologically simplistic to advance the Rawlsian or "realist" view that states act only in their own self-interest, corporations need not be solely committed to maximizing profits as exemplified by B-corporations and others with social responsibility provisions [70]. And finally, the SC model suggests that citizens of industrialized nations bear some moral responsibility by virtue of their benefiting from structural injustices that disadvantage locust-afflicted people, although it must be recognized that their capability of meaningful action is extremely limited.

5. Distribution of Responsibilities

5.1. Four Standards of Global Justice

The C&C and SC models allow us to generate an inventory of potentially responsible agents. However, given both the number of agents, at various scales, who have either the capacity to respond or are implicated by social connections, as well as the complexity of the challenge of desert locust management, such an inventory is necessary but insufficient. We also require a framework for distinguishing among these agents and allocating particular duties to particular agents. Neither the C&C nor the SC model accomplishes this in isolation. As the proponent of the latter model, Young identifies "parameters of reasoning" that can be used to allocate specific duties, including capacity, privilege, interest, and ability [71] (pp. 144–147). However, to better operationalize the standards for fairly distributing the responsibilities for preventative and reactive locust management, we turn to Miller's more perspicuous analysis, which applies to agents ranging from individuals to conglomerate collectives (international organizations, nations, corporations, etc.) [79]. This model provides four concrete standards for allocating responsibility: causal, moral, capacity, and community.

5.1.1. Causal Responsibility

That moral agents who cause harm to others bear a commensurate responsibility to correct the wrong is a commonsense moral intuition. However, as we have argued previously, causal responsibility is neither necessary nor sufficient to generate corresponding moral and political responsibilities, especially in cases of compound and structural injustices. Miller's principle of causal attribution can still be useful in distributing responsibilities to the responsible agents identified by the C&C or SC models of responsibility [79], especially when modified as Shue's (1999 [80] p. 534) "greater contribution principle," which does not imply that the responsible agent is the exclusive cause [80] p. 534. Of the agents who have the capacity to respond to locust plagues or participate and benefit from the global political–economic structures that leave some states more vulnerable to and unable to effectively manage locust plagues, greater burdens should be placed on those who have contributed more to the problem.

Applying causal responsibility to the case of locusts can be challenging, given that swarms (the proximate cause of human suffering) are not moral agents, while the contribution of moral agents—whether individuals, states, or international institutions—are indirect and mediated. However, as Sardo analogously argues, the proper object of responsibility for the structural injustice of climate change is not discrete greenhouse-gas emissions, but the global political–economic structures that both intensify fossil-fuel extraction and unevenly distribute the benefits of such vulnerability and the risks of climate-related hazards [76]. In the case of locust plagues, climate change is aggravating the unpredictability and severity of plagues [35,36]. As such, the industrialized nations with disproportionate CO_2 emissions bear a commensurate, if partial, responsibility for the consequences of erratic and severe weather with regard to pest population dynamics. Additionally, colonial powers played a major role in creating weak governments, unstable societies, aggressive militaries, and dependent economies in their former colonies—all of which undermine both implementing preventive strategies and mounting reactive programs during upsurges and plagues. In an important sense, this complex network of causal responsibility involving locust plagues, climate change, and economic inequality might be best understood as constituting a global, structural injustice. Therefore, the idea of contribution should be expanded to include these climatic, geopolitical, and historical forces that unevenly distribute vulnerability and adaptive capacity for locust plagues.

5.1.2. Moral Responsibility

If an agent's harm is foreseeable, intentional, and unjustified (again, inequities can arise through legitimate processes), there is moral culpability—even without causal responsibility. For example, failure to monitor a risky situation could constitute wrongful neglect [79]. In an international context, even excusable ignorance of how injustices arose does not obviate moral responsibility [81]. In fact, Vanderheiden argues not only that ignorance is not exculpatory, but that agents have a positive responsibility to educate themselves to a reasonable degree about predictable natural disasters and other hazards [82]. What matters is synchronic features (present considerations of an agent's obligations to those who are suffering, limited by the agent's ability to render aid and to bear sacrifice), rather than diachronic considerations (how the suffering came to be). Miller argues that while an agent may have a synchronic moral duty, having participated in the genesis of a harm also can entail moral responsibility [79]. Hence, causal responsibility can engender moral responsibility, but it is not a necessary condition.

Therefore, greater burdens should be assigned to those agents in positions of relative power who foresaw, or should have foreseen, the hazards of locust plagues, but failed to take the necessary preventative action. This would place greater burdens on states and institutional agents than on individuals. Although countries with causal responsibility might have a greater duty to act, standing by while afflicted farmers or pastoralists suffer is not morally defensible even for those countries without a clear diachronic role in the crisis, when taking action could mitigate the harm suffered. Adapting the line of argument provided by Caney [81], our options are to: ascribe mitigation responsibilities to the subsistence agriculturalists, do nothing and thereby allow serious harm, or ascribe duties to more powerful agents. The third option seems the only practical and principled course of action. However, we must keep in mind that "ought implies can" in formulating moral duties, so the pragmatic element of Caney's third option leads to the next standard with respect to global justice.

5.1.3. Capacity to Act

This principle states that those agents with the greatest power to remediate suffering are obligated to do so. In an international context, Miller contends that those with the most resources are expected to shoulder a greater proportion of the burden, a duty that amounts to a kind of progressive taxation [79]. An objection to this standard is that the expectation would be a disincentive for wealthy nations to continue their financially successful practices if doing obligates them to ever greater contributions to poor nations. Miller replies that indefinite growth of production and consumption are not sustainable, regardless of whether global justice requires increased contributions [79]. Moreover, there is no evidence that economic growth depends on unlimited incentives. Another concern is that the synchronic nature of capacity does not require that we consider how an injustice or crisis arose. However, capacity also does not preclude considering causal responsibility.

The application of the capacity principle to locust management is rather straightforward: Wealthier nations are morally obliged to provide greater financial assistance to support both preventative and reactive measures in response to locust plagues. However, it should be noted that money is not the only resource relevant to capacity. Some industrialized nations have greater, relevant scientific and technical expertise. Nor should we focus entirely on nations, as substantial capacity can also be found in universities (e.g., scientific and technical expertise) and corporations (e.g., insecticides, application equipment, and safety supplies). Although we cannot expect a company to donate materials, businesses have a moral obligation to sell their products without profiteering. It should be noted that the capacity to intervene during a plague might be different than that needed to develop and sustain preventive programs. Given the economic efficiency of prevention, an agent's capacity is greatly enhanced by contributing to preventive management. Some might object that capacity is an unrealistic standard in the absence of a sociopolitical relationship between donor and recipient. This brings us to the final principle of global justice.

5.1.4. Communal Responsibility

As our discussion of the scope of global justice suggests, social relationships are morally relevant with regard to the allocation of duties pertaining to global justice. Miller contends that a communitarian principle that recognizes special ties accords with our moral intuitions [79]. For example, we accept, ceteris paribus, that duties to friends are greater than to strangers. If Young's social connections successfully makes the case that responsibility for harm derives from structural injustice, then it follows that an agent's responsibility to provide aid would likewise track structural or communal linkages [77]. However, reducing connections, and hence obligations, to the level of individuals becomes morally impotent given a person's very limited ability to influence structural injustice within the international community (recall that "ought implies can"). The problem with an individualistic approach can be avoided by considering communal relations that exist among conglomerated agents. Nations, agencies, NGOs, and corporations may have particular relations with afflicted countries that entail ethical responsibilities.

With regard to locust management, countries sharing common languages, religions, and cultural practices may often be wealthy nations and their former colonies. Importantly, past relationships of unjust domination created international communities and hence particular moral duties—not by virtue of remediating historical wrongs (although this is relevant to causal responsibility), but by virtue of contemporary connections. So, European nations have particular obligations to their former African and Asian colonies that are afflicted by locusts. In addition, international communities such as the FAO's Regional Commissions for Locust Control generate mutual obligations. When member countries fail to pay their annual dues and the organizations are consequently unable to conduct preventive programs [14], the resulting upsurges and plagues reflect an injustice and moral culpability within these regions. And finally, NGOs, religious denominations, universities, scientific organizations (e.g., the Orthopterists Society, which is an international group devoted to the study of grasshoppers, locusts, and their relatives), and other conglomerate collectives may have communal responsibilities through relationships to locust-afflicted countries, districts, or even villages.

5.2. Resolving the Standards

The principles guiding the just allocation of responsibilities among moral agents might be complementary in some cases (e.g., causal responsibility may align with moral responsibility, or causal responsibility may track communal responsibility). However, the principles may often be conflicting (e.g., an agent with the greatest capacity to act may have no causal responsibility) or unrelated (e.g., being agents within a community has no necessary relationship to those agents having the capacity to act). As such, how to integrate the four standards of global justice is a challenge for which there are at least four possible approaches [79].

5.2.1. Particularism

Few, if any, real-world cases will reduce to a single standard (e.g., a situation with capacity and no elements of causal, moral, or communal responsibility). On the other hand, there is no obvious formula for balancing the four principles, so we might abandon the search for a universal theory and simply deal with every case on its own terms. However, this ad hoc approach provides no guidance as to how the factors should be integrated in a consistent and rational matter.

5.2.2. Master Principle

We might attempt to defend one of the principles and set aside the others. However, arguing for a single consideration seems doomed, as there are sound reasons for each of the standards, and choosing any one of them could result in either solely backward-looking considerations, as with causal responsibility, or forward-looking considerations, as with moral responsibility. The choice of a master principle might also give rise to situations in which nobody would be obligated to help (e.g., an earthquake with the master principle of causal responsibility). There also are surely cases in which justice demands more than a single factor (e.g., a community of nations with radically unequal capacities).

5.2.3. Integration

The standards of global justice could be integrated in two ways. First, the principles could be applied in series (e.g., first moral responsibility, then causal, then communal, then capacity). However, the problem is that in any series, the initial standard will be a matter of degree, and in some cases a lower-order principle might well warrant greater consideration. Moreover, it is not clear how a sequential process would take into account conflicts among the principles (e.g., would there be a simple, ordinal weighting?).

The other integrative approach would be openly pluralistic, applying the principles in parallel. Such a simultaneous methodology would risk degrading into particularism, as there would be no overarching strategy. So, we might propose a heuristic such that immediate suffering is considered first (e.g., hunger from a locust plague) along with the most relevant standards (e.g., morality and capacity), followed by mitigating long-term problems arising from injustice (e.g., preventive programs for locust management) with their relevant standards (e.g., community and causal). However, as stated, even this approach would become idiosyncratic when applied to the complexity of international crises.

5.2.4. Connectivity

Perhaps the most operationally viable approach to finding an optimum between a formulaic integration with a strict "decision rule" and a vague particularism with the risks of ad hoc rationalization is a kind of dynamic compromise providing structural guidance to a pluralistic integration of the standards for global justice. Miller proposes a strategy of applying the principles according to their relative strength of connection to each of the agents [79]. For example, one agent may have a particularly strong causal responsibility, while another might have a highly developed communal obligation, and a third could have a great deal of capacity. Each then would be obligated to act in accordance with its connectivity, assuring that some agent can be held morally accountable and avoiding the problem of the agents waiting for others to act or debating who has the greatest duty. As such, international responsibility is divided among the agents such that they acknowledge connections to the afflicted people that compel action, rather than rationalizing which of the standards are inapplicable and thereby offering excuses for avoiding involvement. The result of connectivity theory is a kind of internal complexity with significant, if not formulaic, clarity—a situation that would appear to mirror real-world cases. Miller points out that this strategy does not preclude sensible ordering of standards reflecting the nature of the problem being addressed. For example, during a humanitarian crisis, capacity may be the highest priority—and whichever agent has the strongest connection to this standard would be compelled to act.

6. Applications to Locust-Management Scenarios

In the following scenarios of desert locust management, we construct simulations using plausible but necessarily idealized moral agents and sociopolitical relationships We considered using actual cases from the last few decades of locust management, but many of the potentially relevant details were not documented, and to generate a tractable analysis within the constraints of this paper, we would be compelled to ignore many of the nuances, organizational layers, and competing demands that were part of the record. Moreover, rather than assigning retrospective blame for past shortcomings, our goal is to illustrate the operationalization of the four principles of global justice using Miller's model of connectivity using realistic, if simplified, conditions [79].

6.1. Scenario 1: A Desert Locust Plague (Reactive Programs)

6.1.1. Scenario

There is a serious invasion of locust swarms during a plague into country X from distant sources in countries Y and Z. All three countries are impoverished with nominal export economies. There are many subsistence farmers in X, a country with a weak government involved in long-term conflicts with Y and civil unrest within its own borders. These military actions have been conducted in political alliance with wealthy country A whose interests align with country X, which is a former colony, while wealthy country B has supported the militaries of countries Y and Z. Country X has provided funding to the regional locust commission (R) that includes its neighbors, but tensions with countries Y and Z have impeded cooperation in terms of developing a strong outbreak-prevention program. All five of the countries (A, B, X, Y, and Z) are members of an international organization that was formed to provide crisis and development assistance on a global basis and whose director (I) determines the funding and logistical support allocated to the regional commission (R). An international NGO (N) is reluctantly accepted by the government of country X, but its work is appreciated by the local people who value the efforts to provide agricultural support and food relief to rural villages.

6.1.2. Analysis

There is an urgent need for assistance in country X to avoid crop loss by subsistence farmers who will suffer hunger and/or loss of land in the short term and perhaps health and education detriments in the long term.

With regard to **causal** responsibility, the swarms originated in countries Y and Z, so their ecosystems are the proximate cause of the swarms in country X. However, only to the extent that the governments of countries Y and Z failed in their duties to guard against outbreaks given available resources, would they have moral (ultimate) responsibility. When considering causal responsibility in terms of contribution rather than exclusive causality, more agents bear responsibilities. The political meddling of countries A and B constitute important, if less-immediate, causes of the situation by having impeded the development of preventive approaches in the region through organization R with support approved by director I. In historical terms, country X was a colony of country A, and this legacy accounts for a continuing lack of internal political capacity. Organization N has done nothing to cause the locust plague.

As for **moral** responsibility, the government of country X has a duty to provide for its citizens. The other countries (A, B, Y, and Z) have a common, humanitarian obligation to the suffering of those in country X. Organization R has no special, moral obligation, but the director (I) has the individual duty to fulfill the charter of the organization to aid nations in peril. Organization N has raised its operating funds through an explicit commitment to relieving hunger.

In terms of **capacity**, countries A and B have the resources needed to mitigate the immanent suffering, while countries X, Y, and Z are without the financial or logistical means to fully address a locust plague. The resources from countries A and B could be provided bilaterally to country X, but director I oversees an organization constructed to receive and distribute funds to needy people, including through commission R. Organization N is overwhelmed by the demand for food in locust-afflicted villages.

Country X is in a geopolitical **community** relationship with its neighbors (mere physical proximity would not constitute a community, although bordering countries typically establish international agreements). All three countries belong to the same regional locust commission (R), although this association is undermined by political hostilities. The colonial history and political bonds form a strong communal relationship between countries X and A. All of the countries belong to the international global community for which director I has responsibility. The director's obligations are diluted by virtue of its expansive membership. Organization N is begrudgingly present in the eyes of country X's government, although its work is valued locally.

In summary, given the urgency of the situation, the moral principles might be ordered such that capacity > community > moral = causal. Then, considering the evident strength of connections among the moral agents, the result would be one in which the responsibility for relieving the human suffering falls very strongly on country A and somewhat less on both country B and the director (I) of the international organization. Still less and relatively equal moral responsibility would accrue to countries X, Y, and Z, as well as organizations R and N.

6.2. Scenario 2: Desert Locust Recession (Preventive Programs)

6.2.1. Scenario

During a recession period, countries U, V, and W comprise a regional commission (organization R) charged with developing outbreak-prevention programs with the support of an international organization (I). While all three countries are impoverished, U and V have functional governments, while W is wracked with corruption by an authoritarian government. Countries U and V share a colonial history with country C, a wealthy nation. Country W has an abundance of oil, but most of the profits flow to the world's largest energy company (E, with stockholder, e) headquartered and taxed in country D, a wealthy nation. Historically, country U has few locust breeding sites; country V has many such habitats, but landmines (provided by country C during hostilities a decade earlier) make access dangerous; country W also has many breeding sites, but very poor roads. Models generated by organization I predict imminent shifts in rainfall patterns as a consequence of climate change attributable to a group of industrialized countries (G, which includes

C and D). These precipitation patterns will expand breeding habitats in country U and foster those in V and W. An international nongovernmental organization (N) pursues agricultural-development projects in the region, but lacks expertise in locust management.

6.2.2. Analysis

A viable preventive program in the region (R) requires the cooperation of countries U, V, and W. However, these agents cannot sustain such a program in light of their depleted economies and competing demands for healthcare, clean water, rural education, etc.

In terms of **causal** responsibility, a failure of prevention could be attributed to the relevant countries (U, V, and W) and the regional organization (R). However, the poverty of these agents can be traced to the colonialism of country C (along with the landmines making access dangerous) and the exploitation of natural resources by company E, which in turn benefits both its stockholder e and country D. The international organization (I) is limited by the funds provided by member nations. The intensification and expansion of locust breeding areas is assignable to those countries, G, responsible for emitting the greatest share of greenhouse gases.

With regard to moral responsibility, the prevention of a humanitarian disaster by a locust plague is arguably most heavily assigned to the international organization (I), given their mandate. In light of the benefits that have accrued to wealthy nations C and D through colonialism and continuing extraction of resources, these agents bear substantial responsibility. A similar case could be made for company E, although the direct linkage to locust-afflicted nations is somewhat weaker, and the connectivity of its stockholder (e) is much less but nonzero. The governments of U, V, and W also have some responsibility to their own people as well as those in the larger area, but their ethical burden is mitigated by poverty (particularly country W, given its weak government). This reasoning also applies to the regional commission R, although this organization has a much narrower duty than the affiliated countries and receives support from organization I. Finally, the NGO has no relevant expertise in locust management and no obligations to do that which is outside of its ability ("ought implies can").

Turning to **capacity**, regional organization (R) and the wealthy countries (C, D, and the collective, G) bear the greatest burden. International organization I has a multiplicity of responsibilities, only one of which is prevention of locust plagues. The impoverished countries (U, V, and W) have broad obligations and limited resources, as does the nongovernmental organization N, while country W further lacks the political ability to contribute. The energy company (E) has assets needed for a preventive program and a duty to provide these while meeting fiduciary obligations to its shareholders, including e, whose capacity is vanishingly small.

And finally, considering **community**, countries U, V, and W form an association geographically and through their collective support of organization R, which represents the locust-management community. There is a strong communal component of country C with countries U and V by virtue of historical relations. Country W's government makes for a weaker communal ability, although it has an economic relationship with company E, while stockholder e participates in the corporate community. Both E and e therefore have an economic connection to country D. The nongovernmental organization (N) has a strong relationship with countries U and V, but lesser so with country W. Both the international organization (I) and the industrialized countries (G) are weak elements of a community concerned with locust outbreak prevention by virtue of diverse global connections.

In summary, given the chronic, rather than acute, nature of this case, the moral principles might be ordered such that causal = moral > community > capacity. Then, taking into account the evident strength of connectivities among the moral agents, the result would be that the greatest obligations to prevent the harm resulting from a locust plague accrue at the country level, including the wealthy (C and D) and impoverished (U and V) nations. The responsibilities of the former arise in consideration of their moral duties and the latter in consideration of their causal efficacies. Of the organizations, only the regional commission (R) has a high obligation, while the larger collectives (G and I) and the company (E) have moderate obligations, largely due to weaker communal and causal considerations (even less for stockholder e). Likewise, country W has moderate obligations in consideration of its weak government, which undermines capacity and community. The nongovernmental organization (N) has high moral and communal factors, but no causal responsibility or capacity, so its overall obligation is low.

6.3. Summary

In both scenarios (preventive and reactive), moral responsibility primarily falls to wealthy countries. However, in the preventive scenario, the allocations are more evenly distributed among countries, and the international and regional organizations bear a greater burden than in the reactive scenario. In the reactive case, the wealthy countries clearly have the greatest responsibility, which reflects the weight put on capacity during a crisis (of course, funds may pass through the international organization).

The allocation of responsibilities for global justice in these cases does not diverge markedly from what we would take to be common, moral intuitions in which causal responsibility and capacity would seem to be important considerations and wealthy countries are thought to bear the greatest burden. Rather, the advantage of adapting and applying Miller's fourfold principles along with network theory is that we have an explicit set of reasons for our ethical inclinations rather than merely doing what "seems right" without being able to explain or defend our decisions to those who fail to share our intuitions [79]. In short, it is desirable to do the right thing, but it is far more intellectually and morally valuable to do the right thing for the right reasons.

That said, we do not possess anything approaching certainty through some sort of moral mathematics. This analysis does not yield artificially precise quantitative results that might be generated by converting the various factors into numerical values for a spreadsheet. As much as such exactitude would seem desirable, it would substitute easy answers for the hard work of democratic deliberation. So, the framework provides a means by which individuals and communities can think through the nature of their responsibilities in the course of negotiating a just approach to developing both preventive programs of locust management and reactive responses during upsurges that may become plagues.

Perhaps even more importantly, with a conceptual formalization, we now have a method for critically analyzing novel cases of locust management that are likely to emerge given rapidly changing political, economic, and environmental conditions. Unfamiliar conditions may reduce the relevancy of our experiences and lead to conflicting moral intuitions. As new situations arise, having a theoretical framework for identifying moral agents and distributing responsibilities to achieve global justice will be important.

7. Conclusions

Our purpose in this paper was to address the rising awareness in the field of locust management that the limiting factor in making substantive progress has become a serious consideration, as well as integration of the social sciences and humanities with the natural sciences [7,21,83]. Oftentimes, we know what strategy and methods to implement from entomological and ecological perspectives to improve the efficacy of management, and so the obstacles are not more or better data about the insects, although greater knowledge is surely an asset, and we would not discourage continued research on locust biology with important implications for sound pest management (e.g., [84]). Rather, the impediments are economic, political, and cultural understanding. As such, we sought to sketch an approach to making socially and ethically sound decisions about desert locust management through the framework of global justice, informed by conventional moral theories. Two findings emerged from our analysis, perhaps neither of which is terribly surprising to those in the field.

First, there is a tremendous diversity of moral agents, including individuals, scientific organizations, nongovernmental organizations, corporations, nations, government agencies, regional commissions, and international bodies. If anything, our analysis likely expanded the inventory of collective, moral agents while providing a sound rationale for inclusion.

Second, justly allocating responsibilities among the agents during a preventive or reactive program of locust management is a complex task. However, there are intellectually compelling principles to provide at least a working model for how these duties ought to be distributed. While natural scientists might be understandably frustrated by the imprecision of our framework, we would point out Aristotle's admonition:

> For it belongs to an educated person to seek out precision in each genus to the extent that the nature of the matter allows: to accept persuasive speech from a skilled mathematician appears comparable to demanding demonstrations from a skilled rhetorician. [85] (pp. 3–4)

While a formula for assigning responsibility is not a sensible endpoint, having the criteria for moral agency, an explicit set of principles, and a plausible, if inexact, theory for justly distributing the burdens of locust management hopefully constitutes a valuable starting point for further research and discourse. Our foray into the nexus of locust management, social science, and humanities represents a step in working toward a philosophically coherent, scientifically sound, conceptually defensible, and operationally viable method.

While our discussion of justice has an anthropocentric focus, it should be noted that many theorists of global justice have argued for expanding the boundaries of justice beyond species membership in addition to political membership. Singer, for example, argues that because all sentient creatures experience pleasure and pain, they are also deserving of moral consideration [86]. Nussbaum also extends her capabilities approach to ground interspecies justice, based on the idea that nonhuman animals are capable of living dignified and flourishing lives, with corresponding needs to promote such flourishing [87]. Schlosberg focuses this approach to the level of the ecosystem, rather than the individual or species, noting that harm to nonhuman animals, in recognition that individual creatures are harmed by habitat disruption and destruction and that ecosystem integrity and functioning contribute to human and nonhuman flourishing [88].

Although a detailed engagement with questions of interspecies justice is beyond the scope of this paper, these arguments raise important questions in the context of locust swarms that demand further interdisciplinary research. Insofar as locust swarms and their management disrupt the integrity and functioning of ecosystems, they may generate obligations of justice on behalf of those ecosystems, especially when such disruptions are caused by anthropogenic climate change. Furthermore, management strategies should be sensitive to ecosystem functioning and integrity, and the long-term ecological consequences must be considered when evaluating the justice of both preventive and reactive strategies. Finally, the moral framework we have proposed does not solve the political problem of actually getting agents (especially powerful states) to bear their responsibilities with regard to locust management. That is a topic for another line of research that must incorporate scholarship in political theory, including Schiff's work on how different ways of narrating injustices can affect how agents acknowledge and act upon their responsibilities [89], and Ackerly's human-rights model of responsibility, which identifies principles-in-practice for agents to leverage their existing social roles and relationships to discharge their responsibilities for combatting injustice [90]. Our goal is that, having opened up space for dialogue between natural scientific research on desert locust management and political and moral philosophies of global justice, future scholarship can engage the questions our analysis generates.

Author Contributions: Conceptualization, J.A.L. and M.C.S.; methodology, J.A.L. and M.C.S.; formal analysis, J.A.L. and M.C.S.; writing—original draft preparation, J.A.L.; writing—review and editing, M.C.S.; investigation, J.A.L. and M.C.S. Both authors have read and agreed to the published version of the manuscript.

Funding: This research received no external funding.

Acknowledgments: The authors are grateful to the editors and reviewers for helpful and constructive feedback.

Conflicts of Interest: The authors declare no conflict of interest.

References

1. Uvarov, B. *Grasshoppers and Locusts: A Handbook of General Acridology*; Cambridge University Press: London, UK, 1966; Volume 1.
2. Uvarov, B. *Grasshoppers and Locusts: A Handbook of General Acridology*; Centre for Overseas Pest Research: London, UK, 1977; Volume 2.
3. Pener, M.P. Endocrine aspects of phase polymorphism in locusts. In *Endocrinology of Insects*; Downer, R.G.H., Laufer, H., Eds.; Alan R. Liss: New York, NY, USA, 1983; pp. 379–394.
4. Cullen, D.A.; Cease, A.V.; Latchininsky, A.V.; Ayali, A.; Berry, K.; Buhl, J.; Keyser, R.D.; Foquet, B.; Hadrich, J.C.; Matheson, T.; et al. From molecules to management: Mechanisms and consequences of locust phase polyphenism. *Adv. Insect Phys.* **2017**, *53*, 165–185.
5. Symmons, P. A critique of "Preventive control and desert locust plagues". *Crop Prot.* **2009**, *28*, 905–907. [CrossRef]
6. Medina, H.; Cease, A.; Trumper, E. The resurgence of the South American locust (*Schistocerca cancellata*). *Metaleptea* **2017**, *37*, 17–21.
7. Zhang, L.; Lecoq, M.; Latchininsky, A.; Hunter, D. Locust and grasshopper management. *Annu. Rev. Entomol.* **2019**, *64*, 15–34. [CrossRef]
8. Symmons, P.M.; Cressman, K.K. *Desert Locust Guidelines*; Food and Agricultural Organization of the United Nations: Rome, Italy, 2001.
9. Lecoq, M. Le criquet pélerin: Enseignements de la dernière invasion et perspectives offertes par la biomodélisation. In *La Lutte Anti-Acridienne*; Essaid, A., Ed.; Faculté des Sciences, CIRUISEF, AUPELF-UREF: Paris, France, 1991; pp. 71–98.
10. Doré, A.; Barbier, M. Maintenir la vigilance. *Rev. Anthropol. Connaiss.* **2015**, *9*, 189–212. [CrossRef]
11. Gay, P.-E.; Lecoq, M.; Piou, C. The limitations of locust preventive management faced with spatial uncertainty: Exploration with a multi-agent model. *Pest Manag. Sci.* **2019**, *76*, 1094–1102. [CrossRef] [PubMed]
12. De Vreyer, P.; Guilbert, N.; Mesple-Sompsa, S. Impact of natural disasters on education outcomes: Evidence from the 1987–89 locust plague in Mali. *J. Afr. Econ.* **2014**, *24*, 57–100. [CrossRef]
13. Manthey, S. Curse of the Good Rains: The Long-Run Impacts of Locust Infestations, Evidence from the 1986–1989 Locust Plague in the Sahel. Master's Thesis, Lund University, Lund, Sweden, 2020.
14. Showler, A.T. The desert locust in Africa and Western Asia: Complexities of war, politics, perilous terrain, and development. In *Radcliffe's IPM World Textbook*; Radcliffe, E.B., Hutchison, W.D., Cancelado, R.E., Eds.; University of Minnesota: St. Paul, MN, USA, 2019.
15. Lecoq, M. Recent progress in desert and migratory locust management in Africa: Are preventative actions possible? *J. Orthop. Res.* **2001**, *10*, 277–291. [CrossRef]
16. Magor, J.I.; Lecoq, M.; Hunter, D.M. Preventive control and desert locust plagues. *Crop Prot.* **2008**, *27*, 1527–1533. [CrossRef]
17. van Huis, A.; Cressman, K.; Magor, J.I. Preventing desert locust plagues: Optimizing management interventions. *Entomol. Exp. Appl.* **2007**, *122*, 191–214. [CrossRef]
18. Abou-Ali, H.; Belhaj, M. Cost Benefit Analysis of Desert Locust Control: A Multicountry Perspective. Economic Research Forum Working Papers. 2008. Available online: https://econpapers.repec.org/paper/ergwpaper/801.htm (accessed on 11 January 2021).
19. Millist, N.; Abdalla, A. *Benefit–Cost Analysis of Australia Plague Locust Control Operations for 2010–2011*; Australian Bureau of Agricultural and Resource Economics and Sciences: Canberra, Australia, 2011.
20. Latchininsky, A.V. Locusts and remote sensing: A review. *J. Appl. Rem. Sens.* **2013**, *7*. [CrossRef]
21. Lecoq, M. Desert locust management: From ecology to anthropology. *J. Orthop. Res.* **2005**, *14*, 179–186. [CrossRef]
22. Toleubayev, K.; Jansen, K.; van Huis, A. Locust control in transition: The loss and reinvention of collective action in post-Soviet Kazakhstan. *Ecol. Soc.* **2007**, *12*, 38. [CrossRef]
23. De Vreyer, P.; Guilbert, N.; Mesplé-Somps, S. *The 1987–1989 Locust Plague in Mali: Evidences of the Heterogeneous Impact of Income Shocks on Education Outcomes*; Working Papers No DT/2012/05; Örebro University School of Business: Örebro, Sweden, 2012.
24. Péloquin, C. Unruly Nature and Technological Authority: Governing Locust Swarms in the Sahel. Ph.D. Thesis, University of Arizona, Tucson, AZ, USA, 2014.
25. Cease, A.J.; Elser, J.J.; Fenichel, E.P.; Hadrich, J.C. Living with locusts: Connecting soil nitrogen, locust outbreaks, livelihoods, and livestock markets. *BioScience* **2015**, *65*, 551–558. [CrossRef]
26. Deshormes, A. *Institutional Study to Enhance the Roles and Responsibilities of the Desert Locust Control Commission Established under Article XIV*; Food & Agriculture Organization of the United Nations: Rome, Italy, 2011.
27. Tom, K. The UN Should Learn That Ideology Won't Stop a Plague of Locusts. Real Clear World Website. Available online: https://www.realclearworld.com/2020/08/06/un_ideology_wont_stop_a_plague_of_locusts_501135.html (accessed on 11 January 2021).
28. Metlerkamp, L.U.S. Agribusiness Takes Aim at Global Food Policy Reform. Available online: https://foodtank.com/news/2020/05/u-s-agribusiness-takes-aim-at-global-food-policy-reform/ (accessed on 11 January 2021).
29. Lecoq, M. Some considerations of the current desert locust situation in East Africa. *Metaleptea* **2020**, *40*, 25–26.

30. Gay, P.-E.; Lecoq, M.; Piou, C. Improving preventive locust management: Insights from a multi-agent model. *Pest Manag. Sci.* **2018**, *74*, 46–58. [CrossRef] [PubMed]
31. Food & Agriculture Organization of the United Nations. Locust watch. *Desert Locust Bull.* **2019**, *483*, 1–8.
32. Bodin, Ö. Collaborative environmental governance: Achieving collective action in social-ecological systems. *Science* **2017**, *357*, eaan1114. [CrossRef]
33. Blair, B.; Lovecraft, A.L.; Hum, R. The disaster chronotope: Spatial and temporal learning in governance of extreme events. In *Governance of Risk, Hazards and Disasters*; Forino, G., Bonati, S., Calandra, L.M., Eds.; Routledge: London, UK, 2018; pp. 43–64.
34. Shi, P.; Li, N.; Ye, Q.; Dong, W.; Han, G.; Fang, W. Research on integrated disaster risk governance in the context of global environmental change. *Int. J. Disaster Risk Sci.* **2010**, *1*, 17–23. [CrossRef]
35. Meynard, C.N.; Lecoq, M.; Chapuis, M.P.; Piou, C. On the relative role of climate change and management in the current Desert Locust outbreak in East Africa. *Glob. Chang. Biol.* **2020**, *26*, 3753–3755. [CrossRef]
36. Salih, A.A.M.; Baraibar, M.; Mwangi, K.K.; Artan, G. Climate change and locust outbreak in East Africa. *Nat. Clim. Chang.* **2020**, *10*, 584–585. [CrossRef]
37. Thomas, N.; Nigam, S. Twentieth-century climate change over Africa: Seasonal hydroclimate trends and Sahara desert expansion. *J. Clim.* **2018**, *31*, 3349–3370. [CrossRef]
38. Food & Agriculture Organization of the United Nations. Desert Locust. Available online: http://www.fao.org/locusts/en/ (accessed on 11 January 2021).
39. Latchininsky, A.; Sword, G.; Sergeev, M.; Cigliano, M.M.; Lecoq, M. Locusts and grasshoppers: Behavior, ecology, and biogeography. *Psyche* **2011**, *2011*, 578327. [CrossRef]
40. Food & Agriculture Organization of the United Nations. Locust Watch. Available online: http://www.fao.org/ag/locusts/en/info/2094/2548/index.html (accessed on 11 January 2021).
41. Belayneh, Y.T. Acridid pest management in the developing world: A challenge to the rural population, a dilemma to the international community. *J. Orthoptera Res.* **2005**, *14*, 187–195. [CrossRef]
42. Brader, L.; Djibo, H.; Faye, F.G.; Ghaout, S.; Lazar, M.; Luzietoso, P.N.; Ould Babah, M.A. *Towards a More Effective Response to Desert Locusts and Their Impacts on Food Security, Livelihoods and Poverty: Multilateral Evaluation of the 2003–2005 Desert Locust Campaign*; Food & Agriculture Organization of the United Nations: Rome, Italy, 2006.
43. Lockwood, J.A.; Showler, A.T.; Latchininsky, A.V. Can we make locust and grasshopper management sustainable? *J. Orthop. Res.* **2001**, *10*, 315–329. [CrossRef]
44. Driver, J. The History of Utilitarianism. In *The Stanford Encyclopedia of Philosophy*; Zalta, E.N., Ed.; 2014. Available online: https://plato.stanford.edu/archives/win2014/entries/utilitarianism-history/ (accessed on 11 January 2021).
45. Alexander, L.; Moore, M. Deontological Ethics. In *The Stanford Encyclopedia of Philosophy*; Zalta, E.N., Ed.; 2016. Available online: https://plato.stanford.edu/archives/win2016/entries/ethics-deontological/ (accessed on 11 January 2021).
46. Hursthouse, R.; Pettigrove, G. Virtue Ethics. In *The Stanford Encyclopedia of Philosophy*; Zalta, E.N., Ed.; 2018. Available online: https://plato.stanford.edu/archives/win2018/entries/ethics-virtue/ (accessed on 11 January 2021).
47. Singer, P. Famine, affluence, and morality. *Philos. Public Aff.* **1972**, *1*, 229–243.
48. O'Neill, O. Kantian approaches to some famine problems. In *Matters of Life and Death*; Reagan, T., Ed.; McGraw-Hill: New York, NY, USA, 1993; pp. 258–270.
49. MacIntyre, A. *After Virtue: A Study in Moral Theory*, 3rd ed.; University of Notre Dame Press: Notre Dame, IN, USA, 2007.
50. Van Hooft, S. Cosmopolitanism as virtue. *J. Glob. Ethics* **2007**, *3*, 303–315. [CrossRef]
51. Hayek, F.A. *The Collected Works of F. A. Hayek Volume XVII: The Constitution of Liberty*; Hamowy, R., Ed.; University of Chicago Press: Chicago, IL, USA, 2011.
52. Sandel, M.J. *Justice: What's the Right Thing to Do?* Farrar, Straus and Giroux: New York, NY, USA, 2009.
53. Rawls, J. *A Theory of Justice*; Harvard University Press: Cambridge, MA, USA, 1971.
54. Young, I.M. *Justice and the Politics of Difference*; Princeton University Press: Princeton, NJ, USA, 1990.
55. Goodhart, M. *Injustice: Political Theory for the Real World*; Oxford University Press: New York, NY, USA, 2018.
56. Lockwood, J.A.; Kemp, W.P.; Onsager, J.A. Long-term, large-scale effects of insecticidal control on rangeland grasshopper populations. *J. Econ. Entomol.* **1988**, *81*, 1258–1264. [CrossRef]
57. Shue, H. *Climate Justice: Vulnerability and Protection*; Oxford University Press: New York, NY, USA, 2014.
58. Acemoglu, D.; Johnson, S.; Robinson, J.A. The colonial origins of comparative development: An empirical investigation. *Am. Econ. Rev.* **2001**, *91*, 1369–1401. [CrossRef]
59. Shklar, J. *The Faces of Injustice*; Yale University Press: New Haven, CT, USA, 1990.
60. Erskine, T. Assigning responsibilities to institutional moral agents: The case of states and quasi-states. *Ethics Int. Aff.* **2001**, *15*, 67–85. [CrossRef]
61. Rawls, J. *The Law of Peoples: With, the Idea of Public Reason Revisited*; Harvard University Press: Cambridge, MA, USA, 1999.
62. Beitz, C.R. Justice and international relations. *Philos. Public Aff.* **1975**, *4*, 360–389.
63. Shue, H. *Basic Rights: Subsistence, Affluence, and U.S. Foreign Policy*; Princeton University Press: Princeton, NJ, USA, 1980.
64. Nussbaum, M. Capabilities and social justice. *Int. Stud. Rev.* **2002**, *4*, 123–135. [CrossRef]
65. Nagel, T. The problem of global justice. *Philos. Public Aff.* **2005**, *33*, 113–147. [CrossRef]
66. Miller, D. Cosmopolitanism: A critique. *Crit. Rev. Int. Soc. Political Philos.* **2002**, *5*, 80–85. [CrossRef]

67. Forst, R. Towards a critical theory of transnational justice. *Metaphilosophy* **2001**, *32*, 160–179. [CrossRef]
68. Kant, I. *The Metaphysics of Morals*; Gregor, M., Translator; Cambridge University Press: Cambridge, UK, 1996.
69. Feinberg, J. Collective responsibility. *J. Philos.* **1968**, *65*, 674–688. [CrossRef]
70. O'Neill, O. Agents of justice. *Metaphilos* **2001**, *32*, 180–195. [CrossRef]
71. Young, I.M. *Responsibility for Justice*; Oxford University Press: New York, NY, USA, 2011.
72. French, P. *Collective and Corporate Responsibility*; Columbia University Press: New York, NY, USA, 1984.
73. Global Locust Initiative, Arizona State University Global Sustainability. Available online: https://sustainability.asu.edu/global-locust-initiative/ (accessed on 11 January 2021).
74. Sen, A. Capability and well-being. In *The Quality of Life*; Nussbaum, M.C., Sen, A., Eds.; Clarendon Press: Oxford, UK, 1993; pp. 30–53.
75. Caney, S. Climate change and the duties of the advantaged. *Crit. Rev. Int. Soc. Political Philos.* **2010**, *13*, 203–228. [CrossRef]
76. Sardo, M.C. Responsibility for climate justice: Political not moral. *Eur. J. Political Theory* **2020**. [CrossRef]
77. Young, I.M. Responsibility and global justice: A social connection model. *Soc. Philos. Policy* **2006**, *23*, 102–130. [CrossRef]
78. van de Walle, N. Economic reform in a democratizing Africa. *Comp. Political* **1999**, *32*, 21–41. [CrossRef]
79. Miller, D. National Responsibility and Global Justice. *Crit. Rev. Int. Soc. Political Philos.* **2008**, *11*, 383–399. [CrossRef]
80. Shue, H. Global environment and international inequality. *Int. Aff.* **1999**, *75*, 531–545. [CrossRef]
81. Caney, S. Cosmopolitan justice, responsibility, and global climate change. *Leiden J. Int. Law* **2005**, *30*, 747–775. [CrossRef]
82. Vanderheiden, S. Knowledge, uncertainty, and responsibility: Responding to climate change. *Public Aff. Q.* **2004**, *18*, 141–158.
83. Sörlin, S. Environmental humanities: Why should biologists interested in the environment take the humanities seriously? *BioScience* **2012**, *62*, 788–789. [CrossRef]
84. Sword, G.A.; Lecoq, M.; Simpson, S.J. Phase polyphenism and preventative locust management. *J. Insect Phys.* **2010**, *56*, 949–957. [CrossRef]
85. Aristotle. *Aristotle's Nicomachean Ethics*; Bartlett, R.C.; Collins, S.D., Translators; University of Chicago Press: Chicago, IL, USA, 2012.
86. Singer, P. *Animal Liberation*; HarperCollins Publishers Inc.: New York, NY, USA, 2002.
87. Nussbaum, M.C. *Frontiers of Justice: Disability, Nationality, Species Membership*; The Belknap Press of Harvard University Press: Cambridge, MA, USA, 2007.
88. Schlosberg, D. *Defining Environmental Justice: Theories, Movements, and Nature*; Oxford University Press: Oxford, UK, 2007.
89. Schiff, J.L. *Burdens of Political Responsibility: Narrative and the Cultivation of Responsiveness*; Cambridge University Press: Cambridge, UK, 2014.
90. Ackerly, B.A. *Just Responsibility: A Human Rights Theory of Global Justice*; Oxford University Press: New York, NY, USA, 2018.

Short Biography of Authors

Jeffrey Alan Lockwood Professor of Natural Sciences and Humanities, Department of Philosophy and Religious Studies, University of Wyoming (2003–present). B.S. in Biology, New Mexico Tech, 1982; Ph.D. in Entomology, Louisiana State University, 1985. Professor of Entomology, University of Wyoming (1986–2003); Visiting Fellow, CSIRO Division of Entomology, Canberra, Australia (1993–1994); Research Fellow, Organization for Economic Cooperation and Development, AgResearch, Mosgiel, New Zealand. Research interests include ecology and management of rangeland grasshoppers, environmental ethics, and philosophy of science. Member of Entomological Society of America, Orthopterists Society, Independent Book Publishers Association, American Society of Composers, Authors and Publishers. Awards include: Western Region Extension Directors Association Award of Excellence, Finalist for Best Contemporary Nonfiction from Western Writers of America, Silver medal for the Benjamin Franklin Award from the Independent Book Publishers Association, John Burroughs Award, Albert Schweitzer Award, Pushcart Prize, and University of Wyoming's: Extraordinary Merit in Research Award and Excellence in Internationalization Award.

Michael Christopher Sardo Non-Tenure Track Assistant Professor, Department of Politics, Occidental College (2019–present). B.A. in Government and Philosophy, The College of William and Mary in Virginia, 2011; M.A. and Ph.D. in Political Science, Northwestern University, 2017. Research interests include political responsibility, global justice, climate change, the anthropocene, and the history of political thought.

agronomy

MDPI

Article

Could Entomophagy Be an Effective Mitigation Measure in Desert Locust Management?

Ahmed Ali Samejo [1], Riffat Sultana [1,*], Santosh Kumar [2] and Samiullah Soomro [1]

1 Department of Zoology, University of Sindh, Jamshoro 75500, Pakistan; samejo_ali7@hotmail.com (A.A.S.); samiullah.soomro@scholars.usindh.edu.pk (S.S.)
2 Department of Zoology, Cholistan University of Veterinary and Animal Sciences, Bahawalpur 75500, Pakistan; santoshkumar@cuvas.edu.pk
* Correspondence: riffat.sultana@usindh.edu.pk

Abstract: The desert locust has been a notorious pest since ancient times. A population upsurge hit Pakistan in 2019 and caused tremendous damage to agriculture and livelihoods. To take advantage of this ongoing upsurge, we conducted a field study to verify whether locust collection could be an interesting control method to protect crops in the event of an invasion, as well as an accepted food resource for poor rural communities. A village in the Thar desertic region was selected as a type-locality. An awareness campaign was launched to promote the collection and consumption of locusts as well as to alert people of their nutritional value. Two large swarms arrived near the village and several other swarms affected places nearby. Around 3033 kg of locusts were collected through handpicking at night. Most of the locusts were eaten and, as a result, hoppers of the next generation did not emerge in the type-locality; however, hopper bands appeared in areas where entomophagy was not practiced. The study area had less locust activity because swarms could not lay eggs due to entomophagy by the villagers. The consumption of desert locusts could be an effective practice to prevent malnutrition and protein deficiency and, to a certain extent, an efficient mitigation measure to help local populations to better protect themselves and their crops against locust outbreaks. Collection and consumption of locusts should be encouraged while remaining realistic about its real impact on locust control. This should also be done in concert with local authorities to take into account the risks to human health and to avoid the consumption of insects treated with pesticides.

Keywords: entomophagy; malnutrition; desert locust; outbreak; food; insect

Citation: Samejo, A.A.; Sultana, R.; Kumar, S.; Soomro, S. Could Entomophagy Be an Effective Mitigation Measure in Desert Locust Management? *Agronomy* **2021**, *11*, 455. https://doi.org/10.3390/agronomy11030455

Academic Editors: Michel Lecoq, Arianne Cease and Imre J. Holb

Received: 26 January 2021
Accepted: 25 February 2021
Published: 28 February 2021

1. Introduction

Among the large number of pest locust species, the desert locust *Schistocerca gregaria* (Forskål, 1775) is the best known and most feared across borders and centuries. Major invasions follow one another at high frequency (eight for the period 1860–1962) [1–3]. These invasions occur when rainfall conditions are favorable to the insects (rainfall well distributed in time and space) in their primary breeding areas. Outbreaks are mainly located in desertic zones that are often difficult to access, but whose total area remains limited (12%) relative to the total area that can be invaded by swarms (up to 31 million km^2) [2,4]. Plagues of desert locusts are disastrous for agriculture and vegetation, destroy cash crops and livestock, and cost billions of dollars to control. The two most recent upsurges in 1986–1989 and 2003–2005 required treatment with insecticides mainly in Africa and the Middle East, 16.8 and 13 million hectares, and costing an estimated 274 and 500 million US dollars, respectively [5–7].

Desert locusts are difficult to combat but can be controlled by proactive monitoring, early action, and the targeted use of appropriate control measures, as needed. Prevention is coordinated internationally by the Food and Agriculture Organization of the United Nations (FAO) [8–11]. Thanks to this strategy, ever-more efficient pesticides, and other control measures, new products that are less environmentally damaging, such as the mycopesticides, and 50 years of hindsight, the invasions are now less frequent, less widespread, and if they cannot be stopped at an early stage, shorter lived and better managed [1,3,10,12–16].

This insect was calamitous in the past [5,17], and dramatic events continue to occur. For various reasons, some outbreaks cannot be stopped at an early stage [12,14,18]. During outbreaks, the swarms disperse quickly over a vast territory; control measures are not used because they are too expensive and difficult to utilize; villagers are overwhelmed in the face of the massive and sudden arrival of the swarms; and crop damage is considerable [3,14]. This is evidenced by the recent upsurge initiated in 2018, which is still ongoing. Originating on the southern Arabian Peninsula, *S. gregaria* has invaded much of East Africa, the Middle East, and Southwest Asia to Pakistan and India and is causing alarm for the food security of many developing and poor countries [9,19]. The economic, social, and environmental consequences of such events may persist or appear several years later [20].

Whether for prevention or in the event of an upsurge, organizing control measures relies mainly on state agencies. Rural communities are often destitute with no means at their disposal. These farmers have developed physical methods that they continue to practice: making noise or burning tires to scare away swarms, burning locusts as they roost at night in the vegetation, digging trenches and burying hopper bands, and plowing the egg-laying fields to destroy the eggs [21]. Harvesting locusts as food can replace other control methods [22]. Manual collection, which may seem derisory, is widely practiced around the world and represents a control method (to reduce the local population), a nutritionally interesting food resource, and a source of income for families to at least partially compensate for crop damage.

The practice of entomophagy is ancient. Locusts have been consumed for centuries and still are in some regions today. In many countries around the world, they are considered a delicacy [23–25]. Entomophagy is an old strategy used to get food after locusts have devastated crops, and the FAO [24] has already documented the importance of edible insects including locusts. When they are swarming, locusts can be collected in large numbers (a single swarm can cover 1200 km^2), relatively quickly and easily during the night when they remain almost motionless. The insects can be collected using hands, bags, and buckets. Nutritionally, locusts are excellent sources of protein and other essential nutrients, both as food for people or feed for other animals [26,27]. To take advantage of the ongoing upsurge in Pakistan, we conducted a field study to verify whether locust collection could indeed control locust populations as well as be an accepted food resource for poor rural communities.

2. Materials and Methods

Selection of Villages and Local People Awareness

Ten villages of the Mahandre-Jo-Par union council (25°35′17.59″ N/70°10′13.16″ E), in the Thar desertic area within Sindh province, were selected as a study site (Figure 1). A sensitization program was conducted in June–July 2019 (1) to raise awareness of the benefits of consuming desert locusts as a source of protein and (2) to determine if collecting locusts on a massive scale could control their proliferation. To remove any reluctance, the villagers were reminded that locusts are halal in the Islamic religion.

Villagers collected locusts at night, by hand picking, in July–October 2019, using light traps (torches or headlights of vehicles) (Figure 2A,B). Locusts were stored in polypropylene bags or other locally available containers (Figure 2C,D). Each bag was weighed (Metis electronic scale with a capacity of 100 kg). The people who consumed the locusts were interviewed about their methods of preparation and cooking, the quantity they consumed each day, and the quality of their taste. To control the locusts locally, we carried out four field trips into their type-locality and other surrounding villages, after the arrival of each of the swarms, to control the emergence of hoppers, the occurrence of hopper bands, and the insects' behavior.

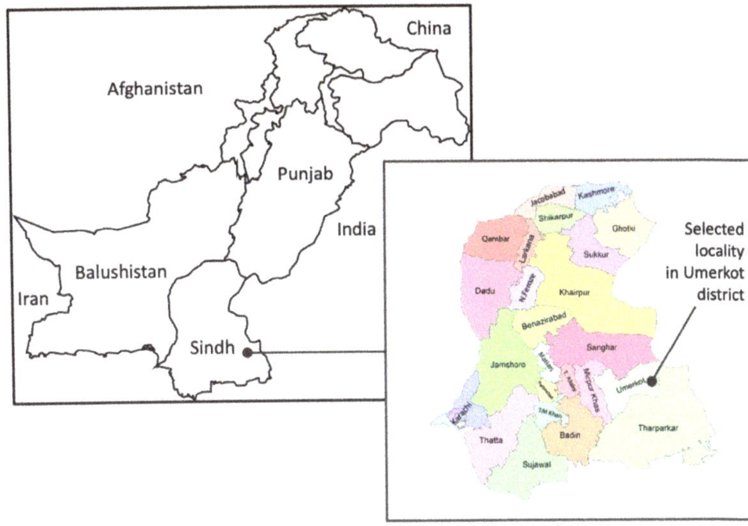

Figure 1. Locality surveyed to inform the local Thari populations about the benefits of locusts as a source of protein.

Figure 2. Collecting desert locusts at night in a field of crops: (**A**) vehicle headlight used as a light trap, (**B**) hand torch for collecting locusts and container for storage, (**C**) polypropylene bag for locusts, (**D**) polypropylene bag containing locusts ready for delivery. Photos by Ahmed Ali Samejo, Santosh Kumar, and Samiullah Soomro.

3. Results

Villagers said swarms of locusts arrived in this area of Pakistan after 22 years of recession. Locusts began to damage all green vegetation, cash crops, and fodder. As a result, people were motivated to collect and eat the locusts both to protect their crops and to improve their diets. On 4 July 2019, a mature swarm of desert locusts landed in the vicinity of Mahandre-Jo-Par village (type-locality) in the evening. Word travelled between villages about the arrival of the swarm, spread over 5 km^2. Residents went to that area at night carrying torches and polypropylene bags for catching locusts. The people knew from local tradition that locusts cannot be caught during the day, but are easily caught at night using light traps. At night, locusts mostly roost on crops and wild shrubs, and some females dig in soil with their ovipositor. For four consecutive nights, the villagers filled bags with locusts. In those four days, around 1275 kilograms of locusts were collected (Table 1). After collection, the bags were closed and brought back to the villagers' homes. Most of the locusts were eaten, although a few fled due to the disturbance. To prepare them for consumption, the locusts were first put in boiling water (Figure 3A), then dried for a day (Figure 3B). The dried locusts were then fried in various ways to make crisp, spicy, and crunchy dishes (Figure 3C–F).

Table 1. Collection of desert locust by villagers after the arrival of the first swarm on 4 July 2019.

Villages	Weight of Desert Locust Collected (kg)				
	1st Day	2nd Day	3rd Day	4th Day	Total
Agrani	22	41	81	45	189
Mahendrani	-	41	74	48	163
Jam-ji-Dhani	17	27	53	24	121
Saadan-j-Dhani	12	21	47	18	98
Jamal-ji-Dhani	-	19	44	23	86
Senhrani	23	31	62	17	133
Ramlani	18	35	52	24	129
Mehlani	-	25	66	26	117
Misryani	-	20	51	27	98
Sekhario	17	33	67	24	141
Total	109	293	597	276	1275

Figure 3. Various dishes based on desert locusts prepared by the Thari. (**A**) Locust boiling in water, (**B**) locusts spread out on a bed for drying, (**C**) cooking of dried locusts, (**D**) locust cooked with lemon, (**E**) "Biryani" (a typical dish of the Indian subcontinent) with locusts, (**F**) crispy locust snacks. Photos by Santosh Kumar, Nawaz Ali Samejo, Muhammad Ibrahim Samejo, Saraj-ud-din Samejo, and Samiullah Soomro.

Another huge mature swarm, covering "a large area" (villager estimate unfortunately imprecise), arrived on 24 September 2019, moving westward. The swarm likely originated from the adjacent breeding areas in Rajasthan, India. The villagers were more motivated to collect these locusts, as they remembered the previous July swarm, and the preparation and consumption of tasty dishes. The villagers formed groups to collect locusts in the field, using vehicle headlights as a light trap. The desert locusts were often in pairs for mating and were easily picked by hand. During this swarm's time in the fields, the people collected about 1758 kg of locusts over five consecutive nights (Table 2).

Table 2. Collection of desert locusts by villagers after the arrival of the second swarm on 24 September 2019.

Villages	Weight of Desert Locust Collected (kg)					
	1st Day	2nd Day	3rd Day	4th Day	5th Day	Total
Agrani	26	61	53	59	19	218
Mahendrani	17	48	57	51	22	195
Jam-ji-Dhani	12	34	29	66	27	168
Saadan-j-Dhani	21	41	33	39	13	147
Jamal-ji-Dhani	15	46	29	33	-	123
Senhrani	27	53	34	57	23	194
Ramlani	31	37	52	47	18	185
Mehlani	25	42	58	39	28	192
Misryani	17	26	57	38	-	138
Sekhario	28	63	48	37	22	198
Total	219	451	450	466	172	1758

In early October 2019, desert locust hoppers emerged from the ground, where the swarms had arrived earlier for laying. However, no emerging hoppers were observed around the village of Mahandre-Jo-Par as most of the locusts have been collected and eaten or had already migrated away. In areas where young hoppers emerged, they were observed marching in bands and damaging moth bean (*Vigna aconitifolia*), a cash crop in the Thar. In Mahandre-Jo-Par, out of 3620 ha cultivated with moth bean, only 4000 rupees per hectare was lost. However, in the three surrounding villages (Khokhrapar, Laplo, and Shekhro), where locusts had not been harvested, losses on the 5888 hectares cultivated with moth bean totaled about 43,000 rupees/hectare. In the fields around the Mahandre-Jo-Par village, the crops were hardly damaged since there were no hoppers there. Sometime later, hopper bands came to these fields from other areas, but the farmers had already harvested their crops. Thus, while the moth bean crop was completely destroyed in other parts of the Thar region, in our study area losses were lower, likely as a result of locust collection by the villagers.

4. Discussion

Our main observational study shows that entomophagy can, to a certain extent, be locally effective in desert locust control. This method cannot prevent locust invasions, but can mitigate the problems caused by locusts to allow the villagers to better cope with the plague. Entomophagy has an important advantage over other control methods: it reduces the number of locusts while providing a diet of better nutritional quality to the local residents.

Collecting locusts had a strong local impact on the insects' population in two important ways: the method greatly reduced the locust population at the time of collection and also reduced the next generation to almost nothing, thereby avoiding most of crop losses. Crop losses continued to take place where no locusts were collected. Unfortunately, we cannot determine the precise impacts of locust collection on the swarm because the characteristics of the swarms were not noted, including their size and approximate density; depending on these parameters, the collection efficiency could be more or less effective. When locust swarms are very dense and large, the local villagers may be overwhelmed and their collection efficiency may be lower, while the nutritional benefit remains the same.

Collection efficiency also depends on the maturity of the locust population. Collection is less efficient and has a less significant impact on population density if the swarm is immature, arrives during or at the end of the day, eats a large part of the available vegetation and crops, and continues its migration the next day. If the swarm is mature, the locusts may remain for several consecutive days and nights for female laying, before flying away. The collection can then be more intense (as in our study), the reduction of the locust population greater, and the impact on local dynamics and crop damage reduced.

Residents of various countries (i.e., Indonesia, M. Lecoq p.c.) have better ways to collect locusts, such as using large nets carried by several people. Of course, the techniques must be within the capacity of the local farmers. Selling locusts can be a significant source of income for poor rural families, and farmers can be encouraged to collect locusts to be processed into feed for poultry. During the most recent invasion in Pakistan's Punjab province, as part of a pilot project led by the Food Security Department, farmers were encouraged to collect locusts and sell them to poultry feed businesses. In addition, researchers from the Pakistan Agriculture Research Council collected dead locusts to make bio-fertilizers for crops; locust-based fertilizers have advantageous N (9%) and P (7%) content [28]. Locusts also can be mixed with other bio-waste to make compost [29].

Finally, we should not expect to control invasions by collecting locusts by hand, even on a large scale. First, there is general agreement that it is impossible to collect enough locusts to end an invasion; there are simply too many insects (a swarms spotted in Kenya in January 2020 covered an area of 40 km by 60 km [30]). It is also impossible to prevent an invasion as the locusts emerge from outbreak areas, which are most often located in desertic and sparsely populated regions. At best, this collection method could be a useful mitigation measure in the event of an upsurge/invasion that could not be controlled at an earlier stage using preventative measures. The collection method also allows local residents, in the absence of other control tools, to protect their own crops, and also to compensate for possible nutritional deficiencies by eating the insects.

Eating swarming locusts is a widespread practice wherever plagues occur. Locusts and grasshoppers have been eaten for centuries and, in some places, still are today. During periods of increased locust activity, piles of dead locusts can be found in the market places [30]. Locusts are rich in protein; about 62% of the dry weight of an adult desert locust is proteins, 17% is fats, and the remainder is inorganic constituents [31]. Thus, entomophagy may have a large positive nutritional impact, and it is an important practice in parts of the desert locust habitat range, especially in certain regions of Pakistan. For a long time, the people of the Thar desert region have faced shortages of food and water, making them extremely vulnerable. Food shortages lead to severe malnutrition (especially to protein deficiency), leading to high maternal mortality rates [32,33]. As an important source of protein, the consumption of desert locusts can help prevent malnutrition. Therefore, collection must, perhaps above all, be considered an interesting food source for poor and undernourished rural populations.

However, using locusts as food has many potential risks [34]. Edible insects are underestimated as a reservoir of human and animal parasites [35]. Using locusts as food for humans or animals must consider health constraints, the health risk being not only microbiological but also chemical [36]. Indeed, outbreaks are currently managed using chemical insecticides [37]. As pesticides do not always kill the locusts immediately, collected locusts may still contain traces of pesticides and be toxic. During an ongoing locust upsurge in East Africa, the FAO [30] strongly advises against eating swarming locusts (living or dead) because it is highly unlikely that the nutritional benefits will outweigh the negative effects of the chemical residues of organophosphate, carbamate, or pyrethroid insecticides [38,39]. The practice of entomophagy, if it is to go beyond its current limits, should therefore be closely supervised by state agencies to avoid health risks and be coordinated with pesticide applications.

5. Conclusions

The consumption of desert locusts could effectively prevent malnutrition and protein deficiency and, to a certain extent, could also be an efficient mitigation measure to help local human populations to better protect themselves against locust outbreaks and the resulting crop damage. Collecting and consuming locusts should be encouraged wherever outbreaks are observed, while remaining realistic about the real impact entomophagy has on locust control. This practice should be coordinated with the local authorities to avoid the consumption of insects treated with pesticides. Sensitization about entomophagy, its health and crop protection benefit, as well as its constraints, is surely required as is a research development on this subject in order to better assess the nutritional and economic benefits in the long-term.

Author Contributions: Conceptualization, A.A.S.; methodology, A.A.S. and R.S.; field surveys, A.A.S., S.S., and S.K.; analysis and drafting, A.A.S., R.S., and S.S. All authors have read and agreed to the published version of the manuscript.

Funding: This study was partially funded by the Higher Education Commission Islamabad, Pakistan (Project No. 6737 SINDH/NRPU/R & D/HEC).

Acknowledgments: The authors are highly thankful to villagers of the village Mahandre-Jo-Par who arranged an awareness program and helped in the collection of data in the field. The authors are also grateful to Michel Lecoq (CIRAD, France), Koutaro Old Maeno (JIRCAS, Japan), and Dana Desonie (The Science Writer, US) for their kind comments and corrections for improvement of the manuscript.

Conflicts of Interest: The authors declare no conflict of interest. The funders had no role in the design of the study; in the collection, analyses, or interpretation of data; in the writing of the manuscript, or in the decision to publish the results.

References

1. Magor, J.I.; Lecoq, M.; Hunter, D.M. Preventive control and desert locust plagues. *Crop Prot.* **2008**, *27*, 152733. [CrossRef]
2. Sword, G.A.; Lecoq, M.; Simpson, S.J. Phase polyphenism and preventative locust management. *J. Insect Physiol.* **2010**, *56*, 949–957. [CrossRef] [PubMed]
3. Brader, L.; Djibo, H.; Faye, F.G.; Ghaout, S.; Lazar, M.; Luzietoso, P.N.; Babah, M.A.O. Towards a more effective response to desert locusts and their impacts on food security, livelihoods and poverty. In *Multilateral Evaluation of the 2003–05 Desert Locust Campaign*; Food and Agriculture Organisation: Rome, Italy, 2006.
4. van Huis, A. Desert locust plagues. *Endeavor* **1995**, *19*, 118–124. [CrossRef]
5. Belayneh, Y.T. Acridid pest management in the developing world: A challenge to the rural population, a dilemma to the international community. *J. Orthoptera Res.* **2005**, *14*, 187–195. Available online: https://www.jstor.org/stable/3657127 (accessed on 26 January 2020). [CrossRef]
6. Jago, N.D. The 1985–1989 grasshopper/locust upsurge in Sub-Saharan Africa: The usage of conventional pesticides and equipment. *Bol. Sanid. Veg. Plagas.* **1990**, *20*, 11–20.
7. Skaf, R.; Popov, G.B.; Roffey, J. The Desert Locust: An international challenge. *Philos. Trans. R. Soc. B* **1990**, *328*, 525–538. [CrossRef]
8. Cressman, K.; Hodson, D. Surveillance, information sharing and early warning systems for transboundary plant pests diseases: The FAO experience. *Arab J. Plant Prot.* **2009**, *27*, 226–232.
9. FAO. *Locust Watch: Desert Locust*; Food and Agricultural Organization of the United Nations: Rome, Italy, 2021; Available online: http://www.fao.org/ag/locusts/en/info/info/index.html (accessed on 20 January 2021).
10. Lecoq, M. Desert Locust Threat to Agricultural Development and Food Security and FAO/International Role in its Control. *Arab J. Plant Prot.* **2003**, *21*, 188–193.
11. Showler, A.T. Desert locust control: The effectiveness of proactive interventions and the goal of outbreak prevention. *Am. Entomol.* **2019**, *65*, 180–191. [CrossRef]
12. Lecoq, M. Recent progress in Desert and Migratory Locust management in Africa. Are preventive actions possible? *J. Orthoptera Res.* **2001**, *10*, 277–291. [CrossRef]
13. Lecoq, M. Vers une solution durable au problème du criquet pèlerin ? *Sécheresse* **2004**, *15*, 217–224.
14. Lecoq, M. Desert locust management: From ecology to anthropology. *J. Orthoptera Res.* **2005**, *14*, 179–186. [CrossRef]
15. Lecoq, M.; Zhang, L. *Encyclopedia of Pest Orthoptera of the World*; China Agricultural University Press: Beijing, China, 2019
16. Showler, A.T. A summary of control strategies for the desert locust, Schistocerca gregaria (Forskål). *Agric. Ecosyst. Environ.* **2002**, *90*, 97–103. [CrossRef]

17. Latchininsky, A.V.; Sword, G.; Sergeev, M.; Cigliano, M.M.; Lecoq, M. Locusts and grasshoppers: Behavior, ecology, and biogeography. *Psyche* **2011**, *2011*, 578327. [CrossRef]
18. Gay, P.-E.; Lecoq, M.; Piou, C. Improving preventive locust management: Insights from a multi-agent model. *Pest Manag. Sci.* **2018**, *74*, 46–58. [CrossRef]
19. Meynard, C.N.; Lecoq, M.; Chapuis, M.P.; Piou, C. On the relative role of climate change and management in the current Desert Locust outbreak in East Africa. *Glob. Chang. Biol.* **2020**, *26*, 3753–3755. [CrossRef]
20. De Vreyer, P.; Guilbert, N.; Mesple-Sompsa, S. Impact of natural disasters on education outcomes: Evidence from the 1987–89 locust plague in Mali. *J. Afr. Econ.* **2014**, *24*, 57–100. [CrossRef]
21. Sharma, A. Locust Control Management: Moving from Traditional to New Technologies—An Empirical Analysis. *Entomol. Ornithol. Herpetol.* **2014**, *4*, 141.
22. van Huis, A. Insect pests as food and feed. *J. Insects Food Feed* **2020**, *6*, 327–331. [CrossRef]
23. van Huis, A. Potential of Insects as Food and Feed in Assuring Food Security. *Annu. Rev. Entomol.* **2011**, *58*, 563–583. [CrossRef]
24. van Huis, A.; Van Itterbeek, J.; Klunder, H.; Mertens, E.; Halloran, A.; Muir, G.; Vantomme, P. *Edible Insects: Future Prospects for Food and Feed Security*; Food and Agriculture Organization of the United Nations: Rome, Italy, 2013; Available online: https://edepot.wur.nl/258042 (accessed on 25 January 2021).
25. DeFoliart, G.R. The human use of insects as food and as animal feed. *Bull. Entomol. Soc. Am.* **1989**, *35*, 22–36. [CrossRef]
26. Straub, P.; Tanga, C.M.; Osuga, I.; Windisch, W.; Subramanian, S. Experimental feeding studies with crickets and locusts on the use of feed mixtures composed of storable feed materials commonly used in livestock production. *Anim. Feed Sci. Technol.* **2019**, *255*, 114215. [CrossRef]
27. DeFoliart, G.R. Insects as a source of protein. *Bull. Entomol. Soc. Am.* **1975**, *21*, 161–163. [CrossRef]
28. Govt Plan to Use Locusts for Developping Bio-Compost. Available online: https://www.dawn.com/news/1567279 (accessed on 11 February 2021).
29. Pakistan Plans to Convert Locust Swarms into Fertilisers. Available online: https://www.outlookindia.com/newsscroll/pakistan-plans-to-convert-locust-swarms-into-fertilisers/1877180 (accessed on 11 February 2021).
30. FAO. *Desert Locust—FAQs*; Food and Agricultural Organization of the United Nations: Rome, Italy, 2020; Available online: http://www.fao.org/locusts/faqs/en/ (accessed on 20 January 2021).
31. Uvarov, B. *Grasshoppers and Locusts. A Handbook of General Acridology*; Cambridge University Press: London, UK, 1966.
32. Shaikh, M.A. Water scarcity in Tharparkar. In Proceedings of the Seventh International Water Technology Conference, Cairo, Egypt, 1–3 April 2003; pp. 63–70. Available online: https://www.iwtc.info/2003_pdf/01-4.pdf (accessed on 26 January 2021).
33. Memon, M.H.; Aamir, N.; Ahmed, N. Climate Change and Drought: Impact of Food Insecurity on Gender Based Vulnerability in District Tharparkar. *Pak. Dev. Rev.* **2018**, *57*, 307–331. [CrossRef]
34. Schlüter, O.; Rumpold, B.; Holzhauser, T.; Roth, A.; Vogel, R.F.; Quasigroch, W.; Vogel, S.; Heinz, V.; Jäger, H.; Bandick, N.; et al. Safety aspects of the production of foods and food ingredients from insects. *Mol. Nutr. Food Res.* **2016**, *61*, 1600520. [CrossRef] [PubMed]
35. Gałęcki, R.; Sokół, R. A parasitological evaluation of edible insects and their role in the transmission of parasitic diseases to humans and animals. *PLoS ONE* **2019**, *14*, e0219303. [CrossRef]
36. Imathiu, S. Benefits and food safety concerns associated with consumption of edible insects. *NFS J.* **2020**, *18*, 1–11. [CrossRef]
37. Zhang, L.; Lecoq, M.; Latchininsky, A.; Hunter, D. Locust and grasshopper management. *Annu. Rev. Entomol.* **2019**, *64*, 15–34. [CrossRef]
38. FAO. Evaluation of Field Trials Data on the Efficacy and Selectivity of Insecticides on Locusts and Grasshoppers. In *Report to FAO by the Pesticide Referee Group, Proceedings of the Tenth Meeting, Gammarth, Tunisia, 10–12 December 2014*; Food and Agricultural Organization of the United Nations: Rome, Italy, 2014.
39. Peveling, R. Environmental conservation and locust control—Possible conflicts and solutions. *J. Orthopt. Res.* **2001**, *10*, 171–187. [CrossRef]

agronomy

Review

Ecological and Human Diet Value of Locusts in a Changing World

Gabriella J. Kietzka [1,*], Michel Lecoq [2] and Michael J. Samways [1]

1 Department of Conservation Ecology and Entomology, Stellenbosch University, Stellenbosch 7600, South Africa; samways@sun.ac.za
2 CIRAD, UMR CBGP, F-34398 Montpellier, France; mlecoq34@gmail.com
* Correspondence: gabikietzka@gmail.com; Tel.: +27-81-030-4491

Abstract: Since ancient times, locusts have been serious pests wreaking havoc on settled agriculture throughout much of the world. Numerous locust practices have been developed to control infestations. This has led to most commentaries portraying locust infestations only in a negative light while focusing on finding best management practices for suppressing locust populations and lessening crop damage caused by swarms. Yet, locusts are also of great ecological significance in being not only an extraordinary natural phenomenon but also major components of ecosystem nutrient cycling, arising long before settled agriculture. Furthermore, for humans, locusts are a nutritious food source, historically and currently being consumed directly. Locust control today should more regularly include their harvesting. This is now more feasible, as environmentally friendly biopesticides can be used to replace harmful organic pesticides. We focus here on the ecological significance of locusts by using calculations based on a 1 km^2 area of swarming and breeding Desert locusts, *Schistocerca gregaria*, and show that the huge biomass of locust individuals contributes greatly to ecosystem processes while also having great potential use in human nutrition, especially where there is an urgent need for improved dietary intake and nutrition.

Keywords: *Schistocerca gregaria*; Desert locust; ecosystem processes; nutrient cycling; nutritional value

Citation: Kietzka, G.J.; Lecoq, M.; Samways, M.J. Ecological and Human Diet Value of Locusts in a Changing World. *Agronomy* **2021**, *11*, 1856. https://doi.org/10.3390/agronomy11091856

Academic Editor: Andrea Sciarretta

Received: 16 August 2021
Accepted: 9 September 2021
Published: 16 September 2021

1. Introduction

"Their (locusts) disappearance ... would not interfere with any other food chain, because locusts are not the basic food of any other bird or beast. So, if we could invoke mystic powers and say, 'All locusts, be gone', I don't think this would make very much difference to world ecology" [1]. This statement by Haskell (1971), during his John Curtis 'Woodstock' Lecture on International Locust Research and Control, whether spoken out of bravado or ignorance, disregards the ecological significance of everything in nature, even locusts.

In the ancient Egyptian texts of the New Kingdom (1070–1550 BCE), locusts are positively referred to as the might of the Egyptian army while negatively as the defeated enemy armies, with 'locusts' being a metaphor for 'multitude' [2]. This double perspective has a biological equivalent: locusts as an extraordinary and ecologically important natural phenomenon yet also a harbinger of human hardship through resource loss and pestilence. A locust outbreak in the northwestern provinces, including Egypt, in the early fourth century is even thought to have led to an outbreak of rats (and their associated fleas) through an abundance of food provided by locust cadavers, which would have caused a devastating plague among the local human population [3].

A locust is a grasshopper (Orthoptera: Acrididae), normally solitarious but periodically displaying a massive increase in abundance, largely in response to changed and favorable weather conditions [4]. The great population increase is often associated with changes in its biochemistry, physiology, morphology, and behavior, known as density-dependent phase polyphenism [5]. The changed form often becomes nomadic, causing major economic damage to crop plants [5,6]. Although other insect species periodically increase to very high numbers and migrate, causing economic crop damage, it is the voracity of the polyphagous locust in both the young and adult stages, combined with gregarization and associated extremely high local biomass, that characterizes a locust plague.

Phylogenetically, phase polyphenism is not basal, with locusts arising from sedentary ancestors and the various phase-like traits, each having evolved in different contexts [7]. These contexts include isolation and climatic conditions. However, there can be loss of gregarization. The Desert locust, *Schistocerca gregaria* (Forskål), has its origin in the Old World [8], with the gregarious *gregaria* subspecies in the north of Africa, Arabia, and southwestern Asia, and the derived subspecies *flaviventris*, isolated in the south of the African continent, less gregarious and very rarely swarming [9]. This is possibly an adaptive response to climatic conditions experienced by the southern subspecies, which found no selective advantage for gregarization [7]. While the focus here is on the Old-World Desert locust, we recognize that there are many locusts also present in the New World, with indications that some of these at least have been eaten by indigenous peoples. Space precludes here the detailing of these other species in human nutrition, we strongly recommend that further studies consider the nutritional opportunities provided by them.

Globally, there are several species of grasshoppers that currently, or at some time in the past, have shown a tendency towards gregarization and caused economic damage (Table S1). Many of these still do. Some species occur as different subspecies at different geographical locations and vary in their behavior and morphology. While we generally view locusts as having a major negative impact on human activities, they are also of great ecological significance, having periodic and localized intense effects on local ecosystems. These effects can ostensibly be negative, such as massive herbivore pressure on crops or even indigenous plants, or be positive, as through deposition of nutrients to new locations. Here, we review the ecological significance of locusts, past and present. We focus on the infamous Desert locust, using it to sketch how locust swarms move nutrients over space and time, specifically the cycling of nitrogen and carbon into the soil ecosystem for plant uptake and their nutritional value as food for humans.

2. Background: Biology and Ecology of Locusts

2.1. Abiotic and Biotic Drivers of Locust Outbreaks

A combination of interacting factors is responsible for locust outbreaks. Winds of a particular strength and direction enable adult locusts to assemble [10,11]. In dry climatic zones, the formation of locust swarms is associated with heavy and extensive rainfall events, as occurred on the horn of Africa and in East Africa (2019–2020). When these adults converge on localized patches of green vegetation appearing after rains, they form large congregations [12]. Adults then select suitable oviposition sites [13], with salinity and soil moisture, up to 5 cm below the surface, being important abiotic variables [14]. In areas where vegetation and oviposition resources are discontinuous, solitarious adults will aggregate where resources are concentrated. This is to a lesser extent when resources are of poor quality and evenly spread across the landscape [15]. Plant community characteristics also play a major role in determining the possibility of an outbreak [16]. High rainfall enables vigorous vegetation growth, which supports high locust population growth in dry areas [17,18]. Once eggs have hatched, and when nymphs are densely crowded, mutual tactile stimulation of the hind legs increases their serotonin levels [19], which leads directly to changes in behavior and, in the following generations, to color and morphological changes, higher appetite, and increased intensity of breeding [20,21].

2.2. Anthropogenic Drivers

Various anthropic actions can promote locust outbreaks, and there is increasing evidence on the impact of human activity [22]. Environmental changes in West Africa appear to be responsible for the increased economic importance of *Zonocerus variegatus* [23]. Overgrazing, by creating favorable conditions for gregariousness, is a common cause of outbreaks of certain species such as the Moroccan locust (*Dociostaurus maroccanus*) [24,25]. For some species, such as the Italian locust, *Calliptamus italicus*, the abandonment of cultivated fields can result in huge outbreaks [26]. In northern China, intensive grazing and subsequent degradation of grasslands favor outbreaks of *Oedaleus asiaticus*, probably by reducing the protein content of plants [27]. In Australia, outbreaks of *Austroicetes cruciata* and *Chortoicetes terminifera* may have resulted from ecological changes following the introduction of European livestock and agriculture [28]. In southeastern Asia, *Locusta migratoria* outbreaks are normally inhibited by the humid tropical environments. However, deforestation in the area is adversely synergistic with drought conditions favoring swarming and greatly elevated population levels [29]. Solitarious individuals of *Schistocerca gregaria* on the Sudan Red Sea coast especially congregate in the sandy, high-nitrogen-containing, moisture-retaining soils of wadies, and where Millet (*Panicum turgidum*) and the intercrop *Heliotropium arbainense* are grown. These agricultural landscapes constitute <5% of the total area but are significant sources of locust outbreaks [16].

2.3. Historical and Current Control Methods

Figure 1 summarizes some commonly used traditional and new control methods. Current evidence suggests that before pesticides, locusts were controlled by mechanical or other physical means. Some techniques include setting fire to roosting adults, trampling or chasing hoppers and burying them in trenches, and plowing known locust egg beds. These traditional methods do not require specialized equipment, are low-cost, and do not adversely impact the environment. However, they are labor intensive, time consuming, and often ineffective [30].

Mechanical and chemical control methods

- Make fires or noise to prevent adult swarms from settling
- Cover roosting sites with straw and burn
- Chase hoppers into trenches and bury, burn or drown them
- Trample or use branches to beat hoppers
- Plough or dig up egg pods at oviposition sites
- Apply neem and linseed oils to prevent locusts from swarming and to damage their antennae
- Use garlic as a deterrant
- Apply a thin layer of flour to crops to gum up locust mouth parts
- Instigate cover crops with nets sprayed with neem oil or garlic
- Baiting: mix insecticide dust with a carrier (bran or maize) and scatter around locust swarms
- Dusting: mix insecticide dust with a powder (chalk or talcom) and scatter, or fill a hesian bag with mixture and beat with a stick to apply
- Spraying: most common method, uses a sprayer to atomize and apply a liquid pesticide
- Ultra-low volume (ULV) spraying: newest technology, small droplets that are oil based
- Drones for ULV spraying

Figure 1. Popular past and present mechanical and chemical control methods used to control locust swarms and bands.

Sodium fluorosilicate and sodium arsenate were the first chemical controls used in India in the 1800s. Dusting and, to a greater extent, baiting were the first means of insecticide application. In the 1940s, the organochlorines benzene hexachloride (BHC) and dieldrin were introduced. These were applied by spraying, initially on the ground, and then in 1951, the first aerial application of dieldrin was trialed on a Desert locust swarm. In the 1970s, these persistent organochlorines and alike were banned, and a global shift towards, mainly, organophosphates, such as fenitrothion and malathion, occurred, which are still used today in addition to a few others (benzoylurea, carbamate, pyrethroid, and phenyl pyrazole) [31].

Later technology allowed for biological control agents, for example the protozoan, *Nosema locustae* [32], and the synthesis of biopesticides, which are developed from various strains of the entomopathogenic fungi, *Metarhizium acridum* [33]. These are not known to be harmful to humans, and to date have not been detected as adverse to biodiversity or environmental health. A major obstacle inhibiting extensive utilization of the fungi is that locusts only begin to die five days post-spray, with maximum mortality occurring one to two weeks post-application. Even so, the credibility of biopesticides for locust management programs is increasingly being recognized [22,34]. While the application of pathogens like entomopathogenic fungi and nematodes from an environmental point of view are preferable to synthetic organic pesticides, there must be some consideration of possible side-effects on other species, especially rare and threatened insects. This requires urgent further assessment and monitoring.

The latest spraying technique uses a much smaller volume of liquid, referred to as ultra-low-volume (ULV) spraying. The formulation of ULV is oil based to prevent the small droplets from evaporating, and specialized sprayers are used to ensure the insecticide is applied efficiently and safely. These have been designed to accommodate all application methods and can be carried or mounted on a motor vehicle or on an airplane or helicopter [35,36]. Recent projects have developed and begun testing *Metarhizium* products in an oil form to make them suitable for ULV application [35–37]. Today, geographical positioning systems (GPS) on ground vehicles are used for precise pesticide application, following drone flyovers for early detection of locust swarm formation and for rapid application of ULV sprays [37,38]. However, much research is still needed to evaluate the effectiveness of these methods under a range of circumstances.

3. Locusts in a Changing World

3.1. The Desert Locust (*Schistocerca gregaria*) as an Example

The Desert locust reproduces rapidly and migrates over very long distances. It has the genetic predisposition to exist in two forms, *solitaria* or *gregaria* [4,39]. During recessions, solitary Desert locusts are usually restricted to the semiarid and arid deserts of Africa and the Near East and Southwest Asia that receive less than 200 mm of rain annually. This covers an area of about 15 million km^2, consisting of about 30 countries [40]. During the invasion periods, approximately 31 million km^2 and over 60 countries are recurrently vulnerable to Desert locust swarms, affecting up to 20% of the world's land area [41]. For these reasons, the Desert locust is considered the most dangerous migratory pest in the world. There is an abundance of information available on this pest, which makes it the ideal species to use as an example for quantifying the ecological significance and nutritional value to humans. Here, characteristics related to the Desert locust and its swarming dynamics were used for the objectives listed in Figure 2.

Table 1 summarizes the characteristics used for calculations, acquired from a variety of sources. It was evident from the literature that there is great variability in Desert locust (and other species) swarm dynamics, biological characteristics, and morphological traits. Therefore, the tables and subsequent calculations provided here serve as examples and are subject to change according to the resources, region, and subspecies used.

For the calculations that follow, a 1 km^2 area of 60 million adult Desert locusts was used. This was based on the average density of a swarm in Kenya in 1955 and resembles average swarm densities from other resources [42,43]. For each objective, unique assumptions are stated, but all calculations assumed: (1) locusts lived for the average (113 days) adult life span; (2) hopper instar stages 1 to 4 lasted 6.5 days each, and stage 5 lasted 10 days;

(3) the swarm comprised a 1:1 ratio of males to females; and (4) hoppers were evenly divided between the five instar stages.

Determine the quantity of food consumed by 1 km² of adult locusts and their progeny

- Food consumed by 1 swarming desert locust during its adult life span
- Number of hoppers produced by 1 km² of adult swarming locusts
- Food consumed by 1 km² of adults, settling density 60 million/km²
- Food consumed by hoppers, accounting for mortality and differences between instar stages

Determine value of locust cadavers and frass for nutrient cycling (nitrogen and carbon)

- Weight of frass produced per adult over adult life span
- Weight of frass produced per hopper at each instar
- Weight of nymph cadavers based on mortality at each instar
- Nitrogen and carbon cycled by 1 km² of adult locusts and their progeny
- Relate nutrients cycled to rice and maize plant intake and production for human consumtion

Determine the nutritional value of 1 km² of adult locusts and their progeny

- Body weight of 1 km² of adults, 60 million individuals
- Body weight of their progeny at various instar stages
- Calculate mineral content based on body weights
- Relate mineral content to human nutritional needs based on adult daily recommended intake

Figure 2. Main objectives and calculations relating to the Desert locust (*S. gregaria*).

Table 1. Main parameters of the Desert locust (*S. gregaria*) in its swarming phase.

Factor	Value
Adults	
Adult body length	F: 5 to 6 cm (5.5 cm) M: 4.5 to 5 cm (4.75 cm) [43]
Adult body weight	F: 3.5 g, M: 2.2 g Avge: 2.85 [43]
Daily food consumption	Own body weight [44]
Adult duration life stage	Min: 75 days, Max: 150 days Avge: 113 days [43]
Swarms	
Locust density settling	Avge: 60 million/km² (patchy vegetation) [42] (Average density settled swarm in Kenya 1955)
Flight speed	Range: 3.8–4.3 m/s and 3 m/s, Avge: 3.5 m/s = 12.6 km/h [45]
Daily flight period	Min: 9–10 h, Max: 13–20 h, Avge: 14.5 h [43]
Eggs	
Number of times female lays	2–3, Avge: 2 [43]
Eggs per pod (1st, 2nd and 3rd laying)	60–80 (1), 50–70 (2), 35–70 (3), Avge: 70, 60, 52.5 [43]
Eggs per generation	140 per female [43]
Egg pod density	200–500/m² in groups, Avge: 350/m² in groups [43]
Pod length	3–4cm [43]
Egg mortality	Avge: 33% [43]
Egg weight	Avge: 5.92 mg [46]
Pod weight	Avge: 536.8 mg [46]
Hoppers	
Number of instars	5 [43]
Nymph body lengths (mm)	7 (S1), 15 (S2), 20 (S3), 33 (S4), 50 (S5) [43]
Nymph body weights (mg)	30–40 (S1), 50–80 (S2), 120–200 (S3), 500–700 (S4), 1000–1200 (S5), Avge: 35, 65, 160, 600, 1100 [13]
Daily food consumption	Own body weight [43]
Duration of stages	S1–S4: 6–7 days, Avge: 6.5 days, S5: 10 days [43]
Mortality	70% (S1), 20% (S2), 10% (S3–S5) [43]

The ecological effects of locust outbreaks are poorly researched. To sustainably manage ecosystems affected by locust outbreaks, a clear understanding of how locusts influence ecosystem function and structure is required. Two of the most obvious but opposing effects are (1) locust swarms rapidly reduce plant aboveground biomass and can suppress plant growth. (2) However, they also play an important role in nutrient cycling [47,48]. During outbreaks, locust frass and cadavers land on the soil, and as organic detritus, they are decomposed by microorganisms. Nutrients, such as nitrogen and carbon, are released into the soil and are available for plant uptake [7,49]. The damage caused by a swarm of locusts as well as the nutrients they give back will not be confined to an area. The daily distance traveled by an adult swarm of Desert locusts is around 183 km/d, which is based on the average flight speed and average daily flight period (Table 1). A quarter degree is equal to 30 km^2, which means a swarm will move approximately 6.1 quarter degrees in a day. This can be used to calculate how long it will take a swarm to reach a certain area on any map.

3.2. *Food Consumed by 1 km^2 of Adult Locusts and Their Progeny*

Calculations adhered to the previously mentioned assumptions. Throughout their adult life stage, female locusts consume more than male locusts (395 g and 249 g, respectively), but the average of the two, 332 g, was used in line with a 1:1 sex ratio (Tables 1 and 2). Considering each instar duration and body weight, an individual nymph was estimated to overall consume 16.59 g of food (Tables 1 and 2). Over the entire life span of a Desert locust, it is estimated to consume roughly 339 g of food. To put these values into perspective, using a locust density of 60 million/km^2, around 2.814 billion hoppers would be produced. This is based on the assumptions that half of the population are female, each female will produce 140 eggs, and egg mortality is 33% (Table 1). In one generation, with mortality at each instar accounted for, and occurring halfway through the stage, hoppers (7,941,799 kg) and adults (18,240,000 kg) would consume 26,182 t of plant material (Table 2).

Table 2. Food consumption estimates by adult Desert locusts and their progeny in 1 km^2 area of 60 million adults.

Variable	Measure
Average food consumed by 1 adult locust	322 g (F: 395 g) (M: 249 g)
Food consumed by 1 nymph (all instars)	16.59 g
Average food consumed by 1 locust	338.59 g = 339 g (F: 411.59 g) (M: 265.59 g)
Number of nymphs produced by adults	2.814 billion hoppers
Food consumed by 1 km^2 of adults	18,240,000 kg
Food consumed by their progeny	7,941,799 kg
Food consumed by adults and progeny	26,181,799 kg ≈ 26,182 t

We now consider the ecological value of the Desert locust in terms of its involvement in nutrient cycling. We do this using two objectives: (1) plant food consumed by both adult and young locusts per unit land area (1 km^2), and (2) the delivery of nutrients to the same land surface area via locust cadavers and frass.

3.3. *The Value of Locust Cadavers and Frass for Nitrogen and Carbon Cycling*

The effect of locust swarms on nitrogen cycling is dependent on the quantity of nitrogen in their frass and cadavers, as well as on the time it takes to become available to plants. Through consuming vegetation, locusts divert nitrogen from plants to the soil, and as a result, soil mineral nitrogen can be greater than in the absence of locust invasions [47]. However, if nitrogen is mineralized too rapidly it can be leached and lost from the ecosystem [48]. Sometimes when locusts infest cultivated crops, such as wheat, they may preferentially feed on broad-leaved weeds in the vicinity, which can result in a net transfer of nitrogen from the weeds to adjacent crop plants [49].

The response of plants to defoliation by locusts depends somewhat on the availability of nitrogen [50]. The speed with which nitrogen in locust cadavers and frass become available affects the plants' ability to regrow after defoliation. If nitrogen is rapidly mineralized, it could be available to plants in the same season and aid in their recovery post locust outbreak. Alternatively, nitrogen will be slowly available if it is immobilized or recalcitrant and will likely favour plants that are more efficient users or better competitors for soil nitrogen [47].

Calculations adhere to the aforementioned assumptions: data on adult cadavers were obtained from Fielding et al. [47] and based on the species *Melanoplus borealis* and *Chorthippus curtipennis*. Importantly, species nitrogen and carbon contents did not differ between species cadavers. Adult locust cadaver weights were assumed to be the average male and female body weights [43] (Table 1). Frass weights for adult Desert locusts were averaged from two different studies, for females Hill et al. [51] and males Norris [52]. Frass nutrient composition is dependent on diet, among other factors, and averages from the data were used and comprised a range of diets from laboratory-raised and free-range specimens [47]. For hoppers, it was assumed they produce the same weight of frass each day equal to their body weight for each instar, and each stage lasted the average number of days (Table 1). For adult locusts, all were assumed to die and contribute to cadaver weight. For hopper frass and cadaver weight calculations, mortality was accounted for at each stage and presumed to occur halfway through the stage to account for the spread of deaths over time, and carbon (C) and nitrogen (N) content and mineralization were the same as for adults [43,47]. Percentage of C in frass and cadavers was calculated using the C/N ratio (Table 3).

Table 3. Percentages of nitrogen (N) and carbon (C) content in locust frass and cadavers and percentages mineralized after 28 days.

Source	%N	C/N	%C	%N Mineralized	%C Mineralized
Frass	2.7	16.99	45.87	10.83	27.17
Cadavers	10.7	4.3	46.01	44	44

Content of N found in locust frass was 2.7% and 10.7% for cadavers, and of that, 10.83% and 44% were mineralized within 28 days, respectively (Table 3). Carbon content of locust frass was 45.87% and 46.01% for cadavers, and of that, 27.17% and 44% were mineralized, respectively (Table 3). Throughout the average adult life stage, 113 days, a female adult produces 32 g of frass and a male 23 g. Their cadavers, which are equal to their average body weights, are 3.5 g and 2.2 g, respectively (Tables 1 and 4). A hopper that survived all instar stages produces 16.59 g of frass (Table 4). The weight of frass and cadavers produced by 60 million adult Desert locusts and their progeny is estimated around 9756 t (Table 4). Of the total weight, around 276,502 kg of N and 44,475,093.50 kg of C would be transferred (Table 4). After incubation in soil for 28 days at 15 °C (near the average growing season soil temperature in central Alaska [47]), approximately 35,758 kg of N and 1,228,562.18 kg of C would be mineralized (Table 4).

Table 4. Weights and nutrient cycling dynamics, nitrogen (N) and carbon (C) of 1 km^2 of adult Desert locusts and their progenies.

Variable	Value
Weight of frass produced by 1 adult locust	F: 32 g, M: 23 g
Weight of adult cadaver	F: 3.5 g, M: 2.2 g
Weight of frass produced by 1 hopper	16.59 g (S1: 0.23 g, S2: 0.42 g, S3: 1.04 g, S4: 3.9 g, S5: 11 g)
Weight of hopper cadavers (all stages)	(S1: 0.035 g, S2: 0.065 g, S3: 0.16 g, S4: 0.6, S5: 1.1)
1 km^2 area of adult locusts	
Weight of frass from adults	1,650,000 kg
N from frass of adults, N mineralized	44,550 kg, mineralized: 4824.77 kg
C from frass of adults, C mineralized	756,855 kg, mineralized: 205,637.50 kg
Weight of cadavers from adults	17,100 kg
N from cadavers of adults, N mineralized	1829.7 kg, mineralized: 805.07 kg
C from cadavers of adults, C mineralized	7867.1 kg, mineralized: 3461.79 kg
Progeny of 1 km^2 area of adult locusts	
Weight of hopper frass	7,941,798 kg
N from frass of hoppers, N mineralized	214,428.55 kg, mineralized: 23,222.61 kg
C from frass of hoppers, C mineralized	3,642,902.74 kg, mineralized: 989,776.68 kg
Weight of hopper cadavers	146,671 kg
N from hopper cadavers, N mineralized	15,693.80 kg, mineralized: 6905.27 kg
C from hopper cadavers, C mineralized	67,468.66 kg, mineralized: 29,686.21 kg
Overall N, overall N mineralized	276,502.05 kg, mineralized: 35,757.72 kg
Overall C, overall C mineralized	4,475,093.50 kg, mineralized: 1,228,562.18 kg

3.4. Nutritional Value of Locusts for Human Food

Locust swarms are mostly viewed as negative events. However, from a human food perspective, they have great potential, with a long history of locusts used as food. Entomophagy is the ancient practice of humans eating insects. As early as 2000 BC, there is evidence of letters written to kings of the Middle East that reported the consumption of locusts, in particular the Desert locust [41]. The Old Testament describes one of the first acceptable uses of insects as food and permits the consumption of 'the locust of any kind, the bald locust of any kind, the cricket of any kind, and the grasshopper of any kind' (Leviticus 11:22; [53]). Historically, locusts were seen as a delicacy consumed by the social elite, for example, on a Neo Assyrian palace relief slab, from the eighth century BC, skewered locusts are being presented to a royal banquet. In William Shakespeare's tragedy Othello, assumed to be written in 1603, locusts are referred to as a delicacy: Othello (1,3), Iago: "The food that to him now is as luscious as locusts shall be to him shortly as bitter as coloquintida." Furthermore, Queen Ranavalona II (1829–1883) of Madagascar indulged in locusts collected by her servants [41]. Even in the New Testament, locusts were depicted as a delicacy, when John the Baptist is 'preaching in the wilderness of Judea,' wearing 'a garment of camel hair and a leather belt around his waist' and eating 'locusts and wild honey' (Matthew 3:4; [53]. Traditionally, the Khoisan of southern Africa also ate grasshoppers and locusts, after roasting them on grills, as depicted in a painting by Samuel Daniell in 1805 (Figure 3a) [54]. In Brazil, the Nambikwara people consume locusts almost daily, particularly the species *Rhammatocerus schistocercoides*, either hoppers in the rainy season or adults in the dry season. The locusts are toasted over charcoal and eaten as such, mixed with fruit juices, or added to cassava flour to make a kind of bread (Figure 3b) [55–57]. The examples could be increased as the practice was common in many countries and civilizations.

Figure 3. (**a**) 'Khoisan engaged in roasting grasshoppers on grills', a painting by Samuel Daniell, 1805. The Khoisan are the early indigenous peoples of southern Africa who traditionally ate insects in their everyday diet. (Reproduction from *Suid-Afrikaanse in Beeld* by A. Preston. Bion Books, South Africa (Copyright Commons)). (**b**) A Nambikwara woman in Mato Grosso (Brazil) toasting locusts (photographed by E. Setz, 1992).

Despite the role of entomophagy in the history and evolution of the human diet, nowadays, use of insects as food varies greatly across the world, yet its potential is great [58]. Many countries struggle with food shortages, with huge numbers of people dying from starvation every day. The FAO has estimated that to feed the global population in 2050, food production must increase by 70% [59]. Locusts can serve as a source of animal protein, and their consumption should be promoted for environmental, health, and livelihood-related reasons, both social and economic [60]. However, there is often abhorrence or fear of entomophagy in many Western societies, a barrier related to cultural factors and history, as well as a lack of information on the possible effects of introducing locusts into food [61]. To conserve what natural habitats and associated biodiversity remain, as well as feed an already struggling human population, a change in attitude towards entomophagy in westernized societies is needed [62]. Nevertheless, locusts are regularly consumed by humans in Asia, Africa, and southern USA. In southeastern Asia, locusts are even being farmed for food and livestock feed [63]. Desert locusts in particular are a nutritious food source for humans, as they contain substantial proportions of proteins, fats, energy, and minerals. Furthermore, they metabolize ingested phytosterols into derivatives with potential health benefits for humans [64].

The combined weight of the 60 million adult locusts was 17,100 kg. Their progenies were equally divided between weight categories of the five instar stages and together weighed 1,103,088 kg. Data from Wahed et al. [65] were used to calculate mineral content and chemical composition of 1 km^2 of adult locusts and their progenies. The locusts they used were mass reared in Egypt; adults and nymphs were continually bred for many generations under laboratory conditions according to Vanden Broeck et al. [66]. These values were compared to three other studies on the nutrient content of adult Desert locusts, one study used specimens from commercial suppliers [67], another collected locusts from Kenya [64], and another from Sudan [63].

According to the mineral analyses of Wahed et al. [65], mineral content was always lower in adults than hoppers for phosphorus (8.744 and 9.135), potassium (673.278 and 782.696), calcium (3.38 and 5.07), and magnesium (1.356 and 1.824) mg/100 g. Compared to other studies, mineral contents of adult locusts were generally lower for most minerals, except for potassium, which varied between studies. The combined mineral weights for 1 km^2 of adult Desert locusts and their progenies were: phosphorus, 102,270 g; potassium, 8,748,960 g, calcium, 56,510 g; and magnesium, 20,350 g (Table 5).

Table 5. Mineral content and chemical composition of 1 km^2 adult Desert locusts and their progenies. Adult human recommended daily intake provided in last column.

Factor (g/km^2 of Adults)	Adults	Hoppers	Combined	Adult Recommended Intake (g/d) (WHO/FAO/UNU, 2004)
Phosphorus (P)	1500	100,770	102,270	0.7
Potassium (K)	115,130	8,633,830	8,748,960	4.7
Calcium (Ca)	580	55,930	56,510	1 to 1.3 (Avge: 1.15)
Magnesium (Mg)	230	20,120	20,350	0.22 to 0.26 (Avge: 0.24)
Protein	9,711,090	727,155,610	736,866,700	46 (F); 56 (M) (Avge: 51)
Fat	4,928,220	167,117,830	172,046,050	44 to 77 (Avge: 60.5)
Total Carbohydrate	509,580	50,631,740	51,141,320	225 to 325 (Avge: 275)
Fiber	1,350,900	87,143,950	88,494,850	25 to 30 (Avge: 27.5)

According to chemical content analyses on the Desert locust by Wahed et al. [65], hoppers contained more protein, total carbohydrate, and ash than adults. For adult locusts, these values are comparable to those reported from other studies [63,64,67]. Based on these values, 60 million swarming adult locusts and their progenies would comprise 736,866,700 g of protein (Table 5). The carbohydrate content of both hoppers (4.59 g/100 g) and adult locusts (2.98 g/100 g) was much lower than the recommended daily intake values (225 to 325 g/d) (Table 5). These low values are to be expected because edible insects generally are not good sources of carbohydrates [64]. Despite this, a 1 km^2 area of adult locusts and their progeny would contain 51,141,320 g of carbohydrates (Table 5). Adults and hoppers had the same crude fiber content (7.9 g/100 g), which was slightly higher than other reported values [64,67]. The combined fiber produced by a 1 km^2 area of adult locusts and their hoppers is 88,494,850 g. Crude fiber probably arises from locust chitin and plays an important role in human digestion [68]. Adult locusts had higher fat content (28.82 g/100 g) than hoppers (15.15 g/100 g), and values were within the range of other reported values [61,62,65]. A 1 km^2 area of adult locusts and their progeny would comprise 172,046,050 g of fat (Table 5). Fat content can vary greatly between individuals, depending on age and diet. For humans, fats contribute to nutrition as an energy source, a supplier of essential fatty acids, and increase the palatability of foods by adding flavor [64,67].

3.5. Putting Things into Perspective

Throughout the world, rice and maize are among the most important crops for human consumption, and both are severely impacted by desert locust swarms [69,70]. Dobermann and Witt [71] conducted a survey of 207 rice farms in China, India, Indonesia, the Philippines, Thailand, and Vietnam. From the obtained values, the average grain yield was 5919 kg/ha, and average nitrogen, phosphorus, and potassium (NPK) uptakes were 117, 18, and 123 kg/ha, respectively. Values were based on one or two crops grown in 1995–1996 or 1997, giving a total sample size of 391. For maize, average plant density is 70,000 plants/ha (mostly varying from 60,000 to 80,000) [72–74], with a yield of around 9500 kg/ha and average NPK uptakes of 609, 357, and 280 kg/ha, respectively [75].

The N mineralized from the frass and cadavers of a 1 km^2 area (100 ha) of locusts and their progeny could meet the N uptake requirements of around 306 ha of rice crops and 59 ha of maize plants. The P and K released from the cadavers of locust adults and their offspring could individually meet the nutrient requirements of about 5682 ha (for P) and 71,130 ha (for K) of rice crops and 286 ha (for P) and 31,246 ha (for K) of maize plants. When considering the uptake requirements of rice and maize plants for all three nutrients, the nutrients supplied by 1 km^2 area of locusts and their progeny could result in a yield of approximately 1,811,214 kg of rice grain and 560,500 kg of maize grain.

Kenya is a country most severely affected by the recent (2019) locust breakout, with around 70,000 ha of agricultural damage. Due to progressive changes in eating habits, rice consumption has been increasing by up to 15% per year since its introduction in 1907 [76]. With an average per capita consumption of 20.6 kg/y, the rice yield resulting from the nutrients mineralized by 1 km^2 area of locusts and their progenies could supply sufficient rice grain for about 87,900 Kenyans for a year. In Kenya, maize is the most important staple for calorie nutrition [77], with per capita consumption estimated at 114 kg/y [78]. The maize yield resulting from the nutrients supplied by 1 km^2 of locusts, and their offspring could provide sufficient maize grain for about 4900 Kenyans for a year [79].

With regards to meeting the dietary requirements of humans from consumption of locusts, 60 million adult locusts and their progeny would meet the following daily intake requirements: for phosphorus, 146,100 adults for a day or 400 adults for a year; for potassium, 1,861,480 adults for a day or 5096 adults for a year; for calcium, 49,145 adults for a day or 134 adults for a year; and for magnesium, 84,791 adults for a day or 232 adults for a year (Table 5).

The protein content of both life stages is greater than many conventional human foods, including chicken and beef [80]. Protein acquired from insects is of high quality, and consequently, the protein from the Desert locust is considered a good source of animal protein for humans [64]. Consumption of 100 adult or hopper Desert locusts will contribute over 100% of the recommended daily intake for adult females (46 g/d) and males (56 g/d) (Table 5). The protein from 1 km^2 of adult locusts and their progeny would provide enough protein for 14,488,366 people for a day or 39,584 people for a year, sufficient carbohydrates to meet the daily intake needs of around 186,000 people or the yearly requirements of approximately 500 people, the daily fiber needs of more than 3 million people or yearly requirements of about 9000 people, and comprise enough fat content for around 3 million adult humans for a day or 7800 people for a year (Table 5).

4. Discussion

Locust plagues have been referred to as the oldest entomological problem. These insects have directly affected settled agriculture from its very beginnings. The earliest known record of a locust was drawn on the wall of an Egyptian tomb, dating back to around 2400 BCE. The historical battle with locust infestations has also been widespread, with evidence from ancient Egyptian, Chinese, Hebrew, Roman, and Greek texts [81]. Despite years of experience, research, and technological advance, and even though invasions are now less frequent and more rapidly controlled [22,39], these locust plagues still wreak havoc, with the world currently facing one of the worst outbreaks in decades [82].

The negative impacts associated with locust swarms are well documented, particularly those affecting agricultural activities. Swarms can have a major impact on the food security and livelihoods of affected communities, particularly the poorest. Losses to crops and pasture can lead to severe food shortages, strong price movements in markets, insufficient availability of grazing areas, sale of animals at very low prices to meet household subsistence needs and to purchase feed for remaining animals, early transhumance of herds and strong tensions between transhumant pastoralists and local farmers, and large human migrations to urban areas [83]. Other economic consequences may occur at harvest, as grain may be contaminated with insect parts and downgraded to a feed grain sold at a lower price [84]. Furthermore, the negative income shock may have a long-term impact on the educational outcomes (school enrollment and completion) of children living in rural areas [85]. And of course, the widespread use of highly toxic chemical insecticides over large areas can have significant effects on the health of human populations, their livestock and food crops, as well as potential damage to local ecosystems and key wildlife [86,87].

However, locusts are not all bad, as they play a major role in nutrient cycling. Here, we focus on the ecological significance of locust swarms for nutrient transfer and cycling, providing quantitative evidence from one of the best investigated locusts, the Desert locust. Cycling of nutrients is a critical factor in the current era of great environmental change and soil deterioration. Today is also a time of high human population density and where means are being sought to procure alternative sources of protein other than traditional meat consumption. Locusts would seem to offer potential benefits.

Locust outbreaks undoubtedly affect ecosystem structure and functioning in many ways. However, while plants experience defoliation, they also gain nutrients, and as locusts move around, the nutrients are being shifted spatially. While the negative impacts of locusts are well documented, especially reduction in aboveground biomass due to feeding, the actual and potential benefits of locust swarms are seldom acknowledged. Here, we show that a 1 km^2 area of Desert locusts could produce around 36,000 kg of mineralized nitrogen and 1,200,000 kg of mineralized carbon for redistribution within an ecosystem and across ecosystems.

Several authors show that locusts, and orthopterans in general, play an important trophic role [88], are stimulators of microbiological processes of decomposition and mineralisation of phytomass [89], can improve the productivity of grasslands in the long term [90], and in case of outbreak, can change considerably fluxes of some chemical elements in the herbaceous ecosystems [26]. Locust frass and cadavers are rich in nutrients, which are transferred to the soil via decomposition by microorganisms and fungi, absorbed by plants, increasing net ecosystem productivity [50] and ecosystem nutrient cycling through rapid mineralization rates of nitrogen and carbon. In one area, nutrients could potentially occur within the same season and promote recovery of the damaged crops or increase soil health to benefit crops for the next growing season [91]. However, as locusts also move far, and over many weeks and months, they are redistributing nutrients over space and time, which may directly benefit local farmers who would have suffered crop loss. Although the damage by locust swarms is devastating, the redistribution of nutrients could be from agricultural areas to ones that are nutrient poor with threatened natural vegetation. In this way, locusts could be viewed as conservationists of our ravaged natural ecosystems, as well as benefiting farmers in other locations at other times.

Locusts are a highly nutritious food source for humans and animals, as they contain significant proportions of proteins, fats, and minerals [64]. A 1 km^2 area of Desert locusts can provide enough protein to support almost 40,000 people for a year. In many countries, locusts are already eaten every time an outbreak occurs, and yet, it has been suggested that rather than eating them, pesticides are used.

Some of the methods used to harvest locusts during outbreaks include collecting by hand, using large sweep nets, and pulling sheets over the vegetation they are roosting on or with machines that can suck them up [92]. After a Desert locust outbreak in Pakistan, 10 villages harvested 1275 kg of locusts in four days from a swarm of 5 km^2. The locusts were first boiled, then dried in the sun for a day, after which they could be stored and used to make various appealing dishes, such as biryani, or eaten as a snack [93].

In the 1970s, the Bombay locust (*Patanga succincta*) was a huge problem in Thailand, feeding on sorghum and corn [94]. After pesticide application attempts were unsuccessful, a campaign was organized from 1978 to 1981 to harvest the locusts for consumption. This proved so successful that the locust no longer seems a problem in Thailand [94], although some argue that the data are inconclusive and that the locust may have been affected by a fungus that killed a large part of the population [1,41,95]. Alternatively, locusts can be used for feed or as fertilizer. An organization in Kenya paid farmers USD 0.45 per kg of locusts, which were dried and crushed into a powder that was used as an organic fertilizer or in animal feed [96]. The only serious risk associated with eating locusts is poisoning when they have been treated with pesticides [97]. However, recent developments have allowed for the synthesis of biopesticides, which use entomopathogenic fungi or protozoa as biocontrol methods. Many of these, such as various commercial strains of *Metarhizium acridum*, are of minimal risk to vertebrates and the environment [98]. Since locust outbreaks are intermittent, their use in human food can only be considered as an occasional food supplement when uncontrolled outbreaks develop into an invasion. This perspective also bears in mind however, that locust carcasses can be dried, ground, and stored for future needs.

We end on a note of caution. If there is extensive and intensive human consumption of locusts, might there be erosion of the ecological value of locusts, especially in terms of redistribution of plant nutrients? This would need to be a major research topic for the future. However, for now, it is better that humans eat locusts than kill them with harmful pesticides.

5. Conclusions

For many years, locusts have been viewed only in a negative light. Although they are responsible for devastating crop losses, they also play an important role in nutrient cycling. Due to the high concentration of nutrients in their bodies and frass, combined with the ability to disperse over long distances over time, locusts are important for the redistribution of nutrients across regions, often in arid nutrient-poor areas. This takes place among many locust and grasshopper species around the world. The effects can be so large that locust swarms can influence landscape-level ecological processes. Locusts can also be beneficial to humans, and their consumption should be used as a mitigation measure and to effectively prevent malnutrition, at least in times of locust outbreaks. Locust meal can be prepared from captured individuals, dried and ground, and then stored for precautionary use in times without locust plagues. However, there must be resolution and coordination among locust control practitioners using pesticides and people harvesting locusts as food to prevent poisoning from eating insects treated with pesticides. Future locust control strategies should consider a combination of biopesticides and collecting locusts for human consumption to the benefit of humanity and our natural ecosystems.

Supplementary Materials: The following are available online at https://www.mdpi.com/article/10.3390/agronomy11091856/s1, Table S1: Main species of pest locusts and grasshoppers around the world today (modified from Lecoq and Zhang 2019).

Author Contributions: Conceptualization, M.J.S. and M.L.; Formal Analysis, G.J.K.; Writing—Original Draft Preparation, G.J.K., M.J.S. and M.L.; Writing—Review and Editing, G.J.K., M.J.S. and M.L. All authors have read and agreed to the published version of the manuscript.

Funding: This research received no external funding.

Conflicts of Interest: The authors declare no conflict of interest.

References

1. Haskell, P.T. International Locust Research and Control. *J. R. Soc. Arts* **1971**, *119*, 249–263.
2. Sayed, M.R. Locust and Its Signification in Ptolemaic Texts. *J. Hist. Archaeol. Anthropol. Sci.* **2018**, *3*, 584–588. [CrossRef]
3. McNeill, W.H. *Plagues and Peoples*; Anchor Books: New York, NY, USA, 1998.
4. Cullen, D.A.; Cease, A.J.; Latchininsky, A.V.; Ayali, A.; Berry, K.; Buhl, J.; De Keyser, R.; Foquet, B.; Hadrich, J.C.; Matheson, T.; et al. From molecules to management: Mechanisms and consequences of locust phase polyphenism. *Adv. Insect Phys.* **2017**, *53*, 167–285.
5. Pener, M.P.; Simpson, S.J. Locust Phase Polyphenism: An Update. *Adv. Insect Physiol.* **2009**, *36*, 1–272.
6. Applebaum, S.W.; Heifetz, Y. Density-Dependent Physiological Phase in Insects. *Annu. Rev. Entomol.* **1999**, *44*, 317–341. [CrossRef] [PubMed]
7. Song, H. Phylogenetic Perspectives on the Evolution of Locust Phase Polyphenism. *J. Orthoptera Res.* **2005**, *14*, 235–245. [CrossRef]
8. Song, H.; Moulton, M.J.; Hiatt, K.D.; Whiting, M.F. Uncovering Historical Signature of Mitochondrial DNA Hidden in the Nuclear Genome: The Biogeography of Schistocerca Revisited. *Cladistics* **2013**, *29*, 643–662. [CrossRef]
9. Chapuis, M.P.; Bazelet, C.S.; Blondin, L.; Foucart, A.; Vitalis, R.; Samways, M.J. Subspecific Taxonomy of the Desert Locust, *Schistocerca gregaria* (Orthoptera: Acrididae), Based on Molecular and Morphological Characters. *Syst. Entomol.* **2016**, *41*, 516–530. [CrossRef]
10. Rainey, R. *Migration and Meteorology. Flight Behaviour and the Atmospheric Environment of Locusts and Other Migrant Pests*; Oxford University Press: Oxford, UK, 1989.
11. Balança, G.; Gay, P.; Rachadi, T.; Lecoq, M. Interpretation of Recent Outbreaks of the Migratory Locust *Locusta migratoria migratorioides* (Reiche and Fairmaire, 1850) [Orthoptera, Acrididae] in Lake Chad Basin According to Rainfall Data. *J. Orthoptera Res.* **1999**, *8*, 83–92. [CrossRef]
12. Kennedy, J.S. The Behaviour of the Desert Locust (*Schistocerca gregaria* (Forsk.))(Orthopt.) in an Outbreak Centre. *Trans. R. Entomol. Soc. Lond.* **1939**, *89*, 385–542. [CrossRef]
13. Roffey, J.; Popov, G. Environmental and Behavioural Processes in a Desert Locust Outbreak. *Nature* **1968**, *219*, 446–450. [CrossRef]
14. Ji, R.; Xie, B.Y.; Li, D.M.; Li, Z.; Zeng, X.C. Relationships between Spatial Pattern of Locusta Migratoria Manilensis Eggpods and Soil Property Variability in Coastal Areas. *Soil Biol. Biochem.* **2007**, *39*, 1865–1869. [CrossRef]
15. Bouaïchi, A.; Simpson, S.J.; Roessingh, P. The Influence of Environmental Microstructure on the Behavioural Phase State and Distribution of the Desert Locust *Schistocerca gregaria*. *Physiol. Entomol.* **1996**, *21*, 247–256. [CrossRef]
16. Van Der Werf, W.; Woldewahid, G.; Van Huis, A.; Butrous, M.; Sykora, K. Plant Communities Can Predict the Distribution of Solitarious Desert Locust *Schistocerca gregaria*. *J. Appl. Ecol.* **2005**, *42*, 989–997. [CrossRef]
17. Uvarov, B. *Grasshoppers and Locusts. A Handbook of General Acridology. Volume 1*; Cambridge University Press: Cambridge, UK, 1966.
18. Uvarov, B. *Grasshoppers and Locusts. A Handbook of General Acridology. Volume 2*; Cambridge University Press: Cambridge, UK, 1977.
19. Hägele, B.F.; Simpson, S.J. The Influence of Mechanical, Visual and Contact Chemical Stimulation on the Behavioural Phase State of Solitarious Desert Locusts (*Schistocerca gregaria*). *J. Insect Physiol.* **2000**, *46*, 1295–1301. [CrossRef]
20. Rogers, S.M.; Matheson, T.; Despland, E.; Dodgson, T.; Burrows, M.; Simpson, S.J. Mechanosensory-Induced Behavioural Gregarization in the Desert Locust *Schistocerca gregaria*. *J. Exp. Biol.* **2003**, *206*, 3991–4002. [CrossRef]
21. Anstey, M.L.; Rogers, S.M.; Ott, S.R.; Burrows, M.; Simpson, S.J. Serotonin Mediates Behavioral Gregarization Underlying Swarm Formation in Desert Locusts. *Science* **2009**, *323*, 627–630. [CrossRef] [PubMed]
22. Zhang, L.; Lecoq, M.; Latchininsky, A.; Hunter, D. Locust and Grasshopper Management. *Annu. Rev. Entomol.* **2019**, *64*, 15–34. [CrossRef] [PubMed]
23. Modder, W. Control of the Variegated Grasshopper *Zonocerus variegatus* (L.) on Cassava. *Afr. Crop. Sci. J.* **1994**, *2*, 391–406.
24. Latchininsky, A.V. Moroccan Locust *Dociostaurus maroccanus* (Thunberg, 1815): A Faunistic Rarity or an Important Economic Pest? *J. Insect Conserv.* **1998**, *2*, 167–178. [CrossRef]
25. Skaf, R.M. Le Criquet marocain au Proche-Orient et sa grégarisation sous l'influence de l'homme. *Bull. Soc. Ecol.* **1972**, *3*, 247–325.
26. Sergeev, M.G. Ups and Downs of the Italian Locust (*Calliptamus italicus* L.) Populations in the Siberian Steppes: On the Horns of Dilemmas. *Agronomy* **2021**, *11*, 746. [CrossRef]
27. Cease, A.; Elser, J.; Ford, C.; Hao, S.; Kang, L.; Harrison, J. Heavy Livestock Grazing Promotes Locust Outbreaks by Lowering Plant Nitrogen Content. *Science* **2012**, *335*, 467–469. [CrossRef]
28. Deveson, E.D. *Naturae Amator* and the Grasshopper Infestations of South Australia's Early Years. *Trans. R. Soc. S. Aust.* **2012**, *136*, 1–15. [CrossRef]
29. Lecoq, M.; Sukirno. Drought and an exceptional outbreak of the oriental migratory locust, *Locusta migratoria manilensis* (Meyen 1835) in Indonesia (Orthoptera: Acrididae). *J. Orthoptera Res.* **1999**, *8*, 153–161. [CrossRef]
30. Dobson, H.M. *Desert Locust Guidelines 4. Control*; Food and Agriculture Organization of the United Nations (FAO): Rome, Italy, 2001.
31. FAO. Evaluation of field trials data on the efficacy and selectivity of insecticides on locusts and grasshoppers. In Proceedings of the Report to FAO by the Pesticide Referee Group Tenth Meeting, Gammarth, Tunisia, 10–12 December 2014; FAO: Rome, Italy, 1996.
32. Zhang, L.; Lecoq, M. *Nosema locustae* (Protozoa, Microsporidia), a Biological Agent for Locust and Grasshopper Control. *Agronomy* **2021**, *11*, 711. [CrossRef]
33. Lomer, C.J.; Bateman, R.P.; Johnson, D.L.; Langewald, J.; Thomas, M. Biological Control of Locusts and Grasshoppers. *Annu. Rev. Entomol.* **2001**, *46*, 667–702. [CrossRef]

34. Hunter, D.M. Credibility of an IPM Approach for Locust and Grasshopper Control: The Australian Example. *J. Orthoptera Res.* **2010**, *19*, 133–137. [CrossRef]
35. Rachadi, T. *Locust Control. Handbook*; EU: Wageningen, The Netherlands, 2010.
36. Sharma, A. Locust Control Management: Moving from Traditional to New Technologies—An Empirical Analysis. *Entomol. Ornithol. Herpetol.* **2015**, *4*, 2161–2183.
37. Cressman, K. The Use of New Technologies in Desert Locust Early Warning. *Outlooks Pest. Manag.* **2008**, *19*, 55–59. [CrossRef]
38. Matthews, G.A. New Technology for Desert Locust Control. *Agronomy* **2021**, *11*, 1052. [CrossRef]
39. Simpson, S.J.; Raubenheimer, D.; Behmer, S.T.; Whitworth, A.; Wright, G.A. A Comparison of Nutritional Regulation in Solitarious- and Gregarious-Phase Nymphs of the Desert Locust *Schistocerca gregaria*. *J. Exp. Biol.* **2002**, *205*, 121–129. [CrossRef] [PubMed]
40. Sword, G.A.; Lecoq, M.; Simpson, S.J. Phase Polyphenism and Preventative Locust Management. *J. Insect Physiol.* **2010**, *56*, 949–957. [CrossRef] [PubMed]
41. Van Huis, A. Harvesting Desert Locusts for Food and Feed May Contribute to Crop Protection but Will Not Suppress Upsurges and Plagues. *J. Insects Food Feed* **2021**, *7*, 245–248. [CrossRef]
42. Rainey, R.C. The Use of Insecticides against the Desert Locust. *J. Sci. Food Agric.* **1958**, *9*, 677–692. [CrossRef]
43. Symmons, P.M.; Cressman, K. *Desert Locust Guidelines 1. Biology and Behaviour*; FAO: Rome, Italy, 2001.
44. Peng, W.; Ma, N.L.; Zhang, D.; Zhou, Q.; Yue, X.; Khoo, S.C.; Yang, H.; Guan, R.; Chen, H.; Zhang, X.; et al. A Review of Historical and Recent Locust Outbreaks: Links to Global Warming, Food Security and Mitigation Strategies. *Environ. Res.* **2020**, *191*, 110046. [CrossRef] [PubMed]
45. Krogh, A.; Weis-Fogh, T. A Roundabout for Studying Sustained Flight of Locusts. *J. Exp. Biol.* **1952**, *29*, 211–219. [CrossRef]
46. Maeno, K.; Tanaka, S. Phase-Specific Responses to Different Qualities of Food in the Desert Locust, *Schistocerca gregaria*: Developmental, Morphological and Reproductive Characteristics. *J. Insect Physiol.* **2011**, *57*, 514–520. [CrossRef]
47. Fielding, D.J.; Trainor, E.; Zhang, M. Diet Influences Rates of Carbon and Nitrogen Mineralization from Decomposing Grasshopper Frass and Cadavers. *Biol. Fertil. Soils* **2013**, *49*, 537–544. [CrossRef]
48. Reynolds, B.C.; Hunter, M.D.; Crossley, D.A., Jr. Effects of canopy herbivory on nutrient cycling in a Northern Hardwood Forest in Western North Carolina. *Selbyana* **2000**, *21*, 74–78.
49. Fielding, D.J.; Conn, J.S. Feeding Preference for and Impact on an Invasive Weed (*Crepis tectorum*) by a Native, Generalist Insect Herbivore, *Melanoplus borealis* (Orthoptera: Acrididae). *Ecol. Popul. Biol. Ann. Entomol. Soc. Am.* **2011**, *104*, 1303–1308. [CrossRef]
50. Lovett, G.M.; Christenson, L.M.; Groffman, P.M.; Jones, C.G.; Hart, J.E.; Mitchell, M.J. Insect Defoliation and Nitrogen Cycling in Forests. *Bioscience* **2002**, *52*, 335–341. [CrossRef]
51. Hill, L.; Mordue, W.; Highnam, K.C. The Endocrine System, Frontal Ganglion, and Feeding during Maturation in the Female Desert Locust. *J. Insect Physiol.* **1966**, *12*, 1197–1208. [CrossRef]
52. Norris, M.J. Group Effects on Feeding in Adult Males of the Desert Locust, *Schistocerca gregaria* (Forsk.), in Relation to Sexual Maturation. *Bull. Entomol. Res.* **1961**, *51*, 731–753. [CrossRef]
53. Evans, J.; Alemu, M.H.; Flore, R.; Frøst, M.B.; Halloran, A.; Jensen, A.B.; Maciel-Vergara, G.; Meyer-Rochow, V.B.; Münke-Svendsen, C.; Olsen, S.B.; et al. "Entomophagy": An Evolving Terminology in Need of Review. *J. Insects Food Feed* **2015**, *1*, 293–305. [CrossRef]
54. Samways, M.J. Approaches and Perspectives in Insect Species Conservation. In *Insect Conservation: A Global Synthesis*; CABI: Wallingford, UK, 2019; pp. 113–151.
55. Lévi-Strauss, C. La vie familiale et sociale des indiens Nambikwara. *J. Soc. Am.* **1948**, *37*, 1–132. [CrossRef]
56. Lecoq, M.; Pierozzi, I. *Rhammatocerus schistocercoides* Locust Outbreaks in Mato Grosso (Brazil): A Long-Standing Phenomenon. *Int. J. Sustain. Dev. World Ecol.* **1995**, *2*, 45–53. [CrossRef]
57. Setz, E.Z.F. Animals in the Nambiquara Diet: Methods of Collection and Processing. *J. Ethnobiol.* **1991**, *11*, 1–22.
58. Raubenheimer, D.; Rothman, J.M. Nutritional Ecology of Entomophagy in Humans and Other Primates. *Annu. Rev. Entomol.* **2013**, *58*, 141–160. [CrossRef]
59. Lensvelt, E.J.S.; Steenbekkers, L.P.A. Exploring Consumer Acceptance of Entomophagy: A Survey and Experiment in Australia and the Netherlands. *Ecol. Food Nutr.* **2014**, *53*, 543–561. [CrossRef] [PubMed]
60. Van Huis, A. Potential of Insects as Food and Feed in Assuring Food Security. *Annu. Rev. Entomol.* **2013**, *58*, 563–583. [CrossRef]
61. Ochiai, M.; Inada, M.; Horiguchi, S. Nutritional and safety evaluation of locust (Caelifera) powder as a novel food material. *J. Food Sci.* **2020**, *85*, 279–288. [CrossRef] [PubMed]
62. Yen, A.L. Edible Insects: Traditional Knowledge or Western Phobia? *Entomol. Res.* **2009**, *39*, 289–298. [CrossRef]
63. Mariod, A.A. *African Edible Insects as Alternative Source of Food, Oil, Protein and Bioactive Components*; Springer Nature: Cham, Switzerland, 2020.
64. Kinyuru, J.N. Nutrient Content and Lipid Characteristics of Desert Locust (*Schistoscerca gregaria*) Swarm in Kenya. *Int. J. Trop. Insect Sci.* **2020**, *41*, 1993–1999. [CrossRef]
65. Wahed, A.E.; Samira, M.N.; Ahmad, A.F. Variations in Chemical Composition Value of Adults and Nymphs Desert Locust, *Schistocerca gregaria* Forskål (Orthoptera: Acrididae). *J. Plant Prot. Pathol.* **2019**, *10*, 677–681.
66. Vanden Broeck, J.; Chiou, S.J.; Schoofs, L.; Hamdaoui, A.; Vandenbussche, F.; Simonet, G.; Wataleb, S.; De Loof, A. Cloning of Two CDNAs Encoding Three Small Serine Protease Inhibiting Peptides from the Desert Locust *Schistocerca gregaria* and Analysis of Tissue-Dependent and Stage-Dependent Expression. *Eur. J. Biochem.* **1998**, *254*, 90–95. [CrossRef]

67. Zielińska, E.; Baraniak, B.; Karaś, M.; Rybczyńska, K.; Jakubczyk, A. Selected Species of Edible Insects as a Source of Nutrient Composition. *Food Res. Int.* **2015**, *77*, 460–466. [CrossRef]
68. Badanaro, F.; Amevoin, K.; Lamboni, C.; Amouzou, K. Edible Cirina Forda (Westwood, 1849) (Lepidoptera: Saturniidae) Caterpillar among Moba People of the Savannah Region in North Togo: From Collector to Consumer. *Asian J. Appl. Sci. Eng.* **2014**, *3*, 13. [CrossRef]
69. Musuna, A.C.Z. Cereal Crop Losses Caused by Locusts in Eastern, Central and Southern Africa Region. *Int. J. Trop. Insect Sci.* **1988**, *9*, 701–707. [CrossRef]
70. Tandzi, L.N.; Mutengwa, C.S. Estimation of Maize (*Zea mays* L.) Yield Per Harvest Area: Appropriate Methods. *Agronomy* **2020**, *10*, 29. [CrossRef]
71. Dobermann, A.; Witt, C. *The Potential Impact of Crop. Intensification on Carbon and Nitrogen Cycling in Intensive Rice Systems*; International Rice Research Institute: Los Baños, Philippines, 2000.
72. Abuzar, M.R.; Sadozai, G.U.; Baloch, M.S.; Baloch, A.A.; Shah, I.H.; Javaid, T.; Hussain, N. Effect of Plant Population Densities on Yield of Maize. *J. Anim. Plant Sci.* **2011**, *21*, 692–695.
73. Hou, P.; Liu, Y.; Liu, W.; Liu, G.; Xie, R.; Wang, K.; Ming, B.; Wang, Y.; Zhao, R.; Zhang, W.; et al. How to Increase Maize Production without Extra Nitrogen Input. *Resour. Conserv. Recycl.* **2020**, *160*, 104913. [CrossRef]
74. Shao, H.; Xia, T.; Wu, D.; Chen, F.; Mi, G. Root Growth and Root System Architecture of Field-Grown Maize in Response to High Planting Density. *Plant Soil* **2018**, *430*, 395–411. [CrossRef]
75. Du Plessis, J. *Maize Production*; Department of Agriculture: Pretoria, South Africa, 2003.
76. FAO. Rice Consumption Per Capita in Kenya. Available online: https://www.helgilibrary.com/indicators/rice-consumption-per-capita/kenya/ (accessed on 26 July 2021).
77. Marenya, P.P.; Wanyama, R.; Alemu, S.; Woyengo, V. Trait Preference Trade-Offs among Maize Farmers in Western Kenya. *Heliyon* **2021**, *7*, e06389. [CrossRef] [PubMed]
78. MoALF. Agricultural Sector Transformation & Growth Strategy 2019–2029. Available online: https://www.kilimodata.developlocal.org (accessed on 26 July 2021).
79. CARD. Rice for Africa–Kenya. Available online: https://riceforafrica.net/card-countries/group-1-countries/kenya/kenya,-march-2021 (accessed on 26 July 2021).
80. Rumpold, B.A.; Schlüter, O.K. Nutritional Composition and Safety Aspects of Edible Insects. *Mol. Nutr. Food Res.* **2013**, *57*, 802–823. [CrossRef] [PubMed]
81. Uvarov, B. The Locust Plague. *J. R. Soc. Arts* **1942**, *91*, 109–118. [CrossRef]
82. Kumar, P.S.; Shukla, G.; Panwar, P.; Chakravarty, S. Locusts Plague: An Emerging Threat to India. *Curr. Sci.* **2021**, *120*, 10.
83. Brader, L.; DJibo, H.; Faya, F.G.; Ghaout, S.; Lazar, M.; Luzietoso, P.N.; Ould Babah, M.A. *Towards a More Effective Response to Desert Locusts and Their Impacts on Food Security, Livelihood and Poverty*; FAO: Rome, Italy, 2006.
84. Onyango, A. The Public Health Implication of the 2020 Locust Invasion in East Africa. Available online: https://globalhealth.euclid.int/the-public-health-implication-of-the-2020-locust-invasion-in-east-africa/ (accessed on 27 July 2021).
85. De Vreyer, P.; Guilbert, N.; Mesple-Somps, S. Impact of Natural Disasters on Education Outcomes: Evidence from the 1987–89 Locust Plague in Mali. *J. Afr. Econ.* **2015**, *24*, 57–100. [CrossRef]
86. Everts, J.W.; Ba, L. Environmental Effects of Locust Control: State of the Art and Perspectives. *New Strateg. Locust Control.* **1997**, 331–336. [CrossRef]
87. Peveling, R. Environmental Conservation and Locust Control-Possible Conflicts and Solutions. *J. Orthoptera Res.* **2001**, *10*, 171–187. [CrossRef]
88. Stebaev, I.V.; Gukasian, A.B. Orthoptera (Tettigonoidea and Acrididae) as stimulators of microbiological processes of decomposition and mineralization of phytomass in the meadow steppes of West Siberia. *Zool. Zh.* **1963**, *42*, 216–221.
89. Gandar, M.V. The dynamics and trophic ecology of grasshoppers (Acridoidea) in a South African savanna. *Oecologia* **1982**, *54*, 370–378. [CrossRef] [PubMed]
90. Belovsky, G.E. Do grasshoppers diminish? In *Grasshoppers and Grassland Health: Managing Grasshopper Outbreaks without Risking Environmental Disaster*; Lockwood, J.A., Latchininsky, A.V., Sergeev, M.G., Eds.; Kluwer Academic Publishers: Dordrecht, The Netherlands; Boston, MA, USA; London, UK, 2000; pp. 7–29.
91. Song, J.; Wu, D.; Shao, P.; Hui, D.; Wan, S. Ecosystem Carbon Exchange in Response to Locust Outbreaks in a Temperate Steppe. *Oecologia* **2015**, *1*, 579–590. [CrossRef] [PubMed]
92. Van Huis, A. Insect Pests as Food and Feed. *J. Insects Food Feed* **2020**, *6*, 327–331. [CrossRef]
93. Samejo, A.A.; Sultana, R.; Kumar, S.; Soomro, S. Could Entomophagy Be an Effective Mitigation Measure in Desert Locust Management? *Agronomy* **2021**, *11*, 455. [CrossRef]
94. Chen, P.; Wongsiri, S.; Jamyanya, T.; Rinderer, T.; Vongsamanode, S.; Matsuka, M.; Sylvester, A.; Oldroyd, B. Honey Bees and Other Edible Insects Used as Human Food in Thailand. *Am. Entomol.* **1998**, *44*, 24–29. [CrossRef]
95. Samšiňáková, A.; Purrini, K. Über Eine Natürliche Infektion Der Heuschrecke, Patanga Succincta Durch Den Pilz *Metarhizium anisopliae* in Thailand. *J. Appl. Entomol.* **1986**, *102*, 273–277. [CrossRef]

96. Ratner, B. Farmers Fight Back: Making Animal Feed from a Locust Plague. Available online: https://widerimage.reuters.com/story/farmersfight-back-making-animal-feed-from-a-locust-plague (accessed on 26 July 2021).
97. Saeed, T.; Abu Dagga, F.; Saraf, M. Analysis of Residual Pesticides Present in Edible Locusts Captured in Kuwait. *Arab Gulf J. Sci. Res.* **1993**, *11*, 1–5.
98. Zimmermann, G. Review on Safety of the Entomopathogenic Fungus *Metarhizium anisopliae*. *Biocontrol Sci. Technol.* **2007**, *17*, 879–920. [CrossRef]

MDPI
St. Alban-Anlage 66
4052 Basel
Switzerland
Tel. +41 61 683 77 34
Fax +41 61 302 89 18
www.mdpi.com

Agronomy Editorial Office
E-mail: agronomy@mdpi.com
www.mdpi.com/journal/agronomy

www.ingramcontent.com/pod-product-compliance
Lightning Source LLC
Chambersburg PA
CBHW061603220926
51850CB00005B/411
9783036557526